Ordinary and Partial Differential Equations

Ordinary and Partial Differential Equations

Edited by
Patrick McCann

www.willfordpress.com

Published by Willford Press,
118-35 Queens Blvd., Suite 400,
Forest Hills, NY 11375, USA

ISBN: 978-1-68285-824-0

Cataloging-in-Publication Data

Ordinary and partial differential equations / edited by Patrick McCann.
p. cm.
Includes bibliographical references and index.
ISBN 978-1-68285-824-0
1. Differential equations. 2. Differential equations, Partial. 3. Calculus. I. McCann, Patrick.
QA372 .O73 2020
515.35--dc23

For information on all Willford Press publications
visit our website at www.willfordpress.com

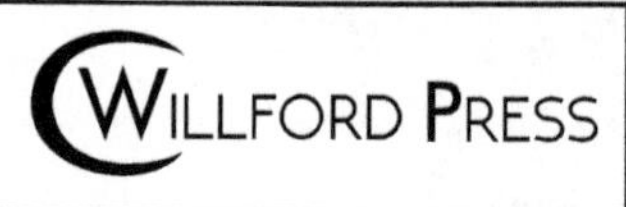

Contents

Preface

This book has been an outcome of determined endeavour from a group of educationists in the field. The primary objective was to involve a broad spectrum of professionals from diverse cultural background involved in the field for developing new researches. The book not only targets students but also scholars pursuing higher research for further enhancement of the theoretical and practical applications of the subject.

The statement which expresses the equality of two expressions is known as an equation. A differential equation is a kind of mathematical equation that shows the connection between a function and its derivatives. Functions represent the physical quantities and derivatives show their rates of change. The differential equation seeks to define the relationship between the two. It can be classified into various types such as ordinary differential equations and partial differential equations. Ordinary differential equation contains one or more than one function of an independent variable. It is related to the derivatives of these functions. Partial differential equations contain unknown multi-variable functions as well as their partial derivatives. These are generally used to formulate problems which contain functions of several variables. The topics included in this book on ordinary and partial differential equations are of utmost significance and bound to provide incredible insights to readers. It presents researches and studies performed by experts across the globe. This book is appropriate for students seeking detailed information in this area as well as for experts.

It was an honour to edit such a profound book and also a challenging task to compile and examine all the relevant data for accuracy and originality. I wish to acknowledge the efforts of the contributors for submitting such brilliant and diverse chapters in the field and for endlessly working for the completion of the book. Last, but not the least; I thank my family for being a constant source of support in all my research endeavours.

Editor

1

Linear θ-Method and Compact θ-Method for Generalised Reaction-Diffusion Equation with Delay

Fengyan Wu,[1,2] Qiong Wang,[2] Xiujun Cheng,[1,2] and Xiaoli Chen[1,2]

[1]*Center for Mathematical Sciences, Huazhong University of Science and Technology, Wuhan 430074, China*
[2]*School of Mathematics and Statistics, Huazhong University of Science and Technology, Wuhan 430074, China*

Correspondence should be addressed to Xiujun Cheng; xiujuncheng@hust.edu.cn

Academic Editor: Jiwei Zhang

This paper is concerned with the analysis of the linear θ-method and compact θ-method for solving delay reaction-diffusion equation. Solvability, consistence, stability, and convergence of the two methods are studied. When $\theta \in [0, 1/2)$, sufficient and necessary conditions are given to show that the two methods are asymptotically stable. When $\theta \in [1/2, 1]$, the two methods are proven to be unconditionally asymptotically stable. Finally, several examples are carried out to confirm the theoretical results.

1. Introduction

Partial functional differential equations (PFDEs) are widely used to model many natural phenomena in various scientific fields [1–9]. In order to gain a better understanding of the complicated dynamics, numerous researchers have investigated PFDEs. For instance, Garrido-Atienza and Real discussed the existence and uniqueness of solutions for delay evolution equations [10]. Mei et al. analysed the stability of travelling waves for nonlocal time-delayed reaction-diffusion equations [11]. Polyanin and Zhurov constructed exact solutions for delay reaction-diffusion equations and more complex nonlinear equations by the functional constraints method [12].

However, the exact solutions are difficult to be obtained [1]. Most researchers have to seek efficient and effective numerical methods to numerically solve PFDEs. Jackiewicz and Zubik-Kowal utilised spectral collocation and waveform relaxation methods to study nonlinear delay partial differential equations [13]. Chen and Wang utilised the variational iteration method to solve a neutral functional-differential equation with proportional delays [14]. Li et al. used the discontinuous Galerkin methods to solve the delay differential equations [15–17]. Bhrawy et al. applied an accurate Chebyshev pseudospectral scheme to study the multidimensional parabolic problems with time delays [18]. Aziz and Amin employed the Haar wavelet to study the numerical solution of a class of delay differential and delay partial differential equations [19].

When it comes to solving PFDEs numerically, here comes one question, that is, whether the numerical solution approximates the exact solution in a stable manner, especially for a long time. In this study, we use the following model as the test equation for analysing stability of the numerical method, which is an extension to the previous work [20–22].

$$\begin{aligned} \frac{\partial}{\partial t}u(x,t) &= r_1\frac{\partial^2}{\partial x^2}u(x,t) + r_2\frac{\partial^2}{\partial x^2}u(x,t-\tau) \\ &\quad + r_3u(x,t) + r_4u(x,t-\tau), \\ &\qquad\qquad t>0,\ 0<x<\pi, \\ u(x,t) &= u_0(x,t), \quad -\tau\le t\le 0,\ 0\le x\le \pi, \\ u(0,t) &= u(\pi,t) = 0, \quad t\ge -\tau. \end{aligned} \tag{1}$$

Here and hereafter parameters $r_1 > 0$ and $r_2 > 0$ denote the diffusion coefficients, $r_3 \in \mathbb{R}$, $r_4 \in \mathbb{R}$, and $\tau > 0$ is the delay term. In particular, when $r_3 = r_4 = 0$, the above model (1) is reduced to the original test equation in [20–22].

For the case where $r_3 = r_4 = 0$, model (1) has been studied by many researchers [2, 20–27]. In this work, we will examine the case where $r_1 > 0$, $r_2 > 0$, $r_3 \in \mathbb{R}$, and $r_4 \in \mathbb{R}$, which is a generalisation of above mentioned work, and analyse the stability condition of the numerical method. The standard second-order central difference method and compact finite difference method are utilised to discrete the diffusion operator, respectively, and the linear θ-method is utilised to discrete the temporal direction. For convenience, we name the standard second-order central difference version as linear θ-method and the compact finite difference method version as compact θ-method. With the spectral radius condition, we consider the stability of the linear θ-method and compact θ-method, respectively.

The rest of this paper is organized as follows. In Section 2, we give a sufficient delay-independent condition for Problem (1) to be asymptotically stable. In Section 3, we propose the linear θ-method for solving Problem (1); solvability, stability, and convergence of the method are discussed. In Section 4, we extend the compact θ-method to solve Problem (1). In Section 5, several numerical tests are performed to validate the theoretical results.

2. Stability of PFDE (1)

In this section, based on Tian's work [20], we give a sufficient condition for the trivial solution of Problem (1) to be asymptotically stable.

Definition 1. The trivial solution $u(x,t) \equiv 0$ of PFDE (1) is called asymptotically stable if its solution $u(x,t)$ corresponding to a sufficiently differentiable function $u_0(x,t)$ with $u_0(0,t) = u_0(\pi,t) = 0$ satisfies

$$\lim_{t\to\infty} u(x,t) = 0. \tag{2}$$

Lemma 2 (cf. [3]). *All the roots of $pe^z + q - ze^z = 0$, where p and q are real, have negative real parts if and only if*

(1) $p < 1$,

(2) $p < -q < \sqrt{a_1^2 + p^2}$,

where a_1 is the root of $g = p\tan(g)$ such that $0 < g < \pi$. If $p = 0$, we take $a_1 = \pi/2$.

Theorem 3. *Assume that the solution of Problem (1) is $u(x,t) = e^{\lambda t}e^{inx}$, where $\lambda \in \mathbb{C}$, $n \in \mathbb{R}$, $x \in [0,\pi]$, and $t \geq 0$. Then the sufficient condition for the trivial solution of Problem (1) to be asymptotically stable is that*

(1) $\tau r_3 < 1 + \tau r_1 n^2$,

(2) $\tau(r_3 - r_1 n^2) < \tau(r_2 n^2 - r_4) < \sqrt{a_1^2 + \tau^2(r_3 - r_1 n^2)^2}$,

where a_1 is the root of $g = \tau(r_3 - r_1 n^2)\tan(g)$ such that $0 < g < \pi$. If $r_3 = r_1 n^2$, we take $a_1 = \pi/2$.

Proof. Let $X = B[0,\pi]$ denote the Banach space equipped with the maximum norm, and $D(\mathscr{A}) = \{y \in X : y'' \in X, y(0) = y(\pi) = 0\}$, and $\mathscr{A}y = y''$ for $y \in D(\mathscr{A})$.

Let $r_1 n^2 (n = 1,2,\cdots)$ be the eigenvalues of $-\mathscr{A}$. According to [1, 28], if all zeros of the characteristic equations

$$f(\lambda) = \lambda + r_1 n^2 - r_3 + \left(r_2 n^2 - r_4\right) e^{-\lambda\tau} \tag{3}$$

have negative real part, then the trivial solution is asymptotically stable. Meanwhile, if at least one zero has positive real part, then it is unstable.

Let $f(\lambda) = 0$; that is, $\lambda = r_3 - r_1 n^2 + (r_4 - r_2 n^2)e^{-\lambda\tau}$. Multiplying by $e^{\lambda\tau}$, we have

$$\lambda e^{\lambda\tau} = \left(r_3 - r_1 n^2\right) e^{\lambda\tau} + \left(r_4 - r_2 n^2\right). \tag{4}$$

Setting $\lambda\tau = z$, we get

$$ze^z = \tau\left(r_3 - r_1 n^2\right) e^z + \tau\left(r_4 - r_2 n^2\right). \tag{5}$$

Denote $p = \tau(r_3 - r_1 n^2)$, and $q = \tau(r_4 - r_2 n^2)$, and rewrite the above equation as

$$ze^z = pe^z + q. \tag{6}$$

Applying Lemma 2, if

(1) $\tau r_3 < 1 + \tau r_1 n^2$,

(2) $\tau(r_3 - r_1 n^2) < \tau(r_2 n^2 - r_4) < \sqrt{a_1^2 + \tau^2(r_3 - r_1 n^2)^2}$,

then the real parts of all zeros of the characteristic equations are negative. Therefore, the trivial solution of Problem (1) is asymptotically stable. Otherwise, there exists a zero λ_0 whose real part is positive such that $f(\lambda_0) = 0$. Hence, the trivial solution is unstable. It completes the proof. □

3. Linear θ-Method

In this section, the linear θ-method is presented to solve Problem (1).

Denote $\Omega_{\Delta t} = \{t_k \mid k = -m, -m+1, \cdots\}$ as a uniform partition on the time interval $[-\tau, \infty)$, where $\Delta t = \tau/m$ is the time step size and $t_k = k\Delta t$. Denote $\Omega_{\Delta x} = \{x_j \mid j = 0, 1, \cdots, N\}$ as a uniform mesh on the space interval $\Omega = [0,\pi]$, where $\Delta x = \pi/N$ is the space step size and $x_j = j\Delta x$. Here m and N are two positive integers. Let u_j^k be the numerical approximation of $u(x_j, t_k)$ and let $\mathscr{V} = \{u_j^k \mid 0 \leq j \leq N,\ k \geq -m\}$ be the grid function defined on $\Omega_{\Delta x} \times \Omega_{\Delta t}$. For any grid function $u \in \mathscr{V}$, we use the following notations:

$$\begin{aligned} \delta_t u_j^k &= \frac{u_j^{k+1} - u_j^k}{\Delta t}, \\ \delta_x^2 u_j^k &= \frac{u_{j+1}^k - 2u_j^k + u_{j-1}^k}{\Delta x^2}, \\ u_j^{k+1/2} &= \frac{1}{2}\left(u_j^k + u_j^{k+1}\right). \end{aligned} \tag{7}$$

Now, applying the standard second-order central difference method to discrete the diffusion operator, we obtain the following linear θ-method:

$$\begin{aligned}\delta_t u_j^k &= r_1\left[(1-\theta)\,\delta_x^2 u_j^k + \theta\delta_x^2 u_j^{k+1}\right]\\ &\quad + r_2\left[(1-\theta)\,\delta_x^2 u_j^{k-m} + \theta\delta_x^2 u_j^{k-m+1}\right]\\ &\quad + r_3\left[(1-\theta)\,u_j^k + \theta u_j^{k+1}\right]\\ &\quad + r_4\left[(1-\theta)\,u_j^{k-m} + \theta u_j^{k-m+1}\right],\\ &\qquad j = 1,2,\cdots,N-1,\ k = 0,1,\cdots,\\ u_j^k &= u_0\left(x_j,t_k\right),\\ &\qquad j = 1,2,\cdots,N-1,\ k = -m,-m+1,\cdots,0,\\ u_0^k &= u_N^k = 0,\quad k = -m,-m+1,\cdots.\end{aligned} \tag{8}$$

We can rewrite the linear θ-method (8) as the following matrix form:

$$\begin{aligned}\phi_0(S)\,U^{k+1} &= \phi_1(S)\,U^k - \phi_m(S)\,U^{k+1-m}\\ &\quad - \phi_{m+1}(S)\,U^{k-m},\end{aligned} \tag{9}$$

where

$$\begin{aligned}U^k &= \left(u_1^k,u_2^k,\cdots,u_{N-1}^k\right)^T,\\ a &= \frac{r_1\Delta t}{\Delta x^2},\\ b &= \frac{r_2\Delta t}{\Delta x^2},\\ c &= r_3\Delta t,\\ d &= r_4\Delta t,\\ \phi_0(\eta) &= 1 + 2a\theta - c\theta - a\theta\eta,\\ \phi_1(\eta) &= 1 - 2a(1-\theta) + c(1-\theta) + a(1-\theta)\,\eta,\\ \phi_m(\eta) &= 2b\theta - d\theta - b\theta\eta,\\ \phi_{m+1}(\eta) &= 2b(1-\theta) - d(1-\theta) - b(1-\theta)\,\eta,\end{aligned} \tag{10}$$

$$S = \begin{pmatrix} 0 & 1 & 0 & \dots & 0 & 0\\ 1 & 0 & 1 & \dots & 0 & 0\\ & \ddots & \ddots & \ddots & \ddots & \\ 0 & 0 & \dots & 1 & 0 & 1\\ 0 & 0 & \dots & 0 & 1 & 0\end{pmatrix}_{(N-1)\times(N-1)},$$

and the eigenvalues of matrix S are $\lambda_j = 2\cos(j\Delta x)$, $j = 1,2,\cdots,N-1$.

3.1. Solvability of Linear θ-Method

Theorem 4. *The linear θ-method (8) is solvable and has a unique solution.*

Proof. The mathematical induction is utilised to prove it. We can obtain the solution of U^1 according to the initial condition. Now, assume that the solution of U^l has been determined. Then we can derive the solution of U^{l+1} with (9). It follows from (9) that the coefficient matrix of the linear system is

$$\phi_0(S) = (1 + 2a\theta - c\theta)\,I - a\theta S. \tag{11}$$

It is easy to verify that the matrix $\phi_0(S)$ is symmetric positive definite. Therefore, the solution of U^{l+1} is determined uniquely. By mathematical induction, the existence and uniqueness of the solution of difference system (8) are obtained immediately. □

3.2. Asymptotic Stability of Linear θ-Method.

Section 2 gives the sufficient condition for the trivial solution of Problem (1) to be asymptotically stable. Next, we will analyse the numerical stability of the linear θ-method (8) under this condition.

Definition 5. A numerical method applied to Problem (1) is called asymptotically stable about the trivial solution if its approximate solution u_j^k corresponding to a sufficiently differentiable function $u_0(x,t)$ with $u_0(0,t) = u_0(\pi,t) = 0$ satisfies

$$\lim_{k\to\infty}\max_{1\le j\le N}\left|u_j^k\right| = 0. \tag{12}$$

In order to prove that a polynomial is a Schur polynomial, the following lemma is needed.

Lemma 6 (cf. [29]). *Let $\gamma_m(z) = \alpha(z)z^m - \beta(z)$ be a polynomial, where $\alpha(z)$ and $\beta(z)$ are polynomials of constant degree. Then, the polynomial $\gamma_m(z)$ is a Schur polynomial for any $m \ge 1$ if and only if the following conditions hold:*

(i) $\alpha(z) = 0 \Rightarrow |z| < 1$,

(ii) $|\alpha(z)| \ge |\beta(z)|$, *for all* $z \in \mathbb{C}$, $|z| = 1$,

(iii) $\gamma_m(z) \ne 0$, *for all* $z \in \mathbb{C}$, $|z| = 1$.

Taking the analytical technique in [20–22], we know that the linear θ-method (8) is asymptotically stable about the trivial solution if and only if

$$\begin{aligned}P_{m,j}^{\theta}(z) \equiv \phi_0\left(\lambda_j\right)z^{m+1} - \phi_1\left(\lambda_j\right)z^m + \phi_m\left(\lambda_j\right)z\\ + \phi_{m+1}\left(\lambda_j\right)\end{aligned} \tag{13}$$

is a Schur polynomial for any $m \ge 1$.

Basic calculations give

$$P_{m,j}^{\theta}(z) = \mu_j(z)\,z^m - \nu_j(z), \tag{14}$$

where

$$\begin{aligned}\mu_j(z) &= \left\{1 + \theta\left[a\left(2-\lambda_j\right) - c\right]\right\}z - 1\\ &\quad + (1-\theta)\left[a\left(2-\lambda_j\right) - c\right],\\ \nu_j(z) &= \left[d - b\left(2-\lambda_j\right)\right]\left[\theta z + (1-\theta)\right].\end{aligned} \tag{15}$$

With the help of Lemma 6, we obtain the following theorem when $\theta \in [0,1/2)$, which offers a sufficient and necessary condition of asymptotic stability for the linear θ-method.

Theorem 7. *Suppose that $a > c/2(1-\cos(\Delta x))$, $b > d/2(1-\cos(\Delta x))$, and $2a(1-\cos(\Delta x)) - c > 2b(1+\cos(\Delta x)) - d$.*

Then the linear θ-method (8) is asymptotically stable about the trivial solution for $\theta \in [0, 1/2)$ if and only if

$$(1-2\theta)\left[a+b-\frac{c+d}{2(1+\cos(\Delta x))}\right] < \frac{1}{1+\cos(\Delta x)}, \quad (16)$$

where $a = r_1\Delta t/\Delta x^2$, $b = r_2\Delta t/\Delta x^2$, $c = r_3\Delta t$, and $d = r_4\Delta t$.

Proof. ($\Rightarrow$) First, let us verify item (*i*) of Lemma 6. According to $\mu_j(z) = 0$, we derive that

$$|z| = \left|1 - \frac{a(2-\lambda_j)-c}{1+\theta[a(2-\lambda_j)-c]}\right|. \quad (17)$$

By the same technique used in [22], one can check that $|z| < 1$.

Next, to verify the rest of items of Lemma 6, that is, $\mu_j(z) > \nu_j(z)$, $j = 1, 2, \cdots, N-1$, for all $z \in \mathbb{C}$, $|z| = 1$, we define the following complex variable function:

$$\begin{aligned} w &= \frac{\mu_j(z)}{\nu_j(z)} \\ &= \frac{\{1+\theta[a(2-\lambda_j)-c]\}z - 1 + (1-\theta)[a(2-\lambda_j)-c]}{[d-b(2-\lambda_j)][\theta z + (1-\theta)]}. \end{aligned} \quad (18)$$

Letting $w = x + yi$ and $|z| = 1$, after some basic calculations (see [20–22]), we find

$$\begin{aligned} \min_{|z|=1,z\in\mathbb{C}} |w| &= \min_{|z|=1,z\in\mathbb{C}} \left|\frac{\mu_j(z)}{\nu_j(z)}\right| \\ &= \min\left\{\left|\frac{a(2-\lambda_j)-c}{b(2-\lambda_j)-d}\right|, \right. \\ &\left. \left|\frac{2-[a(2-\lambda_j)-c](1-2\theta)}{[b(2-\lambda_j)-d](1-2\theta)}\right|\right\}. \end{aligned} \quad (19)$$

The value of $\min_{|z|=1,z\in\mathbb{C}}|w|$ will be discussed in the following two different cases.

Case a ($\min_{|z|=1,z\in\mathbb{C}}|\mu_j(z)/\nu_j(z)| = |(a(2-\lambda_j)-c)/(b(2-\lambda_j)-d)|$). By conditions $a > c/2(1-\cos(\Delta x))$, $b > d/2(1-\cos(\Delta x))$, and $2a(1-\cos(\Delta x)) - c > 2b(1+\cos(\Delta x)) - d$, and noting that $2(1-\cos(\Delta x)) \le (2-\lambda_j) \le 2(1-\cos((N-1)\Delta x)) = 2(1+\cos(\Delta x))$, we have $a(2-\lambda_j) - c > b(2-\lambda_j) - d > 0$. Therefore, for all $z \in \mathbb{C}$, $|z| = 1$, we find

$$\left|\frac{\mu_j(z)}{\nu_j(z)}\right| \ge \min_{|z|=1,z\in\mathbb{C}} \left|\frac{\mu_j(z)}{\nu_j(z)}\right| = \frac{a(2-\lambda_j)-c}{b(2-\lambda_j)-d} > 1. \quad (20)$$

Case b ($\min_{|z|=1,z\in\mathbb{C}}|\mu_j(z)/\nu_j(z)| = |(2-[c+a(2-\lambda_j)](1-2\theta))/[d+b(2-\lambda_j)](1-2\theta)|$). It follows from condition (16) that

$$2 - [a(2-\lambda_j)-c](1-2\theta) > 0. \quad (21)$$

Noticing the fact that $2 - \lambda_j \le 2(1+\cos(\Delta x))$ and condition (16), we have

$$\begin{aligned} \min_{|z|=1,z\in\mathbb{C}} \left|\frac{\mu_j(z)}{\nu_j(z)}\right| &= \frac{2-[a(2-\lambda_j)-c](1-2\theta)}{[b(2-\lambda_j)-d](1-2\theta)} \\ &\ge \frac{2-[2a(1+\cos(\Delta x))-c](1-2\theta)}{[2b(1+\cos(\Delta x))-d](1-2\theta)} > 1. \end{aligned} \quad (22)$$

In brief, combining **Case a** and **Case b**, we conclude that, for all $z \in \mathbb{C}$, $|z| = 1$, $\mu_j(z) > \nu_j(z)$, $j = 1, 2, \cdots, N-1$ holds, which implies that items (*ii*) and (*iii*) of Lemma 6 hold.

Now, with the help of Lemma 6, we derive that the linear θ-method (8) is asymptotically stable about the trivial solution.

($\Leftarrow$) Next, we prove the necessary part from two points by contradiction:

(i) Suppose that $(1-2\theta)[a+b-(c+d)/2(1+\cos(\Delta x))] = 1/(1+\cos(\Delta x))$. Let m be even, $j = N-1$, and $z = -1$, and then, for $|z| = 1$, we get that $P^{\theta}_{m,N-1}(-1) = 0$, which indicates that condition (*iii*) of Lemma 6 does not hold. Thus, the linear θ-method (8) is not asymptotically stable.

(ii) Suppose that $(1-2\theta)[a+b-(c+d)/2(1+\cos(\Delta x))] > 1/(1+\cos(\Delta x))$. Let $j = N-1$ and $z = -1$, and then, after some basic calculations, we arrive at $|\nu_{N-1}(-1)| > |\mu_{N-1}(-1)|$. This signifies that condition (*ii*) of Lemma 6 does not hold. Therefore, the linear θ-method (8) is not asymptotically stable.

Then, we know that (16) is a necessary condition for asymptotic stability. This completes the proof. □

Remark 8. When $r_3 = r_4 = 0$, the sufficient and necessary condition (16) in Theorem 7 is simplified to

$$(1-2\theta)(a+b) < \frac{1}{1+\cos(\Delta x)}, \quad (23)$$

which is consistent with the previous work [20].

Next, when $\theta \in [1/2, 1]$, we will prove that the linear θ-method (8) is unconditionally asymptotically stable with respect to the trivial solution.

Theorem 9. *Suppose that $a > c/2(1-\cos(\Delta x))$, $b > d/2(1-\cos(\Delta x))$, and $2a(1-\cos(\Delta x)) - c > 2b(1+\cos(\Delta x)) - d$. Then the linear θ-method (8) is unconditionally asymptotically stable about the trivial solution for $\theta \in [1/2, 1]$.*

Proof. We will prove the theorem with Lemma 6. First, it follows from $\mu_j(z) = 0$ that

$$|z| = \left|1 - \frac{a(2-\lambda_j)-c}{1+\theta[a(2-\lambda_j)-c]}\right|. \quad (24)$$

Similar to the proof of Theorem 7, we get $|z| < 1$.

Then, we check items (ii) and (iii) of Lemma 6. To do that, we introduce the following complex variable function:

$$\begin{aligned} w &= \frac{\mu_j(z)}{\nu_j(z)} \\ &= \frac{\{1+\theta[a(2-\lambda_j)-c]\}z-1+(1-\theta)[a(2-\lambda_j)-c]}{[d-b(2-\lambda_j)][\theta z+(1-\theta)]}. \end{aligned} \tag{25}$$

(i) $\theta = 1/2$. Set $w = x + yi$ and $|z| = 1$. Basic calculations give

$$\min_{|z|=1,z\in\mathbb{C}} \left|\frac{\mu_j(z)}{\nu_j(z)}\right| = \left|\frac{a(2-\lambda_j)-c}{b(2-\lambda_j)-d}\right|. \tag{26}$$

It follows from assumptions $a > c/2(1-\cos(\Delta x))$, $b > d/2(1-\cos(\Delta x))$, and $2a(1-\cos(\Delta x))-c > 2b(1+\cos(\Delta x))-d$ that $a(2-\lambda_j)-c > b(2-\lambda_j)-d > 0$. Then, for all $z \in \mathbb{C}$, $|z| = 1$, we find that

$$\left|\frac{\mu_j(z)}{\nu_j(z)}\right| \geq \min_{|z|=1,z\in\mathbb{C}} \left|\frac{\mu_j(z)}{\nu_j(z)}\right| = \frac{a(2-\lambda_j)-c}{b(2-\lambda_j)-d} > 1, \tag{27}$$

which indicates that items *(ii)* and *(iii)* of Lemma 6 hold.

(ii) $\theta \in (1/2, 1]$. Similarly, we can get

$$\min_{|z|=1,z\in\mathbb{C}} \left|\frac{\mu_j(z)}{\nu_j(z)}\right| = \left|\frac{a(2-\lambda_j)-c}{b(2-\lambda_j)-d}\right|. \tag{28}$$

In this case, for all $z \in \mathbb{C}$, $|z| = 1$, we also obtain that

$$\left|\frac{\mu_j(z)}{\nu_j(z)}\right| \geq \min_{|z|=1,z\in\mathbb{C}} \left|\frac{\mu_j(z)}{\nu_j(z)}\right| = \frac{a(2-\lambda_j)-c}{b(2-\lambda_j)-d} > 1. \tag{29}$$

According to Lemma 6, we conclude that the linear θ-method (8) is asymptotically stable about the trivial solution. This completes the proof of the theorem. □

3.3. Convergence of Linear θ-Method. Here and below, when we discuss the convergence of numerical methods, we will always assume that the solution $u(x,t)$ of Problem (1) is smooth enough and satisfies

$$\left|\frac{\partial^{j+k}}{\partial x^j \partial t^k} u(x,t)\right| \leq C, \quad 0 \leq j \leq 6,\ 0 \leq k \leq 3,\ C > 0, \tag{30}$$

where C is a constant.

Let $U_j^k = u(x_j, t_k)$, $j = 0, 1, \cdots, k = -m, -m+1, \cdots$. Then, we get

$$\begin{aligned} \delta_t U_j^k &= r_1\left[(1-\theta)\delta_x^2 U_j^k + \theta\delta_x^2 U_j^{k+1}\right] \\ &\quad + r_2\left[(1-\theta)\delta_x^2 U_j^{k-m} + \theta\delta_x^2 U_j^{k-m+1}\right] \\ &\quad + r_3\left[(1-\theta)U_j^k + \theta U_j^{k+1}\right] \\ &\quad + r_4\left[(1-\theta)U_j^{k-m} + \theta U_j^{k-m+1}\right] + R_j^k, \end{aligned} \tag{31}$$

where R_j^k is the local truncation error. Taylor expansion yields that there exists a constant $\overline{C}$ such that, for $j = 1, 2, \cdots, N-1$, $k = 0, 1, \cdots$,

$$|R_j^k| \leq \begin{cases} \overline{C}(\Delta t^2 + \Delta x^2), & \theta = \dfrac{1}{2}, \\ \overline{C}(\Delta t + \Delta x^2), & 0 \leq \theta < \dfrac{1}{2} \text{ or } \dfrac{1}{2} < \theta \leq 1. \end{cases} \tag{32}$$

Thus, the consistence of linear θ-method (8) is obtained. Now, the convergence result is presented in the following theorem.

Theorem 10. *Assume that the assumptions in Theorems 7 and 9 hold. Then, for $k = 1, 2, \cdots$, we have the following convergent result:*

$$\|e^k\| \leq \begin{cases} \widehat{C}(\Delta t^2 + \Delta x^2), & \theta = \dfrac{1}{2}, \\ \widehat{C}(\Delta t + \Delta x^2), & 0 \leq \theta < \dfrac{1}{2} \text{ or } \dfrac{1}{2} < \theta \leq 1, \end{cases} \tag{33}$$

where $e^k = [u_1^k - U_1^k, u_2^k - U_2^k, \cdots, u_{N-1}^k - U_{N-1}^k]^T$ and $\widehat{C}$ is a constant that is independent of $\Delta t, \Delta x$.

Proof. It follows from Theorem 4 that difference system (8) is solvable and has a unique solution. Moreover, the assumptions in Theorems 7 and 9 hold, signifying that the method is stable. Together with the consistence of the method, we derive that (33) holds by the Lax equivalence theorem [30, 31]. □

4. Extension to Compact θ-Method

In this section, we would like to use the compact θ-method with a higher convergence order in space to extend our work. We introduce the compact difference operator,

$$\mathcal{A}_h u_j^k = \begin{cases} \dfrac{u_{j-1}^k + 10u_j^k + u_{j+1}^k}{12}, & j = 1, 2, \cdots, N-1, \\ u_j^k & j = 0, N, \end{cases} \tag{34}$$

and an important lemma below, which will be needed to construct and prove our main results.

Lemma 11 (cf. [32]). *Assume that $v(x) \in C^6[0, \pi]$. Then*

$$\begin{aligned} &\frac{v''(x_{j-1}) + 10v''(x_j) + v''(x_{j+1})}{12} \\ &\quad - \frac{v(x_{j-1}) - 2v(x_j) + v(x_{j+1})}{\Delta x^2} \\ &= \frac{\Delta x^4}{240} v^{(6)}(\omega_j), \end{aligned} \tag{35}$$

where $\omega_j \in (x_{j-1}, x_{j+1})$.

Now, applying the compact difference operator (34) to discrete the diffusion operator, we have the compact θ-method:

TABLE 1: Stability and convergence order of different methods.

	$\theta \in \left[0, \frac{1}{2}\right)$	$\theta \in \left[\frac{1}{2}, 1\right]$	Order
Linear θ-method for problem of [20]	$(1-2\theta)(a+b) < \frac{1}{1+\cos(\Delta x)}$	Unconditionally stable	2
Compact θ-method for problem of [20]	$\frac{1}{6}+(1-2\theta)(a+b) < \frac{1}{1+\cos(\Delta x)}$	Unconditionally stable	4
Linear θ-method for problem (1)	$(1-2\theta)\left[a+b-\frac{c+d}{2(1+\cos(\Delta x))}\right] < \frac{1}{1+\cos(\Delta x)}$	Unconditionally stable	2
Compact θ-method for problem (1)	$\frac{1}{6}+(1-2\theta)\left[a+b-\frac{c+d}{2(1+\cos(\Delta x))}\right] < \frac{1}{1+\cos(\Delta x)}$	Unconditionally stable	4

$$\begin{aligned}
\mathscr{A}_h\delta_t u_j^k &= r_1\left[(1-\theta)\delta_x^2 u_j^k + \theta\delta_x^2 u_j^{k+1}\right] \\
&\quad + r_2\left[(1-\theta)\delta_x^2 u_j^{k-m} + \theta\delta_x^2 u_j^{k-m+1}\right] \\
&\quad + r_3\left[(1-\theta)u_j^k + \theta u_j^{k+1}\right] \\
&\quad + r_4\left[(1-\theta)u_j^{k-m} + \theta u_j^{k-m+1}\right], \\
&\qquad j = 1,2,\cdots,N-1,\ k = 0,1,\cdots, \\
u_j^k &= u_0\left(x_j, t_k\right), \\
&\qquad j = 1,2,\cdots,N-1,\ k = -m,-m+1,\cdots,0, \\
u_0^k &= u_N^k = 0, \quad k = -m,-m+1,\cdots.
\end{aligned} \tag{36}$$

The compact θ-method (36) can be rewritten in the following matrix form:

$$\begin{aligned}
\psi_0(S)U^{k+1} &= \psi_1(S)U^k - \psi_m(S)U^{k+1-m} \\
&\quad - \psi_{m+1}(S)U^{k-m},
\end{aligned} \tag{37}$$

where

$$\begin{aligned}
\psi_0(\eta) &= \frac{5}{6} + 2a\theta - c\theta + \left(\frac{1}{12} - a\theta\right)\eta, \\
\psi_1(\eta) &= \frac{5}{6} - 2a(1-\theta) + c(1-\theta) \\
&\quad + \left(\frac{1}{12} + a(1-\theta)\right)\eta, \\
\psi_m(\eta) &= 2b\theta - d\theta - b\theta\eta, \\
\psi_{m+1}(\eta) &= 2b(1-\theta) - d(1-\theta) - b(1-\theta)\eta.
\end{aligned} \tag{38}$$

Similarly, the solvability, asymptotic stability, and convergence of the compact θ-method (36) can also be obtained. For conciseness, we merely list our main results and omit the details.

Theorem 12. *Suppose that $a > c/2(1-\cos(\Delta x))$, $b > d/2(1-\cos(\Delta x))$, and $2a(1-\cos(\Delta x))-c > 2b(1+\cos(\Delta x))-d$. Then the compact θ-method (36) is asymptotically stable about the trivial solution for $\theta \in [0,1/2)$ if and only if*

$$\begin{aligned}
&\frac{1}{6} + (1-2\theta)\left[a+b-\frac{c+d}{2(1+\cos(\Delta x))}\right] \\
&\quad < \frac{1}{1+\cos(\Delta x)},
\end{aligned} \tag{39}$$

where $a = r_1\Delta t/\Delta x^2$, $b = r_2\Delta t/\Delta x^2$, $c = r_3\Delta t$, and $d = r_4\Delta t$.

Remark 13. When $r_3 = r_4 = 0$, the sufficient and necessary condition (39) in Theorem 12 is reduced to

$$\frac{1}{6} + (1-2\theta)(a+b) < \frac{1}{1+\cos(\Delta x)}, \tag{40}$$

which is consistent with the previous work [22].

Theorem 14. *Suppose that $a > c/2(1-\cos(\Delta x))$, $b > d/2(1-\cos(\Delta x))$, and $2a(1-\cos(\Delta x))-c > 2b(1+\cos(\Delta x))-d$. Then the compact θ-method (36) is unconditionally asymptotically stable about the trivial solution for $\theta \in [1/2,1]$.*

Theorem 15. *Assume that the assumptions in Theorems 12 and 14 hold. Then, for $k = 1,2,\cdots$, we have the following convergent result:*

$$\left\|e^k\right\| \le \begin{cases} \widetilde{C}\left(\Delta t^2 + \Delta x^4\right), & \theta = \frac{1}{2}, \\ \widetilde{C}\left(\Delta t + \Delta x^4\right), & 0 \le \theta < \frac{1}{2} \text{ or } \frac{1}{2} < \theta \le 1, \end{cases} \tag{41}$$

where $\widetilde{C}$ is a constant that is independent of temporal and spatial stepsizes.

Remark 16. The comparison of linear θ-method and compact θ-method applied to problem in [20] and Problem (1) is presented in Table 1.

5. Numerical Tests

In this section, several numerical experiments are carried out to illustrate the theoretical results.

5.1. Stability Tests of Linear θ-Method and Compact θ-Method. We use the following equation to test stability of the proposed method:

$$\begin{aligned}
\frac{\partial}{\partial t}u(x,t) &= \frac{\partial^2}{\partial x^2}u(x,t) + 0.5\frac{\partial^2}{\partial x^2}u(x,t-\tau) \\
&\quad - u(x,t) - 0.5u(x,t-\tau), \\
&\qquad 0 < t \le T,\ 0 < x < \pi, \\
u(x,t) &= \sin(x), \quad -\tau \le t \le 0,\ 0 \le x \le \pi, \\
u(0,t) &= u(\pi,t) = 0, \quad t \ge -\tau.
\end{aligned} \tag{42}$$

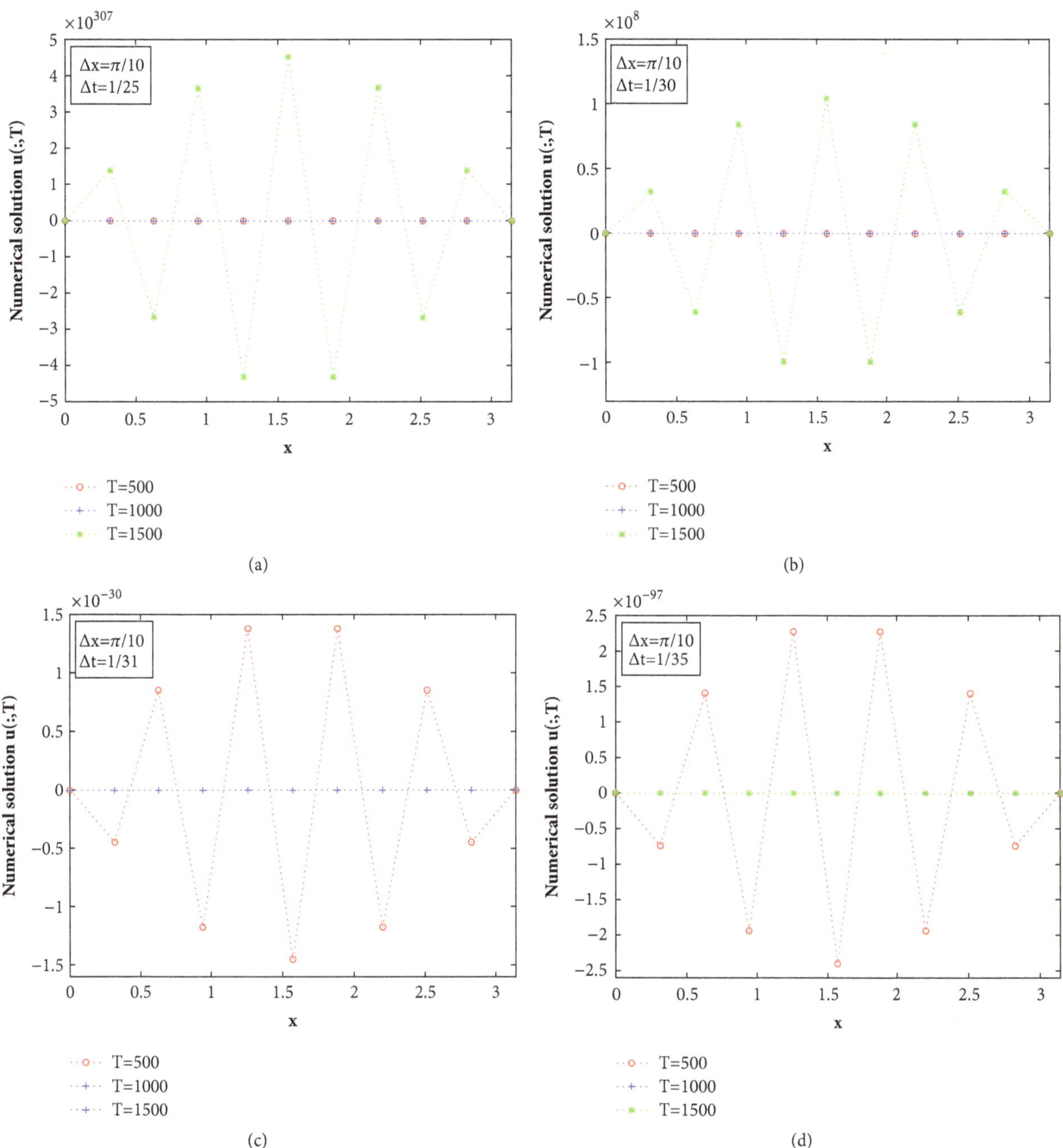

FIGURE 1: **Numerical solution at the different final time T for varying parameter** m ($\theta = 0$ and $\tau = 1$). (a) $m = 25$. (b) $m = 30$. (c) $m = 31$. (d) $m = 35$.

Here, let $\tau = 1$ and $\Delta x = \pi/10$. We set parameters $r_1 = 1.0$, $r_2 = 0.5$, $r_3 = -1.0$, and $r_4 = -0.5$ such that the trivial solution of Problem (1) is asymptotically stable.

5.1.1. Linear θ-Method. First, to verify the effectiveness of the sufficient and necessary condition (16) of the linear θ-method, here we choose the case where $\theta = 0$ to illustrate that. Noting that $\Delta t = \tau/m$, where $m > 0$ is an integer, and substituting parameters $\theta = 0$, $\tau = 1$, $\Delta x = \pi/10$, $r_1 = 1.0$, $r_2 = 0.5$, $r_3 = -1.0$, and $r_4 = -0.5$ into condition (16), we derive that the proposed method is asymptotically stable if $m > 30.4025$. In other words, if $m \geq 31$, then the method is asymptotically stable. Meanwhile, if $m \leq 30$, then the method is not asymptotically stable. In Figure 1, we can get a pictorial understanding of that.

It is easily seen from Figures 1(a) and 1(b) that the numerical solution is unstable as time goes on for $m = 25$ and 30. As shown in Figures 1(c) and 1(d), we know that the numerical solution is asymptotically stable for $m = 31$ and 35. Furthermore, denote the left-hand side of (16) as lhs $= (1-2\theta)[a+b-(c+d)/2(1+\cos(\Delta x))]$, and denote the right-hand side of (16) as rhs $= 1/(1+\cos(\Delta x))$, and Ind $=$ lhs$-$rhs. From above paragraph, we know that Ind > 0 for $m \leq 30$, and Ind < 0 for $m \geq 31$. The relationship between Ind and m is

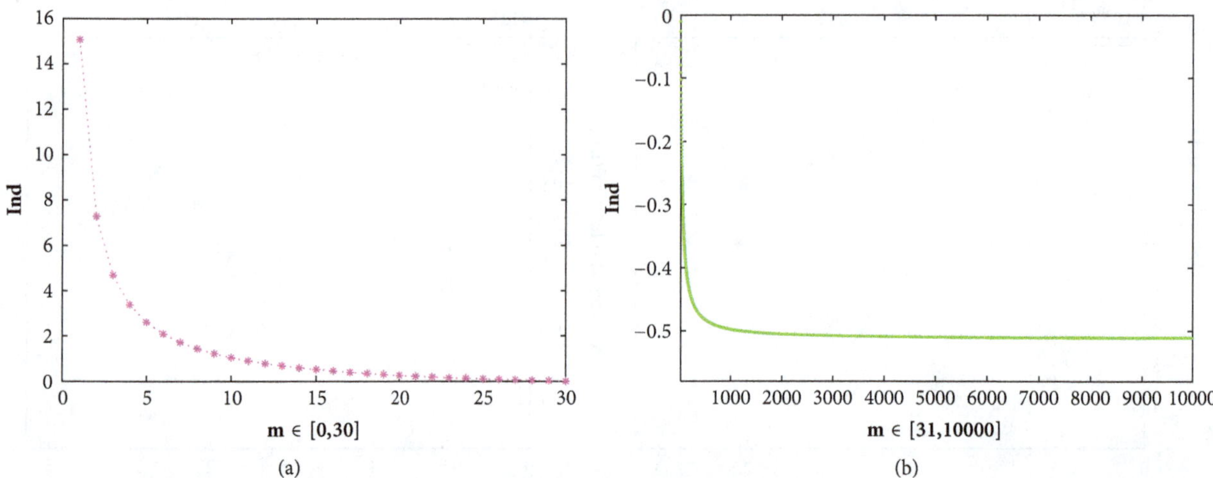

FIGURE 2: **Ind as a function of parameter m** ($\theta = 0$, $\tau = 1$, $\Delta x = \pi/10$, $r_1 = 1.0$, $r_2 = 0.5$, $r_3 = -1.0$, and $r_4 = -0.5$). (a) $m \in [1, 30]$. (b) $m \in [31, 1000]$.

shown in Figures 2(a) and 2(b). All these illustrate the results in Theorem 7.

Next, when $\theta = 1/2$ or 1, we apply the linear θ-method and use different stepsizes to solve problem (42). Theoretically, the numerical solution is asymptotically stable by Theorem 9. Numerically, we know that the numerical solution is asymptotically stable from the plots of Figure 3, which is consistent with the theoretical result.

5.1.2. Compact θ-Method. First, we choose the case $\theta = 0$ to verify the effectiveness of the sufficient and necessary condition (39) of the compact θ-method. Under the case that $\theta = 0$, $\tau = 1$, $\Delta x = \pi/10$, $r_1 = r_3 = 1.0$, and $r_2 = r_4 = 0.5$, and noting that $\Delta t = \tau/m$, by condition (39), the proposed method is asymptotically stable if and only if $m > 45.052$. In other words, if $m \geq 46$, then the proposed method is asymptotically stable. Meanwhile, if $m \leq 45$, then the proposed method is not asymptotically stable. In order to validate it, we give a pictorial understanding of that in Figure 4.

From Figures 4(a) and 4(b), it is easily seen that the numerical solution is unstable as time goes on for $m = 40$ and 45. The numerical solution is asymptotically stable for $m = 46$ and 50 in Figures 4(c) and 4(d), respectively. Furthermore, denote the left-hand side of (39) as LHS $= 1/6+(1-2\theta)[a+b-(c+d)/2(1+\cos(\Delta x))]$, and denote the right-hand side of (39) as RHS $= 1/(1+\cos(\Delta x))$, and Ind = LHS − RHS. According to the analysis of above paragraph, we know that Ind > 0 for $m \leq 45$, and Ind < 0 for $m \geq 46$. The schematic presentation of the relationship between Ind and positive integer m is given in Figures 5(a) and 5(b). All the numerical results agree well with the findings in Theorem 12.

Then, for $\theta = 1/2$ or 1, we apply the compact θ-method and choose different stepsizes to solve problem (42). The numerical results are shown in Figure 6. Theoretically, according to Theorem 14, the numerical solution is asymptotically stable. Numerically, from these figures, we know that the numerical solution is asymptotically stable, which confirms the theoretical result.

5.2. Convergence Tests of Linear θ-Method and Compact θ-Method. We use the following equation to show our convergence results:

$$\begin{aligned}\frac{\partial}{\partial t}u(x,t) &= \frac{\partial^2}{\partial x^2}u(x,t) + 0.5\frac{\partial^2}{\partial x^2}u(x,t-\tau) \\ &\quad - u(x,t) - 0.5u(x,t-\tau) + f(x,t), \\ &\qquad 0 < t \leq T,\ 0 < x < \pi, \\ u(x,t) &= u_0(x,t), \quad -\tau \leq t \leq 0,\ 0 \leq x \leq \pi, \\ u(0,t) &= u(\pi,t) = 0, \quad t \geq -\tau.\end{aligned} \tag{43}$$

Here, the added term $h(x,t)$ and the initial condition $u_0(x,t)$ are specified so that the exact solution is $u(x,t) = e^{-t}\sin(x)$.

We set the parameters $r_1 = 1.0$, $r_2 = 0.5$, $r_3 = -1.0$, $r_4 = -0.5$, $\tau = 0.5$, and $T = 2$ and solve problem (43) with different spatial and temporal stepsizes ($\Delta x = \pi/N$ and $\Delta t = \tau/m$).

For $\theta = 0$, we let $\Delta t = 1e - 5$ to guarantee that the linear θ-method (8) is asymptotically stable. When the compact θ-method (36) is applied to solve problem (43), we let $\Delta t \approx \Delta x^4$. The numerical errors and corresponding orders in different sense of norms are displayed in Table 2. Clearly, these results confirm the convergence of the two methods.

For $\theta = 1/2$, when the linear θ-method (8) is applied to solve problem (43), we let $\Delta t \approx \Delta x$, and for the compact θ-method (36), we let $\Delta t \approx \Delta x^2$. For $\theta = 1$, when method (8) is applied to solve problem (43), we let $\Delta t \approx \Delta x^2$, and for method (36), we let $\Delta t \approx \Delta x^4$. The numerical errors and corresponding orders in different sense of norms are listed in Tables 3 and 4, respectively. It is readily found that these results confirm the convergence of the two methods. Obviously, the compact θ-method gives a better convergence result in the space.

Table 2: Errors and convergence orders when $\theta = 0$, $T = 2$, and $\tau = 0.5$.

Linear θ-method					Compact θ-method				
N	L^2-error	Order	L^∞-error	Order	N	L^2-error	Order	L^∞-error	Order
2	$2.40E-02$	–	$1.91E-02$	–	7	$1.15E-03$	–	$8.95E-04$	–
4	$5.62E-03$	2.09	$4.48E-03$	2.09	14	$7.31E-05$	3.98	$5.83E-05$	3.94
8	$1.38E-03$	2.02	$1.10E-03$	2.02	28	$4.58E-06$	4.00	$3.65E-06$	4.00
16	$3.44E-04$	2.01	$2.75E-04$	2.01	56	$2.86E-07$	4.00	$2.28E-07$	4.00
32	$8.57E-05$	2.01	$6.84E-05$	2.01	112	$1.79E-08$	4.00	$1.43E-08$	4.00

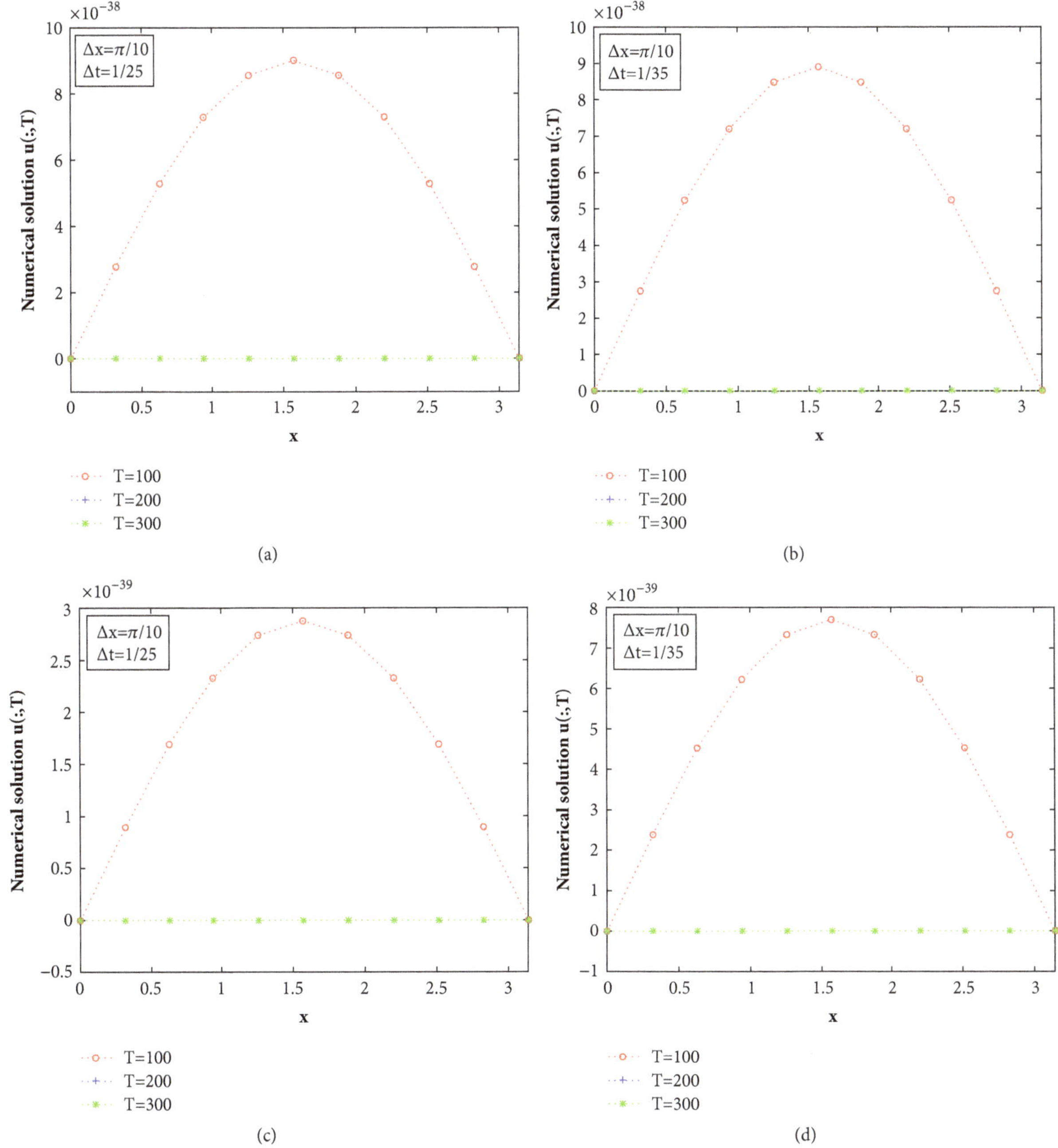

Figure 3: **Numerical solution at the different final time T for varying parameters θ and m** ($\tau = 1$). (a) $\theta = 0.5$ and $m = 25$. (b) $\theta = 0.5$ and $m = 35$. (c) $\theta = 1$ and $m = 25$. (d) $\theta = 1$ and $m = 35$.

TABLE 3: Errors and convergence orders when $\theta = 0.5$, $T = 2$, and $\tau = 0.5$.

	Linear θ-method				Compact θ-method			
N	L^2-error	Order	L^∞-error	Order	L^2-error	Order	L^∞-error	Order
5	$1.40E-03$	–	$1.06E-03$	–	$7.69E-04$	–	$5.83E-04$	–
10	$3.81E-04$	1.87	$3.04E-04$	1.80	$4.46E-05$	4.11	$3.56E-05$	4.03
20	$9.67E-05$	1.98	$7.72E-05$	1.98	$2.74E-06$	4.03	$2.19E-06$	4.03
40	$2.61E-05$	1.89	$2.09E-05$	1.89	$1.69E-07$	4.01	$1.35E-07$	4.01
80	$6.15E-06$	2.09	$4.91E-06$	2.09	$1.06E-08$	4.00	$8.44E-09$	4.00

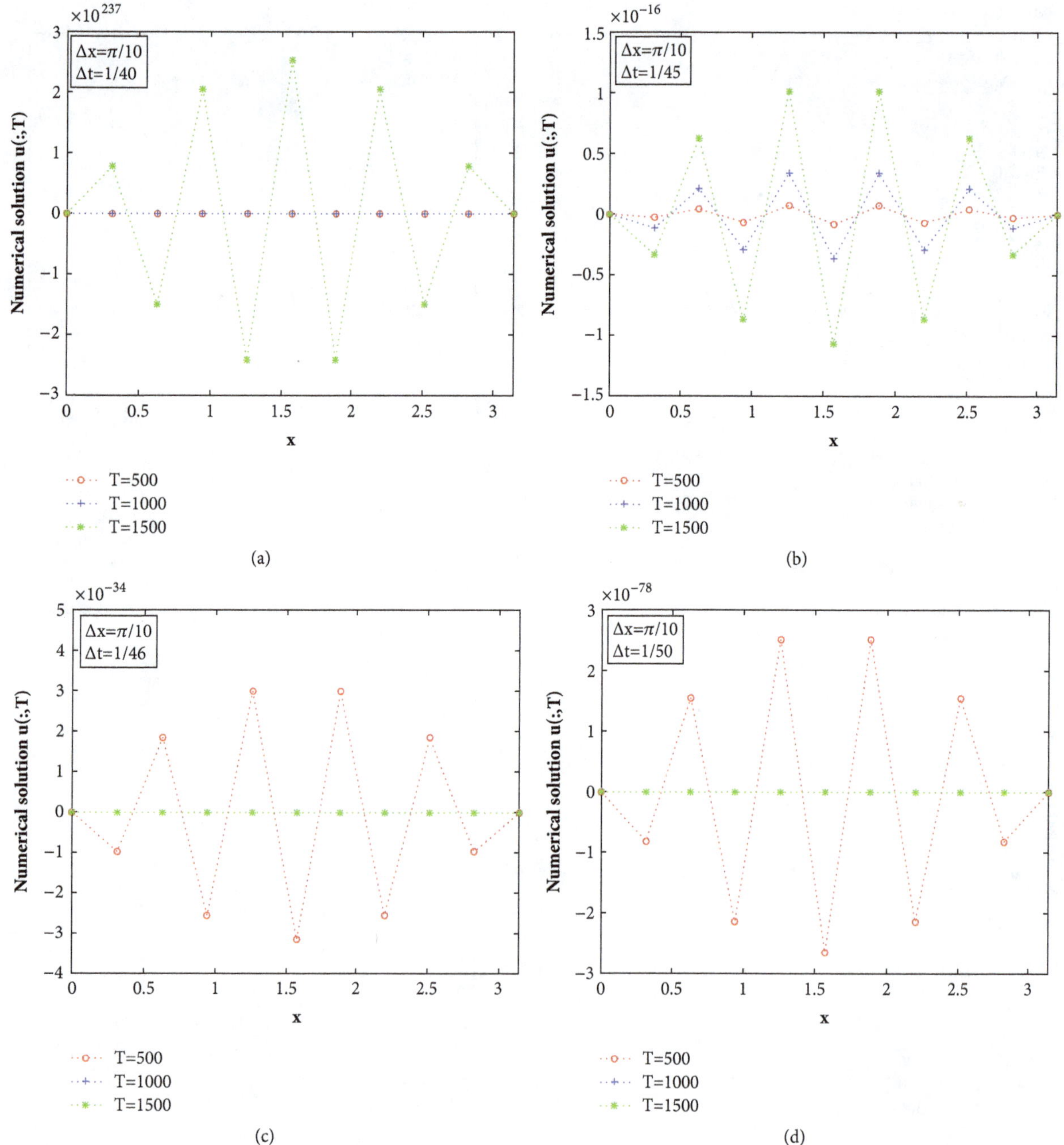

FIGURE 4: **Numerical solution at the different final time T for varying parameter** m ($\theta = 0$ and $\tau = 1$). (a) $m = 40$. (b) $m = 45$. (c) $m = 46$. (d) $m = 50$.

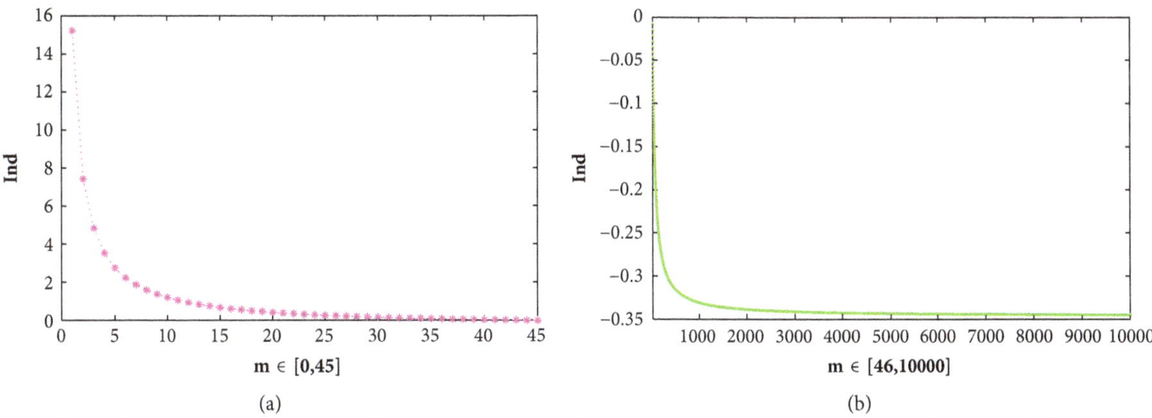

FIGURE 5: **Ind as a function of parameter m** ($\theta = 0$, $\tau = 1$, $\Delta x = \pi/10$, $r_1 = 1.0$, $r_2 = 0.5$, $r_3 = -1.0$, and $r_4 = -0.5$). (a) $m \in [1, 45]$. (b) $m \in [46, 1000]$.

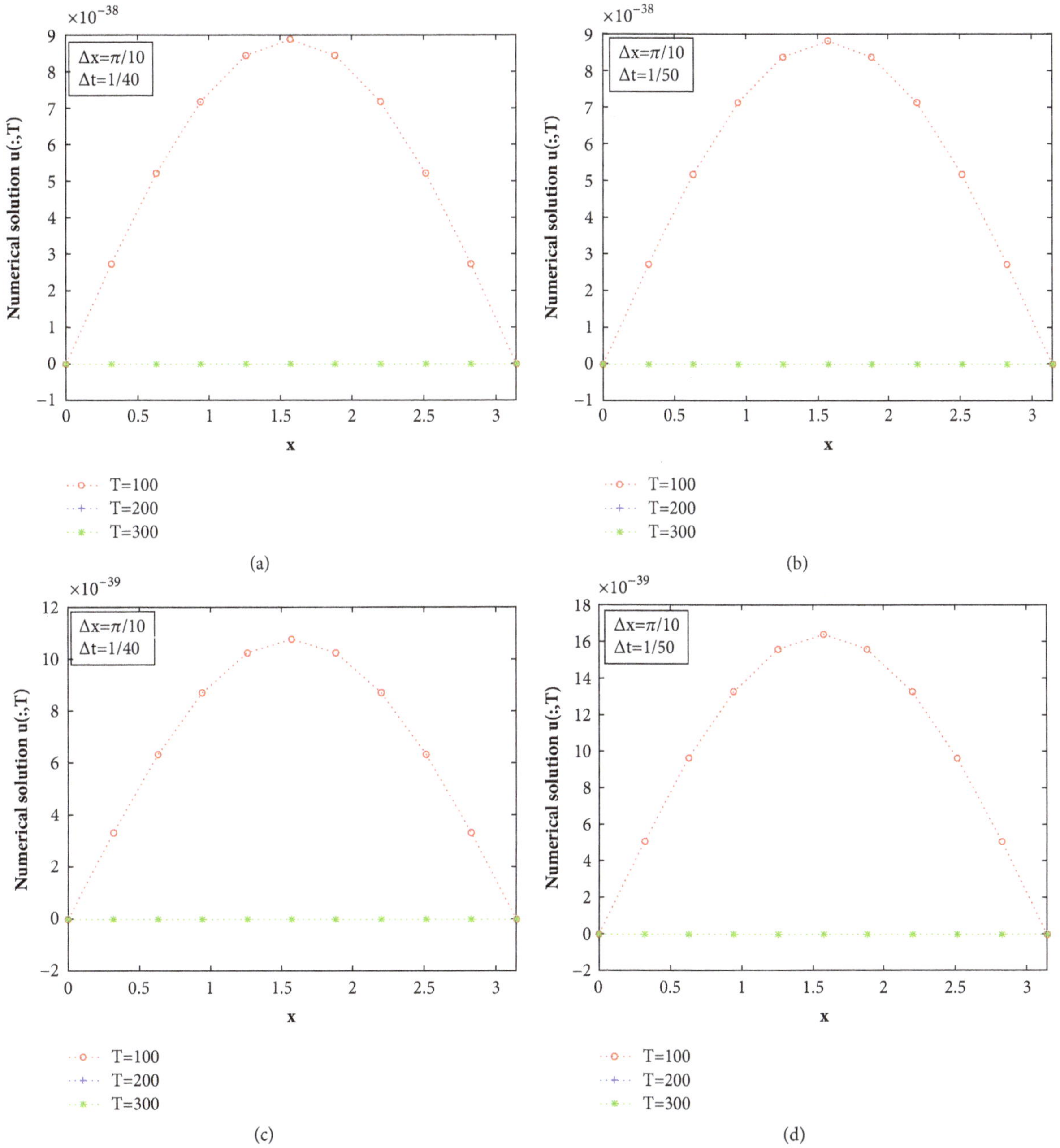

(a) (b)

(c) (d)

FIGURE 6: **Numerical solution at the different final time T for varying parameters θ and m** ($\tau = 1$). (a) $\theta = 0.5$ and $m = 40$. (b) $\theta = 0.5$ and $m = 50$. (c) $\theta = 1$ and $m = 40$. (d) $\theta = 1$ and $m = 50$.

TABLE 4: Errors and convergence orders when $\theta = 1, T = 2$, and $\tau = 0.5$.

	Linear θ-method				Compact θ-method			
N	L^2-error	Order	L^∞-error	Order	L^2-error	Order	L^∞-error	Order
5	$2.16E-02$	–	$1.64E-02$	–	$5.45E-03$	–	$4.14E-03$	–
10	$4.34E-03$	2.31	$3.47E-03$	2.24	$2.96E-04$	4.20	$2.36E-04$	4.13
20	$9.83E-04$	2.14	$7.84E-04$	2.14	$1.82E-05$	4.03	$1.45E-05$	4.03
40	$2.38E-04$	2.04	$1.90E-04$	2.04	$1.13E-06$	4.00	$9.05E-07$	4.00
80	$5.92E-05$	2.01	$4.72E-05$	2.01	$7.09E-08$	4.00	$5.65E-08$	4.00

Conflicts of Interest

The authors declare that they have no conflicts of interest.

Acknowledgments

This work is supported by NSFC (Grants nos. 11571128 and 11771162).

References

[1] J. Wu, "Theory and applications of partial functional-differential equations," in *Applied Mathematical Sciences*, vol. 119, Springer, New York, NY, USA, 1996.

[2] J. Green and H. W. Stech, *Diffusion and Hereditary Effects in a Class of Population Models Differential Equations and Applications in Ecology, Epidemics, and Population Problems*, Academic Press, New York, NY, USA, 1981.

[3] R. E. Bellman and K. L. Cooke, *Differential-Difference Equations*, Academic Press, New York, NY, USA, 1963.

[4] R. V. Culshaw, S. Ruan, and G. Webb, "A mathematical model of cell-to-cell spread of HIV-1 that includes a time delay," *Journal of Mathematical Biology*, vol. 46, pp. 425–444, 2003.

[5] D. Li, J. Zhang, and Z. Zhang, "Unconditionally optimal error estimates of a linearized galerkin method for nonlinear time fractional reaction–subdiffusion equations," *Journal of Scientific Computing*, 2018, https://doi.org/10.1007/s10915-018-0642-9.

[6] D. Li and J. Zhang, "Efficient implementation to numerically solve the nonlinear time fractional parabolic problems on unbounded spatial domain," *Journal of Computational Physics*, vol. 322, pp. 415–428, 2016.

[7] D. Li and J. Wang, "Unconditionally optimal error analysis of Crank-Nicolson Galerkin FEMs for a strongly nonlinear parabolic system," *Journal of Scientific Computing*, vol. 72, no. 2, pp. 892–915, 2017.

[8] J. Kongson and S. Amornsamankul, "A model of the signal transduction process under a delay," *East Asian Journal on Applied Mathematics*, vol. 7, no. 4, pp. 741–751, 2017.

[9] R. Kumar, A. K. Sharma, and K. Agnihotri, "Dynamics of an innovation diffusion model with time delay," *East Asian Journal on Applied Mathematics*, vol. 7, no. 3, pp. 455–481, 2017.

[10] M. Garrido-Atienza and J. Real, "Existence and uniqueness of solutions for delay evolution equations of second order in time," *Journal of Mathematical Analysis and Applications*, vol. 283, no. 2, pp. 582–609, 2003.

[11] M. Mei, C. Ou, and X.-Q. Zhao, "Global stability of monostable traveling waves for nonlocal time-delayed reaction-diffusion equations," *SIAM Journal on Mathematical Analysis*, vol. 42, no. 6, pp. 2762–2790, 2010.

[12] A. D. Polyanin and A. I. Zhurov, "Functional constraints method for constructing exact solutions to delay reaction-diffusion equations and more complex nonlinear equations," *Communications in Nonlinear Science and Numerical Simulation*, vol. 19, no. 3, pp. 417–430, 2014.

[13] Z. Jackiewicz and B. Zubik-Kowal, "Spectral collocation and waveform relaxation methods for nonlinear delay partial differential equations," *Applied Numerical Mathematics*, vol. 56, no. 3-4, pp. 433–443, 2006.

[14] X. Chen and L. Wang, "The variational iteration method for solving a neutral functional-differential equation with proportional delays," *Computers & Mathematics with Applications*, vol. 59, no. 8, pp. 2696–2702, 2010.

[15] D. Li and C. Zhang, "Nonlinear stability of discontinuous Galerkin methods for delay differential equations," *Applied Mathematics Letters*, vol. 23, no. 4, pp. 457–461, 2010.

[16] D. Li and C. Zhang, "L^∞ error estimates of discontinuous Galerkin methods for delay differential equations," *Applied Numerical Mathematics*, vol. 82, pp. 1–10, 2014.

[17] D. Li and C. Zhang, "Superconvergence of a discontinuous Galerkin method for first-order linear delay differential equations," *Journal of Computational Mathematics*, vol. 29, no. 5, pp. 574–588, 2011.

[18] A. H. Bhrawy, M. A. Abdelkawy, and F. Mallawi, "An accurate Chebyshev pseudospectral scheme for multi-dimensional parabolic problems with time delays," *Boundary Value Problems*, vol. 1, pp. 1–20, 2015.

[19] I. Aziz and R. Amin, "Numerical solution of a class of delay differential and delay partial differential equations via Haar wavelet," *Applied Mathematical Modelling: Simulation and Computation for Engineering and Environmental Systems*, vol. 40, no. 23-24, pp. 10286–10299, 2016.

[20] H. Tian, "Asymptotic stability analysis of the linear θ-method for linear parabolic differential equations with delay," *Journal of Difference Equations and Applications*, vol. 15, no. 5, pp. 473–487, 2009.

[21] Q. Zhang, M. Chen, Y. Xu, and D. Xu, "Compact θ-method for the generalized delay diffusion equation," *Applied Mathematics and Computation*, vol. 316, pp. 357–369, 2018.

[22] F. Wu, D. Li, J. Wen, and J. Duan, "Stability and convergence of compact finite difference method for parabolic problems with delay," *Applied Mathematics and Computation*, vol. 322, pp. 129–139, 2018.

[23] J. A. Martín, F. Rodríguez, and R. Company, "Analytic solution of mixed problems for the generalized diffusion equation with delay," *Mathematical and Computer Modelling*, vol. 40, no. 3-4, pp. 361–369, 2004.

[24] P. García, M. A. Castro, J. A. Martín, and A. Sirvent, "Numerical solutions of diffusion mathematical models with delay," *Mathematical and Computer Modelling*, vol. 50, no. 5-6, pp. 860–868, 2009.

[25] P. García, M. A. Castro, J. A. Martín, and A. Sirvent, "Convergence of two implicit numerical schemes for diffusion mathematical models with delay," *Mathematical and Computer Modelling*, vol. 52, no. 7-8, pp. 1279–1287, 2010.

[26] L. Blanco-Cocom and E. Ávila-Vales, "Convergence and stability analysis of the θ-method for delayed diffusion mathematical models," *Applied Mathematics and Computation*, vol. 231, pp. 16–25, 2014.

[27] H. Liang, "Convergence and asymptotic stability of Galerkin methods for linear parabolic equations with delays," *Applied Mathematics and Computation*, vol. 264, pp. 160–178, 2015.

[28] C. C. Travis and G. F. Webb, "Existence and stability for partial functional differential equations," *Transactions of the American Mathematical Society*, vol. 200, pp. 395–418, 1974.

[29] M. Z. Liu and M. N. Spijker, "The stability of the θ-methods in the numerical solution of delay differential equations," *IMA Journal of Numerical Analysis (IMAJNA)*, vol. 10, no. 1, pp. 31–48, 1990.

[30] K. W. Morton and D. F. Mayers, *Numerical Solution of Partial Differential Equation*, Cambridge University Press, Cambridge, UK, 2005.

[31] R. D. Richtmyer and K. W. Morton, *Difference Methods for Initial-Value Problems*, Interscience Tracts in Pure and Applied Mathematics, No. 4, Interscience Publishers, New York, NY, USA, 2nd edition, 1967.

[32] Z.-Z. Sun, "Compact difference schemes for heat equation with Neumann boundary conditions," *Numerical Methods for Partial Differential Equations*, vol. 25, no. 6, pp. 1320–1341, 2009.

Existence and Uniqueness of Solutions for BVP of Nonlinear Fractional Differential Equation

Cheng-Min Su, Jian-Ping Sun, and Ya-Hong Zhao

Department of Applied Mathematics, Lanzhou University of Technology, Lanzhou 730050, China

Correspondence should be addressed to Jian-Ping Sun; jpsun@lut.cn

Academic Editor: Julio D. Rossi

In this paper, we study the existence and uniqueness of solutions for the following boundary value problem of nonlinear fractional differential equation: $({}^{C}D^{q}_{0+}u)(t) = f(t, u(t))$, $t \in (0, 1)$, $u(0) = u''(0) = 0$, $({}^{C}D^{\sigma_1}_{0+}u)(1) = \lambda(I^{\sigma_2}_{0+}u)(1)$, where $2 < q < 3$, $0 < \sigma_1 \leq 1$, $\sigma_2 > 0$, and $\lambda \neq \Gamma(2 + \sigma_2)/\Gamma(2 - \sigma_1)$. The main tools used are nonlinear alternative of Leray-Schauder type and Banach contraction principle.

1. Introduction

Fractional calculus has wide applications in many fields of science and engineering, for example, fluid flow, biosciences, rheology, electrical networks, chemical physics, control theory of dynamical systems, and optics and signal processing [1].

Recently, nonlinear fractional differential equations have been discussed under the following boundary conditions (BCs for short):

(1) Integer derivative BCs:

$$u(0) = u(1) = 0,$$
$$u(0) + u'(0) = 0, u(1) + u'(1) = 0,$$
$$u(0) = u'(1) = u''(0) = 0,$$
$$u(0) = 0, u'(0) + u''(0) = 0, u'(1) + u''(1) = 0,$$
$$u(0) = u_0, u'(0) = u_0^*, u''(T) = u_T,$$
$$u(0) = u'(1) = u''(0) = \cdots = u^{(n-1)}(0) = 0;$$

see papers [2–7], respectively.

(2) Integer derivative and integral BCs:

$$\alpha u(0) - \beta u'(0) = \int_0^1 g(s)u(s)ds, \gamma u(1) + \delta u'(1) = \int_0^1 h(s)u(s)ds,$$
$$u(0) = u'(0) = u''(0) = 0, u(1) = \lambda \int_0^\eta u(s)ds;$$

see papers [8, 9], respectively.

(3) Integer and fractional derivative BCs:

$$u(0) = ({}^{C}D^{\sigma_1}_{0+}u)(1) = 0, u''(0) = ({}^{C}D^{\sigma_2}_{0+}u)(1) = 0,$$
$$u(0) = u''(0) = 0, u'(1) = ({}^{C}D^{\sigma}_{0+}u)(1),$$
$$u(0) = u'(0) = 0, u'(1) = ({}^{C}D^{\sigma}_{0+}u)(1),$$
$$u(0) = 0, (D^{\beta}_{0+}u)(1) = \sum_{i=1}^{m-2} \xi_i (D^{\beta}_{0+}u)(\eta_i),$$
$$u(0) = 0, u(1) + (D^{\beta}_{0+}u)(1) = ku(\xi) + l(D^{\beta}_{0+}u)(\eta),$$
$$u(0) = 0, (D^{\beta}_{0+}u)(1) = a(D^{\beta}_{0+}u)(\xi),$$
$$u(0) = u'(0) = \cdots = u^{(n-2)}(0) = 0, (D^{\alpha}_{0+}u)(1) = 0;$$

see papers [10–16], respectively.

(4) Integer derivative and fractional integral BCs:

$$u(0) = \alpha I^{p}_{0+}u(\eta),$$
$$u(0) = 0, u'(1) = I^{\sigma}_{0+}u(1);$$

see papers [17, 18], respectively.

Besides, there are some other BCs involved in fractional differential equations, such as nonlinear BCs; refer to [19, 20].

Motivated greatly by the above-mentioned works, in this paper, we study the following boundary value problem (BVP for short) of nonlinear fractional differential equation with fractional integral BCs as well as integer and fractional derivative

$$\left({}^{C}D^{q}_{0+}u\right)(t) = f(t, u(t)), \quad t \in (0,1),$$
$$u(0) = u''(0) = 0, \tag{1}$$
$$\left({}^{C}D^{\sigma_1}_{0+}u\right)(1) = \lambda\left(I^{\sigma_2}_{0+}u\right)(1),$$

where ${}^{C}D^{q}_{0+}$ and ${}^{C}D^{\sigma_1}_{0+}$ denote the standard Caputo fractional derivatives and $I^{\sigma_2}_{0+}$ denotes the standard Riemann-Liouville fractional integral. Throughout this paper, we always assume that $2 < q < 3$, $0 < \sigma_1 \le 1$, $\sigma_2 > 0$, $\lambda \ne \Gamma(2+\sigma_2)/\Gamma(2-\sigma_1)$, and $f : [0,1] \times \mathbb{R} \to \mathbb{R}$ is continuous.

In order to prove our main results, the following well-known fixed point theorems are needed.

Theorem 1 (nonlinear alternative of Leray-Schauder type [21]). *Let B be a Banach space with $E \subseteq B$ closed and convex. Assume Ω is a relatively open subset of E with $\theta \in \Omega$ and $T : \overline{\Omega} \to E$ is a continuous and compact map. Then either*

(a) *T has a fixed point in $\overline{\Omega}$ or*

(b) *there exists $u \in \partial\Omega$ and $\eta \in (0,1)$ such that $u = \eta T u$.*

Theorem 2 (Banach contraction principle [22]). *Let (X, d) be a complete metric space and $T: X \to X$ be contractive. Then T has a unique fixed point in X.*

2. Preliminaries

In this section, we always assume that $\mathbb{N} = \{1, 2, 3, \ldots\}$, $\alpha, \beta > 0$, and $[\alpha]$ denotes the integer part of α. Now, for the convenience of the reader, we give the definitions of the Riemann-Liouville fractional integrals and fractional derivatives and the Caputo fractional derivatives on a finite interval of the real line, which may be found in [1].

Definition 3. The Riemann-Liouville fractional integrals $I^{\alpha}_{0+}u$ and $I^{\alpha}_{1-}u$ of order α on $[0,1]$ are defined by

$$\left(I^{\alpha}_{0+}u\right)(t) := \frac{1}{\Gamma(\alpha)}\int_0^t \frac{u(s)\,ds}{(t-s)^{1-\alpha}},$$
$$\left(I^{\alpha}_{1-}u\right)(t) := \frac{1}{\Gamma(\alpha)}\int_t^1 \frac{u(s)\,ds}{(s-t)^{1-\alpha}}, \tag{2}$$

respectively.

Definition 4. The Riemann-Liouville fractional derivatives $D^{\alpha}_{0+}u$ and $D^{\alpha}_{1-}u$ of order α on $[0,1]$ are defined by

$$\left(D^{\alpha}_{0+}u\right)(t) := \left(\frac{d}{dt}\right)^n \left(I^{n-\alpha}_{0+}u\right)(t) = \frac{1}{\Gamma(n-\alpha)}\left(\frac{d}{dt}\right)^n \int_0^t \frac{u(s)\,ds}{(t-s)^{\alpha-n+1}},$$

$$\left(D^{\alpha}_{1-}u\right)(t) := \left(-\frac{d}{dt}\right)^n \left(I^{n-\alpha}_{1-}u\right)(t) = \frac{1}{\Gamma(n-\alpha)}\left(-\frac{d}{dt}\right)^n \int_t^1 \frac{u(s)\,ds}{(s-t)^{\alpha-n+1}}, \tag{3}$$

respectively, where $n = [\alpha] + 1$.

Definition 5. Let $D^{\alpha}_{0+}[u(s)](t) \equiv (D^{\alpha}_{0+}u)(t)$ and $D^{\alpha}_{1-}[u(s)](t) \equiv (D^{\alpha}_{1-}u)(t)$ be the Riemann-Liouville fractional derivatives of order α. Then the Caputo fractional derivatives ${}^{C}D^{\alpha}_{0+}u$ and ${}^{C}D^{\alpha}_{1-}u$ of order α on $[0,1]$ are defined by

$$\left({}^{C}D^{\alpha}_{0+}u\right)(t) := \left(D^{\alpha}_{0+}\left[u(s) - \sum_{k=0}^{n-1}\frac{u^{(k)}(0)}{k!}s^k\right]\right)(t),$$
$$\left({}^{C}D^{\alpha}_{1-}u\right)(t) := \left(D^{\alpha}_{1-}\left[u(s) - \sum_{k=0}^{n-1}\frac{u^{(k)}(1)}{k!}(1-s)^k\right]\right)(t), \tag{4}$$

respectively, where

$$n = \begin{cases} [\alpha] + 1, & \alpha \notin \mathbb{N}, \\ \alpha, & \alpha \in \mathbb{N}. \end{cases} \tag{5}$$

Lemma 6 (see [23]). *If $\alpha + \beta > 1$, then the equation $(I^{\alpha}_{0+}I^{\beta}_{0+}u)(t) = (I^{\alpha+\beta}_{0+}u)(t)$, $t \in [0,1]$, is satisfied for $u \in L_1[0,1]$.*

Lemma 7 (see [23]). *Let $\beta > \alpha$. Then the equation $({}^{C}D^{\alpha}_{0+}I^{\beta}_{0+}u)(t) = (I^{\beta-\alpha}_{0+}u)(t)$, $t \in [0,1]$, is satisfied for $u \in C[0,1]$.*

Lemma 8 (see [1]). *Let n be given by (5). Then the following relations hold:*

(1) *For $k \in \{0, 1, 2, \ldots, n-1\}$, ${}^{C}D^{\alpha}_{0+}t^k = 0$.*

(2) *If $\beta > n$, then ${}^{C}D^{\alpha}_{0+}t^{\beta-1} = (\Gamma(\beta)/\Gamma(\beta-\alpha))t^{\beta-\alpha-1}$.*

Lemma 9 (see [1]). *Let n be given by (5) and $u \in C^n[0,1]$. Then*

$$\left(I^{\alpha}_{0+}\,{}^{C}D^{\alpha}_{0+}u\right)(t) = u(t) + c_0 + c_1 t + c_2 t^2 + \cdots + c_{n-1}t^{n-1}, \tag{6}$$

where $c_i \in \mathbb{R}$, $i = 0, 1, \ldots, n-1$.

For any $x \in L_1[0,1]$, we define

$$\|x\|_{L_1} = \int_0^1 |x(t)|\,dt. \tag{7}$$

Lemma 10. *Let $u \in L_1[0,1]$ be nonnegative. Then $(I^{\alpha+1}_{0+}u)(t) \le \|I^{\alpha}_{0+}u\|_{L_1}$, $t \in [0,1]$.*

Proof. For any $t \in [0,1]$, we have

$$\begin{aligned}\left(I_{0+}^{\alpha+1}u\right)(t) &= \frac{1}{\Gamma(\alpha+1)}\int_0^t \frac{u(s)}{(t-s)^{-\alpha}}ds \\ &= \frac{1}{\alpha\Gamma(\alpha)}\int_0^t u(s)(t-s)^{\alpha}ds \\ &= \frac{1}{\Gamma(\alpha)}\int_0^t u(s)\int_s^t (r-s)^{\alpha-1}dr\,ds \\ &= \frac{1}{\Gamma(\alpha)}\int_0^t\int_0^r \frac{u(s)}{(r-s)^{1-\alpha}}ds\,dr \\ &\le \int_0^1 \frac{1}{\Gamma(\alpha)}\int_0^r \frac{u(s)}{(r-s)^{1-\alpha}}ds\,dr \\ &= \int_0^1 \left(I_{0+}^{\alpha}u\right)(r)\,dr = \left\|I_{0+}^{\alpha}u\right\|_{L_1}.\end{aligned} \tag{8}$$

□

3. Main Results

In the remainder of this paper, for any nonnegative function $g \in L_1[0,1]$, we denote

$$\begin{aligned}M_g &= \left\|I_{0+}^{q-1}g\right\|_{L_1} \\ &+ \frac{\Gamma(2+\sigma_2)\Gamma(2-\sigma_1)}{|\lambda\Gamma(2-\sigma_1)-\Gamma(2+\sigma_2)|}\left[|\lambda|\left(I_{0+}^{q+\sigma_2}g\right)(1)\right. \\ &\left.+\left(I_{0+}^{q-\sigma_1}g\right)(1)\right]\end{aligned} \tag{9}$$

and for any $y \in C[0,1]$, we use the norm

$$\|y\|_{\infty} = \max_{t\in[0,1]}|y(t)|. \tag{10}$$

Lemma 11. *Let $y \in C[0,1]$ be a given function. Then the BVP*

$$\begin{aligned}\left({}^{C}D_{0+}^{q}u\right)(t) &= y(t), \quad t \in (0,1), \\ u(0) &= u''(0) = 0, \\ \left({}^{C}D_{0+}^{\sigma_1}u\right)(1) &= \lambda\left(I_{0+}^{\sigma_2}u\right)(1)\end{aligned} \tag{11}$$

has a unique solution

$$u(t) = \int_0^1 G(t,s)\,y(s)\,ds, \quad t \in [0,1], \tag{12}$$

where

$$\begin{aligned}G(t,s) = &-\frac{\Gamma(2+\sigma_2)\Gamma(2-\sigma_1)}{\lambda\Gamma(2-\sigma_1)-\Gamma(2+\sigma_2)} \\ &\cdot t\left[\frac{\lambda(1-s)^{q+\sigma_2-1}}{\Gamma(q+\sigma_2)}-\frac{(1-s)^{q-\sigma_1-1}}{\Gamma(q-\sigma_1)}\right] \\ &+\begin{cases}\dfrac{(t-s)^{q-1}}{\Gamma(q)}, & 0\le s\le t\le 1, \\ 0, & 0\le t\le s\le 1.\end{cases}\end{aligned} \tag{13}$$

Proof. It follows from the equation in (11) and Lemma 9 that

$$u(t) = \left(I_{0+}^{q}y\right)(t) - c_0 - c_1 t - c_2 t^2, \quad t \in [0,1]. \tag{14}$$

So,

$$u'(t) = \left(I_{0+}^{q-1}y\right)(t) - c_1 - 2c_2 t, \quad t \in [0,1], \tag{15}$$

$$u''(t) = \left(I_{0+}^{q-2}y\right)(t) - 2c_2, \quad t \in [0,1]. \tag{16}$$

In view of (14), (16), and the BCs $u(0) = u''(0) = 0$, we get

$$c_0 = c_2 = 0, \tag{17}$$

and so,

$$u(t) = \left(I_{0+}^{q}y\right)(t) - c_1 t, \quad t \in [0,1]. \tag{18}$$

Then, by using Lemmas 6, 7, and 8, we may obtain

$$\begin{aligned}\left({}^{C}D_{0+}^{\sigma_1}u\right)(t) &= \left(I_{0+}^{q-\sigma_1}y\right)(t) - c_1\frac{\Gamma(2)}{\Gamma(2-\sigma_1)}t^{1-\sigma_1}, \\ &\qquad t \in [0,1], \\ \left(I_{0+}^{\sigma_2}u\right)(t) &= \left(I_{0+}^{q+\sigma_2}y\right)(t) - c_1\frac{\Gamma(2)}{\Gamma(2+\sigma_2)}t^{1+\sigma_2}, \\ &\qquad t \in [0,1],\end{aligned} \tag{19}$$

which together with the BC $({}^{C}D_{0+}^{\sigma_1}u)(1) = \lambda(I_{0+}^{\sigma_2}u)(1)$ implies that

$$\begin{aligned}c_1 = &\frac{\Gamma(2+\sigma_2)\Gamma(2-\sigma_1)}{\lambda\Gamma(2-\sigma_1)-\Gamma(2+\sigma_2)}\left[\lambda\left(I_{0+}^{q+\sigma_2}y\right)(1)\right. \\ &\left.-\left(I_{0+}^{q-\sigma_1}y\right)(1)\right].\end{aligned} \tag{20}$$

Therefore, the BVP (11) has a unique solution

$$\begin{aligned}u(t) &= \left(I_{0+}^{q}y\right)(t) \\ &- \frac{\Gamma(2+\sigma_2)\Gamma(2-\sigma_1)}{\lambda\Gamma(2-\sigma_1)-\Gamma(2+\sigma_2)}\left[\lambda\left(I_{0+}^{q+\sigma_2}y\right)(1)\right. \\ &\left.-\left(I_{0+}^{q-\sigma_1}y\right)(1)\right]t = \int_0^t\left\{-\frac{\Gamma(2+\sigma_2)\Gamma(2-\sigma_1)}{\lambda\Gamma(2-\sigma_1)-\Gamma(2+\sigma_2)}\right. \\ &\left.\cdot t\left[\frac{\lambda(1-s)^{q+\sigma_2-1}}{\Gamma(q+\sigma_2)}-\frac{(1-s)^{q-\sigma_1-1}}{\Gamma(q-\sigma_1)}\right]+\frac{(t-s)^{q-1}}{\Gamma(q)}\right\} \\ &\cdot y(s)\,ds + \int_t^1\left\{-\frac{\Gamma(2+\sigma_2)\Gamma(2-\sigma_1)}{\lambda\Gamma(2-\sigma_1)-\Gamma(2+\sigma_2)}\right. \\ &\left.\cdot t\left[\frac{\lambda(1-s)^{q+\sigma_2-1}}{\Gamma(q+\sigma_2)}-\frac{(1-s)^{q-\sigma_1-1}}{\Gamma(q-\sigma_1)}\right]\right\}y(s)\,ds \\ &= \int_0^1 G(t,s)\,y(s)\,ds, \quad t \in [0,1].\end{aligned} \tag{21}$$

□

Lemma 12. *Let $g \in L_1[0,1]$ be nonnegative. Then*

$$\int_0^1 |G(t,s)| g(s)\, ds \le M_g, \quad t \in [0,1]. \tag{22}$$

Proof. In view of Lemma 10, we have

$$\begin{aligned}
\int_0^1 |G(t,s)| g(s)\, ds &= \int_0^t |G(t,s)| g(s)\, ds \\
&+ \int_t^1 |G(t,s)| g(s)\, ds \\
&\le \int_0^t \left\{ \frac{\Gamma(2+\sigma_2)\Gamma(2-\sigma_1)}{|\lambda\Gamma(2-\sigma_1) - \Gamma(2+\sigma_2)|} \right. \\
&\cdot t \left[\frac{|\lambda|(1-s)^{q+\sigma_2-1}}{\Gamma(q+\sigma_2)} + \frac{(1-s)^{q-\sigma_1-1}}{\Gamma(q-\sigma_1)} \right] \\
&\left. + \frac{(t-s)^{q-1}}{\Gamma(q)} \right\} g(s)\, ds \\
&+ \int_t^1 \left\{ \frac{\Gamma(2+\sigma_2)\Gamma(2-\sigma_1)}{|\lambda\Gamma(2-\sigma_1) - \Gamma(2+\sigma_2)|} \right. \\
&\left. \cdot t \left[\frac{|\lambda|(1-s)^{q+\sigma_2-1}}{\Gamma(q+\sigma_2)} + \frac{(1-s)^{q-\sigma_1-1}}{\Gamma(q-\sigma_1)} \right] \right\} g(s)\, ds \\
&= \frac{1}{\Gamma(q)} \int_0^t \frac{g(s)}{(t-s)^{1-q}} ds \\
&+ \frac{\Gamma(2+\sigma_2)\Gamma(2-\sigma_1)}{|\lambda\Gamma(2-\sigma_1) - \Gamma(2+\sigma_2)|} t \left[\frac{|\lambda|}{\Gamma(q+\sigma_2)} \right. \\
&\cdot \int_0^1 \frac{g(s)}{(1-s)^{1-q-\sigma_2}} ds + \frac{1}{\Gamma(q-\sigma_1)} \\
&\left. \cdot \int_0^1 \frac{g(s)}{(1-s)^{1-q+\sigma_1}} ds \right] = (I_{0+}^q g)(t) \\
&+ \frac{\Gamma(2+\sigma_2)\Gamma(2-\sigma_1)}{|\lambda\Gamma(2-\sigma_1) - \Gamma(2+\sigma_2)|} t \left[|\lambda| \left(I_{0+}^{q+\sigma_2} g \right)(1) \right. \\
&\left. + \left(I_{0+}^{q-\sigma_1} g \right)(1) \right] \le \left\| I_{0+}^{q-1} g \right\|_{L_1} \\
&+ \frac{\Gamma(2+\sigma_2)\Gamma(2-\sigma_1)}{|\lambda\Gamma(2-\sigma_1) - \Gamma(2+\sigma_2)|} \left[|\lambda| \left(I_{0+}^{q+\sigma_2} g \right)(1) \right. \\
&\left. + \left(I_{0+}^{q-\sigma_1} g \right)(1) \right] = M_g, \quad t \in [0,1].
\end{aligned} \tag{23}$$

□

Now, we define an operator $T : C[0,1] \to C[0,1]$ by

$$(Tu)(t) = \int_0^1 G(t,s) f(s, u(s))\, ds, \quad t \in [0,1]. \tag{24}$$

Obviously, u is a solution of the BVP (1) if and only if u is a fixed point of T.

Theorem 13. *Assume that $f(t,0) \not\equiv 0$, $t \in (0,1)$, and there exist nonnegative functions $g_1, g_2 \in L_1[0,1]$, nonnegative increasing continuous function ϕ defined on $[0,+\infty)$, and $r > 0$ such that*

$$\left| f(t,x) \right| \le g_1(t) + g_2(t)\phi(|x|), \quad (t,x) \in [0,1] \times \mathbb{R}, \tag{25}$$

$$M_{g_1} + \phi(r) M_{g_2} < r. \tag{26}$$

Then the BVP (1) has one nontrivial solution.

Proof. Let $\Omega = \{u \in C[0,1] : \|u\|_\infty < r\}$. Since $G(t,s)$ and $f(t,x)$ are continuous on $[0,1] \times [0,1]$ and $[0,1] \times \mathbb{R}$, respectively, we may denote

$$L = \max_{(t,s)\in[0,1]\times[0,1]} |G(t,s)|, \tag{27}$$

$$H = \max_{(t,x)\in[0,1]\times[-r,r]} \left| f(t,x) \right|. \tag{28}$$

First, we prove that $T: \overline{\Omega} \to C[0,1]$ is continuous. Suppose that u_n $(n = 1,2,\ldots)$, $u_0 \in \overline{\Omega}$, and $\|u_n - u_0\|_\infty \to 0$ $(n \to \infty)$. Then for any n and $s \in [0,1]$, we have $|u_n(s)| \le r$. This together with (27) and (28) implies that, for any n and $t \in [0,1]$,

$$\left| G(t,s) f(s, u_n(s)) \right| \le LH, \quad s \in [0,1]. \tag{29}$$

By applying Lebesgue dominated convergence theorem, we get

$$\begin{aligned}
\lim_{n\to\infty} (Tu_n)(t) &= \lim_{n\to\infty} \int_0^1 G(t,s) f(s, u_n(s))\, ds \\
&= \int_0^1 G(t,s) f(s, u_0(s))\, ds \\
&= (Tu_0)(t), \quad t \in [0,1],
\end{aligned} \tag{30}$$

which indicates that $T: \overline{\Omega} \to C[0,1]$ is continuous.

Next, we show that $T: \overline{\Omega} \to C[0,1]$ is compact. Assume that K is a subset of $\overline{\Omega}$. Then for any $u \in K$, we have

$$|u(s)| \le r, \quad s \in [0,1]. \tag{31}$$

In what follows, we will prove that $T(K)$ is relatively compact. On the one hand, for any $y \in T(K)$, there exists $u \in K$ such that $y = Tu$, and so, it follows from (27), (28), and (31) that

$$\begin{aligned}
\left| y(t) \right| = |(Tu)(t)| &= \left| \int_0^1 G(t,s) f(s, u(s))\, ds \right| \\
&\le \int_0^1 |G(t,s)| \left| f(s, u(s)) \right| ds \le LH, \\
&\qquad t \in [0,1],
\end{aligned} \tag{32}$$

which shows that $T(K)$ is uniformly bounded. On the other hand, for any $\varepsilon > 0$, since $G(t,s)$ is uniformly continuous on

$[0,1]\times[0,1]$, there exists $\delta > 0$ such that, for any $t_1, t_2 \in [0,1]$ with $|t_1 - t_2| < \delta$,

$$|G(t_1,s) - G(t_2,s)| < \frac{\varepsilon}{H}, \quad s \in [0,1]. \tag{33}$$

For any $y \in T(K)$, there exists $u \in K$ such that $y = Tu$, and so, for any $t_1, t_2 \in [0,1]$ with $|t_1 - t_2| < \delta$, it follows from (28), (31), and (33) that

$$\begin{aligned} |y(t_1) - y(t_2)| &= |(Tu)(t_1) - (Tu)(t_2)| \\ &= \left|\int_0^1 [G(t_1,s) - G(t_2,s)] f(s,u(s))\,ds\right| \\ &\le \int_0^1 |G(t_1,s) - G(t_2,s)|\,|f(s,u(s))|\,ds \\ &\le H\int_0^1 |G(t_1,s) - G(t_2,s)|\,ds < \varepsilon, \end{aligned} \tag{34}$$

which indicates that $T(K)$ is equicontinuous. By Arzela-Ascoli theorem, we know that $T(K)$ is relatively compact. Therefore, $T: \overline{\Omega} \to C[0,1]$ is compact.

Now, we will prove that (a) of Theorem 1 is fulfilled. Suppose on the contrary that (b) of Theorem 1 is satisfied; that is, there exists $u \in \partial\Omega$ and $\eta \in (0,1)$ such that $u = \eta Tu$. Then, in view of (25), (26), and Lemma 12, we have

$$\begin{aligned} |u(t)| &= |\eta(Tu)(t)| \le |(Tu)(t)| \\ &= \left|\int_0^1 G(t,s) f(s,u(s))\,ds\right| \\ &\le \int_0^1 |G(t,s)|\,|f(s,u(s))|\,ds \\ &\le \int_0^1 |G(t,s)|\,[g_1(s) + g_2(s)\phi(|u(s)|)]\,ds \\ &\le \int_0^1 |G(t,s)|\,g_1(s)\,ds \\ &\quad + \phi(r)\int_0^1 |G(t,s)|\,g_2(s)\,ds \\ &\le M_{g_1} + \phi(r) M_{g_2} < r, \quad t \in [0,1], \end{aligned} \tag{35}$$

which shows that

$$\|u\|_\infty < r. \tag{36}$$

This contradicts the fact $u \in \partial\Omega$.

So, it follows from Theorem 1 that T has a fixed point u^*, which is a desired solution of the BVP (1). At the same time, since $f(t,0) \not\equiv 0$, $t \in (0,1)$, we know that the zero function is not a solution of the BVP (1). Therefore, u^* is a nontrivial solution of the BVP (1). □

Theorem 14. *Assume that there exists a nonnegative function $g_3 \in L_1[0,1]$ such that*

$$|f(t,x) - f(t,y)| \le g_3(t)\,|x - y|, \quad t \in [0,1],\ x, y \in \mathbb{R}, \tag{37}$$

$$M_{g_3} < 1. \tag{38}$$

Then the BVP (1) has a unique solution.

Proof. For any $u, v \in C[0,1]$, in view of (37) and Lemma 12, we have

$$\begin{aligned} &|(Tu)(t) - (Tv)(t)| \\ &\quad = \left|\int_0^1 G(t,s)\,[f(s,u(s)) - f(s,v(s))]\,ds\right| \\ &\quad \le \int_0^1 |G(t,s)|\,|f(s,u(s)) - f(s,v(s))|\,ds \\ &\quad \le \int_0^1 |G(t,s)|\,g_3(s)\,|u(s) - v(s)|\,ds \\ &\quad \le \|u - v\|_\infty \int_0^1 |G(t,s)|\,g_3(s)\,ds \le M_{g_3}\|u - v\|_\infty, \\ &\qquad t \in [0,1]. \end{aligned} \tag{39}$$

This indicates that

$$\|Tu - Tv\|_\infty \le M_{g_3}\|u - v\|_\infty, \tag{40}$$

which together with (38) implies that T is contractive. So, it follows from Theorem 2 that T has a unique fixed point, and so, the BVP (1) has a unique solution. □

Example 15. We consider the BVP

$$\begin{gathered} \left({}^{C}D_{0+}^{5/2}u\right)(t) = t - \frac{t}{2}\sqrt{|u(t)|}, \quad t \in (0,1), \\ u(0) = u''(0) = 0, \\ \left({}^{C}D_{0+}^{1/2}u\right)(1) = \frac{1}{2}\left(I_{0+}^{3/2}u\right)(1). \end{gathered} \tag{41}$$

Let $f(t,x) = t - (t/2)\sqrt{|x|}$, $(t,x) \in [0,1]\times\mathbb{R}$. Then $f: [0,1]\times\mathbb{R} \to \mathbb{R}$ is continuous and $f(t,0) \not\equiv 0$, $t \in (0,1)$.

If we choose $g_1(t) = t$, $g_2(t) = t/2$, $t \in [0,1]$, and $\phi(y) = \sqrt{y}$, $y \in [0,+\infty)$, then it is easy to verify that (25) is satisfied.

Since $q = 5/2$, $\sigma_1 = 1/2$, $\sigma_2 = 3/2$, and $\lambda = 1/2$, a direct calculation shows that

$$\begin{gathered} \frac{\Gamma(2+\sigma_2)}{\Gamma(2-\sigma_1)} = \frac{15}{4}, \\ M_{g_1} = \frac{6656 + 4305\pi}{43680\sqrt{\pi}}, \\ M_{g_2} = \frac{6656 + 4305\pi}{87360\sqrt{\pi}}. \end{gathered} \tag{42}$$

If we choose $r = 1$, then (26) is fulfilled.

Therefore, it follows from Theorem 13 that the BVP (41) has one nontrivial solution.

Example 16. We consider the BVP

$$\begin{aligned} &\left({}^{C}D_{0+}^{5/2}u\right)(t) \\ &\quad = \frac{t}{\pi}\left\{u(t)\arctan u(t) - \frac{1}{2}\ln\left[1+u^{2}(t)\right]\right\}, \\ &\qquad\qquad t \in (0,1), \quad (43) \\ &u(0) = u''(0) = 0, \\ &\left({}^{C}D_{0+}^{1/2}u\right)(1) = \frac{1}{2}\left(I_{0+}^{3/2}u\right)(1). \end{aligned}$$

Let $f(t,x) = (t/\pi)[x\arctan x - (1/2)\ln(1+x^2)]$, $(t,x) \in [0,1]\times\mathbb{R}$. Then $f : [0,1]\times\mathbb{R} \to \mathbb{R}$ is continuous.

If we choose $g_3(t) = t/2$, $t \in [0,1]$, then we may assert that (37) is satisfied. In fact, for any $t \in [0,1]$, if $x = y$, then (37) is obvious. When $x \neq y$, we may suppose that $x < y$. In this case, by Lagrange mean value theorem, there exists $\xi \in (x,y)$ such that, for any $t \in [0,1]$,

$$\begin{aligned} &|f(t,x) - f(t,y)| = \frac{t}{\pi}\left|x\arctan x - \frac{1}{2}\ln\left(1+x^2\right)\right. \\ &\quad \left. - y\arctan y + \frac{1}{2}\ln\left(1+y^2\right)\right| = \frac{t}{\pi}\left|\arctan\xi\right|\left|x\right. \qquad (44) \\ &\quad \left. - y\right| \le g_3(t)\left|x-y\right|; \end{aligned}$$

that is, (37) is satisfied.

On the other hand, in view of $M_{g_3} = M_{g_2} = (6656 + 4305\pi)/87360\sqrt{\pi}$, we know that (38) is fulfilled.

Therefore, it follows from Theorem 14 that the BVP (43) has a unique solution.

Competing Interests

The authors declare that there is no conflict of interests regarding the publication of this paper.

Acknowledgments

This paper is supported by the National Natural Science Foundation of China (11661049).

References

[1] A. A. Kilbas, H. M. Srivastava, and J. J. Trujillo, *Theory and Applications of Fractional Differential Equations*, vol. 204 of *North-Holland Mathematics Studies*, Elsevier Science B.V., Amsterdam, The Netherlands, 2006.

[2] D. Jiang and C. Yuan, "The positive properties of the Green function for Dirichlet-type boundary value problems of nonlinear fractional differential equations and its application," *Nonlinear Analysis: Theory, Methods & Applications*, vol. 72, no. 2, pp. 710–719, 2010.

[3] S. Zhang, "Positive solutions for boundary-value problems of nonlinear fractional differential equations," *Electronic Journal of Differential Equations*, vol. 2006, no. 36, pp. 1–12, 2006.

[4] Z. Bai and T. Qiu, "Existence of positive solution for singular fractional differential equation," *Applied Mathematics and Computation*, vol. 215, no. 7, pp. 2761–2767, 2009.

[5] J.-R. Yue, J.-P. Sun, and S. Zhang, "Existence of positive solution for BVP of nonlinear fractional differential equation," *Discrete Dynamics in Nature and Society*, vol. 2015, Article ID 736108, 6 pages, 2015.

[6] R. P. Agarwal, M. Benchohra, and S. Hamani, "Boundary value problems for fractional differential equations," *Georgian Mathematical Journal*, vol. 16, no. 3, pp. 401–411, 2009.

[7] K. Zhao and P. Gong, "Existence of positive solutions for a class of higher-order Caputo fractional differential equation," *Qualitative Theory of Dynamical Systems*, vol. 14, no. 1, pp. 157–171, 2015.

[8] W. Yang, "Positive solutions for nonlinear Caputo fractional differential equations with integral boundary conditions," *Journal of Applied Mathematics and Computing*, vol. 44, no. 1-2, pp. 39–59, 2014.

[9] X. Zhang, L. Wang, and Q. Sun, "Existence of positive solutions for a class of nonlinear fractional differential equations with integral boundary conditions and a parameter," *Applied Mathematics and Computation*, vol. 226, pp. 708–718, 2014.

[10] A. Guezane-Lakoud and R. Khaldi, "Existence results for a fractional boundary value problem with fractional Lidstone conditions," *Journal of Applied Mathematics and Computing*, vol. 49, no. 1-2, pp. 261–268, 2015.

[11] A. Guezane-Lakoud and S. Bensebaa, "Solvability of a fractional boundary value problem with fractional derivative condition," *Arabian Journal of Mathematics*, vol. 3, no. 1, pp. 39–48, 2014.

[12] R. Li, "Existence of solutions for nonlinear singular fractional differential equations with fractional derivative condition," *Advances in Difference Equations*, vol. 2014, article 292, 2014.

[13] Z.-W. Lv, "Positive solutions of m-point boundary value problems for fractional differential equations," *Advances in Difference Equations*, vol. 2011, Article ID 571804, 2011.

[14] Y. Ji, Y. Guo, J. Qiu, and L. Yang, "Existence of positive solutions for a boundary value problem of nonlinear fractional differential equations," *Advances in Difference Equations*, vol. 2015, article 13, 2015.

[15] C. F. Li, X. N. Luo, and Y. Zhou, "Existence of positive solutions of the boundary value problem for nonlinear fractional differential equations," *Computers & Mathematics with Applications*, vol. 59, no. 3, pp. 1363–1375, 2010.

[16] J. Xu, Z. Wei, and W. Dong, "Uniqueness of positive solutions for a class of fractional boundary value problems," *Applied Mathematics Letters*, vol. 25, no. 3, pp. 590–593, 2012.

[17] S. K. Ntouyas, "Existence results for first order boundary value problems for fractional differential equations and inclusions with fractional integral boundary conditions," *Journal of Fractional Calculus and Applications*, vol. 3, no. 9, pp. 1–14, 2012.

[18] A. Guezane-Lakoud and R. Khaldi, "Solvability of a fractional boundary value problem with fractional integral condition," *Nonlinear Analysis: Theory, Methods & Applications*, vol. 75, no. 4, pp. 2692–2700, 2012.

[19] W. Feng, S. Sun, X. Li, and M. Xu, "Positive solutions to fractional boundary value problems with nonlinear boundary conditions," *Boundary Value Problems*, 2014:225, 15 pages, 2014.

[20] W. Xie, J. Xiao, and Z. Luo, "Existence of extremal solutions for nonlinear fractional differential equation with nonlinear boundary conditions," *Applied Mathematics Letters*, vol. 41, pp. 46–51, 2015.

[21] D. Guo and V. Lakshmikantham, *Nonlinear Problems in Abstract Cones*, Academic Press, San Diego, Calif, USA, 1988.

[22] A. Granas and J. Dugundji, *Fixed Point Theory*, Springer Monographs in Mathematics, Springer, New York, NY, USA, 2003.

[23] S. G. Samko, A. A. Kilbas, and O. I. Marichev, *Fractional Integrals and Derivatives: Theory and Applications*, Gordon and Breach Science, Yverdon, Switzerland, 1993.

3

Solving Nonlinear Fourth-Order Boundary Value Problems Using a Numerical Approach: $(m+1)$th-Step Block Method

Oluwaseun Adeyeye and Zurni Omar

Mathematics Department, School of Quantitative Sciences, Universiti Utara Malaysia, Sintok, Kedah, Malaysia

Correspondence should be addressed to Oluwaseun Adeyeye; adeyeye_oluwaseun@ahsgs.uum.edu.my

Academic Editor: Davood D. Ganji

Nonlinear boundary value problems (BVPs) are more tedious to solve than their linear counterparts. This is observed in the extra computation required when determining the missing conditions in transforming BVPs to initial value problems. Although a number of numerical approaches are already existent in literature to solve nonlinear BVPs, this article presents a new block method with improved accuracy to solve nonlinear BVPs. A $(m+1)$th-step block method is developed using a modified Taylor series approach to directly solve fourth-order nonlinear boundary value problems (BVPs) where m is the order of the differential equation under consideration. The schemes obtained were combined to simultaneously produce solution to the fourth-order nonlinear BVPs at $m+1$ points iteratively. The derived block method showed improved accuracy in comparison to previously existing authors when solving the same problems. In addition, the suitability of the $(m+1)$th-step block method was displayed in the solution for magnetohydrodynamic squeezing flow in porous medium.

1. Introduction

Boundary value problems (BVPs) arise in several branches of science ranging from physical sciences to engineering. There has been commendable progress in solving problems associated with nonlinear ordinary differential equations (ODEs) involving boundary conditions in recent years. These ODEs are sometimes needed to fulfil certain boundary conditions at more than one point of the independent variable which will result in the problem known as two-point boundary value problem. Two-point nonlinear BVPs often cannot be solved by analytical methods and thus finding approximate solutions for these problems becomes essential.

This article considers the following special type of nonlinear boundary value problem:

$$y^{iv} = f(x, y), \tag{1}$$

with the boundary conditions

$$\begin{aligned} y(a) &= \alpha_1, \\ y'(a) &= \alpha_2, \\ y(b) &= \beta_1, \\ y'(b) &= \beta_2, \end{aligned} \tag{2}$$

where f is a continuous function on $[a, b]$ and the parameters α_i and β_i for $i = 1, 2$ are constants.

A variety of methods have been introduced to solve (1) such as shooting methods, splines methods, finite difference methods, finite element methods, differential transform methods, and collocation methods [1–4]. Recently, the adoption of various families of linear multistep method (LMM) for numerically approximating higher order ODEs has been proposed. However, some LMMs cannot directly solve these higher order ODEs and thus require reduction to a system of first-order ODEs. In some cases, the accuracy of LMMs is low such as the case of predictor-corrector methods which incur high computational rigour. This computational rigour involves the derivation of separate predictors for each grid point of the LMM as seen in the work of Kayode and Adeyeye [5] and Kayode and Obarhua [6]. These drawbacks caused for the introduction of block methods which were first proposed by Milne [7] as a means to obtain starting values for predictor-corrector methods. This concept was also further explored by Sarafyan [8].

Block methods differ from alternate approaches such as differential transform method and collocation method. This is because the formulation of block methods is an evaluation of the linear multistep method at different grid points to generate a family of methods that can be applied to produce approximate solutions of ODEs at each grid point simultaneously. This advantage was mentioned by Lambert [9] among other advantages which include being self-starting, permitting easy change of step-length, and being less expensive in terms of function evaluations. Block methods also yield better accuracy when applied to numerical problems.

This article introduces a $(m + 1)$th-step block method where m is the order of the differential equation. The block method is developed using a modification of the conventional Taylor series expansions approach by Lambert [9]. This derivation is shown in Section 2 of this article while Section 3 considers certain numerical examples and their results to show the accuracy of the block method.

2. Methodology

Lambert [9] highlighted three main approaches for developing LMMs. These include interpolation, numerical integration, and Taylor series expansions. This article adopts the Taylor series expansion approach for LMMs to develop the block method. However, certain modifications were introduced since Lambert [9] focused on first-order methods whereas this article develops a block method for fourth-order ODEs. Therefore, the approach was made suitable to develop block methods and not LMMs alone, hence the name modified Taylor series approach.

2.1. Derivation of the $(m + 1)$th-Step Block Method Using Modified Taylor Series Approach. The algorithm described below is used to show the steps involved in deriving the $(m + 1)$th-step block method using the modified Taylor series approach where m is the order of the differential equation.

Algorithm 1.

Start.

Step 1. Obtain the coefficients of the initial multistep scheme:

$$y_{n+(m+1)} = \sum_{j=(m-4)}^{m} \alpha_{j_v} y_{n+j} + \sum_{j=(m-4)}^{(m+1)} \beta_j f_{n+j}, \quad (3)$$

$v = 1, 2, \ldots, m.$

Step 2. Obtain the coefficients of the additional schemes:

$$\begin{aligned} y_{n+j_{m+1}} &= \sum_{j=(m-4)}^{m} \alpha_{j_v} y_{n+j} + \sum_{j=(m-4)}^{(m+1)} \beta_j f_{n+j}, \\ y_{n+j_{m+2}} &= \sum_{j=(m-4)}^{m} \alpha_{j_v} y_{n+j} + \sum_{j=(m-4)}^{(m+1)} \beta_j f_{n+j}, \\ &\vdots \\ y_{n+j_k} &= \sum_{j=(m-4)}^{m} \alpha_{j_v} y_{n+j} + \sum_{j=(m-4)}^{(m+1)} \beta_j f_{n+j}. \end{aligned} \quad (4)$$

Step 3. Derive the coefficients of the (1st, 2nd, ..., $(m - 1)$th) derivative schemes:

$$\begin{aligned} y_n^{(\varrho)} &= \sum_{j=(m-4)}^{m} \alpha_{j_v} y_{n+j} + \sum_{j=(m-4)}^{(m+1)} \beta_j f_{n+j}, \\ y_{n+1}^{(\varrho)} &= \sum_{j=(m-4)}^{m} \alpha_{j_v} y_{n+j} + \sum_{j=(m-4)}^{(m+1)} \beta_j f_{n+j}, \\ &\vdots \\ y_{n+(m+1)}^{(\varrho)} &= \sum_{j=(m-4)}^{m} \alpha_{j_v} y_{n+j} + \sum_{j=(m-4)}^{(m+1)} \beta_j f_{n+j}, \end{aligned} \quad (5)$$

where $\varrho = 1, 2, \ldots, m - 1$.

Step 4. Combine schemes obtained in Steps 1, 2, and 3 above to form a system of equations with matrix form equivalent $Ax = B$ where $x = (N_0, N_1, N_2, \ldots, N_{m-1})^T$ and $N_0 = (y_{n+(m-3)}, y_{n+(m-2)}, \ldots, y_{n+(m+1)})^T$, $N_1 = (y'_{n+(m-3)}, y'_{n+(m-2)}, \ldots, y'_{n+(m+1)})^T$, $N_2 = (y''_{n+(m-3)}, y''_{n+(m-2)}, \ldots, y''_{n+(m+1)})^T$, $N_{m-1} = (y^{m-1}_{n+(m-3)}, y^{m-1}_{n+(m-2)}, \ldots, y^{m-1}_{n+(m+1)})^T$

Step 5. Adopt matrix inverse approach to system of equations in Step 4 to obtain the expected block method.

Stop.

In Algorithm 1, $y_{n+a} = y(x_{n+a}) = y(x_n + ah)$, $f_{n+j} = f(x_{n+j}, y_{n+j})$, and α_{j_v} and β_j are constants with v defined in Step 1 of Algorithm 1.

Note that in Step 1 of Algorithm 1, the expected α_{j_v} are $\alpha_{j_1}, \alpha_{j_2}, \ldots, \alpha_{j_m}$. The α_{j_v}-values can take ${}^kC_v$ forms and j_v-values not chosen will be used as evaluation points when developing the additional methods in Step 2.

Steps 1–3 of Algorithm 1 require expanding individual terms using Taylor Series expansion such as

$$
\begin{aligned}
y_n &= y(x_n),\\
y_{n+1} &= y(x_n+h) = y(x_n) + hy'(x_n) + \frac{h^2}{2!}y''(x_n) + \cdots,\\
&\vdots\\
y_{n+(k-1)} &= y[x_n + ((k-1)h)] = y(x_n) + ((k-1)h)\,y'(x_n) + \frac{((k-1)h)^2}{2!}y''(x_n) + \cdots,\\
y_{n+k} &= y[x_n+kh] = y(x_n) + (kh)\,y'(x_n) + \frac{(kh)^2}{2!}y''(x_n) + \ldots,\\
f_n &= y^{(m)}(x_n),\\
f_{n+1} &= y^{(m)}(x_n+h) = y^{(m)}(x_n) + hy^{(m+1)}(x_n) + \frac{h^2}{2!}y^{(m+2)}(x_n) + \cdots,\\
&\vdots\\
f_{n+(k-1)} &= y^{(m)}[x_n + ((k-1)h)] = y^{(m)}(x_n) + ((k-1)h)\,y^{(m+1)}(x_n) + \frac{((k-1)h)^2}{2!}y^{(m+2)}(x_n) + \cdots,\\
f_{n+k} &= y^{(m)}[x_n+kh] = y^{(m)}(x_n) + (kh)\,y^{(m+1)}(x_n) + \frac{(kh)^2}{2!}y^{(m+2)}(x_n) + \cdots.
\end{aligned}
\tag{6}
$$

Substituting these expansions in individual equations and equating coefficients of $y^{(m)}x_n$ presents the resulting expressions in matrix form

$$A\mathbf{x} = \mathbf{B}, \tag{7}$$

where

$$
A = \begin{pmatrix}
1 & 1 & 1 & \cdots & 1 & 0 & 0 & 0 & \cdots & 0\\
0 & h & 2h & \cdots & (k-1)h & \vdots & \vdots & \vdots & \cdots & \vdots\\
0 & \frac{(h)^2}{2!} & \frac{(2h)^2}{2!} & \cdots & \frac{((k-1)h)^2}{2!} & \vdots & \vdots & \vdots & \cdots & \vdots\\
\vdots & \vdots & \vdots & \cdots & \vdots & \vdots & \vdots & \vdots & \cdots & \vdots\\
0 & \frac{(h)^m}{m!} & \frac{(2h)^m}{m!} & \cdots & \frac{((k-1)h)^m}{m!} & 1 & 1 & 1 & \cdots & 1\\
0 & \frac{(h)^{(m+1)}}{(m+1)!} & \frac{(2h)^{(m+1)}}{(m+1)!} & \cdots & \frac{((k-1)h)^{(m+1)}}{(m+1)!} & 0 & h & 2h & \cdots & kh\\
0 & . & . & \cdots & . & 0 & \frac{(h)^2}{2!} & \frac{(2h)^2}{2!} & \cdots & \frac{(kh)^2}{2!}\\
0 & . & . & \cdots & . & 0 & . & . & \cdots & .\\
0 & . & . & \cdots & . & 0 & . & . & \cdots & .\\
0 & . & . & \cdots & . & 0 & . & . & \cdots & .\\
0 & \frac{(h)^{(2k)}}{(2k)!} & \frac{(2h)^{(2k)}}{(2k)!} & \cdots & \frac{((k-1)h)^{(2k)}}{(2k)!} & 0 & \frac{(h)^{(2k-m)}}{(2k-m)!} & \frac{(2h)^{(2k-m)}}{(2k-m)!} & \cdots & \frac{(kh)^{(2k-m)}}{(2k-m)!}
\end{pmatrix},
$$

$$\mathbf{x} = (\alpha_0, \alpha_1, \alpha_2, \ldots, \alpha_{k-1}, \beta_0, \beta_1, \beta_2, \ldots, \beta_k)^T,$$

$$\mathbf{B} = \left(1, kh, \frac{(kh)^2}{2!}, \frac{(kh)^3}{3!}, \frac{(kh)^4}{4!}, \ldots, \frac{(kh)^{(2k)}}{(2k)!}\right)^T. \tag{8}$$

Note that Algorithm 1 will not successfully obtain the required block method if matrix A is singular. Thus, the nonsingularity of the resulting matrices is discussed.

2.2. Nonsingularity of Resulting Matrices. The matrix A in (7) is a square matrix with $\det(A) \neq 0$ which follows from the theorems below.

Theorem 2. *Suppose that A is a square matrix with a row where every entry is zero, or a column where every entry is zero. Then* $\det(A) = 0$.

Theorem 3. *Suppose that A is a square matrix with two equal rows, or two equal columns. Then* $\det(A) = 0$.

With respect to Theorem 2, since the matrix A does not have a row or column where every entry is zero, then its inverse exists. On the other hand, matrix A has no equal rows or columns which further affirms that its inverse exists.

Theorems 2 and 3 are sufficient conditions to show that the inverse of the resulting matrix will always exist. In addition, the case of linear dependency is considered as defined in the following theorems.

Theorem 4. *If matrix A has linearly dependent columns, then* $\det(A) = 0$.

Theorem 5. *The rank of a matrix A equals the maximum number of linearly independent column vectors. The matrix A has the same number of linearly independent row vectors as it has linearly independent column vectors*

Thus, Theorem 5 is tested for resulting matrices obtained in developing the $(m + 1)$th-step block method to show that A^{-1} exists.

2.3. Specification of the $(m + 1)$th-Step Block Method. Following Algorithm 1, the specification of the $(m + 1)$th-step block method is as follows. From Step 1, the initial multistep scheme for the $(m + 1)$th-step block method in terms of m is

$$\begin{aligned} y_{n+m+1} &= \alpha_{m-4} y_{n+m-4} + \alpha_{m-3} y_{n+m-3} + \alpha_{m-2} y_{n+m-2} \\ &\quad + \alpha_{m-1} y_{n+m-1} + \alpha_m y_{n+m} + \beta_{m-4} f_{n+m-4} \\ &\quad + \beta_{m-3} f_{n+m-3} + \beta_{m-2} f_{n+m-2} + \beta_{m-1} f_{n+m-1} \\ &\quad + \beta_m f_{n+m} + \beta_{m+1} f_{n+m+1}. \end{aligned} \tag{9}$$

Now, considering (9), the individual terms are expanded using Taylor series expansion as defined in (6). The resulting expansions are substituted back in (9) and rewritten in matrix form $A\mathbf{x} = \mathbf{B}$, where

$$A = \begin{pmatrix}
1 & 1 & 1 & 1 & 1 & 0 & 0 & 0 & 0 & 0 & 0 \\
(m-4)h & (m-3)h & (m-2)h & (m-1)h & (m)h & 0 & 0 & 0 & 0 & 0 & 0 \\
\frac{(m-4)h^2}{2!} & \frac{(m-3)h^2}{2!} & \frac{(m-2)h^2}{2!} & \frac{(m-1)h^2}{2!} & \frac{(m)h^2}{2!} & 0 & 0 & 0 & 0 & 0 & 0 \\
\frac{(m-4)h^3}{3!} & \frac{(m-3)h^3}{3!} & \frac{(m-2)h^3}{3!} & \frac{(m-1)h^3}{3!} & \frac{(m)h^3}{3!} & 0 & 0 & 0 & 0 & 0 & 0 \\
\frac{(m-4)h^4}{4!} & \frac{(m-3)h^4}{4!} & \frac{(m-2)h^4}{4!} & \frac{(m-1)h^4}{4!} & \frac{(m)h^4}{4!} & 1 & 1 & 1 & 1 & 1 & 1 \\
\frac{(m-4)h^5}{5!} & \frac{(m-3)h^5}{5!} & \frac{(m-2)h^5}{5!} & \frac{(m-1)h^5}{5!} & \frac{(m)h^5}{5!} & (m-4)h & (m-3)h & (m-2)h & (m-1)h & (m)h & (m+1)h \\
\frac{(m-4)h^6}{6!} & \frac{(m-3)h^6}{6!} & \frac{(m-2)h^6}{6!} & \frac{(m-1)h^6}{6!} & \frac{(m)h^6}{6!} & \frac{(m-4)h^2}{2!} & \frac{(m-3)h^2}{2!} & \frac{(m-2)h^2}{2!} & \frac{(m-1)h^2}{2!} & \frac{(m)h^2}{2!} & \frac{(m+1)h^2}{2!} \\
\frac{(m-4)h^7}{7!} & \frac{(m-3)h^7}{7!} & \frac{(m-2)h^7}{7!} & \frac{(m-1)h^7}{7!} & \frac{(m)h^7}{7!} & \frac{(m-4)h^3}{3!} & \frac{(m-3)h^3}{3!} & \frac{(m-2)h^3}{3!} & \frac{(m-1)h^3}{3!} & \frac{(m)h^3}{3!} & \frac{(m+1)h^3}{3!} \\
\frac{(m-4)h^8}{8!} & \frac{(m-3)h^8}{8!} & \frac{(m-2)h^8}{8!} & \frac{(m-1)h^8}{8!} & \frac{(m)h^8}{8!} & \frac{(m-4)h^4}{4!} & \frac{(m-3)h^4}{4!} & \frac{(m-2)h^4}{4!} & \frac{(m-1)h^4}{4!} & \frac{(m)h^4}{4!} & \frac{(m+1)h^4}{4!} \\
\frac{(m-4)h^9}{9!} & \frac{(m-3)h^9}{9!} & \frac{(m-2)h^9}{9!} & \frac{(m-1)h^9}{9!} & \frac{(m)h^9}{9!} & \frac{(m-4)h^5}{5!} & \frac{(m-3)h^5}{5!} & \frac{(m-2)h^5}{5!} & \frac{(m-1)h^5}{5!} & \frac{(m)h^5}{5!} & \frac{(m+1)h^5}{5!} \\
\frac{(m-4)h^{10}}{10!} & \frac{(m-3)h^{10}}{10!} & \frac{(m-2)h^{10}}{10!} & \frac{(m-1)h^{10}}{10!} & \frac{(m)h^{10}}{10!} & \frac{(m-4)h^6}{6!} & \frac{(m-3)h^6}{6!} & \frac{(m-2)h^6}{6!} & \frac{(m-1)h^6}{6!} & \frac{(m)h^6}{6!} & \frac{(m+1)h^6}{6!}
\end{pmatrix},$$

$$\mathbf{x} = \left(\alpha_{m-4}, \alpha_{m-3}, \alpha_{m-2}, \alpha_{m-1}, \alpha_m, \beta_{m-4}, \beta_{m-3}, \beta_{m-2}, \beta_{m-1}, \beta_m, \beta_{m+1}\right)^T,$$

$$\mathbf{B} = \left(1, (m+1)h, \frac{((m+1)h)^2}{2!}, \frac{((m+1)h)^3}{3!}, \frac{((m+1)h)^4}{4!}, \frac{((m+1)h)^5}{5!}, \frac{((m+1)h)^6}{6!}, \frac{((m+1)h)^7}{7!}, \frac{((m+1)h)^8}{8!}, \frac{((m+1)h)^9}{9!}, \frac{((m+1)h)^{10}}{10!}\right)^T, \tag{10}$$

where matrix A has rank $= 11$ which implies that there are no linearly dependent columns or rows and the inverse exists. This follows from the theorems in Section 2.2 showing that the matrix is nonsingular. Therefore, the scheme in (9) is obtained using matrix inverse method and substituting the value of m as

$$\begin{aligned} y_{n+m+1} &= y_{n+m-4} - 5y_{n+m-3} + 10y_{n+m-2} - 10y_{n+m-1} \\ &\quad + 5y_{n+m} + \frac{h^4}{720}\left(f_{n+m-4} - 125f_{n+m-3} - 350f_{n+m-2}\right. \\ &\quad \left. + 350f_{n+m-1} + 125f_{n+m} + f_{n+m+1}\right). \end{aligned} \tag{11}$$

Following the subsequent steps of Algorithm 1, the specification of the $(m+1)$th-step block method is as follows:

$$
\begin{aligned}
1814400 y_{n+m-3} &= 1814400\left(y_n + h y_n' + \frac{h^2}{2} y_n'' + \frac{h^3}{6} y_n'''\right) + h^4 (49126 f_n + 49045 f_{n+m-3} \\
&\quad - 40160 f_{n+m-2} + 25430 f_{n+m-1} - 9310 f_{n+m} \\
&\quad + 1469 f_{n+m+1}), \\
14175 y_{n+m-2} &= 14175\left(y_n + 2h y_n' + 2h^2 y_n'' + \frac{4h^3}{3} y_n'''\right) + h^4 (4264 f_n + 7960 f_{n+m-3} \\
&\quad - 4910 f_{n+m-2} + 3080 f_{n+m-1} - 1120 f_{n+m} \\
&\quad + 176 f_{n+m+1}), \\
22400 y_{n+m-1} &= 22400\left(y_n + 3h y_n' + \frac{9h^2}{2} y_n'' + \frac{9h^3}{2} y_n'''\right) + h^4 (25488 f_n + 63315 f_{n+m-3} \\
&\quad - 26460 f_{n+m-2} + 19170 f_{n+m-1} - 7020 f_{n+m} \\
&\quad + 1107 f_{n+m+1}), \\
14175 y_{n+m} &= 14175\left(y_n + 4h y_n' + 8h^2 y_n'' + \frac{32h^3}{3} y_n'''\right) + h^4 (40448 f_n + 116480 f_{n+m-3} \\
&\quad - 29440 f_{n+m-2} + 33280 f_{n+m-1} - 11360 f_{n+m} \\
&\quad + 1792 f_{n+m+1}), \\
y_{n+m+1} &= y_n + 5h y_n' + \frac{25h^2}{2} y_n'' + \frac{125h^3}{6} y_n''' \\
&\quad + \frac{h^4}{72576} (418250 f_n + 1315625 f_{n+m-3} \\
&\quad - 175000 f_{n+m-2} + 418750 f_{n+m-1} - 106250 f_{n+m} \\
&\quad + 18625 f_{n+m+1}),
\end{aligned}
\tag{12}
$$

$$
\begin{aligned}
40320 y_{n+m-3}' &= 40320\left(y_n' + h y_n'' + \frac{h^2}{2} y_n'''\right) \\
&\quad + h^3 (3929 f_n + 4975 f_{n+m-3} - 3862 f_{n+m-2} \\
&\quad + 2422 f_{n+m-1} - 883 f_{n+m} + 139 f_{n+m+1}), \\
630 y_{n+m-2}' &= 630\left(y_n' + 2h y_n'' + 2h^2 y_n'''\right) + h^3 (317 f_n \\
&\quad + 734 f_{n+m-3} - 380 f_{n+m-2} + 244 f_{n+m-1} - 89 f_{n+m} \\
&\quad + 14 f_{n+m+1}), \\
4480 y_{n+m-1}' &= 4480\left(y_n' + 3h y_n'' + \frac{9h^2}{2} y_n'''\right) \\
&\quad + h^3 (5481 f_n + 16119 f_{n+m-3} - 4374 f_{n+m-2} \\
&\quad + 4230 f_{n+m-1} - 1539 f_{n+m} + 243 f_{n+m+1}), \\
315 y_{n+m}' &= 315\left(y_n' + 4h y_n'' + 8h^2 y_n'''\right) + h^3 (712 f_n \\
&\quad + 2336 f_{n+m-3} - 224 f_{n+m-2} + 704 f_{n+m-1} \\
&\quad - 200 f_{n+m} + 32 f_{n+m+1}), \\
8064 y_{n+m+1}' &= 8064\left(y_n' + 5h y_n'' + \frac{25h^2}{2} y_n'''\right) \\
&\quad + h^3 (29125 f_n + 101875 f_{n+m-3} + 1250 f_{n+m-2} \\
&\quad + 38750 f_{n+m-1} - 4375 f_{n+m} + 1375 f_{n+m+1}),
\end{aligned}
\tag{13}
$$

$$
\begin{aligned}
10080 y_{n+m-3}'' &= 10080\left(y_n'' + h y_n'''\right) + h^2 (2462 f_n \\
&\quad + 4315 f_{n+m-3} - 3044 f_{n+m-2} + 1882 f_{n+m-1} \\
&\quad - 682 f_{n+m} + 107 f_{n+m+1}), \\
630 y_{n+m-2}'' &= 630\left(y_n'' + 2h y_n'''\right) + h^2 (355 f_n \\
&\quad + 1088 f_{n+m-3} - 370 f_{n+m-2} + 272 f_{n+m-1} \\
&\quad - 101 f_{n+m} + 16 f_{n+m+1}), \\
1120 y_{n+m-1}'' &= 1120\left(y_n'' + 3h y_n'''\right) + h^2 (984 f_n \\
&\quad + 3501 f_{n+m-3} - 72 f_{n+m-2} + 870 f_{n+m-1} - 288 f_{n+m} \\
&\quad + 45 f_{n+m+1}), \\
315 y_{n+m}'' &= 315\left(y_n'' + 4h y_n'''\right) + h^2 (376 f_n \\
&\quad + 1424 f_{n+m-3} + 176 f_{n+m-2} + 608 f_{n+m-1} - 80 f_{n+m} \\
&\quad + 16 f_{n+m+1}), \\
2016 y_{n+m+1}'' &= 2016\left(y_n'' + 5h y_n'''\right) + h^2 (3050 f_n \\
&\quad + 11875 f_{n+m-3} + 2500 f_{n+m-2} + 6250 f_{n+m-1} \\
&\quad + 1250 f_{n+m} + 275 f_{n+m+1}),
\end{aligned}
\tag{14}
$$

$$
\begin{aligned}
1440 y_{n+1}''' &= 1440 y_n''' + h (475 f_n + 1427 f_{n+m-3} \\
&\quad - 798 f_{n+m-2} + 482 f_{n+m-1} - 173 f_{n+m} + 27 f_{n+m+1}), \\
90 y_{n+2}''' &= 90 y_n''' + h (28 f_n + 129 f_{n+m-3} + 14 f_{n+m-2} \\
&\quad + 14 f_{n+m-1} - 6 f_{n+m} + f_{n+m+1}),
\end{aligned}
$$

$$160y'''_{n+3} = 160y'''_n + h(51f_n + 219f_{n+m-3} + 114f_{n+m-2} + 114f_{n+m-1} - 21f_{n+m} + 3f_{n+m+1}),$$
$$45y'''_{n+4} = 45y'''_n + h(14f_n + 64f_{n+m-3} + 24f_{n+m-2} + 64f_{n+m-1} + 14f_{n+m}),$$
$$288y'''_{n+5} = 288y'''_n + h(95f_n + 375f_{n+m-3} + 250f_{n+m-2} + 250f_{n+m-1} + 375f_{n+m} + 95f_{n+m+1}). \quad (15)$$

2.4. Order and Stability Properties of the $(m+1)$th-Step Block Method. To ensure convergence of the block method, its consistency and zero-stability need to be investigated. This follows from Fatunla (1988) which states that a linear multistep method is convergent iff it is consistent and zero-stable.

Starting with the consistency property, a linear multistep method is *consistent* if it has order $p \geq 1$. Thus, the order of the $(m+1)$th-step block method is investigated.

With reference to the definition in Lambert [9], Henrici [11], and Butcher [12], the order and error constant of the $(m+1)$th-step block method follow Definition 6.

Definition 6. The linear operator associated with LMM is defined as

$$L[y(x);h] = \sum_{j=0}^{k} [\alpha_j y_{n+j} - \beta_j f_{n+j}]. \quad (16)$$

On expanding y_{n+j} and f_{n+j} to obtain

$$L[y(x);h] = C_0 y(x_n) + C_1 h y'(x_n) + \cdots + C_q h^q y^{(q)}(x_n) + \cdots, \quad (17)$$

where

$$C_0 = \alpha_0 + \alpha_1 + \alpha_2 + \cdots + \alpha_k,$$
$$C_1 = \alpha_1 + 2\alpha_2 + \cdots + k\alpha_k,$$
$$\vdots$$
$$C_q = \frac{1}{q!}(\alpha_1 + 2^q\alpha_2 + \cdots + k^q\alpha_k) - \frac{1}{(q-m)!}(\beta_1 + 2^{q-m}\beta_2 + \cdots + k^{q-m}\beta_k), \quad (18)$$
$$q = 2, 3, \ldots,$$

the method is said to be of order p if $C_0 = C_1 = \cdots = C_p = C_{p+1} = \cdots = C_{p+(m-1)} = 0$, $C_{p+m} \neq 0$ and C_{p+m} is the error constant.

The integrators of the block method (12) are of order six methods with the error constants, C_{10} obtained as 2323/3628800, 137/14175, 1737/44800, 1408/14175, and 29375/145152, respectively. Having order $p > 1$, the consistency of the block method is affirmed.

Moving on to the second criterion for convergence which is the zero-stability of the block method. Note that this is the most important stability property a good numerical method should possess as it ensures convergence. The key word "zero" is based on the stability phenomenon in terms of convergence in the limit as step-size (h) tends to zero.

Therefore, to test the zero-stability of the $(m+1)$th-step block method, the integrators are normalized to give the first characteristic polynomial $\rho(r)$ as

$$\rho(r) = \det(rA^0 - A^1) = r^m(r-1), \quad (19)$$

with $A^0 = 5 \times 5$ identity matrix

$$A^1 = \begin{pmatrix} 0 & 0 & 0 & 0 & 1 \\ 0 & 0 & 0 & 0 & 1 \\ 0 & 0 & 0 & 0 & 1 \\ 0 & 0 & 0 & 0 & 1 \\ 0 & 0 & 0 & 0 & 1 \end{pmatrix}. \quad (20)$$

The roots of $\rho(r) = 0$ satisfy $|r_j| \leq 1$. Hence, the $(m+1)$th-step block method is zero-stable.

3. Results and Discussion

This section tests the $(m+1)$th-step block method on some nonlinear problems. The numerical results are shown in Tables 1–3 and Figures 1–3.

Example 7. Consider the following nonlinear boundary value problem [10]:

$$y^{iv} - 6\exp(-4y) = -12(1+x)^4, \quad (21)$$

with boundary conditions

$$y(0) = 0,$$
$$y'(0) = 1, \quad (22)$$
$$y(1) = \ln(2) = y'(1) = 0.5.$$

The exact solution of Example 7 is $y = \ln(1+x)$. The obtained numerical results for this problem are presented in Table 1 with $h = 10^{-1}$. The maximum absolute error obtained by the $(m+1)$th-step block method is 3.84959×10^{-7} which is more accurate than the maximum error of 1.78×10^{-3} by Mustafa et al. [10]. The graphical comparison between exact and computed solution is shown in Figure 1.

Example 8. Consider the following nonlinear boundary value problem [10]:

$$y^{iv} = y^2 - x^{10} + -4x^8 - 4x^7 + 8x^6 - 4x^4 + 120x - 48, \quad (23)$$

with boundary conditions

$$y(0) = y'(0) = 0,$$
$$y(1) = y'(1) = 1. \quad (24)$$

Table 1: Comparison of the $(m+1)$th-step block method with Mustafa et al. [10] for solving Example 7.

x	Exact solution	Computed solution	Error [10]	Error ($(m+1)$th-step block method)
0.0	0.00000000000	0.00000000000	$0.000000e+00$	$0.000000e+00$
0.1	0.09531017980	0.09531018728	0.0002954265	$7.472134e-09$
0.2	0.18232155679	0.18232159067	0.0008719341	$3.387990e-08$
0.3	0.26236426447	0.26236434273	0.0014096072	$7.826736e-08$
0.4	0.33647223662	0.33647237832	0.0017352146	$1.417001e-07$
0.5	0.40546510811	0.40546533420	0.0017810699	$2.260955e-07$
0.6	0.47000362925	0.47000394905	0.0015577013	$3.198013e-07$
0.7	0.53062825106	0.53062863602	0.0011349902	$3.849590e-07$
0.8	0.58778666490	0.58778704558	0.0006286279	$3.806745e-07$
0.9	0.64185388617	0.64185415218	0.0001902154	$2.660067e-07$
1.0	0.69314718056	0.69314718056	$0.000000e+00$	$0.000000e+00$

Table 2: Comparison of the exact and computed solution of Example 8.

x	Exact solution	Computed solution	Error [10]	Error ($(m+1)$th-step block method)
0.0	0.00000000000	0.00000000000	$0.000000e+00$	$0.000000e+00$
0.1	0.01981000000	0.01981000000	0.0004095	$0.000000e+00$
0.2	0.07712000000	0.07712000000	0.0025752	$0.000000e+00$
0.3	0.16623000000	0.16623000000	0.0066432	$0.000000e+00$
0.4	0.27904000000	0.27904000000	0.0115595	$0.000000e+00$
0.5	0.40625000000	0.40625000000	0.0156708	$0.000000e+00$
0.6	0.53856000000	0.53856000000	0.0173246	$0.000000e+00$
0.7	0.66787000000	0.66787000000	0.0154706	$0.000000e+00$
0.8	0.78848000000	0.78848000000	0.0102612	$0.000000e+00$
0.9	0.89829000000	0.89829000000	0.0036517	$0.000000e+00$
1.0	1.00000000000	1.00000000000	$0.000000e+00$	$0.000000e+00$

Table 3: Absolute errors of fifth-order HAM and $(m+1)$th-step block method when $R_{mp}=0$.

z	Exact solution	Error [13]	Error ($(m+1)$th-step block method)
0.0	0	0	0
0.1	0.085233703438701791	$7.58785506649317\times10^{-10}$	4.032885×10^{-14}
0.2	0.171320454429454980	$1.39478356642186\times10^{-9}$	2.363387×10^{-13}
0.3	0.259121838110931650	$1.80948145356296\times10^{-9}$	6.848411×10^{-13}
0.4	0.349516600242079760	$1.94815263920844\times10^{-9}$	1.489975×10^{-12}
0.5	0.443409441985037010	$1.81075676675135\times10^{-9}$	2.768896×10^{-12}
0.6	0.541740074458440520	$1.45204859247627\times10^{-9}$	4.502620×10^{-12}
0.7	0.645492623682151550	$9.70818092582703\times10^{-10}$	6.029954×10^{-12}
0.8	0.755705480041236500	$4.90335771985428\times10^{-10}$	6.408873×10^{-12}
0.9	0.873481690845957730	$1.33847599670389\times10^{-10}$	4.708123×10^{-12}
1.0	1.000000000000000000	$2.22044604925031\times10^{-16}$	0.000000×10^{0}

The exact solution of Example 8 is $y = x^5 - 2x^4 + 2x^2$. The obtained numerical results for this problem are presented in Table 2 with $h = 10^{-1}$. The $(m+1)$th-step block method gives precise and accurate results as the exact solution. This is far more encouraging than the maximum error of 1.73×10^{-2} by Mustafa et al. [10]. The graphical comparison between exact and computed solution is also shown in Figure 2.

Examples 7 and 8 considered the solution of nonlinear boundary value problems solved by Mustafa et al. [10]. In their work, the authors adopted a numerical approach based on subdivision schemes. Although their approach gave good results, the $(m+1)$th-step block method gave better results in terms of accuracy. This superiority in accuracy of the $(m+1)$th-step block method is resultant from its self-starting implementation approach instead of the approach of Mustafa et al. [10] requiring choosing different subdivision schemes with certain adjustment of boundary conditions. The self-starting approach of the block method requires no starting values which could reduce the accuracy of the method

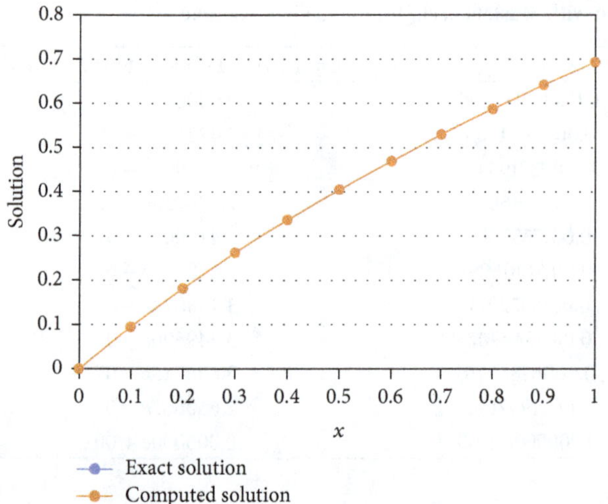

FIGURE 1: Comparison of the exact and computed solution of Example 7.

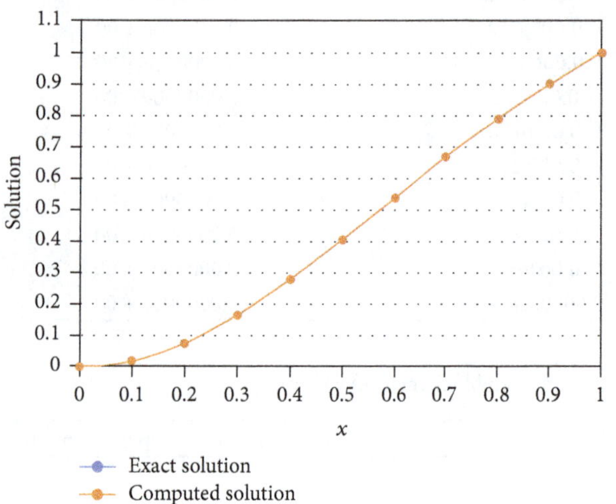

FIGURE 2: Comparison of the exact and computed solution of Example 8.

while also increasing the computational rigour. Rather, the integrators of $(m + 1)$th-step block method were combined as direct simultaneous integrators for the solution of the nonlinear boundary value problems.

In addition, to further show the suitability of the $(m+1)$th-step block method, a physical problem is solved and the results are compared to exiting solutions in literature.

3.1. $(m + 1)$th-Step Block Method Solution for Magnetohydrodynamic Squeezing Flow in Porous Medium. The study of squeezing effect, in addition to other properties such as magnetohydrodynamics (MHD) and porosity, has become one of the most active topics in fluid mechanics. Ullah et al. [13] made an effort to investigate MHD squeezing flow of Newtonian fluid between two parallel plates passing through porous medium by homotopy analysis method (HAM). The authors used similarity transforms to convert the governing

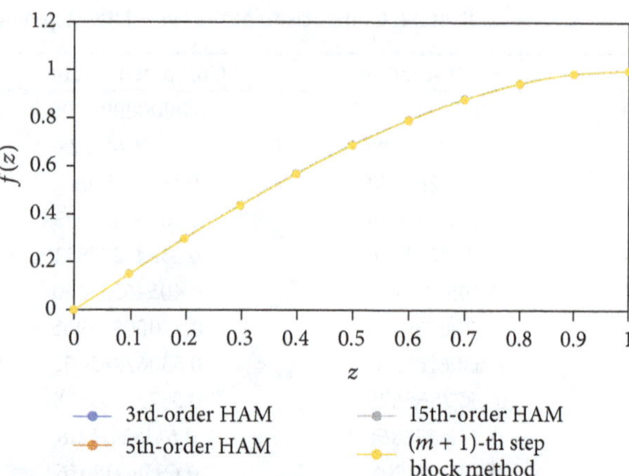

FIGURE 3: Comparison of solution between HAM [13] and $(m+1)$th-step block method.

partial differential equations to equivalent nonlinear boundary value problems and then solved by HAM. This article takes a step further to solve the resulting nonlinear boundary value problems using the $(m + 1)$th-step block method.

The resulting differential equation is a fourth-order nonlinear boundary value problem of the form

$$\frac{d^4}{dz^4}f(z) + R_{mp}f(z)\frac{d^3}{dz^3}f(z) - m_p\frac{d^2}{dz^2}f(z) - m_h\frac{d^2}{dz^2}f(z) = 0, \tag{25}$$

with boundary conditions

$$\begin{aligned}\frac{d^2}{dz^2}f(0) &= 0,\\ \frac{d}{dz}f(1) &= 0,\\ f(0) &= 0,\\ f(1) &= 1,\end{aligned} \tag{26}$$

where R_{mp} is Reynold number and m_h, m_p are Hartmann numbers.

Due to the difficulty to compute an exact solution for (25), Ullah et al. [13] computed varying solutions of (25) for different HAM orders. The $(m + 1)$th-step block method is likewise adopted to solve (25) and convergence is observed to the results proposed by Ullah et al. [13] as seen in Figure 3.

In addition, a special case of (25) is studied when the Reynold number is zero with exact solution obtained using the boundary conditions (26) as

$$f(z) = \frac{e^{2(-\sqrt{M}x)}e^{\sqrt{M}} - e^{\sqrt{M}} + \sqrt{M}xe^{-\sqrt{M}x} + \sqrt{M}xe^{-\sqrt{M}x}e^{2\sqrt{M}}}{e^{-\sqrt{M}x}\left(\sqrt{M}e^{2\sqrt{M}} - e^{2\sqrt{M}} + \sqrt{M} + 1\right)}, \tag{27}$$

where $M = m_h + m_p$.

Comparison is made between the exact solution, fifth-order HAM solution [13], and the $(m + 1)$th-step block method in terms of absolute error.

From Table 3, an improved accuracy was displayed by the $(m + 1)$th-step block method over the homotopy analysis method. This shows the block method is appropriate to evaluate the numerical solution of physical problems modelled as fourth-order nonlinear boundary value problems.

4. Conclusion

This article has introduced a numerical approach based on block methods derived using modified Taylor series approach. The $(m + 1)$th-step block method was adopted for the numerical solution of different nonlinear fourth-order boundary value problems. The numerical results show that the impressive accuracy of the $(m + 1)$th-step block method having the same computations as the exact solution is obtained as shown in Tables 1–3 and Figures 1–3. This grounds the suitability of the $(m + 1)$th-step block method for solving fourth-order nonlinear boundary value problems. In addition, the suitability of the block method in application to physical problems was also investigated by presenting a solution to MHD squeezing flow in a porous medium. Convergence in solution and improved accuracy were properties displayed by the $(m + 1)$th-step block method when solving this fluid model. Thus, the $(m + 1)$th-step block method is appropriate for solving fourth-order nonlinear boundary value problems.

Conflicts of Interest

The authors declare that they have no conflicts of interest.

Authors' Contributions

Both authors contributed equally to this work.

Acknowledgments

This work is supported by Research and Innovation Management Centre (RIMC), Universiti Utara Malaysia.

References

[1] M. Abbas, A. A. Majid, A. I. Ismail, and A. Rashid, "The application of cubic trigonometric B-spline to the numerical solution of the hyperbolic problems," *Applied Mathematics and Computation*, vol. 239, pp. 74–88, 2014.

[2] L. Ahmad Soltani, E. Shivanian, and R. Ezzati, "Shooting homotopy analysis method: a fast method to find multiple solutions of nonlinear boundary value problems arising in fluid mechanics," *Engineering Computations*, vol. 34, no. 2, 2017.

[3] N. Freidoonimehr, M. M. Rashidi, and S. Mahmud, "Unsteady MHD free convective flow past a permeable stretching vertical surface in a nano-fluid," *International Journal of Thermal Sciences*, vol. 87, pp. 136–145, 2015.

[4] Z. Omar and J. O. Kuboye, "New seven-step numerical method for direct solution of fourth order ordinary differential equations," *Journal of Mathematical and Fundamental Sciences*, vol. 48, no. 2, pp. 94–105, 2016.

[5] S. J. Kayode and O. Adeyeye, "A 3-Step hybrid method for direct solution of second order initial value problems," *Australian Journal of Basic and Applied Sciences*, vol. 5, no. 12, pp. 2121–2126, 2011.

[6] S. J. Kayode and F. O. Obarhua, "Continuous y-function hybrid methods for direct solution of differential equations," *International Journal of Differential Equations and Applications*, vol. 12, no. 1, pp. 37–48, 2013.

[7] W. E. Milne, *Numerical Solution of Ordinary Differential Equations*, Wiley, New York, NY, USA, 1953.

[8] D. Sarafyan, "Multistep methods for the numerical solution of ordinary differential equations made self-starting," Mathematics Research Center, MRC-TSR-495, 1965.

[9] J. D. Lambert, *Computational Methods in Ordinary Differential Equations*, Wiley, New York, NY, USA, 1973.

[10] G. Mustafa, M. Abbas, S. T. Ejaz, A. I. M. Ismail, and F. Khan, "A numerical approach based on subdivision schemes for solving non-linear fourth order boundary value problems," *Journal of Computational Analysis and Applications*, vol. 23, no. 4, pp. 607–623, 2017.

[11] P. Henrici, *Discrete Variable Methods in Ordinary Differential Equations*, John Wiley and Sons, New York, NY, USA, 1962.

[12] J. C. Butcher, *Numerical Methods for Ordinary Differential Equations*, Wiley, West Sussex, 2nd edition, 2008.

[13] I. Ullah, M. T. Rahim, H. Khan, and M. Qayyum, "Homotopy analysis solution for magnetohydrodynamic squeezing flow in porous medium," *Advances in Mathematical Physics*, vol. 2016, Article ID 3541512, 9 pages, 2016.

4

Fractional Variational Iteration Method for Solving Fractional Partial Differential Equations with Proportional Delay

Brajesh Kumar Singh and Pramod Kumar

Department of Applied Mathematics, School for Physical Sciences, Babasaheb Bhimrao Ambedkar University, Lucknow 226 025, India

Correspondence should be addressed to Brajesh Kumar Singh; bksingh0584@gmail.com

Academic Editor: Sunil Kumar

This paper deals with an alternative approximate analytic solution to time fractional partial differential equations (TFPDEs) with proportional delay, obtained by using fractional variational iteration method, where the fractional derivative is taken in Caputo sense. The proposed series solutions are found to converge to exact solution rapidly. To confirm the efficiency and validity of FRDTM, the computation of three test problems of TFPDEs with proportional delay was presented. The scheme seems to be very reliable, effective, and efficient powerful technique for solving various types of physical models arising in science and engineering.

1. Introduction

The idea of derivatives of fractional order was described first by great mathematician Newton and Leibnitz in the seventh century and has achieved a great attention due to their numerous applications in nonlinear complex systems arising in various important phenomena in the fluid mechanics, damping laws, electrical networks, signal processing, diffusion-reaction process relaxation processes, mathematical biology, and other fields of science and engineering [1–7]. Fractional derivatives offer more accurate models of real-world problems as compared to integer-order derivatives. The fractional calculus plays a critical role in describing a complex dynamical behavior in tremendous scope of application fields and helps to understand the nature of matter as well as simplifying the controlling design without any loss of hereditary behaviors. Further, the nonlinear oscillation of earthquake can be modeled via fractional derivatives [8]; the fluid-dynamic traffic model with fractional derivatives [9] can eliminate the deficiency arising from the assumption of continuum traffic flow and the fractional nonlinear complex model for seepage flow in porous media [10].

Indeed, it is too tough task to compute an exact solution of a wide class of the differential equations of fractional order. In the past years, different kind of vigorous techniques has been introduced to find an approximate solution of such type of fractional model of differential equations, such as generalized differential transform method [11], Adomian decomposition method [12], homotopy analysis method [13], homotopy analysis transform method [14], modified Laplace transform method [15], and homotopy perturbation transform method (HPTM) [16–18]. FRDTM have been adopted to solve vigorous types of differential equations arising in mathematics, physics, and engineering by Saravanan and Magesh [19], Srivastava et al. [20, 21], Singh and Srivastava [22], Singh and Kumar [23, 24], Singh [25], and Singh and Mahendra [26]. Recently, a fractional model of differential-difference equation model (appeared in nanohydrodynamics, heat conduction in nanoscale, and electronic current that flows through carbon nanotubes) has been studied analytically by adopting homotopy analysis transform method [27], and a hybrid computational approach based on local fractional Sumudu transform with HPM has been employed for numerical study of Klein–Gordon equations on Cantor sets [28]. Atangana and Baleanu [4] proposed a much better version of fractional derivative with a nonsingular and nonlocal kernel, based upon the well-known generalized Mittag-Leffler function, to answer some outstanding questions raised by many researchers within the field of fractional calculus. A relationship of their derivatives with some integral transform operators was presented by Atangana and Koca [29] who show the existence and uniqueness of the system solutions of the fractional system in detail and also obtained a chaotic behavior which was not obtained by local derivative. Goufo [30] adopted newly developed Caputo-Fabrizio fractional derivative without singular kernel to obtain an analytical solution of Korteweg-de Vries-Burgers equation with two perturbations' levels. The two-parameter derivative with non-singular and nonlocal kernel has been introduced by [31] to study the chaotic processes of the fractional system. Goufo [32] used the concept of variable-order derivative to study the stability and convergence analysis of the well-known variable-order replicator-mutator dynamics in a moving medium. For more details in fractional derivatives, we refer the readers to [4, 29–32] and the references therein.

The variational iteration method (VIM) has been developed by Chinese mathematician He [10]. After the seminal work of He, various modification of VIM has been employed to solve various nonlinear problems, such as diffusion and wave equations on cantor sets [33], Riccati differential equation [34], fractional model of coupled Burgers equations [35], and time fractional Fornberg-Whitham equation [36]. For more details, the readers are referred to [33–38] and the references therein.

The partial functional differential equations with proportional delays, a special class of delay partial differential equation, arise specially in the field of biology, medicine, population ecology, control systems, and climate models [39], and complex economic macrodynamics [40].

This paper is concerned with the numerical solution of the initial valued autonomous system of TFPDEs with proportional delay [17] defined by

$$\begin{aligned}\mathscr{D}_t^{\alpha}(u(x,t)) &= f\Big(x, u(a_0x, b_0t), \frac{\partial}{\partial x} \\ &\cdot u(a_1x, b_1t), \ldots, \frac{\partial^m}{\partial x^m}u(a_mx, b_mt)\Big), \\ u^k(x,0) &= \psi_k(x),\end{aligned} \tag{1}$$

where $a_i, b_i \in (0,1)$ for all $i \in N \cup \{0\}$. ψ_k is initial value, f is the differential operator, and the independent variables (x,t) (where t denotes time and x is space variable) denote the position in space or size of cells and maturation level at a time. The solution of (1) may be the voltage, temperature, densities of different particles, form instance, chemicals, cells, and so forth. One significant example of the model, Korteweg-de Vries (KdV) equation, arising in the research of shallow water waves is as follows:

$$\begin{aligned}\mathscr{D}_t^{\alpha}(u(x,t)) &= bu\frac{\partial}{\partial x}u(a_0x, b_0t) + \frac{\partial^3}{\partial x^3}u(a_1x, b_1t), \\ &\qquad 0 < \alpha < 1,\end{aligned} \tag{2}$$

where b is a constant. Another well-known model, time fractional nonlinear Klein–Gordon equation with proportional delay, aries in quantum field theory to describe nonlinear wave interaction:

$$\begin{aligned}\mathscr{D}_t^{\alpha}(u(x,t)) &= u\frac{\partial^2}{\partial x^2}u(a_0x, b_0t) - bu(a_1x, b_1t) \\ &\quad - F(u(a_2x, b_2t)) + h(x,t), \\ &\qquad 1 < \alpha < 2,\end{aligned} \tag{3}$$

where b is a constant, $h(x,t)$ is known analytic function, and F is the nonlinear operator of $u(x,t)$. For details of various types of models, we refer the reader to [17, 39] and the references therein.

To the best of my knowledge, there is little literature of methods to solve TFPDE with delay, such as Chebyshev pseudospectral method for linear differential and differential-functional parabolic equations [41], spectral collocation and waveform relaxation methods [42], and iterated pseudospectral method [43] for nonlinear delay partial differential equations. equations. Abazari and Ganji [44] obtained approximate solutions of PDEs with proportional delay by employing RDTM. Abazari and Kilicman [45] obtained analytical solutions of nonlinear integrodifferential equations with proportional delay by using DTM. Tanthanuch [46] applied group analysis method for nonhomogeneous mucilaginous Burgers equation with proportional delay. The analytical solutions of TFPDE with proportional delay have been obtained by employing homotopy perturbation method by Sakar et al. [17] and Biazar ad Ghanbari [47]. Chena and Wang [48] have adopted variational iteration method (VIM) for solving a neutral functional-differential equation with proportional delays. The main aim of this paper is to propose an alternative approximate solution of the initial valued autonomous system of TFPDE with proportional delay [17] by employing alternative variational iteration method (AVIM).

The paper is sketched into five more sections following Introduction. Specifically, Section 2 deals with the revisit of fractional calculus. Section 3 is devoted to the procedure for the implementation of the AVIM for problem (1). Section 5 is concerned with three test problems with the main aim of establishing the convergency and effectiveness of AVIM. Finally, Section 6 concludes the paper with reference to critical analysis and research perspectives.

2. Preliminaries

Among the various kinds of definitions of fractional derivatives, the definitions mostly applied are due to Riemann-Liouville [1], Caputo [3], Yang [7], He [4], Atangana and Baleanu [4], Caputo-Fabrizio [30], and so forth. This section revisits some basic definitions of fractional calculus due to Liouville [1] which we need to complete the paper.

Definition 1. Let $\mu \in \mathbb{R}$ and $m \in \mathbb{N}$. A function $f : \mathbb{R}^+ \to \mathbb{R}$ belongs to $\mathbb{C}_\mu$ if there exists $k \in \mathbb{R}$, $k > \mu$, and $g \in C[0,\infty)$ such that $f(x) = x^k g(x)$, $\forall x \in \mathbb{R}^+$. Moreover, $f \in \mathbb{C}_\mu^m$ if $f^{(m)} \in \mathbb{C}_\mu$.

Definition 2. Let $\mathcal{J}_t^\alpha$ $(\alpha \geq 0)$ be Riemann-Liouville fractional integral operator and let $f \in \mathbb{C}_\mu$; then

$(*)$ $\mathcal{J}_t^\alpha f(t) = (1/\Gamma(\alpha)) \int_0^t (t-\tau)^{\alpha-1} f(\tau) d\tau$, if $\alpha > 0$,

$(**)$ $\mathcal{J}_t^0 f(t) = f(t)$, where $\Gamma(z) := \int_0^\infty e^{-t} t^{z-1} dt$, $z \in \mathbb{C}$.

For $f \in \mathbb{C}_\mu$, $\mu \geq -1$, $\alpha, \beta \geq 0$, and $\gamma > -1$, the operator $\mathcal{J}_t^\alpha$ satisfies the following properties:

(i) $\mathcal{J}_t^\alpha \mathcal{J}_t^\beta f(x) = \mathcal{J}_t^{\alpha+\beta} f(x) = J_t^\beta \mathcal{J}_t^\alpha f(x)$.

(ii) $\mathcal{J}_t^\alpha x^\gamma = (\Gamma(1+\gamma)/\Gamma(1+\gamma+\alpha)) x^{\alpha+\gamma}$.

It is worth mentioning that Riemann-Liouville derivative exists for any functions that are continuous but Riemann-Liouville derivative has certain disadvantage for describing some natural phenomena; for example, Riemann-Liouville derivative of a constant is not equal to zero. In their work, Caputo and Mainardi [3] proposed a Caputo fractional differentiation operator D_t^α, defined below, which is a modification of definition of Riemann-Liouville to describe the theory of viscoelasticity in order to overcome the discrepancy of Riemann-Liouville derivative [1]. It is worth mentioning that the Caputo fractional derivative allows the utilization of initial and boundary conditions involving integer-order derivatives.

Definition 3. Let $f \in \mathbb{C}_\mu$, $\mu \geq -1$, and $m-1 < \alpha \leq m$, $m \in \mathbb{N}$. Then

$$\begin{aligned} \mathcal{D}_t^\alpha f(t) &= \mathcal{J}_t^{m-\alpha} \mathcal{D}_t^m f(t) \\ &= \frac{1}{\Gamma(m-\alpha)} \int_0^t (t-\tau)^{m-\alpha-1} f^{(m)}(\tau) d\tau. \end{aligned} \tag{4}$$

Moreover, the operator $\mathcal{D}_t^\alpha$ satisfies the following basic properties.

Lemma 4. *Let $m-1 < \alpha \leq m$, $m \in \mathbb{N}$, $f \in \mathbb{C}_\mu^m$, $\mu \geq -1$, and $\gamma > \alpha - 1$, then*

(a) $\mathcal{D}_t^\alpha \mathcal{D}_t^\beta f(x) = \mathcal{D}_t^{\alpha+\beta} f(x)$;

(b) $\mathcal{D}_t^\alpha x^\gamma = (\Gamma(1+\gamma)/\Gamma(1+\gamma-\alpha)) x^{\gamma-\alpha}$,

(c) $\mathcal{D}_t^\alpha \mathcal{J}_t^\alpha f(t) = f(t)$,

(d) $\mathcal{J}_t^\alpha \mathcal{D}_t^\alpha f(t) = f(t) - \sum_{k=0}^m f^{(k)}(0^+)(t^k/k!)$, *for* $t > 0$.

In the present work, Caputo fractional derivative is considered as it deals with traditional initial and boundary conditions in the formulation of the physical problems. For details on fractional derivatives, we refer the interested readers to [1–7].

3. Description of Alternative Variational Iteration Method (AVIM)

Consider an initial valued differential equation:

$$\begin{aligned} &Lu(t) + Nu(t) = g(t), \quad t > 0, \\ &u^{(k)}(0) = c_k, \quad k = 0, 1, \ldots, m-1, \end{aligned} \tag{5}$$

where c_k are real numbers, $L = d^n/dt^n$, and $m \in \mathbb{N}$ is a linear operator; $N \rightarrow$ nonlinear operator and $g(t)$ is a known analytic function.

The correction functional for (5) can be constructed using AVIM as defined in [37] as

$$\begin{aligned} u_{k+1}(t) &= u_k(t) \\ &+ \int_0^t [\lambda(\tau)(Lu_k(\tau) - N\overline{u}_k(\tau) - g(\tau))] d\tau, \end{aligned} \tag{6}$$

where the Lagrange multiplier $\lambda(\tau)$ can be identified optimally by means of variation theory. Generally, the following Lagrange multipliers are used:

$$\lambda(\tau) = \frac{(-1)^m}{(m-1)!} (\tau - t)^{m-1}, \quad m \geq 1. \tag{7}$$

Equations (7) and (6) yield the following iteration formula:

$$u_{k+1}(t) = u_k(t) + A[u_k(t)], \tag{8}$$

where the operator $A[u]$ is defined by

$$\begin{aligned} A(u) &:= \frac{(-1)^m}{(m-1)!} \\ &\cdot \int_0^t \left((\tau - t)^{m-1} (Lu_k(\tau) - N\overline{u}_k(\tau) - g(\tau))\right) d\tau. \end{aligned} \tag{9}$$

Moreover, if we set the components s_k $(k = 0, 1, 2, \ldots)$ as

$$\begin{aligned} s_0 &= u_0, \\ s_1 &= A[s_0], \\ s_2 &= A[s_0 + s_1], \\ &\vdots \\ s_{k+1} &= A[s_0 + s_1 + \cdots + s_k], \end{aligned} \tag{10}$$

then we have $u(t) = \lim_{k\to\infty} u_k(t) = \sum_{k=0}^\infty s_k$. Thus, (9) and (10) yield the series solution of system (5) in the following form:

$$u(t) = \sum_{k=0}^\infty s_k(t). \tag{11}$$

The interested readers are referred to [33, 34, 36–38] for further details.

4. AVIM for FPDEs with Proportional Delay Equations

Consider the initial valued autonomous system of time fractional partial differential equation of order α with $\lceil \alpha \rceil = m \in \mathbb{N}$ as

$$\begin{aligned} &\mathcal{D}_t^\alpha \{u(x,t)\} + Nu(x,t) = g(x,t), \quad t > 0, \\ &u^{(k)}(x,0) = f_k(x), \\ &k = 0, 1, \ldots, m-1, \; x \in R, \end{aligned} \tag{12}$$

where $N \rightarrow$ nonlinear operator, $g = g(x,t) \rightarrow$ known analytic function, and $\mathscr{D}_t^\alpha \rightarrow$ Caputo fractional derivative of order α and $f_k \rightarrow$ a real valued function.

The solution

$$u(x,t) = \lim_{n\to\infty} u_n(x,t) \tag{13}$$

to problem (12) can be derived from the following iteration formula as in [38]:

$$u_{k+1}(t) = u_k(t) - J_t^\alpha \left[D_t^\alpha u_k(x,t) + N u_k(x,t) - g(x,t)\right]. \tag{14}$$

The variational iteration solution, $u(x,t) = \sum_{k=0}^{\infty} v_k(x,t)$, in the present framework is obtained by the following iteration formula for $\lceil \alpha \rceil = m \in \mathbb{N}$ as in [37, 38]:

$$\begin{aligned} v_0 &= \sum_{k=0}^{n-1} \frac{f_k(x)}{k!} t^k, \\ v_{k+1} &= -\frac{\Gamma(\alpha)}{\Gamma(\alpha - m + 1)\Gamma(m)} \mathscr{J}_t^\alpha \left[D_t^\alpha \left[v_0 + \cdots + v_k\right] + N\left[v_0 + \cdots + v_k\right] - g(x,t)\right]. \end{aligned} \tag{15}$$

The iteration formula (15) converges to a solution of problem (12) whenever there exists γ such that $\gamma \in (0,1)$ and $s_{k+1} \le \gamma s_k$ $\forall k \in \mathbb{N} \cup \{0\}$; for details see [37, 38].

5. Application of VIM to TFPDEs with Proportional Delay

This section deals with the effectiveness and validity of VIM, which are demonstrated by means of three test problems of TFPDEs with proportional delay.

Problem 1. Consider initial values system of time fractional order, generalized Burgers equation with proportional delay as given in [17]:

$$\begin{aligned} \mathscr{D}_t^\alpha u(x,t) &= \frac{\partial^2}{\partial x^2} u(x,t) + u\left(\frac{x}{2}, \frac{t}{2}\right) \frac{\partial}{\partial x} u\left(x, \frac{t}{2}\right) + \frac{1}{2} u(x,t), \\ u(x,0) &= x. \end{aligned} \tag{16}$$

Keeping (15) in mind, the iteration formula for (16) can be constructed as

$$\begin{aligned} s_0 &= x, \\ s_{k+1} &= -\mathscr{J}_t^\alpha \left[\mathscr{D}_t^\alpha \left\{\sum_{i=0}^{k} s_i(x,t)\right\} - \sum_{i=0}^{k} \frac{\partial^2 s_i(x,t)}{\partial x^2} - \frac{1}{2} \sum_{i=0}^{k} s_i(x,t) - \left\{\sum_{i=0}^{k} s_i\left(\frac{x}{2}, \frac{t}{2}\right)\right\} \left\{\sum_{i=0}^{k} \frac{\partial s_i(x,t/2)}{\partial x}\right\}\right]. \end{aligned} \tag{17}$$

On simplifying the above relation, we get

$$\begin{aligned} s_0 &= x, \\ s_1 &= \frac{t^\alpha x}{\Gamma[\alpha+1]}, \\ s_2 &= \frac{t^{2\alpha} x \left(2^{1-\alpha} + 1\right)}{2\Gamma[2\alpha+1]}, \\ s_3 &= 2^{-2-3\alpha} t^{3\alpha} x \left[\frac{\left(2+2^\alpha\right)\left(2+4^\alpha\right)\left(\Gamma[1+\alpha]\right)^2 + 2^{1+\alpha}\Gamma[1+2\alpha]}{\left(\Gamma[1+\alpha]\right)^2 \Gamma[1+3\alpha]}\right], \\ &\vdots \end{aligned} \tag{18}$$

The solution of problem (16) is

$$\begin{aligned} u(x,t) &= s_0(x,t) + s_1(x,t) + s_2(x,t) + s_3(x,t) + \cdots \\ &= x\left\{1 + \frac{t^\alpha}{\Gamma[\alpha+1]} + \frac{t^{2\alpha}\left(2^{1-\alpha}+1\right)}{2\Gamma[2\alpha+1]} + \left(\frac{\left(2+2^\alpha\right)\left(2+4^\alpha\right)}{2^{2+3\alpha}\Gamma[1+3\alpha]} + \frac{2^{-1-2\alpha}\Gamma[1+2\alpha]}{\left(\Gamma[1+\alpha]\right)^2\Gamma[1+3\alpha]}\right) \cdot t^{3\alpha} + \cdots\right\} \end{aligned} \tag{19}$$

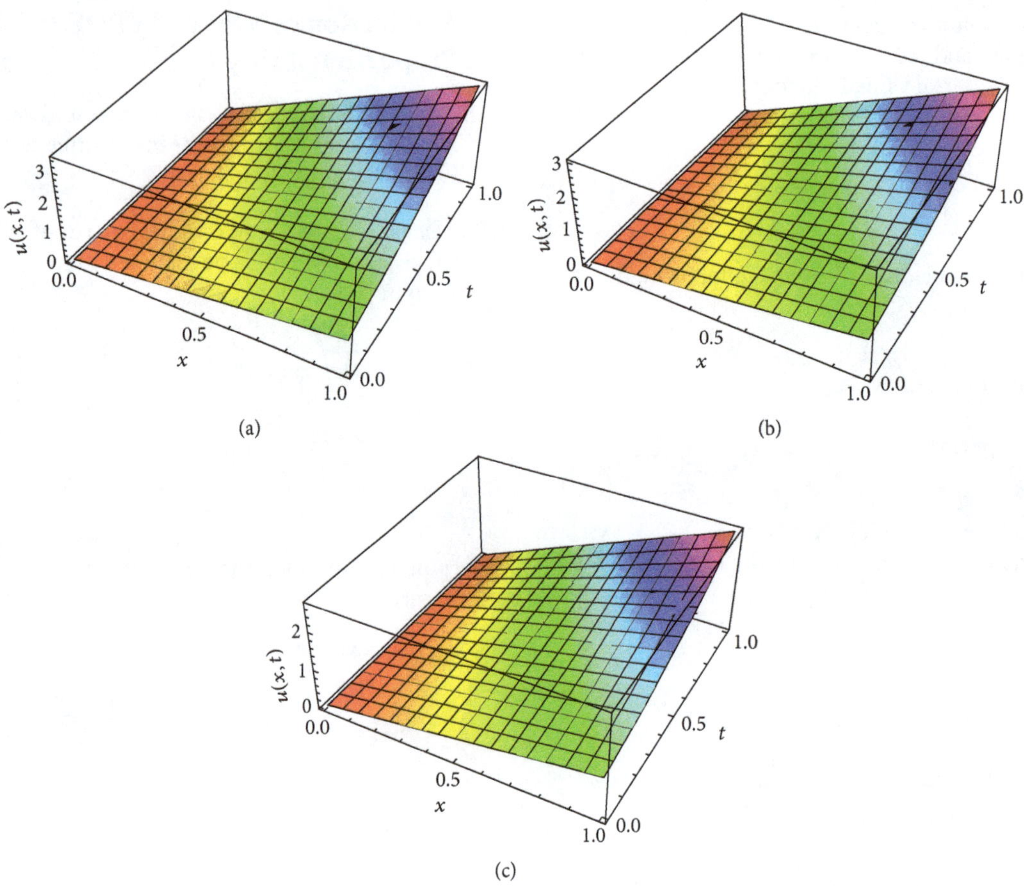

FIGURE 1: The solution behavior of AVIM solution u of Example 1 for (a) $\alpha = 0.8$; (b) $\alpha = 0.9$; (c) $\alpha = 1.0$.

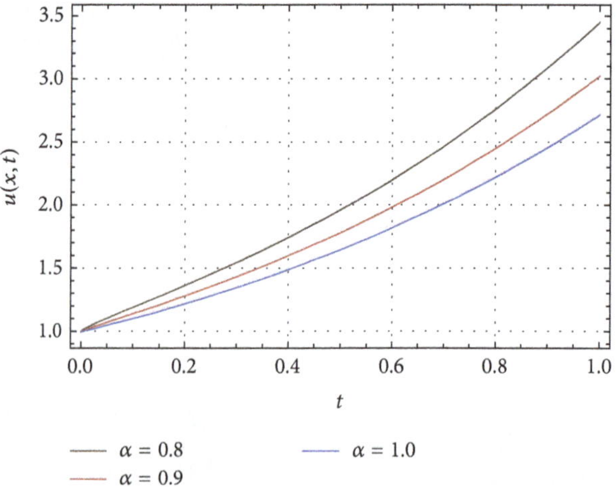

FIGURE 2: Plots of AVIM solution $u(x,t)$ of Example 1 for $\alpha = 0.8, 0.9, 1.0$; $t \in [0,1]$; $x = 1$.

which is closed form to the exact solution and the results due to Sakar et al. [17] and Singh and Kumar [49]. The solution behavior of $u(x,t)$, taking first six terms, for different values of $\alpha = 0.8, 0.9, 1.0$ at different time levels $t \leq 1$ with $x = 1$, is depicted in Figure 1, whereas two dimensional plots are depicted in Figure 2.

For $\alpha = 1$, solution (19) reduces to

$$u(x,t) = x\left(1 + t + \frac{t^2}{2!} + \frac{t^3}{3!} + \frac{t^4}{4!} + \cdots\right) \tag{20}$$

which is same as obtained by DTM and RDTM [44] and HPTM [49] and is a closed form of the exact solution

$u(x,t) = x\exp(t)$. The approximate AVIM solution for $\alpha = 1$ taking first six terms is reported in Table 1; it is mentioned that the results agreed well with solutions obtained by DTM and RDTM [44], HPM [17], and HPTM [49] and approach the exact solutions.

Problem 2. Consider initial value TFPDE with proportional delay as given in [17]:

$$\mathscr{D}_t^\alpha u(x,t) = u\left(x,\frac{t}{2}\right)\frac{\partial^2}{\partial x^2}u\left(x,\frac{t}{2}\right) - u(x,t),$$
$$u(x,0) = x^2. \tag{21}$$

The iteration formula for (21) can be constructed as

$$s_0 = x^2,$$
$$s_{k+1} = -\mathscr{I}_t^\alpha\left[\mathscr{D}_t^\alpha\left\{\sum_{i=0}^k s_i(x,t)\right\} + \sum_{i=0}^k s_i(x,t) - \left\{\sum_{i=0}^k s_i\left(x,\frac{t}{2}\right)\right\}\left\{\sum_{i=0}^k \frac{\partial^2 s_i(x,t/2)}{\partial x^2}\right\}\right]. \tag{22}$$

On solving the above relation, we get

$$s_0 = x^2,$$
$$s_1 = \frac{t^\alpha x^2}{\Gamma[\alpha+1]},$$
$$s_2 = \frac{t^{2\alpha}x^2(-2^{-\alpha}(-4+2^\alpha))}{\Gamma[2\alpha+1]}, \tag{23}$$
$$s_3 = 8^{-\alpha}t^{3\alpha}x^2\left[\frac{(-4+2^\alpha)(-2+2^\alpha)(2+2^\alpha)\Gamma[1+\alpha]^2 + 2^{1+\alpha}\Gamma[1+2\alpha]}{\Gamma[1+\alpha]^2\Gamma[1+3\alpha]}\right],$$
$$\vdots$$

The solution of problem (21) leads to

$$u(x,t) = s_0(x,t) + s_1(x,t) + s_2(x,t) + s_3(x,t) + \cdots \tag{24}$$

which is closed form to the exact solution and the results due to Sakar et al. [17] and Singh and Kumar [49]. The solution behavior of $u(x,t)$ for different values of $\alpha = 0.8, 0.9, 1.0$ at different time levels $t \le 1$ with $x = 1$ is depicted in Figure 3, whereas two dimensional plots are depicted in Figure 4.

In particular, for $\alpha = 1$, solution (24) reduces to

$$u(x,t) = x^2\left(1 + t + \frac{t^2}{2!} + \frac{t^3}{3!} + \frac{t^4}{4!} + \cdots\right) \tag{25}$$

which is same as obtained by DTM and RDTM [44] and HPTM [49] and is a closed form of the exact solution $u(x,t) = x^2\exp(t)$. The approximate AVIM solution for $\alpha = 1$ is reported in Table 2. This confirms that the proposed results agreed well with solutions obtained by DTM and RDTM [44], HPM [17], and HPTM [49] and approach the exact solutions.

Problem 3. Consider initial value TFPDE with proportional delay as given in [17, 44]:

$$\mathscr{D}_t^\alpha u(x,t) = \frac{\partial^2}{\partial x^2}u\left(\frac{x}{2},\frac{t}{2}\right)\frac{\partial}{\partial x}u\left(\frac{x}{2},\frac{t}{2}\right) - \frac{1}{8}\frac{\partial}{\partial x}u(x,t) - u(x,t),$$
$$u(x,0) = x^2. \tag{26}$$

In particular, for $\alpha = 1$, the exact solution is $u(x,t) = x^2\exp(-t)$.

The iteration formula for (26) can be constructed as

$$s_0 = x^2,$$
$$s_{k+1} = -\mathscr{I}_t^\alpha\left[\mathscr{D}_t^\alpha\left\{\sum_{i=0}^k s_i(x,t)\right\} + \sum_{i=0}^k s_i(x,t) + \frac{1}{8}\sum_{i=0}^k \frac{\partial s_i(x,t)}{\partial x} - \left\{\sum_{i=0}^k \frac{\partial s_i(x/2,t/2)}{\partial x}\right\}\left\{\sum_{i=0}^k \frac{\partial^2 s_i(x,t/2)}{\partial x^2}\right\}\right]. \tag{27}$$

Simplification of above relations leads to

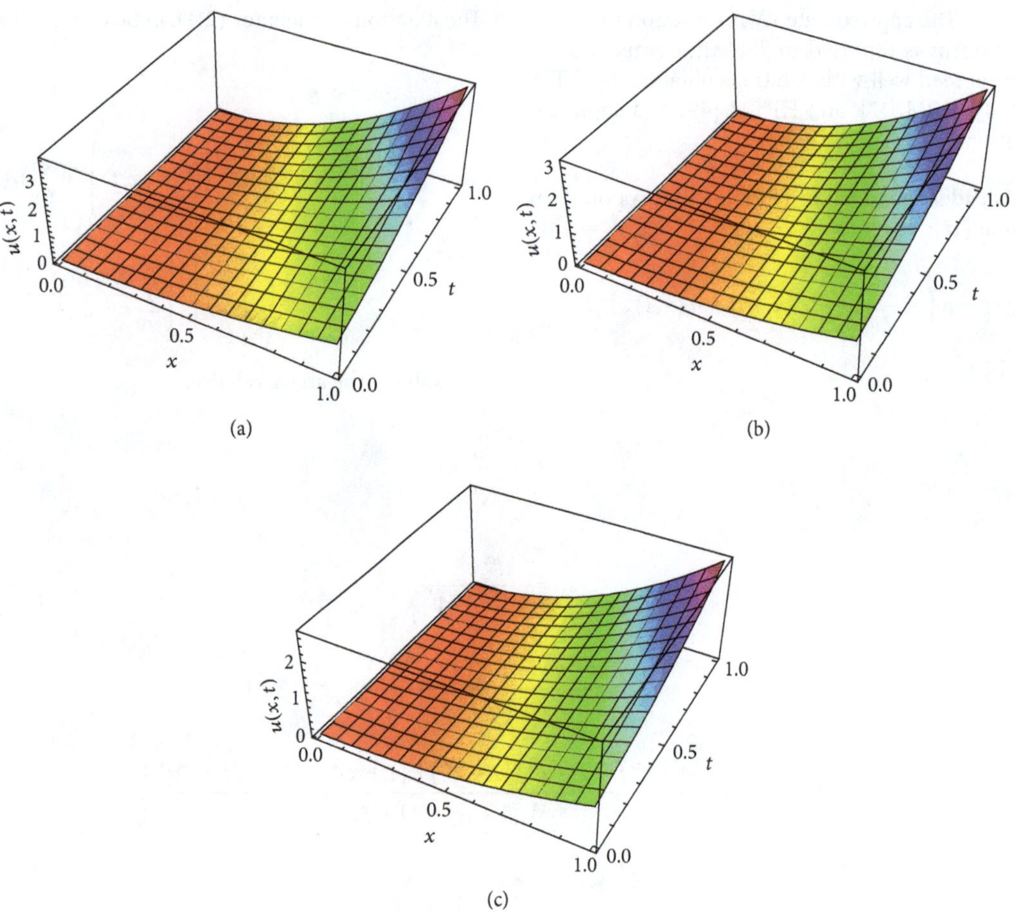

FIGURE 3: The solution behavior of AVIM solution u of Example 2 for (a) $\alpha = 0.8$; (b) $\alpha = 0.9$; (c) $\alpha = 1.0$.

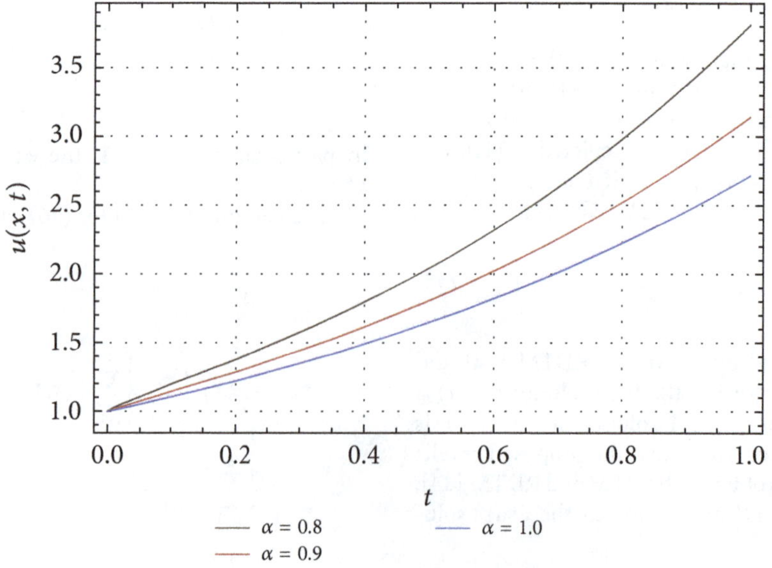

FIGURE 4: Plots of AVIM solution $u(x,t)$ of Example 2 for $\alpha = 0.8, 0.9, 1.0$; $t \in [0,1]$; $x = 1$.

Table 1: Approximate AVIM solution of Example 1 with first six terms for $\alpha = 1$.

x	t	Exact	VIM Approx.	E_{abs}
0.25	0.25	$3.210064E-01$	$3.210063E-01$	$8.789589E-08$
	0.50	$4.121803E-01$	$4.121745E-01$	$5.838508E-06$
	0.75	$5.292500E-01$	$5.291809E-01$	$6.909595E-05$
	1.00	$6.795705E-01$	$6.791667E-01$	$4.037904E-04$
0.5	0.25	$6.420127E-01$	$6.420125E-01$	$1.757918E-07$
	0.50	$8.243606E-01$	$8.243490E-01$	$1.167702E-05$
	0.75	$1.058500E+0$	$1.058362E+0$	$1.381919E-04$
	1.00	$1.359141E+0$	$1.358333E+0$	$8.075809E-04$
0.75	0.25	$9.630191E-01$	$9.630188E-01$	$2.636877E-07$
	0.50	$1.236541E+0$	$1.236523E+0$	$1.751553E-05$
	0.75	$1.587750E+0$	$1.587543E+0$	$2.072879E-04$
	1.00	$2.038711E+0$	$2.037500E+0$	$1.211371E-03$

Table 2: Approximate AVIM solution of Example 2 for $\alpha = 1$ with first six terms.

x	t	Exact	VIM Approx.	E_{abs}
0.25	0.25	$8.025159E-02$	$8.025157E-02$	$2.197397E-08$
	0.50	$1.030451E-01$	$1.030436E-01$	$1.459627E-06$
	0.75	$1.323125E-01$	$1.322952E-01$	$1.727399E-05$
	1.00	$1.698926E-01$	$1.697917E-01$	$1.009476E-04$
0.50	0.25	$3.210064E-01$	$3.210063E-01$	$8.789589E-08$
	0.50	$4.121803E-01$	$4.121745E-01$	$5.838508E-06$
	0.75	$5.292500E-01$	$5.291809E-01$	$6.909595E-05$
	1.00	$6.795705E-01$	$6.791667E-01$	$4.037904E-04$
0.75	0.25	$7.222643E-01$	$7.222641E-01$	$1.977658E-07$
	0.50	$9.274057E-01$	$9.273926E-01$	$1.313664E-05$
	0.75	$1.190813E+0$	$1.190657E+0$	$1.554659E-04$
	1.00	$1.529034E+0$	$1.528125E+0$	$9.085285E-04$

$$
\begin{aligned}
s_0 &= x^2, \\
s_1 &= -\frac{t^\alpha x^2}{\Gamma[\alpha+1]}, \\
s_2 &= \frac{2^{-2+\alpha} t^{2\alpha} x(-2+2[\alpha](1+4x))}{\Gamma[1+2\alpha]}, \\
s_3 &= \left\{ \frac{8^{1+\alpha} x \Gamma[(1/2)\alpha] - \sqrt{\pi}\left(\left(-2+2^\alpha\right)\left(-2+4^\alpha\right) + 2^{4+\alpha} x\left(-1-2^\alpha+4^\alpha(1+2x)\right)\Gamma[1+\alpha]\right)}{2^{5+3\alpha}\sqrt{\pi}\Gamma[1+\alpha]\Gamma[1+3\alpha]} \right\} t^{3\alpha}, \\
&\vdots
\end{aligned}
\tag{28}
$$

Therefore the solution for (26) is

$$u(x,t) = s_0(x,t) + s_1(x,t) + s_2(x,t) + s_3(x,t) + \cdots \tag{29}$$

which is the required solution. The similar solution behavior has been obtained in [17, 49]. The solution behavior of $u(x,t)$ for different values of $\alpha = 0.8, 0.9, 1.0$, at different time

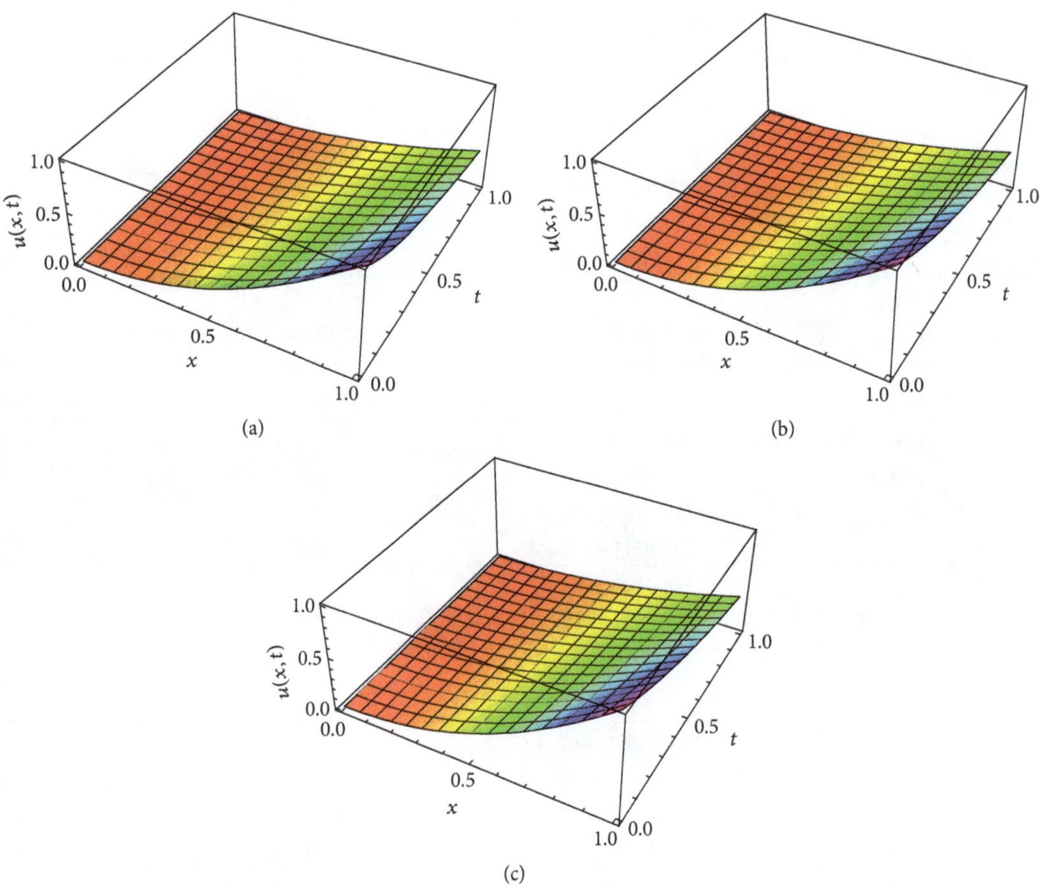

FIGURE 5: The solution behavior of VIM solution u of Example 3 for (a) $\alpha = 0.8$; (b) $\alpha = 0.9$; (c) $\alpha = 1.0$.

levels $t \leq 1$ with $x = 1$, is depicted in Figure 5, whereas two dimensional plots are depicted in Figure 6. Thus, the proposed results agreed well with solutions obtained by HPM [17] and HPTM [49].

For $\alpha = 1$, solution (29) reduces to

$$u(x,t) = \left(1 - t + \frac{t^2}{3} - \frac{t^3}{6} + \frac{t^4}{24} - \frac{t^5}{120} + \frac{t^6}{720} - \cdots\right)x^2 \tag{30}$$

which is same as obtained by DTM and RDTM [44] and HPTM [49] and is a closed form of the exact solution $u(x,t) = x^2 \exp(-t)$. The approximate AVIM solution for $\alpha = 1$ is reported in Table 3. The proposed solution converges to the exact solution.

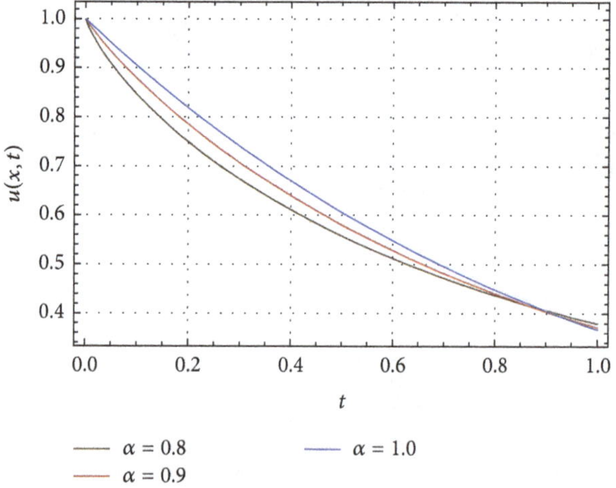

FIGURE 6: Plots of VIM solution $u(x,t)$ of Example 3 for $\alpha = 0.8$, 0.9, 1.0; $t \in [0,1]$; $x = 1$.

6. Conclusion

In this paper, `alternative variation iteration method` is successfully implemented for the numerical computation of initial valued autonomous system of time fractional model of TFPDE with proportional delay, where we use the fractional derivative in Caputo sense. The analytical results have been given in terms of a power series which converges to the exact solutions.

Three test problems are carried out in order to validate and illustrate the efficiency of the method. The proposed solutions agreed excellently with HPM [17], HPTM [49], and DTM [44]. These approximate solutions are obtained without any discretization, perturbation, or restrictive conditions.

TABLE 3: Approximate AVIM solution of Example 3 with first six terms for $\alpha = 1$.

x	t	Exact	VIM Approx.	E_{abs}
0.25	0.25	$4.867505E-02$	$4.867503E-02$	$2.045889E-08$
	0.50	$3.790817E-02$	$3.790690E-02$	$1.265190E-06$
	0.75	$2.952291E-02$	$2.950897E-02$	$1.393738E-05$
	1.00	$2.299247E-02$	$2.291667E-02$	$7.579841E-05$
0.50	0.25	$1.947002E-01$	$1.947001E-01$	$8.183556E-08$
	0.50	$1.516327E-01$	$1.516276E-01$	$5.060761E-06$
	0.75	$1.180916E-01$	$1.180359E-01$	$5.574951E-05$
	1.00	$9.196986E-02$	$9.166667E-02$	$3.031936E-04$
0.75	0.25	$4.380754E-01$	$4.380753E-01$	$1.841300E-07$
	0.50	$3.411735E-01$	$3.411621E-01$	$1.138671E-05$
	0.75	$2.657062E-01$	$2.655807E-01$	$1.254364E-04$
	1.00	$2.069322E-01$	$2.062500E-01$	$6.821857E-04$

Competing Interests

The authors declare that there are no competing interests regarding the publication of this article.

Acknowledgments

Pramod Kumar is thankful to Babasaheb Bhimrao Ambedkar University, Lucknow, India, for financial assistance to carry out the work.

References

[1] K. S. Miller and B. Ross, *An Introduction to the Fractional Calculus and Fractional Differential Equations*, Wiley-Interscience, New York, NY, USA, 1993.

[2] A. A. Kilbas, H. M. Srivastava, and J. J. Trujillo, *Theory and Applications of Fractional Differential Equations*, Elsevier, Amsterdam, The Netherlands, 2006.

[3] M. Caputo and F. Mainardi, "Linear models of dissipation in anelastic solids," *La Rivista del Nuovo Cimento*, vol. 1, no. 2, pp. 161–198, 1971.

[4] A. Atangana and D. Baleanu, "New fractional derivatives with nonlocal and non-singular kernel: theory and application to heat transfer model," *Thermal Science*, vol. 20, no. 2, pp. 763–769, 2016.

[5] J.-H. He, "A tutorial review on fractal spacetime and fractional calculus," *International Journal of Theoretical Physics*, vol. 53, no. 11, pp. 3698–3718, 2014.

[6] E. Goldfain, "Fractional dynamics, cantorian space-time and the gauge hierarchy problem," *Chaos, Solitons and Fractals*, vol. 22, no. 3, pp. 513–520, 2004.

[7] X.-J. Yang, *Advanced Local Fractional Calculus and its Applications*, World Science, New York, NY, USA, 2012.

[8] J. H. He, "Nonlinear oscillation with fractional derivative and its applications," in *Proceedings of the International Conference on Vibrating Engineering 98*, vol. 9, Dalian, China, 1998.

[9] J.-H. He, "Homotopy perturbation technique," *Computer Methods in Applied Mechanics and Engineering*, vol. 178, no. 3-4, pp. 257–262, 1999.

[10] J.-H. He, "Approximate analytical solution for seepage flow with fractional derivatives in porous media," *Computer Methods in Applied Mechanics and Engineering*, vol. 167, no. 1-2, pp. 57–68, 1998.

[11] J. Liu and G. Hou, "Numerical solutions of the space- and time-fractional coupled Burgers equations by generalized differential transform method," *Applied Mathematics and Computation*, vol. 217, no. 16, pp. 7001–7008, 2011.

[12] S. Momani and Z. Odibat, "Analytical solution of a time-fractional Navier–Stokes equation by Adomian decomposition method," *Applied Mathematics and Computation*, vol. 177, no. 2, pp. 488–494, 2006.

[13] M. G. Sakar and F. Erdogan, "The homotopy analysis method for solving the time-fractional Fornberg-Whitham equation and comparison with Adomian's decomposition method," *Applied Mathematical Modelling*, vol. 37, no. 20-21, pp. 8876–8885, 2013.

[14] S. Kumar and D. Kumar, "Fractional modelling for BBM-Burger equation by using new homotopy analysis transform method," *Journal of the Association of Arab Universities for Basic and Applied Sciences*, vol. 16, pp. 16–20, 2014.

[15] S. Kumar, D. Kumar, S. Abbasbandy, and M. M. Rashidi, "Analytical solution of fractional Navier-Stokes equation by using modified Laplace decomposition method," *Ain Shams Engineering Journal*, vol. 5, no. 2, pp. 569–574, 2014.

[16] S. Momani, G. H. Erjaee, and M. H. Alnasr, "The modified homotopy perturbation method for solving strongly nonlinear oscillators," *Computers and Mathematics with Applications*, vol. 58, no. 11-12, pp. 2209–2220, 2009.

[17] M. G. Sakar, F. Uludag, and F. Erdogan, "Numerical solution of time-fractional nonlinear PDEs with proportional delays by homotopy perturbation method," *Applied Mathematical Modelling*, vol. 40, no. 13-14, pp. 6639–6649, 2016.

[18] J. Singh, D. Kumar, and A. Kılıçman, "Homotopy perturbation method for fractional gas dynamics equation using Sumudu transform," *Abstract and Applied Analysis*, vol. 2013, Article ID 934060, 8 pages, 2013.

[19] A. Saravanan and N. Magesh, "An efficient computational technique for solving the Fokker-Planck equation with space and time fractional derivatives," *Journal of King Saud University—Science*, vol. 28, no. 2, pp. 160–166, 2016.

[20] V. K. Srivastava, N. Mishra, S. Kumar, B. K. Singh, and M. K. Awasthi, "Reduced differential transform method for solving $(1 + n)$—dimensional Burgers' equation," *Egyptian Journal of Basic and Applied Sciences*, vol. 1, no. 2, pp. 115–119, 2014.

[21] V. K. Srivastava, S. Kumar, M. K. Awasthi, and B. K. Singh, "Two-dimensional time fractional-order biological population model and its analytical solution," *Egyptian Journal of Basic and Applied Sciences*, vol. 1, no. 1, pp. 71–76, 2014.

[22] B. K. Singh and V. K. Srivastava, "Approximate series solution of multi-dimensional, time fractional-order (heat like) diffusion equations using FRDTM," *Royal Society Open Science*, vol. 2, no. 4, Article ID 140511, 2015.

[23] B. K. Singh and P. Kumar, "FRDTM for numerical simulation of multi-dimensional, time-fractional model of Navier–Stokes equation," *Ain Shams Engineering Journal*, 2016.

[24] B. K. Singh and P. Kumar, "Numerical computation for time-fractional gas dynamics equations by fractional reduced differential transforms method," *Journal of Mathematics and System Science*, vol. 6, no. 6, pp. 248–259, 2016.

[25] B. K. Singh, "Fractional reduced differential transform method for numerical computation of a system of linear and nonlinear fractional partial differential equations," *International Journal of Open Problems in Computer Science and Mathematics*, vol. 9, no. 3, 2016.

[26] B. K. Singh and Mahendra, "A numerical computation of a system of linear and nonlinear time dependent partial differential equations using reduced differential transform method," *International Journal of Differential Equations*, vol. 2016, Article ID 4275389, 8 pages, 2016.

[27] D. Kumar, J. Singh, and D. Baleanu, "Numerical computation of a fractional model of differential-difference equation," *Journal of Computational and Nonlinear Dynamics*, vol. 11, no. 6, Article ID 061004, 2016.

[28] D. Kumar, J. Singh, and D. Baleanu, "A hybrid computational approach for Klein–Gordon equations on Cantor sets," *Nonlinear Dynamics*, vol. 87, no. 1, pp. 511–517, 2017.

[29] A. Atangana and I. Koca, "Chaos in a simple nonlinear system with Atangana-Baleanu derivatives with fractional order," *Chaos, Solitons & Fractals*, vol. 89, pp. 447–454, 2016.

[30] E. F. D. Goufo, "Application of the Caputo-Fabrizio fractional derivative without singular kernel to Korteweg-de Vries-Burgers equation," *Mathematical Modelling and Analysis*, vol. 21, no. 2, pp. 188–198, 2016.

[31] E. F. D. Goufo, "Chaotic processes using the two-parameter derivative with non-singular and non-local kernel: basic theory and applications," *Chaos*, vol. 26, no. 8, Article ID 084305, 2016.

[32] E. F. D. Goufo, "Stability and convergence analysis of a variable order replicator-mutator process in a moving medium," *Journal of Theoretical Biology*, vol. 403, pp. 178–187, 2016.

[33] X.-J. Yang, D. Baleanu, Y. Khan, and S. T. Mohyud-Din, "Local fractional variational iteration method for diffusion and wave equations on Cantor sets," *Romanian Journal of Physics*, vol. 59, no. 1-2, pp. 36–48, 2014.

[34] F. Geng, Y. Lin, and M. Cui, "A piecewise variational iteration method for Riccati differential equations," *Computers and Mathematics with Applications*, vol. 58, no. 11-12, pp. 2518–2522, 2009.

[35] A. Prakash, M. Kumar, and K. K. Sharma, "Numerical method for solving fractional coupled Burgers equations," *Applied Mathematics and Computation*, vol. 260, pp. 314–320, 2015.

[36] M. G. Sakar and H. Ergören, "Alternative variational iteration method for solving the time-fractional Fornberg-Whitham equation," *Applied Mathematical Modelling*, vol. 39, no. 14, pp. 3972–3979, 2015.

[37] Z. M. Odibat, "A study on the convergence of variational iteration method," *Mathematical and Computer Modelling*, vol. 51, no. 9-10, pp. 1181–1192, 2010.

[38] Z. Odibat and S. Momani, "The variational iteration method: an efficient scheme for handling fractional partial differential equations in fluid mechanics," *Computers & Mathematics with Applications*, vol. 58, no. 11-12, pp. 2199–2208, 2009.

[39] J. Wu, *Theory and Applications of Partial Functional Differential Equations*, Springer, New York, NY, USA, 1996.

[40] A. A. Keller, "Contribution of the delay differential equations to the complex economic macrodynamics," *WSEAS Transactions on Systems*, vol. 9, no. 4, pp. 358–371, 2010.

[41] B. Zubik-Kowal, "Chebyshev pseudospectral method and waveform relaxation for differential and differential-functional parabolic equations," *Applied Numerical Mathematics*, vol. 34, no. 2-3, pp. 309–328, 2000.

[42] B. Zubik-Kowal and Z. Jackiewicz, "Spectral collocation and waveform relaxation methods for nonlinear delay partial differential equations," *Applied Numerical Mathematics*, vol. 56, no. 3-4, pp. 433–443, 2006.

[43] J. Mead and B. Zubik-Kowal, "An iterated pseudospectral method for delay partial differential equations," *Applied Numerical Mathematics*, vol. 55, no. 2, pp. 227–250, 2005.

[44] R. Abazari and M. Ganji, "Extended two-dimensional DTM and its application on nonlinear PDEs with proportional delay," *International Journal of Computer Mathematics*, vol. 88, no. 8, pp. 1749–1762, 2011.

[45] R. Abazari and A. Kılıcman, "Application of differential transform method on nonlinear integro–differential equations with proportional delay," *Neural Computing & Applications*, vol. 24, no. 2, pp. 391–397, 2014.

[46] J. Tanthanuch, "Symmetry analysis of the nonhomogeneous inviscid Burgers equation with delay," *Communications in Nonlinear Science and Numerical Simulation*, vol. 17, no. 12, pp. 4978–4987, 2012.

[47] J. Biazar and B. Ghanbari, "The homotopy perturbation method for solving neutral functional-differential equations with proportional delays," *Journal of King Saud University—Science*, vol. 24, no. 1, pp. 33–37, 2012.

[48] X. Chen and L. Wang, "The variational iteration method for solving a neutral functional-differential equation with proportional delays," *Computers & Mathematics with Applications*, vol. 59, no. 8, pp. 2696–2702, 2010.

[49] B. K. Singh and P. Kumar, "Homotopy perturbation transform method for solving fractional partial differential equations with proportional delay," https://arxiv.org/abs/1611.06488v1.

5

Finite Time Synchronization of Extended Nonlinear Dynamical Systems Using Local Coupling

A. Acosta,[1] P. García,[2] H. Leiva,[1] and A. Merlitti[3]

[1]*School of Mathematical Sciences and Information Technology, Department of Mathematics, Yachay Tech, Urcuqui, Ecuador*
[2]*Facultad de Ingeniería en Ciencias Aplicadas, Universidad Técnica del Norte, Ibarra, Ecuador*
[3]*Departamento de Estadística, Facultad de Ciencias Económicas y Sociales, Universidad Central de Venezuela, Caracas, Venezuela*

Correspondence should be addressed to P. García; pgarcia@utn.edu.ec

Academic Editor: Peiguang Wang

We consider two reaction-diffusion equations connected by one-directional coupling function and study the synchronization problem in the case where the coupling function affects the driven system in some specific regions. We derive conditions that ensure that the evolution of the driven system closely tracks the evolution of the driver system at least for a finite time. The framework built to achieve our results is based on the study of an abstract ordinary differential equation in a suitable Hilbert space. As a specific application we consider the Gray-Scott equations and perform numerical simulations that are consistent with our main theoretical results.

1. Introduction

The synchronization of the evolution of systems that are sensitive to changes in the initial condition is a phenomenon that occurs spontaneously in systems ranging from biology to physics. As a matter of fact, starting from publications by Fujisaka and Yamada [1] and later by Pecora and Carroll [2], there have been many explanations about the occurrence of this phenomenon as well as new practical applications [3]; among these we highlight [4]. For localized systems (ODE) the problem is well understood; see, for example, [4, 5] and the references therein. On the other hand, a much smaller number of results are available for extended systems represented by partial differential equations (PDE). Among these in [6, 7], the authors have considered a pair of unidirectionally coupled systems with a linear term that penalizes the separation between the actual states of the systems. When the coupling function is linear, the synchronization problem has been addressed through different approaches like invariant manifold method via Galerkin's approximations [8], via an abstract formulation using semigroup theory [7, 9], or numerically [6].

With the exception of works [6, 10], in the rest of the references [7–9, 11, 12] the coupling function disturbs the system in its entirety. In contrast, in [6, 10], the authors propose a synchronization scheme that does need to disturb the whole driven system. Moreover, the subset of sites in the driven system is chosen arbitrarily.

In this work we present a general procedure for two reaction-diffusion equations connected through a one-directional coupling function. We study the synchronization problem in the case where the coupling function affects the driven system in some specific regions and our approach, which is based in an abstract formulation coming from semigroup theory, allows establishing a relation between the conditions to obtain synchronization in finite time and the intensity of the coupling. To illustrate the theoretical results we consider a pair of equations of Gray-Scott [13].

The paper is organized as follows: in Section 2 we set the problem, in Section 3 we give an abstract representation of the problem in a suitable Hilbert space, in Section 4 we give the main theoretical results, as the existence of bounded solutions of the abstract equation, in Section 5 we give an example and numerical simulations of the performance of the strategy, and finally in Section 6 we give some final remarks.

2. Setting of the Problem

We consider the following system with boundary Dirichlet conditions:

$$u_t = Du_{xx} + f(u), \tag{1}$$

$$v_t = Dv_{xx} + f(v) + p(x)(v - u), \tag{2}$$

where $0 < x < l, t > 0$. $D = \text{diag}(d_1, d_2, \ldots, d_n)$ is a diagonal matrix with positive entries, the function $f : \mathbb{R}^n \to \mathbb{R}^n$ is a continuous locally Lipschitz function, and $p(x)$ is defined as

$$p(x) = \sum_{i=1}^{m} \nu_i \mathscr{X}_{[a_i,b_i]}(x), \tag{3}$$

where, for $i \in \{1, 2, \ldots, m\}$, each $\nu_i \in \mathbb{R}$ and $\mathscr{X}_{[a_i,b_i]}$ is the characteristic function of the interval $[a_i, b_i]$, with $0 < a_1 < b_1 < \cdots < a_m < b_m < l$.

We show that the evolution of (2) closely tracks the evolution of (1) which means v behaves, in some sense, like u. To set precisely our problem and consider it in an abstract framework we start with a bounded solution $\overline{u}(x,t)$ of (1). Thus, there exists $N > 0$ such that

$$|\overline{u}(x,t)| := \sqrt{\sum_{i=1}^{n} \overline{u}_i^2(x,t)} \le N; \quad 0 < x < l,\ t > 0, \tag{4}$$

where $\overline{u}_i$ are the components of the vector valued function $\overline{u}$. Also, we assume that for any interval $J = [a,b] \subset (0, +\infty)$ there exists a constant $K > 0$, depending on J, such that

$$|\overline{u}_t(x,t)| \le K; \quad 0 < x < l,\ t \in J. \tag{5}$$

Let us define a function $g : (0,l) \times (0,\infty) \times \mathbb{R}^n \to \mathbb{R}^n$ by

$$g(x,t,e) := f(e + \overline{u}(x,t)) - f(\overline{u}(x,t)) + p(x)e \tag{6}$$

and consider the transformation

$$e(x,t) = v(x,t) - \overline{u}(x,t). \tag{7}$$

If v is a solution of (2), with input $\overline{u}(x,t)$, then (6) and (7) lead us to the equation

$$e_t = De_{xx} + g(x,t,e). \tag{8}$$

Now, we consider (8) together with Dirichlet boundary conditions:

$$\begin{aligned} e(0,t) &= 0, \\ e(l,t) &= 0, \\ t &> 0. \end{aligned} \tag{9}$$

Our efforts will focus on problem (8)-(9). Concretely, we are interested in solutions such that the driven systems closely track the evolution of the driver systems at least for a finite time interval.

3. Preliminaries and Abstract Formulation of the Problem

In this section, by choosing an appropriate Hilbert space, we discuss some preliminaries and set our problem as an abstract ordinary differential equation. Let us start considering the Hilbert space $H := L^2((0,l), \mathbb{R}^n)$ with the usual inner product; that is, if $\Phi = (\Phi^1, \ldots, \Phi^n)^T$, $\Psi = (\Psi^1, \ldots, \Psi^n)^T \in H$, then

$$\langle \Phi, \Psi \rangle = \int_0^l \left(\sum_{i=1}^{n} \Phi^i(x) \Psi^i(x) \right) dx \tag{10}$$

and the induced norm is given by

$$\|\Phi\|^2 = \int_0^l \left(\sum_{i=1}^{n} \left[\Phi^i(x)\right]^2 \right) dx. \tag{11}$$

Next, we consider the linear unbounded operator $A : D(A) \subset H \to H$ defined by

$$A\Phi := -D\frac{d^2}{dx^2}\Phi, \tag{12}$$

where

$$D(A) = H_0^1((0,l), \mathbb{R}^n) \cap H^2((0,l), \mathbb{R}^n). \tag{13}$$

We summarize some very well-known important properties related to the operator A:

(i) A is a sectorial operator. As a consequence $-A$ generates an analytic semigroup, e^{-At}, which is, for each $t > 0$, compact.

(ii) The spectrum $\sigma(A)$, of A, consists of just eigenvalues $\lambda_{i,j} = d_j(i\pi/l)^2$, with $i = 1, 2, \ldots$ and $j = 1, 2, \ldots, n$. We order the set of eigenvalues $\{\lambda_{i,j}\}$ according to the sequence $0 < \lambda_1 \le \lambda_2 \le \cdots \to \infty$, where

$$\lambda_1 = \min\{d_1, d_2, \ldots, d_n\} \left(\frac{\pi}{l}\right)^2. \tag{14}$$

(iii) There exists a complete orthonormal set $\{\Phi_i\}_{i=1}^{\infty}$, of eigenvectors of A, such that

$$A\Phi = \sum_{i=1}^{\infty} \lambda_i \langle \Phi, \Phi_i \rangle \Phi_i, \quad \Phi \in D(A). \tag{15}$$

(iv) e^{-At} is given by

$$e^{-At}\Phi = \sum_{i=1}^{\infty} e^{-\lambda_i t} \langle \Phi, \Phi_i \rangle \Phi_i, \quad \Phi \in H. \tag{16}$$

In the remainder of this section we mainly follow [14, 15] and the notations used come from [15]. In order to study the nonlinear part of the abstract equation corresponding to (8)-(9), we consider the fractional power spaces and the interpolation spaces associated with the operator A. For any

$\alpha \geq 0$, the domain $D(A^\alpha)$ of the fractional power operator A^α is defined by

$$V^{2\alpha} := D(A^\alpha) := \left\{\Phi \in H : \sum_{i=1}^{\infty} \lambda_i^{2\alpha} |\langle \Phi, \Phi_i \rangle|^2 < \infty \right\}, \tag{17}$$

and the operator A^α is given by

$$A^\alpha \Phi = \sum_{i=1}^{\infty} \lambda_i^\alpha \langle \Phi, \Phi_i \rangle \Phi_i, \quad \forall \Phi \in D(A^\alpha). \tag{18}$$

V^α itself becomes a Hilbert space with the V^α-inner product given by $\langle \Phi, \Psi \rangle_\alpha := \sum_{i=1}^{\infty} \lambda_i^\alpha \langle \Phi, \Phi_i \rangle \langle \Psi, \Phi_i \rangle$ and the V^α-norm is the graph norm associated with A^α; that is, $\|\Phi\|_\alpha = \|A^\alpha \Phi\|$. Moreover, if $\alpha \geq \beta$, then V^α is a continuous embedding into V^β that verifies the estimate $\|\Phi\|_\beta^2 \leq \lambda_1^{\beta-\alpha} \|\Phi\|_\alpha^2$, for all $\Phi \in V^\alpha$. In particular,

$$\|\Phi\|^2 \leq \lambda_1^{-\alpha} \|\Phi\|_\alpha^2, \quad \forall \Phi \in V^\alpha. \tag{19}$$

Also, according to Theorem 1.6.1 in [14] and the discussion given there, for $1/4 < \alpha \leq 1$ we have that $V^{2\alpha}$ is a continuous embedding into $C((0,l), \mathbb{R}^n)$. Thus, there exists a positive constant C such that

$$\sup_{x \in (0,l)} |\Phi(x)| \leq C \|\Phi\|_{2\alpha}, \quad \forall \Phi \in V^{2\alpha}. \tag{20}$$

The next proposition, whose proof is similar to the one given in [7], contains estimates relating the semigroup $\{e^{-At}\}$ with norms $\|\cdot\|_\alpha$ and $\|\cdot\|$. Also, it will play an important role in the discussion of our main theoretical results.

Proposition 1. *For each $\Phi \in V^\alpha$, $\alpha > 0$, one has the following estimates:*

$$\begin{aligned} \left\|e^{-At}\Phi\right\|_\alpha^2 &\leq e^{-2\lambda_1 t} \|\Phi\|_\alpha^2, \quad t \geq 0, \\ \left\|e^{-At}\Phi\right\|_\alpha^2 &\leq t^{-\alpha} \alpha^\alpha e^{-\alpha} e^{-\lambda_1 t} \|\Phi\|^2, \quad t > 0. \end{aligned} \tag{21}$$

Now, we associate with system (8)-(9) an abstract ordinary differential equation on H with an initial condition

$$\begin{aligned} \dot{\Phi} + A\Phi &= F(t, \Phi), \quad t > 0; \\ \Phi(0) &= \Phi_0, \end{aligned} \tag{22}$$

where F, acting on $[0,\infty) \times V^{2\alpha}$, is defined by

$$F(t, \Phi)(x) := g(x, t, \Phi(x)). \tag{23}$$

For some $r > 0$ and $1/4 < \alpha < 1$, we assume F maps $[0,\infty) \times U_r$ into H, where $U_r = \{\Phi \in V^{2\alpha} : \|\Phi\|_{2\alpha} \leq r\}$.

The following lemma establishes that F is Lipschitz continuous in the second variable on U_r.

Lemma 2. *There exists a constant $L = L(U_r)$ such that for $\Phi_1, \Phi_2 \in U_r$, $t > 0$,*

$$\|F(t, \Phi_1) - F(t, \Phi_2)\| \leq L \|\Phi_1 - \Phi_2\|_{2\alpha}. \tag{24}$$

Proof. Given a ball $B_{\bar{r}}(0)$ of radius $\bar{r}$ and center 0 in $\mathbb{R}^n$, there exists a positive constant $\bar{L} = \bar{L}(\bar{r})$ such that $|f(z_2) - f(z_1)| \leq \bar{L}|z_2 - z_1|$ for all $z_1, z_2 \in B_{\bar{r}}(0)$.

For any $\Phi_1, \Phi_2 \in U_r$, we consider $\Delta := |F(t,\Phi_1)(x) - F(t,\Phi_2)(x)|$, $0 < x < l$, and $t > 0$. Now, let us consider N and C as in (4) and (20), respectively. If we choose $\bar{r} = Cr + N$, then there exists $\bar{L} = \bar{L}(\bar{r})$ such that

$$\begin{aligned} \Delta &\leq \bar{L}|\Phi_1(x) - \Phi_2(x)| + |p(x)| |\Phi_1(x) - \Phi_2(x)| \\ &\leq (\bar{L} + |p(x)|) \sup_{x \in (0,l)} |\Phi_1(x) - \Phi_2(x)| \\ &\leq (\bar{L} + |p(x)|) C \|\Phi_1 - \Phi_2\|_{2\alpha}. \end{aligned} \tag{25}$$

Therefore,

$$\begin{aligned} \int_0^l \Delta^2 dx &\leq \int_0^l C^2 (\bar{L} + |p(x)|)^2 \|\Phi_1 - \Phi_2\|_{2\alpha}^2 dx \\ &\leq 2C^2 \left(l\bar{L}^2 + \|p\|^2\right) \|\Phi_1 - \Phi_2\|_{2\alpha}^2. \end{aligned} \tag{26}$$

Thus,

$$\|F(t, \Phi_1) - F(t, \Phi_2)\| \leq L \|\Phi_1 - \Phi_2\|_{2\alpha} \tag{27}$$

with

$$L = \sqrt{2} C \left(l\bar{L}^2 + \|p\|^2\right)^{1/2}. \tag{28}$$

□

We finish this section with a lemma that will be used to obtain our main theoretical results. It can be established as an application of Lemma 3.3.2 in [14].

Lemma 3. *A continuous function $\Phi : (0, t_1) \to V^{2\alpha}$ is a solution of the integral equation*

$$\Phi(t) = e^{-At}\Phi_0 + \int_0^t e^{-A(t-s)} F(s, \Phi(s)) ds, \quad t \in (0, t_1), \tag{29}$$

if and only if Φ is a solution of (22).

4. Main Theoretical Results

Theorem 4. *For any $\Phi_0 \in \text{int}(U_r)$ there exists $t_1 = t_1(\Phi_0) > 0$ such that (22) has a unique solution Φ on $(0, t_1)$ with initial condition $\Phi(0) = \Phi_0$.*

Proof. By Lemma 3, it suffices to prove the corresponding result for integral equation (29).

Choose $\rho > 0$, with $\rho + \|\Phi_0\|_{2\alpha} < r$, such that the set

$$V = \left\{\Phi \in V^{2\alpha} : \|\Phi - \Phi_0\|_{2\alpha} \leq \rho\right\} \tag{30}$$

is contained in U_r. We have, applying Lemma 2, that F is Lipschitz continuous, in the second variable, on V. Moreover, for the estimate

$$\|F(t, \Phi_1) - F(t, \Phi_2)\| \leq L \|\Phi_1 - \Phi_2\|_{2\alpha}, \quad \text{for } t > 0, \ \Phi_1, \Phi_2 \in V, \tag{31}$$

we choose $L = \sqrt{2}C(l\overline{L}^2+\|p\|^2)^{1/2}$, with $\overline{L} = \overline{L}(C(\rho+\|\Phi_0\|_{2\alpha})+N)$. Next, we select $t_1 > 0$ such that

$$\left\|\left(e^{-At} - I\right)\Phi_0\right\|_{2\alpha} \leq \frac{\rho}{2}, \quad 0 \leq t \leq t_1, \tag{32}$$

$$\left(\frac{2\alpha}{e}\right)^{\alpha} L\left(\rho + \|\Phi_0\|_{2\alpha}\right)\int_0^{t_1} u^{-\alpha}e^{-(\lambda_1/2)u}du \leq \frac{\rho}{2}. \tag{33}$$

Let us define S as the set of continuous functions $\Psi : [0,t_1] \to V^{2\alpha}$ such that $\|\Psi(t) - \Phi_0\|_{2\alpha} \leq \rho$ on $[0,t_1]$. If S is endowed with the supreme norm $\|\Psi\|^{t_1} := \sup_{0\leq t\leq t_1}\|\Psi(t)\|_{2\alpha}$, then it is a complete metric space.

Now, for $\Psi \in S$ we define $T(\Psi)$ acting on $[0,t_1]$ as

$$T(\Psi)(t) = e^{-At}\Phi_0 + \int_0^t e^{-A(t-s)}F(s,\Psi(s))\,ds. \tag{34}$$

First, we show that T maps S into itself. In fact,

$$\begin{aligned}\|T(\Psi)(t) - \Phi_0\|_{2\alpha} &\leq \left\|\left(e^{-At} - I\right)\Phi_0\right\|_{2\alpha} \\ &+ \int_0^t \left\|e^{-A(t-s)}F(s,\Psi(s))\right\|_{2\alpha} ds \leq \frac{\rho}{2} \\ &+ \int_0^t \left\|e^{-A(t-s)}F(s,\Psi(s))\right\|_{2\alpha} ds \leq \frac{\rho}{2} + \left(\frac{2\alpha}{e}\right)^{\alpha} \\ &\cdot \int_0^t (t-s)^{-\alpha}e^{-(\lambda_1/2)(t-s)}\|F(s,\Psi(s))\|\,ds \leq \frac{\rho}{2} \\ &+ \left(\frac{2\alpha}{e}\right)^{\alpha}\int_0^t (t-s)^{-\alpha}e^{-(\lambda_1/2)(t-s)}L\|\Psi(s)\|_{2\alpha}\,ds \\ &\leq \frac{\rho}{2} + \left(\frac{2\alpha}{e}\right)^{\alpha} L\left(\rho + \|\Phi_0\|_{2\alpha}\right) \\ &\cdot \int_0^{t_1} (t-s)^{-\alpha}e^{-(\lambda_1/2)(t-s)}ds \leq \rho, \quad \text{for } 0 \leq t \leq t_1.\end{aligned} \tag{35}$$

The fact that $T(\Psi)$ is continuous from $[0,t_1]$ to $V^{2\alpha}$ is easily proved.

Next, we shall prove that T is a contraction. In fact, if $\Psi_1, \Psi_2 \in S$ and $0 \leq t \leq t_1$, then for $\Delta := \|T(\Psi_1)(t) - T(\Psi_2)(t)\|_{2\alpha}$ we have that

$$\begin{aligned}\Delta &\leq \int_0^t \left\|e^{-A(t-s)}\left(F(s,\Psi_1(s)) - F(s,\Psi_2(s))\right)\right\|_{2\alpha} ds \\ &\leq \left(\frac{2\alpha}{e}\right)^{\alpha}\int_0^t (t-s)^{-\alpha} \\ &\cdot e^{-(\lambda_1/2)(t-s)}\|F(s,\Psi_1(s)) - F(s,\Psi_2(s))\|\,ds \\ &\leq \left(\frac{2\alpha}{e}\right)^{\alpha}\int_0^t L(t-s)^{-\alpha} \\ &\cdot e^{-(\lambda_1/2)(t-s)}\left\|\Psi_1(s) - \Psi_2(s)\right\|_{2\alpha} ds \leq \left(\frac{2\alpha}{e}\right)^{\alpha} \\ &\cdot L\left(\int_0^t (t-s)^{-\alpha}e^{-(\lambda_1/2)(t-s)}ds\right)\|\Psi_1 - \Psi_2\|^{t_1} \\ &\leq \frac{\rho}{2\left(\rho + \|\Phi_0\|_{2\alpha}\right)}\|\Psi_1 - \Psi_2\|^{t_1}.\end{aligned} \tag{36}$$

Therefore, $\|T(\Psi_1)-T(\Psi_2)\|^{t_1} \leq (1/2)\|\Psi_1-\Psi_2\|^{t_1}$ for all $\Psi_1, \Psi_2 \in S$.

Finally, by the Banach fixed point theorem, T has a unique fixed point Φ in S, which is a continuous solution of integral equation (29). By Lemma 3, this is the unique solution of (22) on $(0,t_1)$ with initial value $\Phi(0) = \Phi_0$. □

The previous theorem does not tell anything about the maximal interval where Φ is defined. In this regard we have the following.

Theorem 5. *Assume that for every closed set $B \subset \operatorname{int}(U_r)$, $F([0,\infty) \times B)$ is bounded in H. If Φ is a solution of (22) on $(0,t_1)$ and t_1 is maximal, then either $t_1 = +\infty$ or else there exists a sequence $t_n \to t_1^-$ as $n \to \infty$ such that $\Phi(t_n) \to \partial U_r$.*

Proof. Suppose $t_1 < \infty$ and there is not neighborhood N of ∂U_r such that $\Phi(t)$ enters N for t in an interval $[t_1-\epsilon,t_1)$, with ϵ small enough. We may take N of the form $N = U_r - B$ where B is a closed subset of $\operatorname{int}(U_r)$, and $\Phi(t) \in B$ for $t \in [t_1-\epsilon,t_1)$.

We are going to prove that there exists $\Phi_1 \in B$ such that $\Phi(t) \to \Phi_1$ in $V^{2\alpha}$ as $t \to t_1^-$, and this implies that the solution may be extended beyond time t_1 (with $\Phi(t_1) = \Phi_1$), contradicting maximality of t_1.

Now let $M := \sup\{\|F(t,\Phi)\| : t \geq 0,\ \Phi \in B\}$. We first show that $\|\Phi(t)\|_{2\alpha}$ remains bounded on the interval $(0,t_1)$; in fact

$$\begin{aligned}&\|\Phi(t)\|_{2\alpha} \\ &\leq \left\|e^{-At}\Phi_0\right\|_{2\alpha} + \int_0^t \left\|e^{-A(t-s)}F(s,\Phi(s))\right\|_{2\alpha} ds \\ &\leq e^{-\lambda_1 t}\left\|\Phi_0\right\|_{2\alpha} \\ &\quad + \left(\frac{2\alpha}{e}\right)^{\alpha}\int_0^t (t-s)^{-\alpha}e^{-(\lambda_1/2)(t-s)}\|F(s,\Phi(s))\|\,ds \\ &\leq e^{-\lambda_1 t}\left\|\Phi_0\right\|_{2\alpha} + M\left(\frac{2\alpha}{e}\right)^{\alpha}\int_0^t (t-s)^{-\alpha}ds \\ &= e^{-\lambda_1 t}\left\|\Phi_0\right\|_{2\alpha} + M\left(\frac{2\alpha}{e}\right)^{\alpha}\frac{t^{1-\alpha}}{1-\alpha}.\end{aligned} \tag{37}$$

Now we consider the difference $\Phi(t) - \Phi(\tau)$ with t and τ such that $t_1 - \epsilon \leq \tau < t < t_1$. It is obtained that

$$\begin{aligned}&\Phi(t) - \Phi(\tau) \\ &\quad = \left(e^{-At} - e^{-A\tau}\right)\Phi_0 + \int_\tau^t e^{-A(t-s)}F(s,\Phi(s))\,ds \\ &\quad + \int_0^\tau \left(e^{-A(t-s)} - e^{-A(\tau-s)}\right)F(s,\Phi(s))\,ds.\end{aligned} \tag{38}$$

For each term in the right hand side we get an estimate. Let us call $I_1 := \|(e^{-At}-e^{-A\tau})\Phi_0\|_{2\alpha}$, $I_2 := \|\int_\tau^t e^{-A(t-s)}F(s,\Phi(s))ds\|_{2\alpha}$, and $I_3 := \|\int_0^\tau (e^{-A(t-s)} - e^{-A(\tau-s)})F(s,\Phi(s))ds\|_{2\alpha}$; now we have

$$\begin{aligned}
I_1 &= \left\| \int_\tau^t A e^{-As}\Phi_0 ds \right\|_{2\alpha} \le \int_\tau^t \left\| A e^{-As}\Phi_0 \right\|_{2\alpha} ds \\
&= \int_\tau^t \left\| e^{-As}A\Phi_0 \right\|_{2\alpha} ds \le \int_\tau^t \left(\frac{2\alpha}{e}\right)^\alpha \\
&\quad \cdot s^{-\alpha} e^{-(\lambda_1/2)s} \left\| A\Phi_0 \right\| ds \le \left(\frac{2\alpha}{e}\right)^\alpha \left\| A\Phi_0 \right\| \\
&\quad \cdot \int_\tau^t s^{-\alpha} ds = \left(\frac{2\alpha}{e}\right)^\alpha \left\| A\Phi_0 \right\| \frac{t^{1-\alpha} - \tau^{1-\alpha}}{1-\alpha}, \\
I_2 &\le \int_\tau^t \left\| e^{-A(t-s)} F(s,\Phi(s)) \right\|_{2\alpha} ds \le \int_\tau^t \left(\frac{2\alpha}{e}\right)^\alpha (t \\
&\quad - s)^{-\alpha} e^{-(\lambda_1/2)(t-s)} \left\| F(s,\Phi(s)) \right\| ds \le M \left(\frac{2\alpha}{e}\right)^\alpha \\
&\quad \cdot \int_\tau^t (t-s)^{-\alpha} ds = M \left(\frac{2\alpha}{e}\right)^\alpha \frac{(t-\tau)^{1-\alpha}}{1-\alpha}, \\
I_3 &\le \int_0^{\tau-\epsilon} \left\| e^{-A(\tau-s-\epsilon)} \left(e^{-A(t-\tau+\epsilon)} - e^{-A\epsilon} \right) \right. \\
&\quad \left. \cdot F(s,\Phi(s)) \right\|_{2\alpha} ds \\
&\quad + \int_{\tau-\epsilon}^\tau \left\| \left(e^{-A(t-s)} - e^{-A(\tau-s)} \right) F(s,\Phi(s)) \right\|_{2\alpha} ds \\
&\le M \left(\frac{2\alpha}{e}\right)^\alpha \left\| e^{-A(t-\tau+\epsilon)} - e^{-A\epsilon} \right\| \int_0^{\tau-\epsilon} (\tau - s \\
&\quad - \epsilon)^{-\alpha} ds + \int_{\tau-\epsilon}^\tau \left\| e^{-A(t-s)} F(s,\Phi(s)) \right\|_{2\alpha} ds \\
&\quad + \int_{\tau-\epsilon}^\tau \left\| e^{-A(\tau-s)} F(s,\Phi(s)) \right\|_{2\alpha} ds \le M \left(\frac{2\alpha}{e}\right)^\alpha \\
&\quad \cdot \left\| e^{-A(t-\tau+\epsilon)} - e^{-A\epsilon} \right\| \frac{(\tau-\epsilon)^{1-\alpha}}{1-\alpha} + M \left(\frac{2\alpha}{e}\right)^\alpha \\
&\quad \cdot \int_{\tau-\epsilon}^\tau \left((t-s)^{-\alpha} + (\tau-s)^{-\alpha} \right) ds = M \left(\frac{2\alpha}{e}\right)^\alpha \\
&\quad \cdot \left\| e^{-A(t-\tau+\epsilon)} - e^{-A\epsilon} \right\| \frac{(\tau-\epsilon)^{1-\alpha}}{1-\alpha} + M \left(\frac{2\alpha}{e}\right)^\alpha \\
&\quad \cdot \frac{(t-\tau+\epsilon)^{1-\alpha} - (t-\tau)^{1-\alpha} + \epsilon^{1-\alpha}}{1-\alpha}.
\end{aligned} \tag{39}$$

Since e^{-At} is compact for $t > 0$, then $\{e^{-At}\}$ is a uniformly continuous semigroup, which implies that $\|e^{-A(t-\tau+\epsilon)} - e^{-A\epsilon}\| \to 0$ as $t \to \tau$.

Finally from the estimates given for I_1, I_2, and I_3 we conclude that there exists $\Phi_1 \in B$ such that $\lim_{t\to t_1^-}\Phi(t) = \Phi_1$, and the proof is complete. □

Corollary 6. *There exists $\epsilon > 0$ such that the solution Φ, of problem (22), satisfies the estimate*

$$\|\Phi(t)\|_{2\alpha} \le \|\Phi_0\|_{2\alpha} \tag{40}$$

for all t belonging to the interval $[0,\epsilon)$.

Proof. Let B a closed subset of $V^{2\alpha}$ that contains the initial condition Φ_0 in its interior and $M := \sup\{\|F(t,\Phi)\| : t \ge 0,\ \Phi \in B\}$. There exists $\tilde{t}_1 > 0$ such that

$$\|\Phi(t)\|_{2\alpha} \le e^{-\lambda_1 t}\|\Phi_0\|_{2\alpha} + M\left(\frac{2\alpha}{e}\right)^\alpha \frac{t^{1-\alpha}}{1-\alpha}, \quad \text{for } 0 < t < \tilde{t}_1. \tag{41}$$

Therefore

$$\|\Phi(t)\|_{2\alpha} < \|\Phi_0\|_{2\alpha} + M\left(\frac{2\alpha}{e}\right)^\alpha \frac{t^{1-\alpha}}{1-\alpha}, \tag{42}$$

and the result follows due to the fact that $\lim_{t\to 0^+} M(2\alpha/e)^\alpha (t^{1-\alpha}/(1-\alpha))e^{\lambda_1 t} = 0$. □

5. Example and Numerical Simulations

To illustrate our theoretical results we consider the particular case of system (1)-(2):

$$\begin{aligned}
\frac{\partial u_1}{\partial t} &= d_1 \frac{\partial^2 u_1}{\partial x^2} - u_1 u_2^2 + a(1-u_1), \\
\frac{\partial u_2}{\partial t} &= d_2 \frac{\partial^2 u_2}{\partial x^2} + u_1 u_2^2 - (a+b)u_2,
\end{aligned} \tag{43}$$

$$\begin{aligned}
\frac{\partial v_1}{\partial t} &= d_1 \frac{\partial^2 v_1}{\partial x^2} - v_1 v_2^2 + a(1-v_1) + p(x)(v_1 - u_1), \\
\frac{\partial v_2}{\partial t} &= d_2 \frac{\partial^2 v_2}{\partial x^2} + v_1 v_2^2 - (a+b)v_2 + p(x)(v_2 - u_2),
\end{aligned} \tag{44}$$

where $0 < x < l$, $t > 0$. System (43) corresponds to the Gray-Scott cubic autocatalysis model [13] which is related to two irreversible chemical reactions and exhibits mixed mode spatiotemporal chaos. Here a, b, d_1, and d_2 are dimensionless constants, where b corresponds to the rate of conversion of a component into another, a is the rate of the process that feeds a component and drains another, and d_i, $i = 1, 2$, are the diffusion rates. In the context of (1) the function f is defined as $f\binom{u_1}{u_2} = \binom{-u_1u_2^2 + a(1-u_1)}{u_1u_2^2 - (a+b)u_2}$ and the (8), that is, $e_t = De_{xx} + g(x,t,e)$, becomes in

$$\begin{aligned}
\begin{pmatrix} e_1 \\ e_2 \end{pmatrix}_t &= \begin{pmatrix} d_1 & 0 \\ 0 & d_2 \end{pmatrix} \begin{pmatrix} e_1 \\ e_2 \end{pmatrix}_{xx} + \left(e_1 e_2^2 + 2\overline{u}_2(x,t) e_1 e_2 \right. \\
&\quad \left. + \overline{u}_2^2(x,t) e_1 + \overline{u}_1 e_2^2 + 2\overline{u}_1(x,t)\overline{u}_2(x,t) e_2 \right) \begin{pmatrix} -1 \\ 1 \end{pmatrix} \\
&\quad + \begin{pmatrix} -a + p(x) & 0 \\ 0 & -a - b + p(x) \end{pmatrix} \begin{pmatrix} e_1 \\ e_2 \end{pmatrix}.
\end{aligned} \tag{45}$$

Proposition 7. *There exists a real value function h, continuous and increasing on the interval* $(0, \infty)$ *such that* $|g(x,t,e)| \leq h(|e|)$ *for all* (x,t,e) *in* $(0,l) \times (0,\infty) \times \mathbb{R}^2$.

Proof. The estimates

$$\begin{aligned}
\left|e_1 e_2^2\right| &\leq |e|^3, \\
\left|2\bar{u}_2(x,t) e_1 e_2\right| &\leq 2N|e|^2, \\
\left|\bar{u}_2^2(x,t) e_1\right| &\leq N^2|e|, \\
\left|\bar{u}_1 e_2^2\right| &\leq N|e|^2, \\
\left|2\bar{u}_1(x,t)\bar{u}_2(x,t) e_2\right| &\leq 2N^2|e|, \\
\left|\begin{pmatrix} -a+p(x) & 0 \\ 0 & -a-b+p(x) \end{pmatrix}\begin{pmatrix} e_1 \\ e_2 \end{pmatrix}\right| &\leq \mathscr{C}|e|,
\end{aligned} \tag{46}$$

where $\mathscr{C}$ is a constant that depends on a, b, and the function p, imply that $|g(x,t,e)| \leq \sqrt{2}(|e|^3 + 3N|e|^2 + 3N^2|e|) + \mathscr{C}|e|$. Thus, h could be defined by $h(s) = \sqrt{2}(s^3 + 3Ns^2 + 3N^2 s) + \mathscr{C}s$. □

To apply Theorem 4, we observe that for the abstract problem the extended function given in (23) becomes in

$$\begin{aligned}
F(t,\Phi)(x) &= g(x,t,\Phi) \\
&= (\Psi_1(x,t), \Psi_2(x,t))^T + p(x)\Phi(x),
\end{aligned} \tag{47}$$

where

$$\begin{aligned}
\Psi_1(x,t) &= -\Phi_1(x)\Phi_2^2(x) - 2\Phi_1(x)\Phi_2(x)\bar{u}_2(x,t) \\
&\quad - \Phi_1(x)\bar{u}_2^2(x,t) - \Phi_2^2(x)\bar{u}_1(x,t) \\
&\quad - 2\Phi_2(x)\bar{u}_1(x,t)\bar{u}_2(x,t) - a\Phi_1(x), \\
\Psi_2(x,t) &= \Phi_1(x)\Phi_2^2(x) + 2\Phi_1(x)\Phi_2(x)\bar{u}_2(x,t) \\
&\quad + \Phi_1(x)\bar{u}_2^2(x,t) + \Phi_2^2(x)\bar{u}_1(x,t) \\
&\quad + 2\Phi_2(x)\bar{u}_1(x,t)\bar{u}_2(x,t) \\
&\quad - (a+b)\Phi_2(x),
\end{aligned} \tag{48}$$

being $\Phi = (\Phi_1, \Phi_2)^T$, $\bar{u}(x,t) = (\bar{u}_1(x,t), \bar{u}_2(x,t))^T$.

Now, for $\Phi \in V^{2\alpha}$ using (20) and Proposition 7 we obtain that

$$\|F(t,\Phi)\| = \|g(\cdot,t,\Phi)\| \leq h(C\|\Phi\|_{2\alpha}). \tag{49}$$

Hence F maps bounded sets in $[0,\infty) \times V^{2\alpha}$ into bounded sets in H.

In order to realize a numerical implementation to illustrate the main result, the values for the constants d_1, d_2, a, and b appearing in system (43)-(44) are chosen as $d_1 = 5 \times 10^{-3}$, $d_2 = 5 \times 10^{-4}$, $a = 0.028$, and $b = 0.053$, and initial conditions are given by

$$\begin{aligned}
u_1(0,x) &= \sin\left(\frac{\pi x}{l}\right), \\
u_2(0,x) &= \left(e^{-1000(x-2l/3)^2} + e^{-10(x-l/3)^2}\right)\sin\left(\frac{\pi x}{l}\right), \\
v_1(0,x) &= e^{-10(x-l/2)^2}\sin\left(\frac{\pi x}{l}\right), \\
v_2(0,x) &= e^{-10(x-l/2)^2}\sin\left(\frac{\pi x}{l}\right).
\end{aligned} \tag{50}$$

In this case the Lipschitz constant L, appearing in (33), is given by

$$L = \sqrt{2}C\left(l\left(3\sqrt{2}\left(C\left(\rho + \|\Phi_0\|_{2\alpha}\right) + N\right)^2 + a + b\right)^2 + \|p\|^2\right)^{1/2}. \tag{51}$$

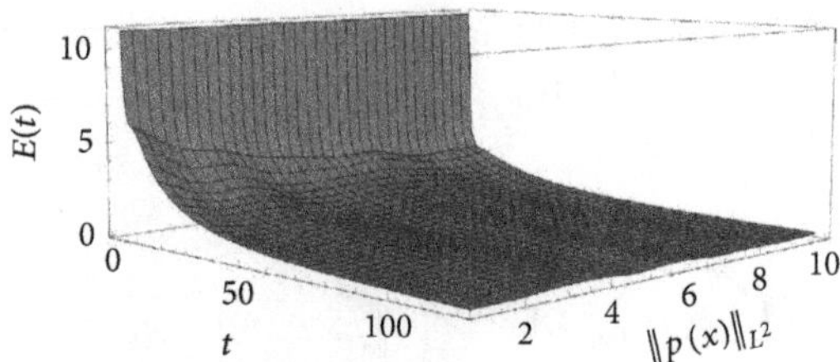

Figure 1: Synchronization error as function of the time and the intensity of the perturbation defined as $\|p(x)\|_{L^2}$.

Figure 1 shows a qualitative result of the synchronization error as function of the time, defined as

$$E(t) = \left[\frac{1}{l}\int_0^l \sum_{i=1}^2 (u_i(x,t) - v_i(x,t))^2 dx\right]^{1/2}, \tag{52}$$

and the intensity of the perturbation defined as $\|p(x)\|_{L^2}$. There, for a fixed time, the error always is minor compared with the initial error, consistent with our main result.

6. Concluding Remarks

We present a synchronization scheme of reaction-diffusion equations connected by a localized one-directional coupling function and give conditions that ensure the synchronization at least for a finite time. Conditions for synchronization depend on a sort of coupling intensity given by the L^2 norm of the coupling function. This norm is related to the intensity of local perturbation and its spatial extension, suggesting that this relation can be optimized in order to improve the synchronization or design of a control scheme.

Finally, although we have proven that the synchronization occurs in an interval of time, the numerical simulations suggest that this interval can be extended.

Disclosure

P. García's permanent address is Laboratorio de Sistemas Complejos, Departamento de Física Aplicada, Facultad de Ingeniería, Universidad Central de Venezuela.

Conflicts of Interest

The authors declare that there are no conflicts of interest regarding the publication of this paper.

References

[1] H. Fujisaka and T. Yamada, "Stability theory of synchronized motion in coupled-oscillator systems," *Progress of Theoretical and Experimental Physics*, vol. 69, no. 1, pp. 32–47, 1983.

[2] L. M. Pecora and T. L. Carroll, "Synchronization in chaotic systems," *Physical Review Letters*, vol. 64, no. 8, pp. 821–824, 1990.

[3] B. Jovic, *Synchronization Techniques for Chaotic Communication Systems*, Springer, Berlin, Germany, 2001.

[4] S. Boccaletti, J. Kurths, G. Osipov, D. L. Valladares, and C. S. Zhou, "The synchronization of chaotic systems," *Physics Reports*, vol. 366, no. 1-2, pp. 1–101, 2002.

[5] L. Pecora, T. Carroll, G. Johnson, D. Mar, and K. S. Fink, "Synchronization stability in coupled oscillator arrays: solution for arbitrary configurations," *International Journal of Bifurcation and Chaos*, vol. 10, no. 2, pp. 273–290, 2000.

[6] L. Kocarev, Z. Tasev, and U. Parlitz, "Synchronizing spatiotemporal chaos of partial differential equations," *Physical Review Letters*, vol. 79, no. 1, pp. 51–54, 1997.

[7] A. Acosta, P. Garca, and H. Leiva, "Synchronization of non-identical extended chaotic systems," *Applicable Analysis: An International Journal*, vol. 92, no. 4, pp. 740–751, 2013.

[8] L. Xie and Y. Zhao, "Synchronization of some kind of PDE chaotic systems by invariant manifold method," *International Journal of Bifurcation and Chaos*, vol. 5, no. 7, pp. 2303–2309, 2005.

[9] P. García, A. Acosta, and H. Leiva, "Synchronization conditions for master-slave reaction diffusion systems," *Europhysics Letters*, vol. 88, no. 6, pp. 60006-1–60006-6, 2009.

[10] Z. Xu and W. Jiangsum, "Synchronization of two discrete Ginzburg-Landau equations using local coupling," *International Journal of Nonlinear Science*, vol. 1, no. 1, pp. 19–29, 2006.

[11] R. O. Grigoriev and A. Handel, "Non-normality and the localized control of extended systems," *Physical Review E: Statistical, Nonlinear, and Soft Matter Physics*, vol. 66, no. 6, Article ID 067201, pp. 067201/1–067201/4, 2002.

[12] K. Wu and B.-S. Chen, "Synchronization of partial differential systems via diffusion coupling," *IEEE Transactions on Circuits and Systems I: Regular Papers*, vol. 59, no. 11, pp. 2655–2668, 2012.

[13] P. Gray and S. K. Scott, "Sustained oscillations and other exotic patterns of behavior in isothermal reactions," *The Journal of Physical Chemistry*, vol. 89, no. 1, pp. 22–32, 1985.

[14] D. Henry, *Geometric Theory of Semilinear Parabolic Equations*, vol. 840 of *Lecture Notes in Mathematics*, Springer, New York, NY, USA, 1981.

[15] G. R. Sell and Y. You, *Dynamics of evolutionary equations*, vol. 143 of *Applied Mathematical Sciences*, Springer, 2002.

6

An Analytical and Approximate Solution for Nonlinear Volterra Partial Integro-Differential Equations with a Weakly Singular Kernel Using the Fractional Differential Transform Method

Rezvan Ghoochani-Shirvan,[1] Jafar Saberi-Nadjafi,[2] and Morteza Gachpazan[2]

[1]*Department of Applied Mathematics, Ferdowsi University of Mashhad, International Campus, Mashhad, Iran*
[2]*Department of Applied Mathematics, School of Mathematical Sciences, Ferdowsi University of Mashhad, Mashhad, Iran*

Correspondence should be addressed to Jafar Saberi-Nadjafi; najafi141@gmail.com

Academic Editor: Jaume Giné

An analytical-approximate method is proposed for a type of nonlinear Volterra partial integro-differential equations with a weakly singular kernel. This method is based on the fractional differential transform method (FDTM). The approximate solutions of these equations are calculated in the form of a finite series with easily computable terms. The analytic solution is represented by an infinite series. We state and prove a theorem regarding an integral equation with a weak kernel by using the fractional differential transform method. The result of the theorem will be used to solve a weakly singular Volterra integral equation later on.

1. Introduction

Many engineering and physical problems result in the analysis of the nonlinear weakly singular Volterra integral equations (WSVIEs). These equations are applied in many areas [1] such as reaction-diffusion problems in small cells [2], theory of elasticity, heat conductions, hydrodynamics, stereology [3], the radiation of heat from semi-infinite solids [4], and other applications. Such equations have been studied by several authors [5–14].

The aim of this paper is applying the fractional differential transform method (FDTM) for solving WSVIE. The fractional differential transform method has recently been developed for solving the differential and integral equations. For example, in [15], FDTM is applied for fractional differential equations and in [16] it is used for fractional integro-differential equations. This method is applied to nonlinear fractional partial differential equations in [17]. The use of the differential transform method (DTM) in electric circuit analysis was first proposed by Zhou [18].

The main challenge of partial integro-differential equations (PIDEs) with a weakly singular kernel is faced when we are looking for an analytical solution. By applying the differential transform method, the result mostly obtained is an analytical solution in the form of a polynomial. The differential transform method is different from the traditional high order Taylor series method, which requires symbolic competition of the necessary derivatives of the data functions. Making use of this method enables us to obtain highly accurate results or exact solutions for a partial integro-differential equation. The use of application of DTM and FDTM does not require linearization, discretization, or perturbation in contrast to the methods discussed in the literature [8, 9, 19].

The form of WSVIE that we will consider in this paper with FDTM is

$$\phi_t = \phi(x,t) + f(x,t) + \int_0^t \int_0^x (x-\xi)^{p-1} N[\phi(\xi,\eta)]\, d\xi\, d\eta, \tag{1}$$

where $0 < p < 1$, $0 \le \xi \le x$, $0 \le \eta \le t$, and $(x,t) \in [0,1] \times [0,1]$ with the initial condition

$$\phi(x,0) = \phi_0(x), \tag{2}$$

where ϕ is an unknown function in $\Lambda(= [0,1]\times[0,1])$ which should be determined and $f(x,t)$, $\phi_0(x)$ are known functions and N is a nonlinear operator. The given functions $f(x,t)$ and $\phi(x,t)$ are assumed to be sufficiently smooth in order to guarantee the existence and uniqueness of a solution $\phi \in C(\Lambda)$. It is assumed that the nonlinear term $N[\phi]$ satisfies the Lipschitz condition in $L^2(\Lambda)$.

The numerical treatment of (1) is not simple because the solutions of WSVIEs usually have a weak singularity at $x = 0$. Different numerical techniques have been developed for the solution of PIDEs [8, 10, 20–25]. In this article, FDTM is applied to solve (1) and the main theorem is proved on the two-dimensional FDTM, while the one-dimensional FDTM has been applied in [26].

The paper is organized as follows: In Section 2, Caputo and Riemann-Liouville fractional derivatives are introduced. In Section 3, the theorems of the fractional differential transform method, preliminaries, and notations are explained. In Section 4, we have proposed the main theorem, for which a WSVIE can be considered as a series of FDT. Further, some examples of the application of FDTM are demonstrated, which show the accuracy of the method, in Section 5. We conclude our discussion in Section 6.

2. Riemann-Liouville and Caputo Fractional Derivatives

There are different kinds of definitions for the fractional derivative of order $q > 0$; among various definitions of fractional derivatives of order $q > 0$, the Riemann-Liouville and Caputo formulas are the most common [27]. The Riemann-Liouville fractional integration of order q is defined as

$$J_{x_0}^{q} f(x) = \frac{1}{\Gamma(q)} \int_{x_0}^{x} (x-t)^{q-1} f(t)\,dt, \qquad q>0,\ x>x_0. \tag{3}$$

The following equations define Riemann-Liouville and Caputo fractional derivatives of order q, respectively:

$$D_{x_0}^{q} f(x) = \frac{d^m}{dx^m}\left[J_{x_0}^{m-q} f(x)\right], \tag{4}$$

$$D_{*x_0}^{q} f(x) = J_{x_0}^{m-q}\left[\frac{d^m}{dx^m} f(x)\right], \tag{5}$$

where $m-1 \le q < m$ and $m \in N$. From (4) and (5), we have

$$D_{x_0}^{q} f(x) = \frac{1}{\Gamma(m-q)} \frac{d^m}{dx^m} \int_{x_0}^{x} (x-t)^{m-q-1} f(t)\,dt, \qquad x>x_0. \tag{6}$$

3. The Fractional Differential Transform Method (FDTM)

There are some approaches to the generalization of the notion of differentiation to fractional orders. According to the Riemann-Liouville formula, the fractional differentiation is defined by (6). The analytical and continuous function $f(x)$ is expended in terms of a fractional power series as follows:

$$f(x) = \sum_{k=0}^{\infty} F(k)(x-x_0)^{k/\alpha}, \tag{7}$$

where α is the order of fraction and $F(k)$ is the fractional differential transform of $f(x)$ [15, 28–30]. In Caputo sense [31], (6) is modified to handle integer-order initial conditions as follows:

$$D^q x_0\left[f(x) - \sum_{k=0}^{m-1}\frac{1}{k!}(x-x_0)^k f^{(k)}(x_0)\right] = \frac{1}{\Gamma(m-q)} \cdot \frac{d^m}{dx^m}\left\{\int_0^x\left[\frac{f(t)-\sum_{k=0}^{m-1}(1/k!)(t-x_0)^k f^{(k)}(x_0)}{(x-t)^{1+q-m}}\right]dt\right\}. \tag{8}$$

Since the initial conditions are implemented to the integer-order derivatives, the transformation of the initial conditions also can be represented as follows:

$$F(k) = \begin{cases} \dfrac{1}{(k/\alpha)!}\left[\dfrac{d^{k/\alpha} f(x)}{dx^{k/\alpha}}\right]_{x=x_0} & \text{if } \dfrac{k}{\alpha} \in N, \\ 0 & \text{if } \dfrac{k}{\alpha} \notin N, \end{cases} \tag{9}$$

for $k = 0, 1, 2, \ldots, (n\alpha - 1)$, where n is the order of FDE that is considered.

Consider a function $u(x,t)$ of two variables, and assume that it can be expressed as a product of two single-variable functions as $u(x,t) = f(x)g(t)$. The expansion of the function $u(x,t)$ in a Taylor series around a point (x_0, t_0) is as follows:

$$u(x,t) = \sum_{k=0}^{\infty}\sum_{h=0}^{\infty} U(k,h)(x-x_0)^{k/\beta}(t-t_0)^{h/\alpha}. \tag{10}$$

If we take (x_0, t_0) as $(0,0)$, then (10) can be illustrated as

$$u(x,t) = \sum_{k=0}^{\infty}\sum_{h=0}^{\infty} U(k,h)(x)^{k/\beta}(t)^{h/\alpha}, \tag{11}$$

where α, β are the order of the fractions, $\alpha, \beta \in N$, and $U(k,h) = F(k)G(h)$ is called the spectrum of $u(x,t)$ and defined by

$$U(k,h) = \frac{1}{\Gamma(k/\beta+1)\Gamma(h/\alpha+1)}\left[D^{k/\beta}D^{h/\alpha}u(x,t)\right]_{(x_0,t_0)}. \tag{12}$$

If we choose $\alpha = 1$ and $\beta = 1$, the fractional two-dimensional differential transform reduces to the classical two-dimensional differential transform. Using (11) and (12), the theorems of FDTM are introduced as follows. The proofs of these theorems can be found in [15, 17].

Theorem 1. *Suppose that $W(k,h)$, $U(k,h)$, and $V(k,h)$ are the differential transformations of the functions $w(x,t)$, $u(x,t)$, and $v(x,t)$, respectively, with order of fraction α and β; then*

(1) *if* $w(x,t) = u(x,t) \pm v(x,t)$, *then* $W(k,h) = U(k,h) \pm V(k,h)$;

(2) *if* $w(x,t) = \lambda u(x,t)$, *then* $W(k,h) = \lambda U(k,h)$;

(3) *if* $w(x,t) = (x-x_0)^p(t-t_0)^q$, *then*

$$W(k,h) = \delta(k-\beta p)\,\delta(h-\alpha q) = \begin{cases} 1 & k=\beta p,\ h=\alpha q, \\ 0 & \text{otherwise}; \end{cases} \tag{13}$$

(4) *if* $w(x,t) = u(x,t)v(x,t)$, *then*

$$W(k,h) = U(k,h)V(k,h) = \sum_{r=0}^{k}\sum_{s=0}^{h} U(r,h-s)V(k-r,s); \tag{14}$$

(5) *if* $w(x,t) = x^p\sin(at+b)$, *then*

$$W(k,h) = \frac{a^h}{h!}\delta(k-p)\sin\left(\frac{h\pi}{2}+b\right); \tag{15}$$

(6) *if* $w(x,t) = x^p\cos(at+b)$, *then*

$$W(k,h) = \frac{a^h}{h!}\delta(k-p)\cos\left(\frac{h\pi}{2}+b\right); \tag{16}$$

(7) *if* $w(x,t) = x^p\exp(\lambda t)$, *then*

$$W(k,h) = \frac{\delta(k-p)\left(\lambda^h\right)}{h!}, \tag{17}$$

where p *and* q *are positive and* λ, a, b *are scalars.*

Theorem 2. *If* $u(x,t) = D_x^q v(x,t)$ *and* $\beta \in N$ *is order of fractional, then*

$$U(k,h) = \frac{\Gamma(q+1+k/\beta)}{\Gamma(1+k/\beta)}V(k+\beta q,h). \tag{18}$$

Definition 3. The Beta function $B(a,b)$ of two variables is defined by

$$B(a,b) = \int_0^1 (1-x)^{a-1}x^{b-1}dx. \tag{19}$$

The following is proved easily:

$$B(a,b) = \frac{\Gamma(a)\Gamma(b)}{\Gamma(a+b)}. \tag{20}$$

Definition 4. The Kronecker delta function is given by

$$\delta(k-m) = \begin{cases} 1, & k=m, \\ 0, & k\neq m. \end{cases} \tag{21}$$

4. Main Theorem

Now, we represent the main theorem of this study, through which a weakly singular Volterra integral equation can be expressed as a series of fractional differential transform for $w(x,t) = \int_0^t\int_0^x (x-\xi)^{p-1}N[\phi(\xi,\eta)]d\xi\,d\eta$, $0<p<1$, $p\in Q$.

Theorem 5. *Suppose that* $\Phi(k,h)$ *and* $W(k,h)$ *are the fractional differential transforms of the functions* $\phi(x,t)$ *and* $w(x,t)$, *respectively, such that*

$$w(x,t) = \int_0^t\int_0^x (x-\xi)^{p-1}N[\phi(\xi,\eta)]\,d\xi\,d\eta, \quad 0<p<1,\ p\in Q. \tag{22}$$

Then, by choosing a suitable $\beta \in z^+$ *such that* $\beta p \in z^+$, *we have*

$$W(k,h) = \frac{1}{((h-q)/\alpha)+1}\sum_{l=0}^{k}\delta(l-\beta p) \cdot \sum_{q=0}^{h}\delta(q-\alpha)B\left(\frac{k-l}{\beta}+1,p\right) \cdot \Phi(k-l,h-q), \tag{23}$$

where $B(\cdot,\cdot)$ *is the Beta function and* δ *is the Kronecker delta function.*

Proof. By putting $N[\phi(x,t)] = \sum_{k=0}^{\infty}\sum_{h=0}^{\infty}\Phi(k,h)x^{k/\beta}t^{h/\alpha}$ in (22), it will change into

$$\begin{aligned} w(x,t) &= \int_0^t\int_0^x (x-\xi)^{p-1}N[\phi(\xi,\eta)]\,d\xi\,d\eta \\ &= \int_0^t\left(\int_0^x (x-\xi)^{p-1}\sum_{k=0}^{\infty}\sum_{h=0}^{\infty}\Phi(k,h)\xi^{k/\beta}\eta^{h/\alpha}d\xi\right)d\eta \\ &= \sum_{k=0}^{\infty}\sum_{h=0}^{\infty}\Phi(k,h)\int_0^t\left(\int_0^x (x-\xi)^{p-1}\xi^{k/\beta}d\xi\right)\eta^{h/\alpha}d\eta. \end{aligned} \tag{24}$$

To calculate $\int_0^x (x-\xi)^{p-1}\xi^{k/\beta}d\xi$, we change variable $x/\xi = v$ and according to Definition 3 we get

$$\int_0^x (x-\xi)^{p-1}\xi^{k/\beta}d\xi = B\left(\frac{k}{\beta}+1,p\right)x^{p+k/\beta}. \tag{25}$$

The following equation is obtained by using Theorem 1 and replacing (25) into (24):

$$w(x,t) = \sum_{k=0}^{\infty}\sum_{h=0}^{\infty}\Phi(k,h)B\left(\frac{k}{\beta}+1,p\right)x^{p+k/\beta}\int_0^t \eta^{h/\alpha}d\eta. \tag{26}$$

Hence

$$w(x,t) = \frac{1}{(h/\alpha)+1}x^p\sum_{k=0}^{\infty}\sum_{h=0}^{\infty}B\left(\frac{k}{\beta}+1,p\right)\Phi(k,h) \cdot x^{k/\beta}t^{h/\alpha+1} = \frac{1}{(h/\alpha)+1}\sum_{l=0}^{\infty}\delta(l-\beta p)$$

$$\cdot x^{l/\beta} \sum_{q=0}^{\infty} \delta(q-\alpha) t^{q/\alpha}$$

$$\cdot \sum_{k=0}^{\infty} \sum_{h=0}^{\infty} B\left(\frac{k}{\beta}+1, p\right) \Phi(k,h) x^{k/\beta} t^{h/\alpha}$$

$$= \frac{1}{(h/\alpha)+1}$$

$$\cdot \sum_{k=0}^{\infty} \sum_{h=0}^{\infty} \sum_{l=0}^{\infty} \delta(l-\beta p) \sum_{q=0}^{\infty} \delta(q-\alpha) B\left(\frac{k}{\beta}+1, p\right)$$

$$\cdot \Phi(k,h) x^{(k+l)/\beta} t^{(h+q)/\alpha} = \frac{1}{((h-q)/\alpha)+1}$$

$$\cdot \sum_{k=0}^{\infty} \sum_{h=0}^{\infty} \sum_{l=0}^{k} \delta(l-\beta p) \sum_{q=0}^{h} \delta(q-\alpha) B\left(\frac{k-l}{\beta}+1, p\right)$$

$$\cdot \Phi(k-l, h-q) x^{k/\beta} t^{h/\alpha}. \tag{27}$$

According to $w(x,t) = \sum_{k=0}^{\infty} \sum_{h=0}^{\infty} W(k,h) x^{k/\beta} t^{h/\alpha}$, one can conclude that

$$W(k,h) = \frac{1}{((h-q)/\alpha)+1} \sum_{l=0}^{k} \delta(l-\beta p) \cdot \sum_{q=0}^{h} \delta(q-\alpha) B\left(\frac{k-l}{\beta}+1, p\right) \Phi(k-l, h-q). \tag{28}$$

The proof is completed. □

5. Description of Method

In this section, we try to describe the FDTM for (1) and initial condition (2). Based on Theorems 1, 2, and 5, FDTM for (1) and (2) would result as follows:

$$(1+h)\Phi(k,h+1) = \Phi(k,h) + F(k,h) + \frac{1}{((h-q)/\alpha)+1} \sum_{l=0}^{k} \delta(l-\beta p) \cdot \sum_{q=0}^{h} \delta(q-\alpha) B\left(\frac{k-l}{\beta}+1, p\right) \Phi(k-l, h-q). \tag{29}$$

$$\Phi(k,h) = \Phi(k,0), \quad k = 0,1,2,3,\ldots,$$

in which $\Phi(k,h)$ and $F(k,h)$ are the FDTM of $\phi(x,t)$ and $f(x,t)$, respectively. Then according to the recurrence relation (11) the unknown function would result.

6. Applications

In this section, we take some examples to clarify the advantages and the accuracy of the fractional differential transform method (FDTM) for solving a kind of nonlinear partial integro-differential equation with a weakly singular kernel. For each of these examples, we obtain a recurrence relation. In all of the examples, we choose $\alpha = 1$ and β is chosen in a way where $\beta p \in z^{+}$.

Example 1. Consider the following nonlinear partial integro-differential equation with a weakly singular kernel with $p = 1/2$ [23]:

$$\phi_t = \phi(x,t) + f(x,t) + \int_0^t \int_0^x (x-\xi)^{p-1} [\phi(\xi,\eta)]^2 \, d\xi \, d\eta, \tag{30}$$

with the initial condition $\phi(x,0) = x^2$ and $f(x,t) = 2t - x^2 - t^2 - (2/315)x^{1/2}t(128x^4 + 112x^2t^2 + 63t^4)$.

For solving (30), we employ the FDTM, to get

$$\begin{aligned}(1+h)\Phi(k,h+1) &= \Phi(k,h) + 2\delta(h-1)\delta(k) \\ &\quad - \delta(h)\delta(k-4) - \delta(h-2)\delta(k) - \frac{256}{315}\delta(h-1) \\ &\quad \cdot \delta(k-9) - \frac{224}{315}\delta(h-3)\delta(k-5) - \frac{126}{315}\delta(h-5) \\ &\quad \cdot \delta(k-1) + \frac{1}{h} B\left(\frac{k-l}{2}+1, \frac{1}{2}\right) \\ &\quad \cdot \sum_{s=0}^{k-1} \sum_{r=0}^{h-1} \Phi(s, h-1-r) \Phi(k-1-s, r),\end{aligned} \tag{31}$$

where $h, k \geq 1$ in the upper bound of the sigmas and the differential transform of initial condition is as follows:

$$\Phi(k,0) = \delta(k-4) = \begin{cases} 1 & k=4 \\ 0 & k \neq 4 \end{cases} \longrightarrow \tag{32}$$

$$\Phi(4,0) = 1.$$

Also we have

$$\begin{aligned}\phi(x,t) &= \sum_{k=0}^{\infty} \sum_{h=0}^{\infty} \Phi(k,h) x^{k/2} t^h \\ &= \Phi(0,0) + \Phi(0,1)t + \Phi(0,2)t^2 + \cdots \\ &\quad + \Phi(1,0)x^{1/2} + \Phi(1,1)x^{1/2}t \\ &\quad + \Phi(1,2)x^{1/2}t^2 + \Phi(1,3)x^{1/2}t^3 + \cdots \\ &\quad + \Phi(2,0)x + \Phi(2,1)xt + \Phi(2,2)xt^2 \\ &\quad + \Phi(2,3)xt^3 + \cdots + \Phi(3,0)x^{3/2} \\ &\quad + \Phi(3,1)x^{3/2}t + \Phi(3,2)x^{3/2}t^2 \\ &\quad + \Phi(3,3)x^{3/2}t^3 + \cdots + \Phi(4,0)x^2 \\ &\quad + \Phi(4,1)x^2t + \Phi(4,2)x^2t^2 + \Phi(4,3)xt^3 \\ &\quad + \cdots.\end{aligned} \tag{33}$$

By using the recurrence relation (31) and the transform initial condition (32), we get the following:

$$
\begin{gathered}
k=0 \\
h=0 \\
\downarrow \\
\Phi(0,1)=\Phi(0,0)+0 \\
\downarrow \\
\Phi(0,1)=0; \\
k=1 \\
h=0 \\
\downarrow \\
\Phi(1,1)=\Phi(1,0)+0 \\
\downarrow \\
\Phi(1,1)=0; \\
k=2 \\
h=0 \\
\downarrow \\
\Phi(2,1)=\Phi(2,0)+0 \\
\downarrow \\
\Phi(2,1)=0; \\
k=3 \\
h=0 \\
\downarrow \\
\Phi(3,1)=\Phi(3,0)+0 \\
\downarrow \\
\Phi(3,1)=0; \\
k=4 \\
h=0 \\
\downarrow \\
\Phi(4,1)=\Phi(4,0)-1 \\
\downarrow \\
\Phi(4,1)=0,
\end{gathered}
\tag{34}
$$

and by applying the same calculations, the following can be concluded:

$$\Phi(k,1)=0 \quad k\geq 5. \tag{35}$$

Also we put

$$
\begin{gathered}
k=0 \\
h=1 \\
\downarrow \\
2\Phi(0,2)=\Phi(0,1)+2 \\
\downarrow \\
\Phi(0,2)=1; \\
k=1 \\
h=1 \\
\downarrow \\
2\Phi(1,2)=\Phi(1,1)+0 \\
\downarrow \\
\Phi(1,2)=0; \\
k=2 \\
h=1 \\
\downarrow \\
2\Phi(2,2)=\Phi(2,1)+0 \\
\downarrow \\
\Phi(2,2)=0 \\
\vdots \\
k=9 \\
h=1 \\
\downarrow \\
2\Phi(9,2)=\Phi(9,1)-\frac{2\times 128}{315} \\
+B\left(4+1,\frac{1}{2}\right)\sum_{s=0}^{9-1}\sum_{r=0}^{0}\Phi(s,0)\,\Phi(8-s,0).
\end{gathered}
\tag{36}
$$

By Definition 3, we have

$$B\left(5,\frac{1}{2}\right)=\frac{2\times 128}{315}. \tag{37}$$

So $\Phi(9,2)=0$.

And by applying the same calculations, we can conclude that

$$\Phi(k,2)=0 \quad k>9. \tag{38}$$

By continuing this process, we can also conclude the following:

$$\Phi(k,h)=0 \quad k\geq 0,\ h\geq 3. \tag{39}$$

Therefore, by substituting the above values into (33), the exact solution is obtained in the following form:

$$\phi(x,t) = x^2 + t^2 \tag{40}$$

which is the particular solution obtained in [23].

Example 2. Consider the following nonlinear partial integro-differential equation with a weakly singular kernel:

$$\phi_t = \phi(x,t) + f(x,t) + \int_0^t \int_0^x (x-\xi)^{-1/4} [\phi(\xi,\eta)]^2 \, d\xi \, d\eta \tag{41}$$

with the initial condition $\phi(x,0) = x$ and $f(x,t) = -x\sin t - x\cos t - (64/231)x^{11/4}t - (32/231)x^{11/4}\sin 2t$.

Taking into consideration the two-dimensional transform for (41) and the related theorems, we have

$$\begin{aligned}(h+1)\Phi(k,h+1) &= \Phi(k,h) - \delta(k-4)\frac{1}{h!}\sin\frac{h\pi}{2} \\ &- \delta(k-4)\frac{1}{h!}\cos\frac{h\pi}{2} - \frac{64}{231}\delta(k-11)\delta(h-1) \\ &- \frac{32}{231}\left(\frac{2^h}{h!}\right)\sin\frac{h\pi}{2}\delta(k-11) + \frac{1}{h} \\ &\cdot B\left(\frac{k-3}{4}+1, \frac{3}{4}\right) \\ &\cdot \sum_{s=0}^{k-3}\sum_{r=0}^{h-1}\Phi(s,h-1-r)\Phi(k-3-s,r).\end{aligned} \tag{42}$$

The differential transform of the initial condition is as follows:

$$\Phi(k,0) = \delta(k-4) = \begin{cases} 1 & k=4 \\ 0 & k \neq 4 \end{cases} \longrightarrow \Phi(4,0) = 1. \tag{43}$$

Also we have

$$\begin{aligned}\phi(x,t) &= \sum_{k=0}^{\infty}\sum_{h=0}^{\infty}\Phi(k,h)x^{k/4}t^h \\ &= \Phi(0,0) + \Phi(0,1)t + \Phi(0,2)t^2 + \cdots \\ &+ \Phi(1,0)x^{1/4} + \Phi(1,1)x^{1/4}t \\ &+ \Phi(1,2)x^{1/4}t^2 + \Phi(1,3)x^{1/4}t^3 + \cdots \\ &+ \Phi(2,0)x^{2/4} + \Phi(2,1)x^{2/4}t \\ &+ \Phi(2,2)x^{2/4}t^2 + \Phi(2,3)x^{2/4}t^3 + \cdots \\ &+ \Phi(3,0)x^{3/4} + \Phi(3,1)x^{3/4}t \\ &+ \Phi(3,2)x^{3/4}t^2 + \Phi(3,3)x^{3/4}t^3 + \cdots \\ &+ \Phi(4,0)x + \Phi(4,1)xt + \Phi(4,2)xt^2 \\ &+ \Phi(4,3)xt^3 + \cdots.\end{aligned} \tag{44}$$

By using the recurrence relation (42) and the differential transform of initial condition (43), we get

$$\begin{gathered} k=0 \\ h=0 \\ \downarrow \\ \Phi(0,1) = \Phi(0,0) - 0 \\ \downarrow \\ \Phi(0,1) = 0; \\ k=1 \\ h=0 \\ \downarrow \\ \Phi(1,1) = \Phi(1,0) - 0 \\ \downarrow \\ \Phi(1,1) = 0; \\ k=2 \\ h=0 \\ \downarrow \\ \Phi(2,1) = \Phi(2,0) - 0 \\ \downarrow \\ \Phi(2,1) = 0; \\ k=3 \\ h=0 \\ \downarrow \\ \Phi(3,1) = \Phi(3,0) - 0 \\ \downarrow \\ \Phi(3,1) = 0; \\ k=4 \\ h=0 \\ \downarrow \\ \Phi(4,1) = \Phi(4,0) - 1 \\ \downarrow \\ \Phi(4,1) = 0 \end{gathered} \tag{45}$$

and by applying the same calculations, it can be concluded that

$$\Phi(k,1) = 0 \quad k \geq 5;$$

$$k = 0$$

$h = 1$

$\downarrow$

$2\Phi(0,2) = \Phi(0,1) - 0$

$\downarrow$

$\Phi(0,2) = 0;$

$k = 1$

$h = 1$

$\downarrow$

$2\Phi(1,2) = \Phi(1,1) - 0$

$\downarrow$

$\Phi(1,2) = 0;$

$k = 2$

$h = 1$

$\downarrow$

$2\Phi(2,2) = \Phi(2,1) - 0$

$\downarrow$

$\Phi(2,2) = 0;$

$k = 3$

$h = 1$

$\downarrow$

$2\Phi(3,2) = \Phi(3,1) - 0$

$\downarrow$

$\Phi(3,2) = 0;$

$k = 4$

$h = 1$

$\downarrow$

$2\Phi(4,2) = \Phi(4,1) - 1$

$\downarrow$

$\Phi(4,2) = \frac{-1}{2}$

$\vdots$

$k = 11$

$h = 1$

$\downarrow$

$$2\Phi(11,2) = \Phi(11,1) - \frac{64}{231} - \frac{64}{231}$$

$$+ B\left(\frac{3}{4}, 3\right) \sum_{S=0}^{8} \Phi(s,0)\,\Phi(8-s,0)$$

$\downarrow$

$\Phi(11,2) = 0;$

$k = 0$

$h = 2$

$\downarrow$

$3\Phi(0,3) = \Phi(0,2) - 0$

$\downarrow$

$\Phi(0,3) = 0;$

$k = 1$

$h = 2$

$\downarrow$

$3\Phi(1,3) = \Phi(1,2) - 0$

$\downarrow$

$\Phi(1,3) = 0;$

$k = 2$

$h = 2$

$\downarrow$

$3\Phi(2,3) = \Phi(2,2) - 0$

$\downarrow$

$\Phi(2,3) = 0;$

$k = 3$

$h = 2$

$\downarrow$

$3\Phi(3,3) = \Phi(3,2) - 0$

$\downarrow$

$\Phi(3,3) = 0;$

$k = 4$

$h = 2$

$\downarrow$

$3\Phi(4,3) = \Phi(4,2) + \frac{1}{2!}$

$\downarrow$

$$\Phi(4,3) = -\frac{1}{2!} + \frac{1}{2!} = 0, \qquad (46)$$

and by applying the same calculations, we conclude the following:

$$\begin{gathered}\Phi(k,3)=0 \quad k\geq 5;\\ k=0\\ h=3\\ \downarrow\\ 4\Phi(0,4)=\Phi(0,3)-0\\ \downarrow\\ \Phi(0,4)=0;\\ k=1\\ h=3\\ \downarrow\\ 4\Phi(1,4)=\Phi(1,3)-0\\ \downarrow\\ \Phi(1,4)=0;\\ k=2\\ h=3\\ \downarrow\\ 4\Phi(2,4)=\Phi(2,3)-0\\ \downarrow\\ \Phi(2,4)=0;\\ k=3\\ h=3\\ \downarrow\\ 4\Phi(3,4)=\Phi(3,3)-0\\ \downarrow\\ \Phi(3,4)=0;\\ k=4\\ h=3\\ \downarrow\\ 4\Phi(4,4)=\Phi(4,3)+\frac{1}{3!}\\ \downarrow\\ \Phi(4,4)=\frac{1}{4!};\\ \Phi(k,4)=0 \quad k\geq 5.\end{gathered} \tag{47}$$

Therefore, by substituting the above values into (44), the exact solution is obtained in the following form:

$$\begin{aligned}\phi(x,t) &= \Phi(4,0)\,x+\Phi(4,1)\,xt+\Phi(4,2)\,xt^2\\ &\quad+\Phi(4,3)\,xt^3+\cdots\\ &= x-\frac{1}{2!}xt^2+\frac{1}{4!}xt^4+\cdots = x\cos t.\end{aligned} \tag{48}$$

Example 3. Consider the following nonlinear partial integro-differential equation with a weakly singular kernel:

$$\begin{aligned}\phi_t &= \phi(x,t)+f(x,t)\\ &\quad+\int_0^t\int_0^x (x-\xi)^{-3/5}\,[\phi(\xi,\eta)]^2\,d\xi\,d\eta\end{aligned} \tag{49}$$

with the initial condition $\phi(x,0) = x$ and $f(x,t) = (125/168)x^{12/5}e^{2t}$.

To solve (49), by applying FDTM, we have

$$\begin{aligned}(h+1)\,\Phi(k,h+1) &= \Phi(k,h)-\frac{125}{168}\delta(k-12)\,\frac{2^h}{h!}\\ &\quad+\frac{1}{h}B\left(\frac{k-2}{5}+1,\frac{2}{5}\right)\\ &\quad\cdot\sum_{s=0}^{k-2}\sum_{r=0}^{h-1}\Phi(s,h-1-r)\,\Phi(k-2-s,r),\end{aligned} \tag{50}$$

where $k \geq 2, h \geq 1$ in the upper bound of the sigmas and differential transform of initial condition is as follows:

$$\Phi(k,0)=\delta(k-5)=\begin{cases}1 & k=5\\ 0 & k\neq 5\end{cases} \longrightarrow \Phi(5,0)=1. \tag{51}$$

Also we have

$$\begin{aligned}\phi(x,t) &= \sum_{k=0}^{\infty}\sum_{h=0}^{\infty}\Phi(k,h)\,x^{k/5}t^h\\ &= \Phi(0,0)+\Phi(0,1)\,t+\Phi(0,2)\,t^2+\cdots\\ &\quad+\Phi(1,0)\,x^{1/5}+\Phi(1,1)\,x^{1/5}t\\ &\quad+\Phi(1,2)\,x^{1/5}t^2+\Phi(1,3)\,x^{1/5}t^3+\cdots\\ &\quad+\Phi(2,0)\,x^{2/5}+\Phi(2,1)\,x^{2/5}t\\ &\quad+\Phi(2,2)\,x^{2/5}t^2+\Phi(2,3)\,x^{2/5}t^3+\cdots\\ &\quad+\Phi(3,0)\,x^{3/5}+\Phi(3,1)\,x^{3/5}t\\ &\quad+\Phi(3,2)\,x^{3/5}t^2+\Phi(3,3)\,x^{3/5}t^3+\cdots\\ &\quad+\Phi(4,0)\,x^{4/5}+\Phi(4,1)\,x^{4/5}t\\ &\quad+\Phi(4,2)\,x^{4/5}t^2+\Phi(4,3)\,x^{4/5}t^3+\cdots\\ &\quad+\Phi(5,0)\,x+\Phi(5,1)\,xt+\Phi(5,2)\,xt^2\\ &\quad+\Phi(5,3)\,xt^3+\cdots.\end{aligned} \tag{52}$$

By using the recurrence relation (50), the differential transform of initial condition (51), and the same calculations of the above-mentioned examples, it is concluded that

$$\phi(x,t) = x + xt + \frac{1}{2!}xt^2 + \frac{1}{3!}xt^3 + \cdots - \frac{125}{168}x^{12/5}\left(t + \frac{t^2}{2!} + \frac{t^3}{3!} + \cdots\right). \quad (53)$$

Of course this solution is an analytical solution.

7. Conclusion

In this paper, we have described the definition and operation of two-dimensional fractional differential transform; fractional derivatives have been considered in the Caputo and Riemann-Liouville sense and the main theorem on fractional differential transform method. Using the fractional differential transform method, a kind of nonlinear partial integro-differential equation with a singular kernel was solved approximately and analytically. We have used FDTM in this paper to solve (30) which was solved by operational matrices in [23]. The advantages of this method are that one obtains satisfactory results in less time, there is no need to calculate any repeated integral, and there is no discretization.

Conflicts of Interest

The authors declare that there are no conflicts of interest regarding the publication of this paper.

References

[1] G. Capobianco, D. Conte, and I. Del Prete, "High performance parallel numerical methods for Volterra equations with weakly singular kernels," *Journal of Computational and Applied Mathematics*, vol. 228, no. 2, pp. 571–579, 2009.

[2] J. A. Dixon, "A nonlinear weakly singular Volterra integro-differential equation arising from a reaction-diffusion study in a small cell," *Journal of Computational and Applied Mathematics*, vol. 18, no. 3, pp. 289–305, 1987.

[3] P. Linz, *Analytical and Numerical Methods for Volterra Equations*, vol. 7 of *SIAM Studies in Applied Mathematics*, Society for Industrial and Applied Mathematics (SIAM), Philadelphia, Pa, USA, 1985.

[4] J. B. Keller and W. E. Olmstead, "Temperature of a nonlinearly radiating semi-infinite solid," *Quarterly of Applied Mathematics*, vol. 29, pp. 559–566, 1971/72.

[5] A. Palamara Orsi, "Product integration for Volterra integral equations of the second kind with weakly singular kernels," *Mathematics of Computation*, vol. 65, no. 215, pp. 1201–1212, 1996.

[6] Y. Chen and T. Tang, "Spectral methods for weakly singular Volterra integral equations with smooth solutions," *Journal of Computational and Applied Mathematics*, vol. 242, pp. 53–69, 2013.

[7] H. Brunner, A. Pedas, and G. Vainikko, "The piecewise polynomial collocation method for nonlinear weakly singular Volterra equations," *Mathematics of Computation*, vol. 68, no. 227, pp. 1079–1095, 1999.

[8] J. C. Lopez Marcos, "A difference scheme for a nonlinear partial integrodifferential equation," *SIAM Journal on Numerical Analysis*, vol. 27, no. 1, pp. 20–31, 1990.

[9] L. Bougoffa, R. C. Rach, and A. Mennouni, "An approximate method for solving a class of weakly-singular Volterra integro-differential equations," *Applied Mathematics and Computation*, vol. 217, no. 22, pp. 8907–8913, 2011.

[10] M. Dehghan, "Solution of a partial integro-differential equation arising from viscoelasticity," *International Journal of Computer Mathematics*, vol. 83, no. 1, pp. 123–129, 2006.

[11] P. Baratella and A. P. Orsi, "A new approach to the numerical solution of weakly singular Volterra integral equations," *Journal of Computational and Applied Mathematics*, vol. 163, no. 2, pp. 401–418, 2004.

[12] T. Diogo and P. Lima, "Superconvergence of collocation methods for a class of weakly singular Volterra integral equations," *Journal of Computational and Applied Mathematics*, vol. 218, no. 2, pp. 307–316, 2008.

[13] T. Tang, "A finite difference scheme for partial integro-differential equations with a weakly singular kernel," *Applied Numerical Mathematics*, vol. 11, no. 4, pp. 309–319, 1993.

[14] Y. Chen and T. Tang, "Spectral methods for weakly singular Volterra integral equations with smooth solutions," *Journal of Computational and Applied Mathematics*, vol. 233, no. 4, pp. 938–950, 2009.

[15] A. Arikoglu and I. Ozkol, "Solution of fractional differential equations by using differential transform method," *Chaos, Solitons & Fractals*, vol. 34, no. 5, pp. 1473–1481, 2007.

[16] A. Arikoglu and I. Ozkol, "Solution of fractional integro-differential equations by using fractional differential transform method," *Chaos, Solitons & Fractals*, vol. 40, no. 2, pp. 521–529, 2009.

[17] S. Momani and Z. Odibat, "A novel method for nonlinear fractional partial differential equations: combination of DTM and generalized Taylor's formula," *Journal of Computational and Applied Mathematics*, vol. 220, no. 1-2, pp. 85–95, 2008.

[18] J. K. Zhou, *differential transformation and its applications for electrical circuits*, Huazhong University Press, Wuha, China, 1986.

[19] R. Hilfer, *Applications of Fractional Calculus in Physics*, World Scientific, Singapore, 2000.

[20] E. Tohidi and F. Toutounian, "Convergence analysis of Bernoulli matrix approach for one-dimensional matrix hyperbolic equations of the first order," *Computers & Mathematics with Applications*, vol. 68, no. 1-2, pp. 1–12, 2014.

[21] F. Toutounian and E. Tohidi, "A new Bernoulli matrix method for solving second order linear partial differential equations with the convergence analysis," *Applied Mathematics and Computation*, vol. 223, pp. 298–310, 2013.

[22] G. Fairweather, "Spline collocation methods for a class of hyperbolic partial integro-differential equations," *SIAM Journal on Numerical Analysis*, vol. 31, no. 2, pp. 444–460, 1994.

[23] S. Singh, V. K. Patel, V. K. Singh, and E. Tohidi, "Numerical solution of nonlinear weakly singular partial integro-differential equation via operational matrices," *Applied Mathematics and Computation*, vol. 298, pp. 310–321, 2017.

[24] S. Singh, V. K. Patel, and V. K. Singh, "Operational matrix approach for the solution of partial integro-differential equation," *Applied Mathematics and Computation*, vol. 283, pp. 195–207, 2016.

[25] W. McLean, I. H. Sloan, and V. Thomee, "Time discretization via Laplace transformation of an integro-differential equation of parabolic type," *Numerische Mathematik*, vol. 102, no. 3, pp. 497–522, 2006.

[26] E. Rahimi, H. Taghvafard, and G. H. Erjaee, "Fractional differential transform method for solving a class of weakly singular Volterra integral equations," *Iranian Journal of Science & Technology*, vol. 38, no. 1, pp. 69–73, 2014.

[27] D. Nazari and S. Shahmorad, "Application of the fractional differential transform method to fractional-order integro-differential equations with nonlocal boundary conditions," *Journal of Computational and Applied Mathematics*, vol. 234, no. 3, pp. 883–891, 2010.

[28] A. A. Kilbas, H. M. Srivastava, and J. J. Trujillo, *Theory and Applications of Fractional Differential Equations*, New York, NY, USA, Elsevier, 2006.

[29] M. Sen, *Introduction to fractional-order operators and their engineering applications*, Department of Aerospace and Mechanical Engineering University of Notre Dame, 2014.

[30] Z. M. Odibat and N. T. Shawagfeh, "Generalized Taylor's formula," *Applied Mathematics and Computation*, vol. 186, no. 1, pp. 286–293, 2007.

[31] M. Caputo, "Linear models of dissipation whose Q is almost frequency independent-II," *The Geophysical Journal of the Royal Astronomical Society*, vol. 13, no. 5, pp. 529–539, 1967.

7

Global and Local Structures of Bifurcation Curves of ODE with Nonlinear Diffusion

Tetsutaro Shibata

Laboratory of Mathematics, Graduate School of Engineering, Hiroshima University, Higashi-Hiroshima, 739-8527, Japan

Correspondence should be addressed to Tetsutaro Shibata; shibata@amath.hiroshima-u.ac.jp

Academic Editor: Patricia J. Y. Wong

We consider the nonlinear eigenvalue problem $[D(u)u']' + \lambda f(u) = 0$, $u(t) > 0$, $t \in I := (0,1)$, $u(0) = u(1) = 0$, where $D(u) = u^k$, $f(u) = u^{2n-k-1} + \sin u$, and $\lambda > 0$ is a bifurcation parameter. Here, $n \in \mathbb{N}$ and k $(0 \le k < 2n-1)$ are constants. This equation is related to the mathematical model of animal dispersal and invasion, and λ is parameterized by the maximum norm $\alpha = \|u_\lambda\|_\infty$ of the solution u_λ associated with λ and is written as $\lambda = \lambda(\alpha)$. Since $f(u)$ contains both power nonlinear term u^{2n-k-1} and oscillatory term $\sin u$, it seems interesting to investigate how the shape of $\lambda(\alpha)$ is affected by $f(u)$. The purpose of this paper is to characterize the total shape of $\lambda(\alpha)$ by n and k. Precisely, we establish three types of shape of $\lambda(\alpha)$, which seem to be new.

1. Introduction

This paper is concerned with the following nonlinear eigenvalue problems:

$$\left[D(u(t))\,u(t)'\right]' + \lambda f(u(t)) = 0, \quad t \in I := (0,1), \tag{1}$$

$$u(t) > 0, \quad t \in I, \tag{2}$$

$$u(0) = u(1) = 0, \tag{3}$$

where $D(u) = u^k$, $f(u) = u^{2n-k-1} + \sin u$, and $\lambda > 0$ is a bifurcation parameter. Here, $n \in \mathbb{N}$ and k $(0 \le k < 2n-1)$ are constants. Bifurcation problems have a long history and there are so many results concerning the asymptotic properties of bifurcation diagrams. We refer to [1–8] and the references therein. Moreover, bifurcation problems with nonlinear diffusion have been proposed in the field of population biology, and several model equations of logistic type have been considered. We refer to [9] and the references therein. In particular, the case $D(u) = u^k$ $(k > 0)$ has been derived from a model equation of animal dispersal and invasion in [10, 11]. In this situation, λ is a parameter which represents the habitat size and diffusion rate. On the other hand, there are several papers which treat the asymptotic behavior of oscillatory bifurcation curves. We refer to [7, 12–19] and the references therein. Our equation (1) contains both nonlinear diffusion term and oscillatory nonlinear terms. The purpose of this paper is to find the difference between the structures of bifurcation curves of the equations with only oscillatory term and those with both nonlinear diffusion term and the oscillatory term in (1). To clarify our intention, let $k = 2$ and $n = 2$. Then (1) is given as

$$\left(u^2 u'\right)' + \lambda (u + \sin u) = 0. \tag{4}$$

The corresponding equation without nonlinear diffusion is the case $k = 0$ and $n = 1$, namely,

$$u'' + \lambda (u + \sin u) = 0. \tag{5}$$

It should be mentioned that, by using a generalized time-map argument in [9], for any given $\alpha > 0$, there exists a unique classical solution pair (λ, u_α) of (1)–(3) satisfying $\alpha = \|u_\alpha\|_\infty$. Furthermore, λ is parameterized by α as $\lambda = \lambda(\alpha)$ and is continuous in $\alpha > 0$. For (5), the following asymptotic formula for $\lambda(\alpha)$ as $\alpha \longrightarrow \infty$ has been obtained.

Theorem 1 (see [12]). *Consider (5) with (2)–(3). Then as* $\alpha \longrightarrow \infty$,

$$\lambda(\alpha) = \pi^2 - 4\frac{\pi}{\alpha}\sqrt{\frac{\pi}{2\alpha}}\sin\left(\alpha - \frac{\pi}{4}\right) + o\left(\alpha^{-3/2}\right). \tag{6}$$

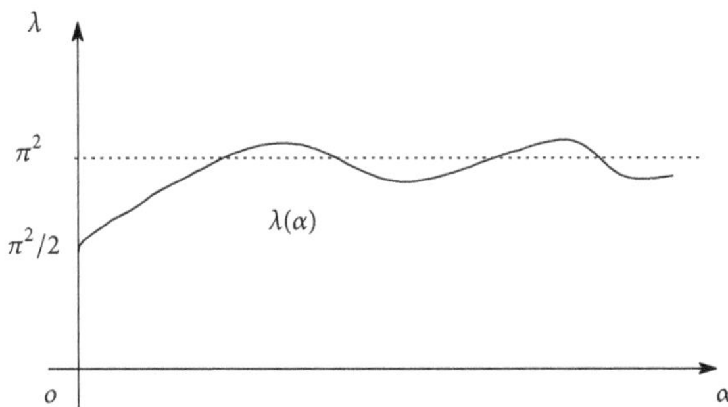

FIGURE 1: The graph of $\lambda(\alpha)$ for (5) ($k = 0, n = 1$).

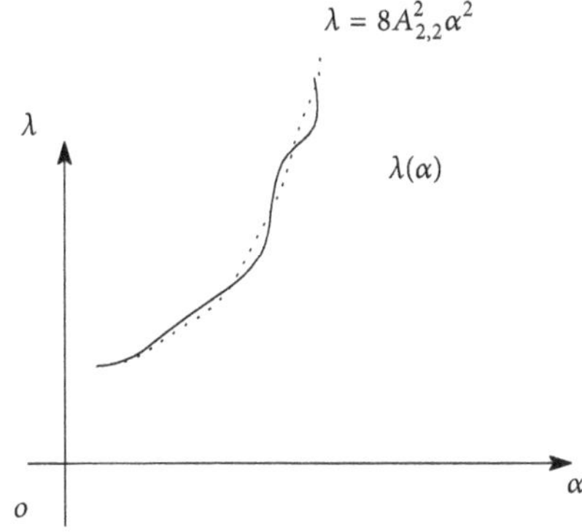

FIGURE 2: The graph of $\lambda(\alpha)$ for $k = n = 2$.

For (5) with (2)-(3), the asymptotic behavior of $\lambda(\alpha)$ as $\alpha \longrightarrow 0$ is as follows. For a solution pair $(\lambda(\alpha), u_\alpha)$ satisfying $\|u_\alpha\|_\infty = \alpha$, put $v_\alpha(t) := u_\alpha(t)/\alpha$ and let $\alpha \longrightarrow 0$. Then we easily obtain the function $v_0 \in C^2(I)$ which satisfies $-v_0''(t) = 2\lambda(0)v_0(t)$, $v_0(t) > 0$ for $t \in I$ with $v_0(0) = v_0(1) = 0$. This implies $\lambda(0) = \pi^2/2$. By this fact and Theorem 1, the bifurcation curve $\lambda(\alpha)$ starts from $\pi^2/2$ and tends to π^2 with oscillation and intersects the line $\lambda = \pi^2$ infinitely many times for $\alpha \gg 1$.

Since (4) includes both the nonlinear diffusion function and oscillatory term, it seems interesting how the nonlinear diffusion functions give effect to the structures of bifurcation curves.

Now we state our main results.

Theorem 2. *Consider (1) with (2)-(3). Then as $\alpha \longrightarrow \infty$,*

$$\lambda(\alpha) = 4n\alpha^{2k+2-2n}\left\{A_{k,n}^2 - 2A_{k,n}\sqrt{\frac{\pi}{2n}}\alpha^{k+(1/2)-2n}\sin\left(\alpha - \frac{\pi}{4}\right) + o\left(\alpha^{k+(1/2)-2n}\right)\right\}, \tag{7}$$

where

$$A_{k,n} = \int_0^1 \frac{s^k}{\sqrt{1-s^{2n}}}ds. \tag{8}$$

By Theorem 2, we obtain the global behavior of $\lambda(\alpha)$ as $\alpha \longrightarrow \infty$ for $n = k = 2$ and see that the asymptotic behavior of $\lambda(\alpha)$ is completely different from that for $k = 0, n = 1$ by comparing Figures 1 and 2.

Now we establish the asymptotic behavior of $\lambda(\alpha)$ as $\alpha \longrightarrow 0$ to obtain a complete understanding of the structure of $\lambda(\alpha)$. Let

$$B_0 := \int_0^1 \frac{s^k}{\sqrt{1-s^{k+2}}}ds, \tag{9}$$

$$B_1 := \frac{k+2}{12(k+4)}\int_0^1 \frac{s^k\left(1-s^{k+4}\right)}{\left(1-s^{k+2}\right)^{3/2}}ds, \tag{10}$$

$$B_2 = \frac{k+2}{2n}\int_0^1 \frac{s^k\left(1-s^{2n}\right)}{\left(1-s^{k+2}\right)^{3/2}}ds, \tag{11}$$

$$B_3 = \frac{n}{k+2}\int_0^1 \frac{s^k\left(1-s^{k+2}\right)}{\left(1-s^{2n}\right)^{3/2}}ds. \tag{12}$$

Theorem 3. *Consider (1) with (2)-(3). Then the following asymptotic formulas hold as $\alpha \longrightarrow 0$.*

(i) *Assume that $k + 4 < 2n$. Then*

$$\lambda(\alpha) = 2(k+2)\alpha^k\left\{B_0^2 + 2B_0B_1\alpha^2 + o\left(\alpha^2\right)\right\}. \tag{13}$$

(ii) *Assume that $2n = k + 4$. Then*

$$\lambda(\alpha) = 2(k+2)\alpha^k\left\{B_0^2 - 10B_0B_1\alpha^2 + o\left(\alpha^2\right)\right\}. \tag{14}$$

(iii) *Assume that $k + 2 < 2n < k + 4$. Then*

$$\lambda(\alpha) = 2(k+2)\alpha^k\left\{B_0^2 - B_0B_2\alpha^{2n-k-2} + o\left(\alpha^{2n-k-2}\right)\right\}. \tag{15}$$

(iv) *Assume that $2n = k + 2$. Then*

$$\lambda(\alpha) = (k+2)\alpha^k\left\{B_0^2 + B_0B_1\alpha^2 + o\left(\alpha^2\right)\right\}. \tag{16}$$

(v) *Assume that $k + 1 < 2n < k + 2$. Then*

$$\lambda(\alpha) = 4n\alpha^{2(k+1-n)}\left\{A_{k,n}^2 - 2A_{k,n}B_3\alpha^{k+2-2n} + o\left(\alpha^{k+2-2n}\right)\right\}. \tag{17}$$

The rough images of the graphs of $\lambda(\alpha)$ for $k = 1, n = 2$, $n = k = 2$, and $k = 1, n = 3$ are given in Figures 3, 4, and 5.

The proofs depend on the generalized time-map argument in [9] and stationary phase method (cf. Lemma 4). It should be mentioned that if we apply Lemma 4 to our situation, careful consideration about the regularity of the functions is necessary.

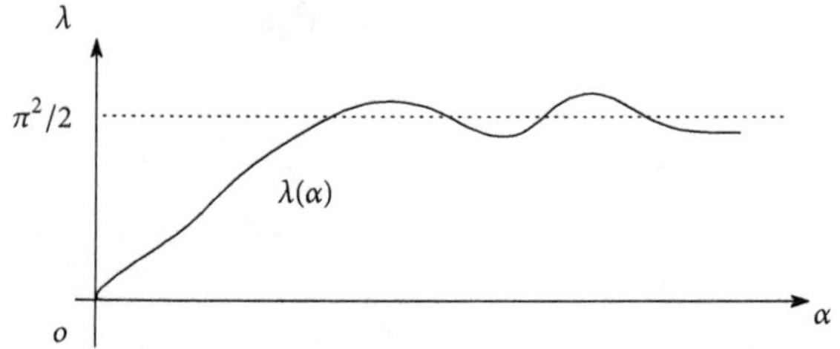

FIGURE 3: The graph of $\lambda(\alpha)$ for $k = 1, n = 2$.

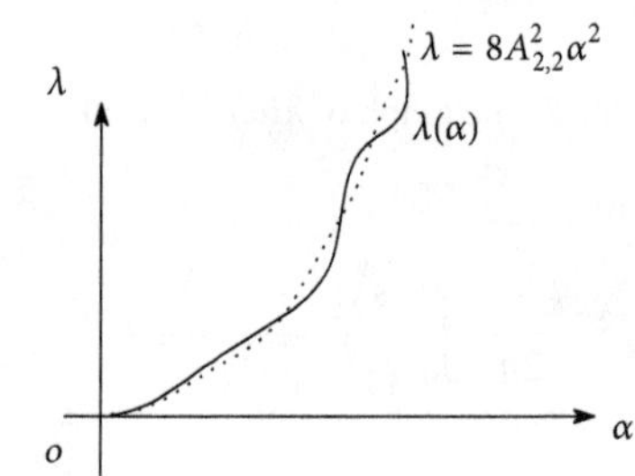

FIGURE 4: The graph of $\lambda(\alpha)$ for $k = n = 2$.

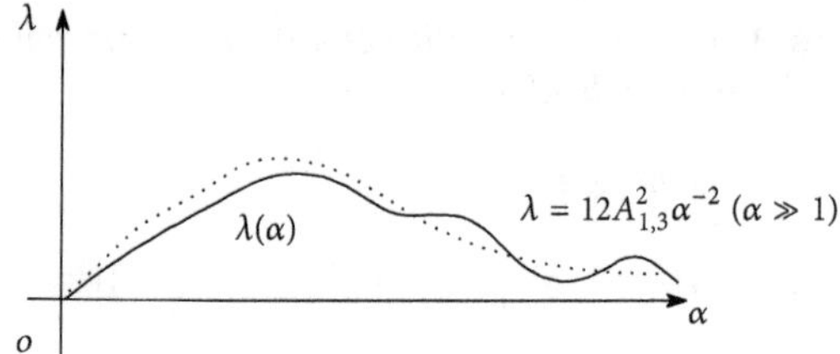

FIGURE 5: The graph of $\lambda(\alpha)$ for $k = 1, n = 3$.

2. Proof of Theorem 2

We put

$$\Lambda := \left\{\alpha > 0 \mid f(\alpha) > 0, \int_u^\alpha f(t)\,D(t)\,dt > 0 \text{ for all } u \in [0,\alpha)\right\}. \tag{18}$$

It was shown in [9, (2.7)] that if $\alpha \in \Lambda$, then $\lambda(\alpha)$ is well defined. In our situation, it is clear that, for $t > 0$, $D(t) > 0$, $f(t) > 0$, so $f(t)D(t) > 0$. Therefore, $\Lambda \equiv \mathbb{R}_+$. By this and the generalized time-map obtained in [9] (cf. (24)) and the time-map argument in [8, Theorem 2.1], we see that, for any given $\alpha > 0$, there exists a unique classical solution pair (λ, u_α) of (1)–(3) satisfying $\alpha = \|u_\alpha\|_\infty$. Furthermore, λ is parameterized by α as $\lambda = \lambda(\alpha)$ and is continuous in $\alpha > 0$. For $u \geq 0$, we put

$$G(u) := \int_0^u f(y)\,D(y)\,dy = \frac{1}{2n}u^{2n} + G_1(u) := \frac{1}{2n}u^{2n} + \int_0^u y^k \sin y\,dy. \tag{19}$$

It is known from [9] that if $(u_\alpha, \lambda(\alpha)) \in C^2(\bar{I}) \times \mathbb{R}_+$ satisfies (1)–(3), then

$$u_\alpha(t) = u_\alpha(1-t), \quad 0 \leq t \leq 1, \tag{20}$$

$$u_\alpha\left(\frac{1}{2}\right) = \max_{0\leq t\leq 1} u_\alpha(t) = \alpha, \tag{21}$$

$$u_\alpha'(t) > 0, \quad 0 < t < \frac{1}{2}. \tag{22}$$

In what follows, we denote by C various positive constants independent of $\alpha \gg 1$. For $0 \leq s \leq 1$ and $\alpha \gg 1$, we have

$$\left|\frac{G_1(\alpha) - G_1(\alpha s)}{\alpha^{2n}(1-s^{2n})}\right| = \left|\frac{\int_{\alpha s}^\alpha w^k \sin w\,dw}{\alpha^{2n}(1-s^{2n})}\right| \leq C\frac{\alpha^{k+1}(1-s^{k+1})}{\alpha^{2n}(1-s^{2n})} \leq C\alpha^{k+1-2n} \ll 1. \tag{23}$$

By this, (19), and Taylor expansion, we have from [9, (2.5)] that

$$\begin{aligned}\sqrt{\frac{\lambda(\alpha)}{2}} &= \int_0^\alpha \frac{D(u)}{\sqrt{G(\alpha) - G(u)}}du \\ &= \int_0^\alpha \frac{u^k}{\sqrt{(1/2n)(\alpha^{2n} - u^{2n}) + G_1(\alpha) - G_1(u)}}du \\ &= \sqrt{2n}\alpha^{k+1-n}\int_0^1 \frac{s^k}{\sqrt{1 - s^{2n} + (2n/\alpha^{2n})(G_1(\alpha) - G_1(\alpha s))}}ds \\ &= \sqrt{2n}\alpha^{k+1-n}\int_0^1 \frac{s^k}{\sqrt{1-s^{2n}}}\left\{1 - \frac{n}{\alpha^{2n}}\frac{G_1(\alpha) - G_1(\alpha s)}{(1-s^{2n})}(1+o(1))\right\}ds \\ &= \sqrt{2n}\alpha^{k+1-n}\left\{\int_0^1 \frac{s^k}{\sqrt{1-s^{2n}}}ds - \frac{n}{\alpha^{2n}}L(\alpha)(1+o(1))\right\},\end{aligned} \tag{24}$$

where

$$L(\alpha) := \int_0^1 \frac{s^k}{(1-s^{2n})^{3/2}}(G_1(\alpha) - G_1(\alpha s))\,ds. \tag{25}$$

We see from (24) and (25) that if we obtain the precise asymptotic formula for $L(\alpha)$ as $\alpha \longrightarrow \infty$, then we obtain Theorem 2. To do this, we apply the stationary phase method to our situation. By combining [13, Lemma 2] and [7, Lemmas 2.24], we have the following equality.

Lemma 4 (see [13, Lemma 2 and 10, Lemma 2.24]). *Assume that the function $f(r) \in C^2[0,1]$, $w(r) \in C^3[0,1]$, and*

$$\begin{aligned} &w'(r) < 0, \quad r \in (0,1], \\ &w'(0) = 0, \\ &w''(0) < 0. \end{aligned} \tag{26}$$

Then as $\mu \longrightarrow \infty$

$$\int_0^1 f(r)\, e^{i\mu w(r)} dr = \frac{1}{2} e^{i(\mu w(0)-(\pi/4))} \sqrt{\frac{2\pi}{\mu |w''(0)|}} f(0) + O\left(\frac{1}{\mu}\right). \tag{27}$$

In particular, by taking the imaginary part of (27), as $\mu \longrightarrow \infty$,

$$\int_0^1 f(r) \sin(\mu w(r))\, dr = \frac{1}{2}\sqrt{\frac{2\pi}{\mu |w''(0)|}} f(0) \sin\left(w(0)\mu - \frac{\pi}{4}\right) + O\left(\frac{1}{\mu}\right). \tag{28}$$

We note that, to obtain (27), we have to be careful about the regularity of f and w.

Lemma 5. *As* $\alpha \longrightarrow \infty$,

$$L(\alpha) = \sqrt{\frac{\pi}{2}} \frac{1}{n^{3/2}} \alpha^{k+(1/2)} \sin\left(\alpha - \frac{\pi}{4}\right) + O\left(\alpha^k\right). \tag{29}$$

Proof. We put $s = \sin\theta$ and

$$Y(\theta) := Y_1(\theta)(G_1(\alpha) - G_1(\alpha\sin\theta)) := \frac{\sin^k\theta}{(1+\sin^2\theta + \cdots + \sin^{2n-2}\theta)^{3/2}} (G_1(\alpha) - G_1(\alpha\sin\theta)). \tag{30}$$

By integration by parts, (25) and (30), we have

$$\begin{aligned} L(\alpha) &= \int_0^1 \frac{s^k (G_1(\alpha) - G_1(\alpha s))}{(1-s^2)^{3/2}(1+s^2+\cdots+s^{2n-2})^{3/2}} ds \\ &= \int_0^{\pi/2} \frac{1}{\cos^2\theta} \frac{\sin^k\theta (G_1(\alpha) - G_1(\alpha\sin\theta))}{(1+\sin^2\theta+\cdots+\sin^{2n-2}\theta)^{3/2}} d\theta \\ &:= L_1(\alpha) - L_2(\alpha) \\ &= [\tan\theta Y(\theta)]_0^{\pi/2} - \int_0^{\pi/2} \tan\theta \{Y_1(\theta)(G_1(\alpha) - G_1(\alpha\sin\theta))\}' d\theta. \end{aligned} \tag{31}$$

By l'Hôpital's rule, we obtain

$$\lim_{\theta\longrightarrow\pi/2} \frac{G_1(\alpha) - G_1(\alpha\sin\theta)}{\cos\theta} = \lim_{\theta\longrightarrow\pi/2} \frac{\alpha\cos\theta(\alpha\sin\theta)^k \sin(\alpha\sin\theta)}{\sin\theta} = 0. \tag{32}$$

This implies that $L_1(\alpha) = 0$. Next,

$$\begin{aligned} L_2(\alpha) &= \int_0^{\pi/2} \tan\theta \{Y_1'(\theta)(G_1(\alpha) - G_1(\alpha\sin\theta)) - Y_1(\theta)\alpha\cos\theta(\alpha\sin\theta)^k \sin(\alpha\sin\theta)\} d\theta. \\ &:= L_{21}(\alpha) - L_{22}(\alpha). \end{aligned} \tag{33}$$

We first calculate $L_{21}(\alpha)$. Assume that $k > 0$. Then

$$Y_1'(\theta) = \frac{\sin^{k-1}\theta\cos\theta}{(1+\sin^2\theta+\cdots+\sin^{2n-2}\theta)^{3/2}} \times \left[k - \frac{3(\sin^2\theta + 2\sin^4\theta + \cdots + (n-1)\sin^{2n-2}\theta)}{1+\sin^2\theta+\cdots+\sin^{2n-2}\theta}\right]. \tag{34}$$

This implies that, for $\alpha \gg 1$,

$$\left|\tan\theta Y_1'(\theta)\right| \le C\left|\sin^k\theta\right| \le C. \tag{35}$$

By direct calculation, we also obtain (35) for the case where $k = 0$. By integration by parts, we obtain

$$\begin{aligned} \left|G_1(\alpha) - G_1(\alpha\sin\theta)\right| &= \left|\int_{\alpha\sin\theta}^{\alpha} w^k \sin w\, dw\right| \\ &\le \left|\left[-w^k\cos w\right]_{\alpha\sin\theta}^{\alpha}\right| + \left|\int_{\alpha\sin\theta}^{\alpha} k w^{k-1}\cos w\, dw\right| \\ &\le C\alpha^k. \end{aligned} \tag{36}$$

By (35) and (36), for $\alpha \gg 1$, we obtain

$$\begin{aligned} \left|L_{21}(\alpha)\right| &= \left|\tan\theta Y_1'(\theta)(G_1(\alpha) - G_1(\alpha\sin\theta))\right| \\ &\le C\alpha^k. \end{aligned} \tag{37}$$

Since

$$L_{22}(\alpha) = \alpha^{k+1}\int_0^{\pi/2} Y_1(\alpha)\sin^{k+1}\theta\sin(\alpha\sin\theta)\, d\theta, \tag{38}$$

by putting $\theta = (\pi/2)(1-r)$, we obtain

$$L_{22}(\alpha) = \frac{\pi}{2}\alpha^{k+1}\int_0^1 \frac{\cos^{2k+1}(\pi/2)r}{(1+\cos^2(\pi/2)r + \cdots + \cos^{2n-2}(\pi/2)r)^{3/2}} \sin\left(\alpha\cos\frac{\pi}{2}r\right) dr. \tag{39}$$

Let

$$f(r) := \frac{\cos^{2k+1}(\pi/2)r}{\left(1+\cos^2(\pi/2)r+\cdots+\cos^{2n-2}(\pi/2)r\right)^{3/2}},$$
$$w(r) := \cos\frac{\pi}{2}r, \tag{40}$$
$$\mu := \alpha.$$

Case 1. Assume that $k > 1/2$ or $k = 0$. Then clearly $f(r) \in C^2[0,1]$, and we are able to apply Lemma 4 to (39). Then we obtain

$$L_{22}(\alpha) = \sqrt{\frac{\pi}{2}}\frac{1}{n^{3/2}}\alpha^{k+(1/2)}\sin\left(\alpha - \frac{\pi}{4}\right) + O\left(\alpha^k\right). \tag{41}$$

By this, (33), and (37), we obtain (29).

Case 2. Assume that $0 < k < 1/2$. Then $f(r) \in C^{1+2k}[0,1]$ with $0 < 2k < 1$. Therefore, $f(r)$ does not satisfy the condition in Lemma 4. However, we found in [14] that we can still apply Lemma 4 to (39) in this situation and obtain (41). For completeness, the reason will be explained in the Appendix. By this, (33), and (41), we obtain (29). Thus the proof is complete. □

By (24) and Lemma 5, we obtain Theorem 2 immediately. Thus the proof is complete.

3. Proof of Theorem 3

In this section, let $0 < \alpha \ll 1$. The proofs of Theorem 3 (i)-(v) are similar. Therefore, we only prove (i) and (iv).

Proof of Theorem 3 (i). We assume that $2n > k + 4$. Then by Taylor expansion, for $0 \le s \le 1$, we have

$$\begin{aligned} G(\alpha) - G(\alpha s) &= \frac{1}{2n}\alpha^{2n}\left(1-s^{2n}\right) + \frac{1}{k+2}\alpha^{k+2}\left(1-s^{k+2}\right) \\ &\quad - \frac{1}{6(k+4)}\alpha^{k+4}\left(1-s^{k+4}\right)(1+o(1)). \end{aligned} \tag{42}$$

By this, (24), Taylor expansion, and putting $u = \alpha s$, we obtain

$$\begin{aligned} \sqrt{\frac{\lambda(\alpha)}{2}} &= \int_0^\alpha \frac{u^k}{\sqrt{(1/2n)\left(\alpha^{2n}-u^{2n}\right)+(1/(k+2))\left(\alpha^{k+2}-u^{k+2}\right)-(1/6(k+4))\left(\alpha^{k+4}-u^{k+4}\right)(1+o(1))}}du \\ &= \sqrt{k+2}\alpha^{k/2}\int_0^1 \frac{s^k}{\sqrt{1-s^{k+2}}\sqrt{1-((k+2)/6(k+4))\left(\left(1-s^{k+4}\right)/\left(1-s^{k+2}\right)\right)\alpha^2+o\left(\alpha^2\right)}}ds \\ &= \sqrt{k+2}\alpha^{k/2}\int_0^1 \frac{s^k}{\sqrt{1-s^{k+2}}}\left(1+\frac{k+2}{12(k+4)}\frac{1-s^{k+4}}{1-s^{k+2}}\alpha^2+o\left(\alpha^2\right)\right)ds = \sqrt{k+2}\alpha^{k/2}\left\{B_0+B_1\alpha^2+o\left(\alpha^2\right)\right\}. \end{aligned} \tag{43}$$

This implies (13). Thus the proof is complete. □

Proof of Theorem 3 (iv). We assume that $2n = k + 2$. Then by (42), for $0 \le s \le 1$, we have

$$\begin{aligned} G(\alpha) - G(w) &= \frac{2}{k+2}\left(\alpha^{k+2}-w^{k+2}\right) \\ &\quad - \frac{1}{6(k+4)}\left(\alpha^{k+4}-w^{k+4}\right)(1+o(1)). \end{aligned} \tag{44}$$

By this, (24), and putting $w = \alpha s$, we obtain

$$\begin{aligned} \sqrt{\frac{\lambda(\alpha)}{2}} &= \int_0^\alpha \frac{w^k}{\sqrt{(2/(k+2))\left(\alpha^{k+2}-w^{k+2}\right)-(1/6(k+4))\left(\alpha^{k+4}-w^{k+4}\right)(1+o(1))}}dw \\ &= \sqrt{\frac{k+2}{2}}\alpha^{k/2}\int_0^1 \frac{s^k}{\sqrt{1-s^{k+2}}}\left\{1-\frac{k+2}{12(k+4)}\frac{1-s^{k+4}}{1-s^{k+2}}\alpha^2+o\left(\alpha^2\right)\right\}^{-1/2}ds \\ &= \sqrt{\frac{k+2}{2}}\alpha^{k/2}\int_0^1 \frac{s^k}{\sqrt{1-s^{k+2}}}\left\{1+\frac{k+2}{24(k+4)}\frac{1-s^{k+4}}{1-s^{k+2}}\alpha^2+o\left(\alpha^2\right)\right\}ds. \end{aligned} \tag{45}$$

This implies

$$\sqrt{\lambda} = \sqrt{k+2}\alpha^{k/2}\left\{B_0 + \frac{1}{2}B_1\alpha^2 + o\left(\alpha^2\right)\right\}. \quad (46)$$

This implies (16). Thus the proof is complete. □

Appendix

In this section, by following the argument in [14], we show that Case 2 in Lemma 5 holds for completeness. We put

$$f(x) = f_1(x) f_2(x) := \cos^{2k+1}\frac{\pi}{2}x \cdot \frac{1}{\left(1+\cos^2(\pi/2)x + \cdots + \cos^{2n-2}(\pi/2)x\right)^{3/2}}. \quad (A.1)$$

Note that $0 < 2k < 1$. We see that $f_2(x) \in C^2[0,1]$. The essential point of the proof of (27) in this case is to show Lemma 2.24 in [7] (see also [7, Lemma 2.25]). Namely, as $\mu \longrightarrow \infty$,

$$\Phi(\mu) := \int_0^1 f(x)e^{-i\mu x^2}dx = \frac{1}{2}\sqrt{\frac{\pi}{\mu}}e^{-i(\pi/4)}f(0) + O\left(\frac{1}{\mu}\right). \quad (A.2)$$

We put $h(x) = (f(x) - f(0))/x$. Then we have $f(x) = f(0) + xh(x)$. We know from [7, Lemma 2.24] that, for $\mu \gg 1$,

$$\int_0^1 e^{-i\mu x^2}dx = \frac{1}{2}\sqrt{\frac{\pi}{\mu}}e^{-i\pi/4} + O\left(\frac{1}{\mu}\right). \quad (A.3)$$

By (A.2) and (A.3), we obtain

$$\begin{aligned}\Phi(\mu) &= f(0)\int_0^1 e^{-i\mu x^2}dx + \int_0^1 xe^{-i\mu x^2}h(x)\,dx \\ &= \frac{1}{2}f(0)\sqrt{\frac{\pi}{\mu}}e^{-i\pi/4} + O\left(\frac{1}{\mu}\right) \\ &\quad + \int_0^1 xe^{-i\mu x^2}h(x)\,dx.\end{aligned} \quad (A.4)$$

We put

$$\Phi_1(\mu) := \int_0^1 xe^{-i\mu x^2}h(x)\,dx. \quad (A.5)$$

Now we prove that $h(x) \in C^1[0,1]$, because if it is proved, then by integration by parts, we easily show that $\Phi_1(\mu) = O(1/\mu)$ and our conclusion (A.2) follows immediately from (A.4) and (A.5). For $0 \le x \le 1$, we have

$$\begin{aligned}h(x) &= \frac{f(x)-f(0)}{x} \\ &= f_2(x)\frac{f_1(x)-f_1(0)}{x} + f_1(0)\frac{f_2(x)-f_2(0)}{x} \\ &:= f_2(x)h_1(x) + f_1(0)h_2(x).\end{aligned} \quad (A.6)$$

Then we have $h_2(x) \in C^1[0,1]$. Furthermore, by direct calculation, we can show that $h_1(x) \in C^1[0,1]$. It is reasonable, because by Taylor expansion, for $0 < x \ll 1$, we have

$$h_1(x) = -\frac{(2k+1)\pi^2}{8}x + O\left(x^3\right). \quad (A.7)$$

Thus the proof is complete.

Conflicts of Interest

The author declares that there are no conflicts of interest.

Acknowledgments

This work was supported by JSPS KAKENHI Grant Number JP17K05330.

References

[1] A. Ambrosetti, H. Brézis, and G. Cerami, "Combined effects of concave and convex nonlinearities in some elliptic problems," *Journal of Functional Analysis*, vol. 122, no. 2, pp. 519–543, 1994.

[2] S. Cano-Casanova and J. López-Gómez, "Existence, uniqueness and blow-up rate of large solutions for a canonical class of one-dimensional problems on the half-line," *Journal of Differential Equations*, vol. 244, no. 12, pp. 3180–3203, 2008.

[3] Y. J. Cheng, "On an open problem of Ambrosetti, Brezis and Cerami," *Differential and Integral Equations*, vol. 15, pp. 1025–1044, 2002.

[4] R. Chiappinelli, D. G. De Figueiredo, and P. Hess, "Bifurcation from infinity and multiple solutions for an elliptic system," *Differential and Integral Equations*, vol. 6, no. 4, pp. 757–771, 1993.

[5] R. Chiappinelli, "Upper and lower bounds for higher order eigenvalues of some semilinear elliptic equations," *Applied Mathematics and Computation*, vol. 216, no. 12, pp. 3772–3777, 2010.

[6] R. Chiappinelli, "Approximation and convergence rate of nonlinear eigenvalues: Lipschitz perturbations of a bounded self-adjoint operator," *Journal of Mathematical Analysis and Applications*, vol. 455, no. 2, pp. 1720–1732, 2017.

[7] P. Korman, *Global solution curves for semilinear elliptic equations*, World Scientific Publishing Co. Pte. Ltd., Hackensack, NJ, USA, 2012.

[8] T. Laetsch, "The number of solutions of a nonlinear two point boundary value problem," *Indiana University Mathematics Journal*, vol. 20, pp. 1–13, 1971.

[9] Y. H. Lee, L. Sherbakov, J. Taber, and J. Shi, "Bifurcation diagrams of population models with nonlinear, diffusion,"

Journal of Computational and Applied Mathematics, vol. 194, no. 2, pp. 357–367, 2006.

[10] J. D. Murray, "Mathematical biology. I. An introduction," in *Interdisciplinary Applied Mathematics*, vol. 17, Springer-Verlag, New York, NY, 3rd edition, 2002.

[11] P. Turchin, "Population consequences of aggregative movement," *Journal of Animal Ecology*, vol. 58, no. 1, pp. 75–100, 1989.

[12] T. Shibata, "Asymptotic length of bifurcation curves related to inverse bifurcation problems," *Journal of Mathematical Analysis and Applications*, vol. 438, no. 2, pp. 629–642, 2016.

[13] P. Korman and Y. Li, "Infinitely many solutions at a resonance," *Electronic Journal of Differential Equations*, pp. 105–111, 2000, Presented at the Differential Equations Conference 5.

[14] T. Shibata, "Global and local structures of oscillatory bifurcation curves," *Journal of Spectral Theory*.

[15] A. Galstian, P. Korman, and Y. Li, "On the oscillations of the solution curve for a class of semilinear equations," *Journal of Mathematical Analysis and Applications*, vol. 321, no. 2, pp. 576–588, 2006.

[16] P. Korman, "An oscillatory bifurcation from infinity, and from zero," *Nonlinear Differential Equations and Applications NoDEA*, vol. 15, no. 3, pp. 335–345, 2008.

[17] T. Shibata, "Oscillatory bifurcation for semilinear ordinary differential equations," *Electronic Journal of Qualitative Theory of Differential Equations*, no. 44, pp. 1–13, 2016.

[18] T. Shibata, "Global and local structures of oscillatory bifurcation curves with application to inverse bifurcation problem," *Topological Methods in Nonlinear Analysis*, vol. 50, no. 2, pp. 603–622, 2017.

[19] T. Shibata, "Global behavior of bifurcation curves for the nonlinear eigenvalue problems with periodic nonlinear terms," *Communications on Pure & Applied Analysis*, vol. 17, no. 5, pp. 2139–2147, 2018.

8

On the Control of Coefficient Function in a Hyperbolic Problem with Dirichlet Conditions

Seda İğret Araz[1] **and Murat Subaşi**[2]

[1]Siirt University, Department of Elementary Mathematics Education, Siirt 56100, Turkey
[2]Atatürk University, Department of Mathematic, Erzurum 25240, Turkey

Correspondence should be addressed to Seda İğret Araz; sedaaraz@siirt.edu.tr

Academic Editor: Omar Abu Arqub

This paper presents theoretical results about control of the coefficient function in a hyperbolic problem with Dirichlet conditions. The existence and uniqueness of the optimal solution for optimal control problem are proved and adjoint problem is used to obtain gradient of the functional. However, a second adjoint problem is given to calculate the gradient on the space $W_2^1(0,l)$. After calculating gradient of the cost functional and proving the Lipschitz continuity of the gradient, necessary condition for optimal solution is constructed.

1. Introduction

Hyperbolic boundary value problems have appeared as mathematical modelling of physical phenomena like small vibration of a string, in the fields of science and engineering. There has been much attention to studies related to optimal control problems involving hyperbolic problems [1]. There have been many studies about optimal control for hyperbolic systems which are considered [2–4].

Some of these important studies can be summarized as follows.

Hasanov [5] has considered problem of controlling the function $w := \{F(x,t); f(t)\}$ for the following problem:

$$u_{tt} = \left(k(x) u_x\right)_x + F(x,t), \quad (x,t) \in (0,l) \times (0,T)$$
$$u(x,0) = u_0(x), \quad u_t(x,0) = u_1(x), \quad x \in (0,l)$$
$$-k(0) u_x(0,t) = f(t), \quad k(l) u_x(l,t) = 0 \quad t \in (0,T) \tag{1}$$

with the conditions

$$u(x,T) = \mu(x), \quad u_t(x,T) = \nu(x), \tag{2}$$

using the functionals

$$J_1(w) = \int_0^l [u(x,T;w) - \mu(x)]^2 dx,$$
$$J_2(w) = \int_0^l [u_t(x,T;w) - \nu(x)]^2 dx, \tag{3}$$
$$J_3(w) = J_1(w) + J_2(w).$$

Majewski [6] has controlled the function $u(x, y) \in L_2(P, \mathbb{R}^M)$ for hyperbolic equation:

$$\frac{\partial^2 z}{\partial x \partial y}(x, y) = \tilde{f}\left(x, y, z(x, y), \frac{\partial z}{\partial x}(x, y), \frac{\partial z}{\partial y}(x, y), u(x, y)\right) \tag{4}$$

$$z(x, 0) = z(0, y) = 0, \quad \forall x, y \in [0, 1]$$

using the functional

$$J^k(z(\cdot), u(\cdot)) = \int_P F^k(x, y, z(x, y), u(x, y))\, dx\, dy. \quad k = 0, 1, 2, \ldots \tag{5}$$

Yeloğlu and Subaşı [7] have dealt with determination pair $w := \{f(x, t), h(x)\}$ in the following problem:

$$p(x) u_{tt} = (k(x) u_x)_x + f(x, t), \quad (x, t) \in \Omega_T$$

$$u(x, 0) = g(x), \quad u_t(x, 0) = h(x), \quad x \in (0, l) \tag{6}$$

$$u(0, t) = 0, \quad u(l, t) = 0, \quad t \in (0, T]$$

for the functional

$$J_\alpha(w) = \int_0^l [u(x, T; w) - y(x)]^2 dx + \alpha \|w\|_W^2. \tag{7}$$

Kröner [8] has specified the function $u(t) \in L_2(0, T)$ for nonlinear hyperbolic equation:

$$y_{tt} - A(u, y) = f$$

$$y(0) = y_0(u), \tag{8}$$

$$y_t(0) = y_1(u)$$

using the functional

$$J(u, y^1) = \int_0^T J_1(y^1(t))\, dt + J_2(y^1(T)) + \frac{\alpha}{2}\|u\|_U^2. \tag{9}$$

Tagiyev [9] has studied the problem of controlling the coefficients $v = (k(x), q(x, t)) \in L_\infty(\Omega) \times L_\infty(Q_T)$ for linear hyperbolic equation:

$$\frac{\partial^2 u}{\partial t^2} - \sum_{i=1}^n \frac{\partial}{\partial x_i}\left(k_i(x) \frac{\partial u}{\partial x_i}\right) + q(x, t) u = f(x, t), \quad (x, t) \in Q_T$$

$$u|_{t=0} = \varphi_0(x), \tag{10}$$

$$\left.\frac{\partial u}{\partial t}\right|_{t=0} = \varphi_1(x),$$

$$u|_{S_T} = 0$$

using the functional

$$J(v) = \alpha_0 \int_{Q_T} |u(x, t, v) - z_0(x, t)|^2 dx\, dt + \alpha_1 \int_\Omega |u(x, T, v) - z_1(x)|^2 dx. \tag{11}$$

2. Statement of the Problem

In this study, we deal with the process of vibration in finite homogeneous string, occupying the interval $(0, l)$. As the control function, we take the transverse elastic force which is in the coefficient of the vibration problem. Also, we propose the usage of a more regular space than the space of square integrable functions in the cost functional. In general, this process exposes some difficulties in the stage of acquiring the gradient. This study offers a second adjoint problem to overcome this case.

In the domain $\Omega := (x, t) \in (0, l) \times (0, T)$, we consider the functional

$$J_\alpha(q) = \int_0^l [u(x, T; q) - y(x)]^2 dx + \alpha \|q - r\|_{W_2^1(0,l)}^2 \tag{12}$$

on the set

$$Q = \left\{q(x) : q(x) \in W_2^1(0, l),\ 0 < q_1 \le q(x) \le q_2,\ \|q(x)\|_{W_2^1(0,l)} \le q_3\right\} \tag{13}$$

subject to the hyperbolic problem

$$u_{tt} - u_{xx} + q(x) u = 0, \quad (x, t) \in \Omega \tag{14}$$

$$u(x, 0) = \varphi_1(x), \quad u_t(x, 0) = \varphi_2(x), \quad x \in (0, l) \tag{15}$$

$$u(0, t) = 0, \quad u(l, t) = 0, \quad t \in (0, T). \tag{16}$$

Here $y(x) \in L_2(0, l)$ is the desired target function to which $u(x, T)$ must be close enough. The function $r(x) \in W_2^1(0, l)$ is an initial guess for optimal solution. $\alpha > 0$ is regularization parameter. $q_1, q_2, q_3 > 0$ are given positive numbers.

The initial status functions are in the following spaces:

$$\varphi_1(x) \in W_2^1(0, l), \quad \varphi_2(x) \in L_2(0, l). \tag{17}$$

The aim of this study is to deal with the problem of

$$J_{\alpha^*} = \inf_{q \in Q} J_\alpha(q) = J_\alpha(q^*) \tag{18}$$

under conditions (12)-(17).

Namely, we want to control the transverse elastic force on the space $W_2^1(0,l)$ and the solution $u(x,T)$ corresponding to this control function must be close enough to $y(x)$ in $L_2(0,l)$. In order to get a stable solution, we choose the space $W_2^1(0,l)$ which is more regular than $L_2(0,l)$.

The inner product and norm in $W_2^1(0,l)$ are defined, respectively, as

$$(f,g)_{W_2^1(0,l)} = \int_0^l \left[f(x).g(x) + \frac{df(x)}{dx}.\frac{dg(x)}{dx} \right] dx,$$
$$\|f\|^2_{W_2^1(0,l)} = \int_0^l \left([f(x)]^2 + \left[\frac{df(x)}{dx}\right]^2 \right) dx. \quad (19)$$

The paper is organized as follows: in Section 3, we obtain the generalized solution for hyperbolic problem. In Section 4, we prove the existence and uniqueness of the optimal solution. In Section 5, we obtain the adjoint problem for the optimal control problem and find the gradient of the functional. The main contribution of this paper is executed in this section. Because the controls are chosen in the space $W_2^1(0,l)$, getting the gradient of the functional necessitates finding a second adjoint problem. In the last section, we demonstrate the Lipschitz continuity of the gradient and state the necessary condition for optimal solution.

3. Solvability of the Problem

In this section, we first give the definition of the generalized solution for hyperbolic problem.

The generalized solution of problem (14)-(15) is the function $u \in \overset{o}{W}{}_2^{1,1}(\Omega)$ satisfying the following integral equality:

$$\int_\Omega [-u_t\eta_t + u_x\eta_x + q(x)u\eta]\, dx\, dt = \int_0^l \varphi_2(x)\eta(x,0)\, dx \quad (20)$$

for $\forall \eta \in \overset{o}{W}{}_2^{1,1}(\Omega), \eta(x,T) = 0$.

It can be seen in [10] that solution in the sense of (20) exists, is unique, and satisfies the following inequality:

$$\max_{0\le t\le T} \left(\|u(\cdot,t)\|^2_{L_2(0,l)} + \|u_t(\cdot,t)\|^2_{L_2(0,l)} + \|u_x(\cdot,t)\|^2_{L_2(0,l)} \right) \le c_1 \left(\|\varphi_1\|^2_{W_2^1(0,l)} + \|\varphi_2\|^2_{L_2(0,l)} \right) \quad (21)$$

where $c_1 = \max\{3c_0, 3c_0/q_1\}$ and $c_0 = \max\{1, q_2\}$ or

$$\|u\|^2_{W_2^{1,1}(\Omega)} \le c_2 \left(\|\varphi_1\|^2_{W_2^1(0,l)} + \|\varphi_2\|^2_{L_2(0,l)} \right). \quad (c_2 = c_1T) \quad (22)$$

Since φ_1 and φ_2 are given functions, it can be written as follows:

$$\|u\|^2_{W_2^{1,1}(\Omega)} \le c_3. \quad (23)$$

Now, we give an increment $\delta q(x) \in W_2^1(0,l)$ to the control function $q(x)$ such as $q+\delta q \in Q$. Then the difference function $\delta u = \delta u(x,t) = u(x,t;q+\delta q) - u(x,t;q)$ is the solution of the following difference initial-boundary problem:

$$\delta u_{tt} - \delta u_{xx} + [q(x) + \delta q(x)]\delta u + \delta q(x) u = 0 \quad (24)$$

$$\delta u(x,0) = 0, \quad \delta u_t(x,0) = 0 \quad (25)$$

$$\delta u(0,t) = 0, \quad \delta u(l,t) = 0. \quad (26)$$

By considering (23), we obtain that the solution of above difference initial-boundary problem holds the following inequality:

$$\max_{0\le t\le T} \left(\|\delta u(.,t)\|^2_{L_2(0,l)} \right) \le c_4 \|\delta q\|^2_{W_2^1(0,l)}. \quad (27)$$

Here $c_4 = (t^3 2l/3)c_3$ is independent from δq.

4. Existence and Uniqueness of the Optimal Solution

To demonstrate the existence and the uniqueness of optimal solution for problem (12)-(17), it is enough to show that conditions of the following theorem given by Goebel [11] hold.

Theorem 1. *Let H be a uniformly convex Banach space and the set Q be a closed, bounded, and convex subset of H. If $\alpha > 0$ and $\beta \ge 1$ are given numbers and the functional $J(q)$ is lower semicontinuous and bounded from below on the set Q, then there is a dense set G of H that the functional*

$$J_\alpha(q) = J(q) + \alpha \|q - r\|^\beta_H \quad (28)$$

takes its minimum on the set Q for $\forall r \in G$. If $\beta > 1$ then minimum is unique.

Before showing that these conditions have been satisfied, we prove that the functional

$$J(q) = \int_0^l [u(x,T;q) - y(x)]^2 dx \quad (29)$$

is continuous. For this, we write the following increment of the functional:

$$\delta J(q) = J(q+\delta q) - J(q) = \int_0^l 2[u(x,T) - y(x)][\delta u(x,T)]\, dx + \int_0^l [\delta u(x,T)]^2 dx. \quad (30)$$

Since $y(x) \in L_2(0,l)$, if we consider inequalities (22) and (27), we conclude that this increment satisfies the following continuity inequality on the set Q:

$$|\delta J(q)| \le c_5 \left(\|\delta q\|_{W_2^1(0,l)} + \|\delta q\|^2_{W_2^1(0,l)} \right). \quad (31)$$

Here c_5 is independent of δq.

Thanks to this inequality, we can say that this functional is also lower semicontinuous and bounded from below on the set Q.

On the other hand, the set $W_2^1(0,l)$ is a uniformly convex Banach space [12], the set Q is a closed, bounded, and convex subset of $W_2^1(0,l)$, and $\beta = 2$.

Therefore the conditions of above theorem hold and optimal solution to the problem (18) is unique.

5. Adjoint Problem and Gradient of the Functional

In this section, we write the Lagrange functional used for finding adjoint problem, before we show the Frechet differentiability of the functional $J_\alpha(q)$ on the set Q. Lagrange functional to the problem is

$$\begin{aligned} L(u,q,\eta) &= \int_0^l [u(x,T;q) - y(x)]^2 dx \\ &+ \alpha \|q - r\|^2_{W_2^1(0,l)} \\ &+ \int_0^T \int_0^l [u_{tt} - u_{xx} + q(x) u] \eta \, dx \, dt. \end{aligned} \quad (32)$$

The first variation of this functional according to the function u is obtained such as

$$\begin{aligned} \delta L &= \int_0^l 2[u(x,T) - y(x) - \eta_t(x,T)] \delta u(x,T) dx \\ &+ \int_0^l \int_0^T (\eta_{tt} - \eta_{xx} + q(x)\eta) \delta u \, dt \, dx = 0. \end{aligned} \quad (33)$$

By means of stationary condition $\delta L = 0$, the following adjoint boundary value problem is found:

$$\eta_{tt} - \eta_{xx} + q(x)\eta = 0 \quad (34)$$

$$\begin{aligned} &\eta(x,T) = 0, \\ &\eta_t(x,T) = 2[u(x,T) - y(x)] \end{aligned} \quad (35)$$

$$\begin{aligned} &\eta(0,t) = 0, \\ &\eta(l,t) = 0. \end{aligned} \quad (36)$$

For $\forall \gamma \in \overset{o}{W}{}_2^{1,1}(\Omega)$, the function $\eta \in C^1([0,T], L_2(0,l)) \cap C^0([0,T], W_2^1(0,l))$ which satisfies the following equality

$$\begin{aligned} &\int_0^T \int_0^l [-\eta_t \gamma_t + \eta_x \gamma_x + q(x)\eta\gamma] \, dx \, dt \\ &= \int_0^l \eta_t(x,0) \gamma(x,0) dx \\ &- \int_0^l 2[u(x,T) - y(x)] \gamma(x,T) dx \end{aligned} \quad (37)$$

is the solution of adjoint boundary value problem (34)-(36).

This solution satisfies the following inequality:

$$\|\eta\|_{L_2(0,l)} \le c_6 \|u(x,T) - y(x)\|_{L_2(0,l)}, \quad \forall t \in [0,T]. \quad (38)$$

Now, we can pass the calculation of the gradient. In order to do this, we must evaluate the increment of the functional $J_\alpha(q)$. The increment can be written such as

$$\begin{aligned} \delta J_\alpha(q) &= J_\alpha(q + \delta q) - J_\alpha(q) \\ &= \int_0^l 2[u(x,T) - y(x)](\delta u) dx \\ &+ \int_0^l (\delta u)^2 dx + 2\alpha \langle q - r, \delta q \rangle_{W_2^1(0,l)}. \end{aligned} \quad (39)$$

The difference problem (24)-(26) and the adjoint problem (34)-(36) give together the equality of

$$\begin{aligned} &2\int_0^l [u(x,T) - y(x)](\delta u) dx \\ &= \int_0^T \int_0^l [\delta q \delta u \eta + \delta q u \eta] \, dx \, dt. \end{aligned} \quad (40)$$

Inserting (40) in (39), we have

$$\begin{aligned} \delta J_\alpha(q) &= \int_0^T \int_0^l u\eta\delta q \, dx \, dt + \int_0^T \int_0^l \eta \delta u \delta q \, dx \, dt \\ &+ \int_0^l (\delta u(x,T))^2 dx + 2\alpha \langle q - r, \delta q \rangle_{W_2^1(0,l)}. \end{aligned} \quad (41)$$

By (27) and (38), the second and third integrals of the above equality give the following inequality:

$$\begin{aligned} &\int_0^T \int_0^l \eta \delta u \delta q \, dx \, dt + \int_0^l (\delta u(x,T))^2 dx \\ &\le c_7 \|\delta q\|^2_{W_2^1(0,l)}. \end{aligned} \quad (42)$$

The statement (41) can be rewritten as

$$\begin{aligned} \delta J_\alpha(q) &= \langle u\eta, \delta q \rangle_{L_2(\Omega)} + 2\alpha \langle q - r, \delta q \rangle_{W_2^1(0,l)} \\ &+ o\left(\|\delta q\|^2_{W_2^1(0,l)} \right) \end{aligned} \quad (43)$$

or

$$\delta J_\alpha(q) = \left\langle \int_0^T u\eta\, dt, \delta q \right\rangle_{L_2(0,l)} + 2\alpha \langle q - r, \delta q\rangle_{W_2^1(0,l)} + o\left(\|\delta q\|^2_{W_2^1(0,l)}\right). \tag{44}$$

In order to pass the inner product in $W_2^1(0,l)$, we rearrange (44) such as

$$\delta J_\alpha(q) = \langle \xi + 2\alpha(q-r), \delta q\rangle_{W_2^1(0,l)} + o\left(\|\delta q\|^2_{W_2^1(0,l)}\right). \tag{45}$$

Here function $\xi(x)$ is the solution of the second adjoint problem:

$$\begin{aligned} -\xi'' + \xi &= \int_0^T u\eta\, dt \\ \xi'(0) &= 0, \\ \xi'(l) &= 0. \end{aligned} \tag{46}$$

Therefore, we have the following gradient:

$$J'_\alpha(q) = \xi + 2\alpha(q-r). \tag{47}$$

6. Lipschitz Continuity of the Gradient

In this section, we introduce a theorem about Lipschitz continuity of the gradient. By this means, we can express the necessary condition for optimal solution.

Theorem 2. *Gradient $J'_\alpha(q)$ satisfies the following Lipschitz inequality:*

$$\left\|J'_\alpha(q+\delta q) - J'_\alpha(q)\right\|^2_{W_2^1(0,l)} \le c_8 \|\delta q\|^2_{W_2^1(0,l)}. \tag{48}$$

Here c_8 is independent from δq.

Hence, it has been proven that the gradient $J'_\alpha(q)$ is continuous on the set Q and it can be seen that it holds the Lipschitz condition with constant $c_8 > 0$.

Proof. Increment of the functional $J'_\alpha(q)$ by giving the increment of δq to the control $q \in Q$ is obtained:

$$\begin{aligned} J'_\alpha(q+\delta q) - J'_\alpha(q) &= \xi_\delta + 2\alpha(q+\delta q - r) - \xi \\ &\quad + 2\alpha(q-r) = \delta\xi + 2\alpha\delta q \end{aligned} \tag{49}$$

where the function $\delta\xi(x)$ is the solution of the increment problem:

$$\delta\xi''(x) - \delta\xi(x) = \int_0^T (u_\delta \delta\eta + \delta u\eta)\, dt. \tag{50}$$

Taking the norm of (49) in the space $W_2^1(0,l)$, we acquire the following inequality belonging to the functional $\delta J'_\alpha(q)$:

$$\left\|\delta J'_\alpha(q)\right\|^2_{W_2^1(0,l)} \le 2\|\delta\xi\|^2_{W_2^1(0,l)} + 8\alpha^2 \|\delta q\|^2_{W_2^1(0,l)}. \tag{51}$$

There is a solution of problem (50) in $W_2^1(0,l)$ and this solution satisfies the following inequality:

$$\|\delta\xi(x)\|_{W_2^1(0,l)} \le \left\|\int_0^T (u_\delta\delta\eta + \delta u\eta)\, dt\right\|_{L_2(0,l)}. \tag{52}$$

The function

$$\begin{aligned} \delta\eta(x,t) &= \eta_\delta(x,t) - \eta(x,t) \\ &= \eta(x,t;q+\delta q) - \eta(x,t;q) \end{aligned} \tag{53}$$

in the right hand side of inequality (52) is the solution of the following problem:

$$\begin{aligned} &\frac{\partial^2\delta\eta}{\partial t^2} - \frac{\partial^2\delta\eta}{\partial x^2} + (q(x) + \delta q(x))\delta\eta + \delta q(x)\eta = 0, \\ &\qquad\qquad (x,t) \in \Omega \\ &\delta\eta(x,T) = 0, \\ &\delta\eta_t(x,T) = 2\delta u(x,T), \\ &\delta\eta(0,t) = \delta\eta(l,t) = 0 \end{aligned} \tag{54}$$

and this function holds the following inequality:

$$\max_{0\le t\le T}\left(\|\delta\eta(.,t)\|^2_{L_2(0,l)}\right) \le c_9 \|\delta q\|^2_{W_2^1(0,l)}. \tag{55}$$

Here c_9 is independent of δq.

So, the function u_δ that takes place in the right hand side of (52) holds the same inequality given as follows:

$$\|u_\delta\|^2_{L_2(\Omega)} \le c_3. \tag{56}$$

Hence inequality (52) has the following property:

$$\begin{aligned} \|\delta\xi(x)\|^2_{W_2^1(0,l)} &\le 2\|u_\delta\|^2_{L_2(\Omega)} \max_{0\le t\le T}\left(\|\delta\eta(.,t)\|^2_{L_2(0,l)}\right) \\ &\quad + 2\|\eta\|^2_{L_2(\Omega)} \max_{0\le t\le T}\left(\|\delta u(.,t)\|^2_{L_2(0,l)}\right). \end{aligned} \tag{57}$$

If inequalities (27), (38), (55), and (56) about functions u_δ, $\delta\eta(.,t)$, η, and $\delta u(.,t)$ are written in (57), then the following assessment is obtained:

$$\|\delta\xi(x)\|^2_{W_2^1(0,l)} \le c_{10}\|\delta q\|^2_{W_2^1(0,l)}. \tag{58}$$

Here c_{10} is independent of δq.

Considering inequality (58), the following is written:

$$\begin{aligned} \left\|\delta J'_\alpha(q)\right\|^2_{W_2^1(0,l)} &\le 2\|\delta\xi(x)\|^2_{W_2^1(0,l)} \\ &\quad + 8\alpha^2\|\delta q(x)\|^2_{W_2^1(0,l)} \\ &\le 2c_{10}\|\delta q\|^2_{W_2^1(0,l)} + 8\alpha^2\|\delta q\|^2_{W_2^1(0,l)} \\ &\le c_{11}\|\delta q\|^2_{W_2^1(0,l)}. \end{aligned} \tag{59}$$

So the following inequality for the gradient $J'_\alpha(q)$ is obtained:

$$\left\|J'_\alpha(q+\delta q) - J'_\alpha(q)\right\|^2_{W_2^1(0,l)} \le c_{11}\|\delta q\|^2_{W_2^1(0,l)}. \tag{60}$$

Once we take as $c_8 = c_{11}$, then the proof is obtained. □

7. The Necessary Condition for Optimal Solution

After showing Lipschitz continuity of the gradient, it can be said that the gradient $J'_{\alpha}(q)$ is continuous on the set Q and it holds the Lipschitz constant $c_8 > 0$. The fact that the functional $J_{\alpha}(q)$ is continuously differentiable on the set Q and the set Q is convex, in that case the following inequality is valid according to theorem in [13]:

$$\left\langle J'_{\alpha}(q^*), q - q^* \right\rangle_{W_2^1(0,l)} \geq 0, \quad \forall q \in Q. \tag{61}$$

Therefore, the following inequality is written for optimal control problem:

$$\langle \xi + 2\alpha (q^* - r), q - q^* \rangle_{W_2^1(0,l)} \geq 0, \quad \forall q \in Q. \tag{62}$$

Conflicts of Interest

The authors declare that they have no conflicts of interest.

References

[1] H. F. Guliyev and K. S. Jabbarova, "An optimal control problem for weakly nonlinear hyperbolic equations," *Acta Mathematica Hungarica*, vol. 131, no. 3, pp. 197–207, 2011.

[2] G. M. Bahaa, "Boundary control problem of infinite order distributed hyperbolic systems involving time lags," *Intelligent Control and Automation*, vol. 3, pp. 211–221, 2012.

[3] A. Kowalewski, I. Lasiecka, and J. Sokołowski, "Sensitivity analysis of hyperbolic optimal control problems," *Computational Optimization and Applications*, vol. 52, no. 1, pp. 147–179, 2012.

[4] S. Serovajsky, "Optimal control for systems described by hyperbolic equation with strong nonlinearity," *Journal of Applied Analysis and Computation*, vol. 3, no. 2, pp. 183–195, 2013.

[5] A. Hasanov, "Simultaneous determination of the source terms in a linear hyperbolic problem from the final overdetermination: weak solution approach," *IMA Journal of Applied Mathematics*, vol. 74, no. 1, pp. 1–19, 2008.

[6] M. Majewski, "Stability analysis of an optimal control problem for a hyperbolic equation," *Journal of Optimization Theory and Applications*, vol. 141, no. 1, pp. 127–146, 2009.

[7] T. Yeloğlu and M. Subaşı, "Simultaneous control of the source terms in a vibrational string problem," *Iranian Journal of Science & Technology, Transaction A*, vol. 34, no. A1, 2010.

[8] A. Kröner, "Adaptive finite element methods for optimal control of second order hyperbolic equations," *Computational Methods in Applied Mathematics*, vol. 11, no. 2, pp. 214–240, 2011.

[9] R. K. Tagiyev, "On optimal control of the hyperbolic equation coefficients," *Automation and Remote Control*, vol. 73, no. 7, pp. 1145–1155, 2012.

[10] O. A. Ladyzhenskaya, *Boundary Value Problems in Mathematical Physics*, Springer-Verlag, New York, USA, 1985.

[11] M. Goebel, "On existence of optimal control," *Mathematische Nachrichten*, vol. 93, pp. 67–73, 1979.

[12] K. Yosida, *Functional Analysis*, 624, Springer-Verlag, New York, USA, 1980.

[13] F. P. Vasilyev, *Ekstremal Problemlerin Çözüm Metotları*, 400, Nauka, 1981.

9

Finite Volume Element Approximation for the Elliptic Equation with Distributed Control

Quanxiang Wang,[1] **Tengjin Zhao**,[2] **and Zhiyue Zhang**[2]

[1]*College of Engineering, Nanjing Agricultural University, Nanjing 210031, China*
[2]*Jiangsu Key Laboratory for NSLSCS, School of Mathematical Sciences, Nanjing Normal University, Nanjing 210023, China*

Correspondence should be addressed to Zhiyue Zhang; zhangzhiyue@njnu.edu.cn

Guest Editor: Omar Abu Arqub

In this paper, we consider a priori error estimates for the finite volume element schemes of optimal control problems, which are governed by linear elliptic partial differential equation. The variational discretization approach is used to deal with the control. The error estimation shows that the combination of variational discretization and finite volume element formulation allows optimal convergence. Numerical results are provided to support our theoretical analysis.

1. Introduction

In recent years, the optimization with partial differential equation constraints (PDEs) has received a significant impulse. Because of wide applicability of the field, a lot of theoretical results have been developed. Generally, it is difficult to obtain the analytical solutions for optimal control problems with PDEs. Factually, only approximate solutions or numerical solutions can be expected. Therefore, many numerical methods have been proposed to solve the problems.

Finite element method is an important numerical method for the problems of partial differential equations and widely used in the numerical solution of optimal control problems. There are extensive studies in convergence of finite element approximation for optimal control problems. For example, priori error estimates for finite element discretization of optimal control problems governed by elliptic equations are discussed in many publications. In [1], a new approach to error control and mesh adaptivity is described for the discretization of the optimal control problems governed by elliptic partial differential equations. In [2], the error estimates for semilinear elliptic optimal controls in the maximum norm are presented. Chen and Liu present a priori error analysis for mixed finite element approximation of quadratic optimal control problems [3]. In [4], a priori error analysis for the finite element discretization of the optimal control problems governed by elliptic state equations is considered. Hou and Li investigate the error estimates of mixed finite element methods for optimal control problems governed by general elliptic equations and derive L^2 and H^1 error estimates for both the control and state variables [5].

The finite volume element method has been one of the most commonly used numerical methods for solving partial differential equations. The advantages of the method are that the computational cost is less than finite element method, and the mass conservation law is maintained. So it has been extensively used in computational fluid dynamics [6–12]. However, there are only a few published results on the finite volume element method for the optimal control problems. In [13], the authors discussed distributed optimal control problems governed by elliptic equations by using the finite volume element methods. The variational discretization approach is used to deal with the control and the error estimates are obtained in some norms. In [14], the authors considered the convergence analysis of discontinuous finite volume methods applied to distributed optimal control problems governed by a class of second-order linear elliptic equations.

In this paper, we will investigate the finite volume element method for the general elliptic optimal control problem with Dirichlet or Neumann boundary conditions. The variational discretization approach is used to deal with the control, which can avoid explicit discretization of the control and improve the approximation. In addition, we discuss the optimal control problems in polygonal domains with corner singularities. In this situation, the solution does not admit integrable second derivatives. The desired convergence results of finite volume element schemes cannot be expected. Two effective methods are proposed to compensate the negative effects of the corner singularities. The corresponding results will be reported in the future.

The rest of the paper is organized as follows. In Section 2, the model problem and the finite volume element schemes are introduced. Section 3 presents the error estimates of the finite volume element schemes. In Section 4, numerical results are supplied to justify the theoretical analysis. Brief conclusions are given in Section 5.

2. Problem Statement and Discretization

2.1. Model Problem. In this paper, we consider the following second-order elliptic partial differential equation:

$$-\nabla\cdot(A\nabla y)+c_0 y=Bu+f,\quad \text{in }\Omega,\tag{1}$$

where $\Omega\subset R^2$ is a bounded convex polygon with boundary $\partial\Omega$, $A=\{a_{ij}(x)\}$ is a 2×2 symmetric and uniformly positive definite matrix, $c_0>0$ is a sufficient smooth function defined on Ω, B denotes the linear and continuous control operator, $Bu\in L^2(\Omega)$, and u and f have enough regularity so that this problem has a unique solution when we combine either homogeneous Dirichlet or Neumann boundary conditions on $\partial\Omega$.

In addition, we use the following notations for the inner products and norms on $L^2(\Omega)$, $H^1(\Omega)$, and $L^\infty(\Omega)$:

$$\begin{aligned}(v,w)&=(v,w)_{L^2(\Omega)},\\ \|v\|&=\|v\|_{L^2(\Omega)},\\ \|v\|_1&=\|v\|_{H^1(\Omega)},\\ \|v\|_\infty&=\|v\|_{L^\infty(\Omega)}.\end{aligned}\tag{2}$$

The corresponding weak formulation for (1) is

$$\text{Find } y\in H \text{ such that } a(y,v)=(Bu+f,v),\quad \forall v\in H,\tag{3}$$

where

$$a(y,v)=\int_\Omega\left(\sum_{i,j=1}^{2}a_{ij}\frac{\partial y}{\partial x_j}\frac{\partial v}{\partial x_i}+c_0yv\right)dx,\tag{4}$$

and

$$(Bu+f,v)=\int_\Omega(Bu+f)\,v\,dx;\tag{5}$$

H denotes either depending on the prescribed type of boundary conditions (homogeneous Neumann or Dirichlet).

Now, we consider the following optimal control problem for state variable y and the control variable u:

$$\min\ J(y,u)=\frac{1}{2}\int_\Omega|y-y_\Omega|^2dx+\frac{\lambda}{2}\int_\Omega|u|^2dx,\tag{6}$$

over all $H\times L^2(\Omega)$ subject to elliptic state problem (3) and the control constraints

$$u_a(x)\le u(x)\le u_b(x),\tag{7}$$

where $y_\Omega\in L^2(\Omega)$ is a given desired state and $\lambda\ge 0$ is a regularization parameter. We define the set of admissible control by

$$U_{ad}=\{u\in L^2(\Omega):u_a(x)\le u\le u_b(x)\},\tag{8}$$

where U_{ad} is a nonempty, closed, and convex subset of $L^2(\Omega)$,$u_a(x)\le u_b(x)$.

From standard arguments for elliptic equations, we can obtain the following propositions.

Proposition 1. *For fixed control $u\in L^2(\Omega)$, the state equation (3) admits a unique solution $y\in H$. Moreover, there is a constant C, which does not depend on $Bu+f$, such that*

$$\|u\|_1\le C\|Bu+f\|.\tag{9}$$

Proposition 2. *Let U_{ad} be a nonempty, closed, bounded, and convex set, y_Ω in $L^2(\Omega)$ and $\lambda>0$; then the optimal control problem (6) admits a unique solution $(\overline{y},\overline{u})$.*

This proof follows standard techniques [15].

The adjoint state equation for $\overline{z}\in H$ is given by

$$a(\overline{z},w)=(\overline{y}-y_\Omega,w),\quad\forall w\in H,\tag{10}$$

where the equation is the weak formulation of the following elliptic problem:

$$-\nabla\cdot(A\nabla\overline{z})+c_0\overline{z}=\overline{y}-y_\Omega,\quad\text{in }\Omega,\tag{11}$$

with homogeneous Neumann or Dirichlet boundary conditions.

Proposition 3. *The necessary and sufficient optimality conditions for (6) and (7) can be expressed as the variational inequality*

$$(\lambda\overline{u}+B^*\overline{z},u-\overline{u})\ge 0,\quad\forall u\in U_{ad}.\tag{12}$$

Further, the variational inequality is equivalent to

$$\overline{u}=P_{[u_a(x),u_b(x)]}\left(-\frac{B^*\overline{z}}{\lambda}\right),\tag{13}$$

where $P_{[u_a(x),u_b(x)]}(\cdot)=\min\{u_b(x),\max\{u_a(x),\cdot\}\}$ denotes the orthogonal projection in $L^2(\Omega)$ onto the admissible set of the control and B^* is the adjoint operator of B.

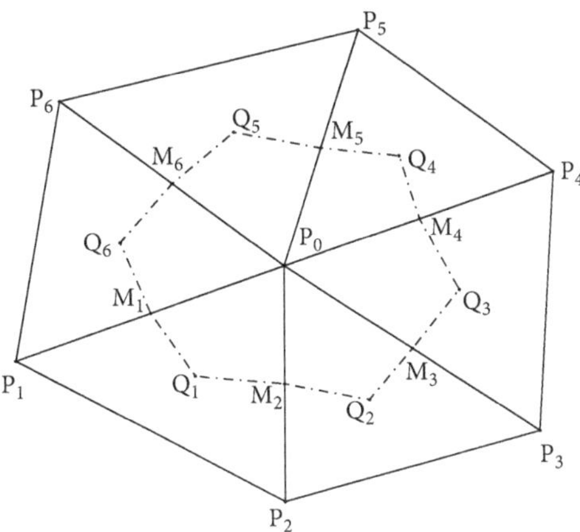

FIGURE 1: Control volume with barycenter as internal point.

2.2. Discretization. Now we describe the finite volume element discretization of the optimal control problem (6).

We consider a quasi-uniform triangulation T_h. Divide $\overline{\Omega}$ into a sum of finite number of small triangles K such that they have no overlapping internal region and a vertex of any triangle does not belong to a side of any other triangle. At last, we can obtain a triangulation such that $\overline{\Omega} = \bigcup_{K\in T_h} K$.

We then construct a dual mesh T_h^* related to T_h. Let P_0 be a node of a triangle, P_i $(i = 1, 2, \ldots, 6)$ the adjacent nodes of P_0, and M_i the midpoint of $\overline{P_0P_i}$. Choose the barycenter Q_i of triangle $\triangle P_0P_iP_{i+1}$ $(P_7 = P_1)$ as the node of the dual mesh. Connect successively $M_1, Q_1, \ldots, M_6, Q_6, M_1$ to form a polygonal region V, called a control volume. Figure 1 presents a sketch of a control volume.

Let U_h be the trial function space defined on the triangulation T_h,

$$U_h = \{v \in C(\Omega) : v|_K \text{ is linear for all } K \in T_h\}, \quad (14)$$

and V_h be the test function space defined on the dual mesh T_h^*,

$$V_h = \{v \in L^2(\Omega) : v|_V \text{ is constant for all } V \in T_h^*\}. \quad (15)$$

In this way, we have

$$\begin{aligned} U_h &= \operatorname{span}\{\phi_1, \phi_2, \ldots, \phi_{N_{node}}\}, \\ V_h &= \operatorname{span}\{\psi_1, \psi_2, \ldots, \psi_{N_{node}}\}, \end{aligned} \quad (16)$$

where ϕ_i are the standard node basis functions with the nodes x_i and ψ_i are the characteristic functions of the control volume V_i.

Let I_h and I_h^* be the interpolation projections onto the trial function space U_h and test function space V_h, respectively. By the interpolation theory, we have for $w \in U_h \cap H^2$

$$\begin{aligned} &|w - I_h w|_m \le Ch^{2-m} |w|_2, \quad m = 0, 1; \\ &\|w - I_h^* w\| \le Ch |w|_1. \end{aligned} \quad (17)$$

Then the finite volume element schemes for (3), (10), and (13) are defined as follows:

$$a_h(\overline{y}_h, I_h^* v) = (B\overline{u}_h + f, I_h^* v), \quad \forall v \in U_h, \quad (18)$$

$$a_h(\overline{z}_h, I_h^* w) = (\overline{y}_h - y_\Omega, I_h^* w), \quad \forall w \in U_h, \quad (19)$$

$$\begin{aligned} &(\lambda \overline{u}_h + B^* \overline{z}_h, u - \overline{u}_h) \ge 0 \quad \forall u \in U_{ad}, \\ &\text{or } \overline{u}_h = P_{[u_a(x), u_b(x)]}\left(-\frac{B^*\overline{z}_h}{\lambda}\right), \end{aligned} \quad (20)$$

where

$$\begin{aligned} &a_h(\overline{y}_h, I_h^* v) \\ &= -\sum_{V_i} \left[I_h^* v \int_{\partial V_i} A\nabla \overline{y}_h \cdot \mathrm{n} ds - \int_{V_i} c_0 \overline{y}_h I_h^* v dx \right]. \end{aligned} \quad (21)$$

3. Error Estimates

In order to present the error estimates, we first introduce some lemmas in preparation of the proof for the main convergence theorem.

3.1. Some Lemmas. According to [16], we have the following lemma, which indicates that the bilinear form $a_h(\cdot, I_h^* \cdot)$ is coercive on U_h.

Lemma 4. *$a_h(\cdot, I_h^* \cdot)$ is positive definite for small enough h; namely, there exist $h_0 > 0$, $\alpha > 0$ such that for $0 < h \le h_0$*

$$a_h(v, I_h^* v) \ge \alpha \|v\|_1^2, \quad \forall v \in U_h. \quad (22)$$

We seldom have a symmetric bilinear form $a_h(\cdot, I_h^* \cdot)$ even though $a(\cdot, \cdot)$ is symmetric. The following lemma is used to measure how far the bilinear form $a_h(\cdot, I_h^* \cdot)$ is from being symmetric [17].

Lemma 5. *There exist positive constants C, h_0 such that, for $u, w \in U_h$ and $0 < h \le h_0$, we have*

$$|a_h(u, I_h^* w) - a_h(w, I_h^* u)| \le Ch \|u\|_1 \|w\|_1. \quad (23)$$

Furthermore, we introduce the auxiliary functions $\overline{y}^h \in U_h$ and $\overline{z}^h \in U_h$ which are the solutions of the following problems:

$$\begin{aligned} a_h\left(\overline{y}^h, I_h^* v\right) &= \left(B\overline{u} + f, I_h^* v\right), \quad \forall v \in U_h, \\ a_h\left(\overline{z}^h, I_h^* w\right) &= \left(\overline{y} - y_\Omega, I_h^* w\right), \quad \forall w \in U_h. \end{aligned} \tag{24}$$

For the problems, we can obtain the following results.

Lemma 6. *Let $\overline{y}_h$ and $\overline{z}_h$ be the solution of (18) and (19) and $\overline{y}^h$, $\overline{z}^h$ be the solution of (24). Then, we have*

$$\left\|\overline{y}_h - \overline{y}^h\right\|_1 \le C\left\|\overline{u}_h - \overline{u}\right\|, \tag{25}$$

$$\left\|\overline{z}_h - \overline{z}^h\right\|_1 \le C\left\|\overline{y}_h - \overline{y}\right\|. \tag{26}$$

Proof. Combining (18) and (24), we have

$$a_h\left(\overline{y}^h - \overline{y}_h, I_h^* v\right) = \left(B\left(\overline{u} - \overline{u}_h\right), I_h^* v\right). \tag{27}$$

By taking $v = \overline{y}^h - \overline{y}_h$ and using Lemma 4, we have

$$\alpha\left\|\overline{y}^h - \overline{y}_h\right\|_1^2 \le \left(B\left(\overline{u} - \overline{u}_h\right), I_h^*\left(\overline{y}^h - \overline{y}_h\right)\right), \tag{28}$$

where Lemma 4 is used. At last, we can obtain (25) with Cauchy-Schwarz inequality. Equation (26) can be obtained similarly. □

The results in [18] can easily be extended to cover the elliptic equations with homogeneous Neumann boundary conditions. Now we list the useful theoretical results in the following lemma.

Lemma 7. *Let $\overline{y}$ and $\overline{z}$ be the solution of (4) and (10), respectively, and $\overline{y}^h$, $\overline{z}^h$ be the solution of (24), $\overline{u}$, f, $y_\Omega \in H^1(\Omega)$, and $A \in W^{2,\infty}$. Then there exists a positive constant $C > 0$ and $h_0 > 0$ such that for $0 < h \le h_0$*

$$\begin{aligned} \left\|\overline{y} - \overline{y}^h\right\| &\le Ch^2, \\ \left\|\overline{y} - \overline{y}^h\right\|_1 &\le Ch, \\ \left\|\overline{y} - \overline{y}^h\right\|_\infty &\le Ch^2 \log\frac{1}{h}, \\ \left\|\overline{z} - \overline{z}^h\right\| &\le Ch^2, \\ \left\|\overline{z} - \overline{z}^h\right\|_1 &\le Ch, \\ \left\|\overline{z} - \overline{z}^h\right\|_\infty &\le Ch^2 \log\frac{1}{h}. \end{aligned} \tag{29}$$

3.2. L^2 Error Estimate

Theorem 8. *Assume that $\overline{u}$ and $\overline{u}_h$ are the solutions of (6) and (20), respectively, $\overline{u}$, f, $y_\Omega \in H^1(\Omega)$, and $A \in W^{2,\infty}$. Then there exists a positive constant $C > 0$ and $h_0 > 0$ such that for $0 < h \le h_0$*

$$\left\|\overline{u} - \overline{u}_h\right\| \le Ch^2. \tag{30}$$

Proof. Let us test (12) with $\overline{u}_h$, and (20) with $\overline{u}$, and sum up the two inequalities; we have

$$\left(\lambda\left(\overline{u} - \overline{u}_h\right) + B^*\left(\overline{z} - \overline{z}_h\right), \overline{u}_h - \overline{u}\right) \ge 0. \tag{31}$$

We further get

$$\begin{aligned} \lambda\left\|\overline{u} - \overline{u}_h\right\|^2 &\le \left(B^*\left(\overline{z} - \overline{z}_h\right), \overline{u}_h - \overline{u}\right) \\ &= \left(\overline{z} - \overline{z}_h, B\left(\overline{u}_h - \overline{u}\right)\right) \\ &= \left(\overline{z} - \overline{z}^h, B\left(\overline{u}_h - \overline{u}\right)\right) \\ &\quad + \left(\overline{z}^h - \overline{z}_h, B\left(\overline{u}_h - \overline{u}\right)\right) \\ &\le \frac{1}{2\lambda}\left\|\overline{z} - \overline{z}^h\right\|^2 + \frac{\lambda}{2}\left\|\overline{u}_h - \overline{u}\right\|^2 \\ &\quad + \left(\overline{z}^h - \overline{z}_h - I_h^*\left(\overline{z}^h - \overline{z}_h\right), B\left(\overline{u}_h - \overline{u}\right)\right) \\ &\quad + \left(I_h^*\left(\overline{z}^h - \overline{z}_h\right), B\left(\overline{u}_h - \overline{u}\right)\right) \\ &= \frac{1}{2\lambda}\left\|\overline{z} - \overline{z}^h\right\|^2 + \frac{\lambda}{2}\left\|\overline{u}_h - \overline{u}\right\|^2 \\ &\quad + \left(\overline{z}^h - \overline{z}_h - I_h^*\left(\overline{z}^h - \overline{z}_h\right), B\left(\overline{u}_h - \overline{u}\right)\right) \\ &\quad + a_h\left(\overline{y}_h - \overline{y}^h, I_h^*\left(\overline{z}^h - \overline{z}_h\right)\right), \end{aligned} \tag{32}$$

where

$$\begin{aligned} &a_h\left(\overline{y}_h - \overline{y}^h, I_h^*\left(\overline{z}^h - \overline{z}_h\right)\right) \\ &\quad = a_h\left(\overline{y}_h - \overline{y}^h, I_h^*\left(\overline{z}^h - \overline{z}_h\right)\right) \\ &\qquad - a_h\left(\overline{z}^h - \overline{z}_h, I_h^*\left(\overline{y}_h - \overline{y}^h\right)\right) \\ &\qquad + a_h\left(\overline{z}^h - \overline{z}_h, I_h^*\left(\overline{y}_h - \overline{y}^h\right)\right) \\ &\quad = a_h\left(\overline{y}_h - \overline{y}^h, I_h^*\left(\overline{z}^h - \overline{z}_h\right)\right) \\ &\qquad - a_h\left(\overline{z}^h - \overline{z}_h, I_h^*\left(\overline{y}_h - \overline{y}^h\right)\right) \\ &\qquad + \left(\overline{y} - \overline{y}_h, I_h^*\left(\overline{y}_h - \overline{y}^h\right)\right) \\ &\quad = a_h\left(\overline{y}_h - \overline{y}^h, I_h^*\left(\overline{z}^h - \overline{z}_h\right)\right) \\ &\qquad - a_h\left(\overline{z}^h - \overline{z}_h, I_h^*\left(\overline{y}_h - \overline{y}^h\right)\right) \\ &\qquad + \left(\overline{y} - \overline{y}^h, I_h^*\left(\overline{y}_h - \overline{y}^h\right)\right) \\ &\qquad - \left(\overline{y}_h - \overline{y}^h, I_h^*\left(\overline{y}_h - \overline{y}^h\right)\right) \\ &\quad \le a_h\left(\overline{y}_h - \overline{y}^h, I_h^*\left(\overline{z}^h - \overline{z}_h\right)\right) \\ &\qquad - a_h\left(\overline{z}^h - \overline{z}_h, I_h^*\left(\overline{y}_h - \overline{y}^h\right)\right) \\ &\qquad + \left(\overline{y} - \overline{y}^h, I_h^*\left(\overline{y}_h - \overline{y}^h\right)\right). \end{aligned} \tag{33}$$

Combining the above equations, we can obtain

$$\begin{aligned}\lambda \|\bar{u}-\bar{u}_h\|^2 &\le \frac{1}{2\lambda}\left\|\bar{z}-\bar{z}^h\right\|^2+\frac{\lambda}{2}\|\bar{u}_h-\bar{u}\|^2 \\ &\quad+\left(\bar{z}^h-\bar{z}_h-I_h^*\left(\bar{z}^h-\bar{z}_h\right),B\left(\bar{u}_h-\bar{u}\right)\right) \\ &\quad+a_h\left(\bar{y}_h-\bar{y}^h,I_h^*\left(\bar{z}^h-\bar{z}_h\right)\right) \\ &\quad-a_h\left(\bar{z}^h-\bar{z}_h,I_h^*\left(\bar{y}_h-\bar{y}^h\right)\right) \\ &\quad+\left(\bar{y}-\bar{y}^h,I_h^*\left(\bar{y}_h-\bar{y}^h\right)\right) \\ &=\frac{1}{2\lambda}\left\|\bar{z}-\bar{z}^h\right\|^2+\frac{\lambda}{2}\|\bar{u}_h-\bar{u}\|^2+E_1+E_2 \\ &\quad+E_3.\end{aligned} \tag{34}$$

According to Lemmas 5, 6, and 7, we have

$$\begin{aligned}E_1&=\left(\bar{z}^h-\bar{z}_h-I_h^*\left(\bar{z}^h-\bar{z}_h\right),B\left(\bar{u}_h-\bar{u}\right)\right) \\ &\le Ch\left\|\bar{z}^h-\bar{z}_h\right\|_1\|\bar{u}_h-\bar{u}\|\le Ch\|\bar{y}-\bar{y}_h\|\|\bar{u}_h-\bar{u}\| \\ &\le Ch\|\bar{u}-\bar{u}_h\|^2.\end{aligned} \tag{35}$$

$$\begin{aligned}E_2&=a_h\left(\bar{y}_h-\bar{y}^h,I_h^*\left(\bar{z}^h-\bar{z}_h\right)\right) \\ &\quad-a_h\left(\bar{z}^h-\bar{z}_h,I_h^*\left(\bar{y}_h-\bar{y}^h\right)\right) \\ &\le Ch\left\|\bar{y}_h-\bar{y}^h\right\|_1\left\|\bar{z}^h-\bar{z}_h\right\|_1 \\ &\le Ch\|\bar{u}_h-\bar{u}\|\|\bar{y}-\bar{y}_h\|\le Ch\|\bar{u}-\bar{u}_h\|^2.\end{aligned} \tag{36}$$

Using Lemmas 5 and 6, we conclude

$$\begin{aligned}E_3&=\left(\bar{y}-\bar{y}^h,I_h^*\left(\bar{y}_h-\bar{y}^h\right)\right)\le\left\|\bar{y}-\bar{y}^h\right\|\left\|\bar{y}_h-\bar{y}^h\right\| \\ &\le\left\|\bar{y}-\bar{y}^h\right\|\|\bar{u}-\bar{u}_h\|\le Ch^2\|\bar{u}-\bar{u}_h\|.\end{aligned} \tag{37}$$

Combining (34)–(37) and using Lemma 7, we can obtain the desirable result

$$\|\bar{u}-\bar{u}_h\|\le Ch^2. \tag{38}$$

□

Theorem 9. *Assume that $\bar{y},\bar{z}$ are the solutions of (6) and (11), respectively, and $\bar{y}_h,\bar{z}_h$ are the solutions of (18) and (19), respectively, $\bar{u}$, f, $y_\Omega\in H^1(\Omega)$, and $A\in W^{2,\infty}$. Then there exists a positive constant $C>0$ such that*

$$\begin{aligned}\|\bar{y}-\bar{y}_h\|&\le Ch^2, \\ \|\bar{z}-\bar{z}_h\|&\le Ch^2.\end{aligned} \tag{39}$$

Proof. Using the triangle inequality, we have

$$\|\bar{y}-\bar{y}_h\|\le\left\|\bar{y}-\bar{y}^h\right\|+\left\|\bar{y}^h-\bar{y}_h\right\|. \tag{40}$$

From Lemma 6 and Theorem 8, we can obtain

$$\left\|\bar{y}^h-\bar{y}_h\right\|\le C\|\bar{u}-\bar{u}_h\|\le Ch^2. \tag{41}$$

Using Lemma 7, we can obtain the desired result

$$\|\bar{y}-\bar{y}_h\|\le Ch^2. \tag{42}$$

Similarly, we have

$$\|\bar{z}-\bar{z}_h\|\le Ch^2. \tag{43}$$

□

3.3. H^1 Error Estimate

Theorem 10. *Assume that $\bar{y},\bar{z}$ are the solutions of (6) and (11), respectively, and $\bar{y}_h,\bar{z}_h$ are the solutions of (18) and (19), respectively, $\bar{u}$, f, $y_\Omega\in L^2(\Omega)$, and $A\in W^{1,\infty}$. Then there exists a positive constant $C>0$ such that*

$$\begin{aligned}\|\bar{y}-\bar{y}_h\|_1&\le Ch, \\ \|\bar{z}-\bar{z}_h\|_1&\le Ch.\end{aligned} \tag{44}$$

Proof. Using the triangle inequality, we have

$$\|\bar{y}-\bar{y}_h\|_1\le\left\|\bar{y}-\bar{y}^h\right\|_1+\left\|\bar{y}^h-\bar{y}_h\right\|_1. \tag{45}$$

From Lemma 4, we can obtain

$$\begin{aligned}\alpha\left\|\bar{y}^h-\bar{y}_h\right\|_1^2&\le a_h\left(\bar{y}^h-\bar{y}_h,I_h^*\left(\bar{y}^h-\bar{y}_h\right)\right) \\ &=\left(B\left(\bar{u}-\bar{u}_h\right),I_h^*\left(\bar{y}^h-\bar{y}_h\right)\right) \\ &\le\frac{1}{2\alpha}\|\bar{u}-\bar{u}_h\|^2+\frac{\alpha}{2}\left\|\bar{y}^h-\bar{y}_h\right\|^2.\end{aligned} \tag{46}$$

By using Lemma 7 and Theorems 8 and 9, we can obtain the desired result

$$\|\bar{y}-\bar{y}_h\|_1\le Ch. \tag{47}$$

Similarly, we have

$$\|\bar{z}-\bar{z}_h\|_1\le Ch. \tag{48}$$

□

Remark 11. In the case $U_{ad}=L^2(\Omega)$, the projection equations (13) and (20) become $\bar{u}=-B^*z/\lambda$ and $\bar{u}_h=-B^*\bar{z}_h/\lambda$, respectively. Using the above theorem, we can obtain the following error estimate:

$$\|\bar{u}-\bar{u}_h\|_1\le Ch. \tag{49}$$

Table 1: Errors of the control for different error norms.

h	L^∞ error	r	L^2 error	r	H^1 error	r
1/8	1.7412E-01	-	4.4468E-02	-	2.3006	-
1/16	4.1870E-02	2.05	1.0374E-02	2.09	1.1313	1.02
1/32	1.0476E-02	2.00	2.5558E-03	2.02	5.6371E-01	1.00
1/64	2.6171E-03	2.00	6.3685E-04	2.00	2.8162E-01	1.00

Table 2: Errors of the state for different error norms.

h	L^∞ error	r	L^2 error	r	H^1 error	r
1/8	1.1429E-03	-	3.7634E-04	-	5.3988E-03	-
1/16	2.1927E-04	2.38	8.5181E-05	2.14	1.1332E-03	2.25
1/32	5.1326E-05	2.09	2.0833E-05	2.03	2.7140E-04	2.06
1/64	1.2609E-05	2.03	5.1751E-06	2.01	6.7061E-05	2.02

3.4. L^∞ Error Estimate

Theorem 12. *Assume that $\overline{y}_h, \overline{z}_h, \overline{u}_h$ are the solutions of (18), (19), and (20), respectively, $\overline{u}$, f, $y_\Omega \in H^1(\Omega)$, and $A \in W^{2,\infty}$. Then there exists a positive constant $C > 0$ such that*

$$\begin{aligned} \left\|\overline{u}-\overline{u}_h\right\|_\infty &\le Ch^2 \log\frac{1}{h}, \\ \left\|\overline{z}-\overline{z}_h\right\|_\infty &\le Ch^2 \log\frac{1}{h}, \\ \left\|\overline{y}-\overline{y}_h\right\|_\infty &\le Ch^2 \log\frac{1}{h}. \end{aligned} \tag{50}$$

Proof. Using the projection equations (13) and (20), we have

$$\begin{aligned} \left\|\overline{u}-\overline{u}_h\right\|_\infty &\le C\left\|\overline{z}-\overline{z}_h\right\|_\infty \\ &\le C\left(\left\|\overline{z}-\overline{z}^h\right\|_\infty + \left\|\overline{z}^h-\overline{z}_h\right\|_\infty\right) \\ &\le C\left\|\overline{z}-\overline{z}^h\right\|_\infty + C\left(\log\frac{1}{h}\right)^{1/2}\left\|\overline{z}^h-\overline{z}_h\right\|_1 \\ &\le C\left\|\overline{z}-\overline{z}^h\right\|_\infty + C\left(\log\frac{1}{h}\right)^{1/2}\left\|\overline{y}-\overline{y}_h\right\| \\ &\le Ch^2\log\frac{1}{h}. \end{aligned} \tag{51}$$

Similarly, we have

$$\left\|\overline{z}-\overline{z}_h\right\|_\infty \le Ch^2 \log\frac{1}{h}. \tag{52}$$

□

4. Numerical Experiments

In this section, we report some numerical results of finite volume element schemes for the elliptic optimal control problems. To illustrate the theoretical analysis, the following rate of convergence r is defined:

$$r = \log_2\left(\frac{\left\|u_{2h}-u\right\|}{\left\|u_h-u\right\|}\right), \tag{53}$$

where u_h is the numerical solution with space step size h and u the analytical solution. The rate approaching the number 2 would indicate second-order accuracy in space.

4.1. Experiment 1. To validate the finite volume element schemes for the solution of elliptic optimal control problems, test example is needed for which the exact solutions are known in advance [15]. We consider the problems with homogeneous Neumann boundary condition,

$$\min \quad J(y,u) = \frac{1}{2}\int_\Omega \left|y-y_\Omega\right|^2 dx + \frac{1}{2}\int_\Omega |u|^2\, dx, \tag{54}$$

subject to

$$\begin{aligned} -\Delta y + y &= u + f, \quad \text{in } \Omega, \\ \nabla y \cdot \mathrm{n} &= 0, \quad \text{on } \partial\Omega, \end{aligned} \tag{55}$$

where Ω denotes unit square $[0,1]\times[0,1]$, $U_{ad}=L^2(\Omega)$, n is the outer unit normal vector, and $f = 1 - \sin^2(2\pi x_1)\sin^2(2\pi x_2)$. Under these settings, the optimal control is

$$\overline{u}(x) = \sin^2(2\pi x_1)\sin^2(2\pi x_2). \tag{56}$$

The adjoint state is

$$\overline{z}(x) = -\sin^2(2\pi x_1)\sin^2(2\pi x_2), \tag{57}$$

and the associated state is

$$\overline{y}(x) = 1. \tag{58}$$

Then we can determine the function y_Ω accordingly.

Errors of finite volume element schemes in L^∞, L^2, and H^1 norm are computed. Data are listed in Tables 1–3. In Tables 1 and 3, errors in H^1 norm have optimal convergence order for both control and adjoint state. These results confirm our theoretical error analysis (44). In Table 2, due to additional smoothness of the state, the H^1 error is $O(h^2)$. The convergence results in Tables 1–3 demonstrate second-order accuracy in L^∞ and L^2 norm for the control, state, and adjoint state.

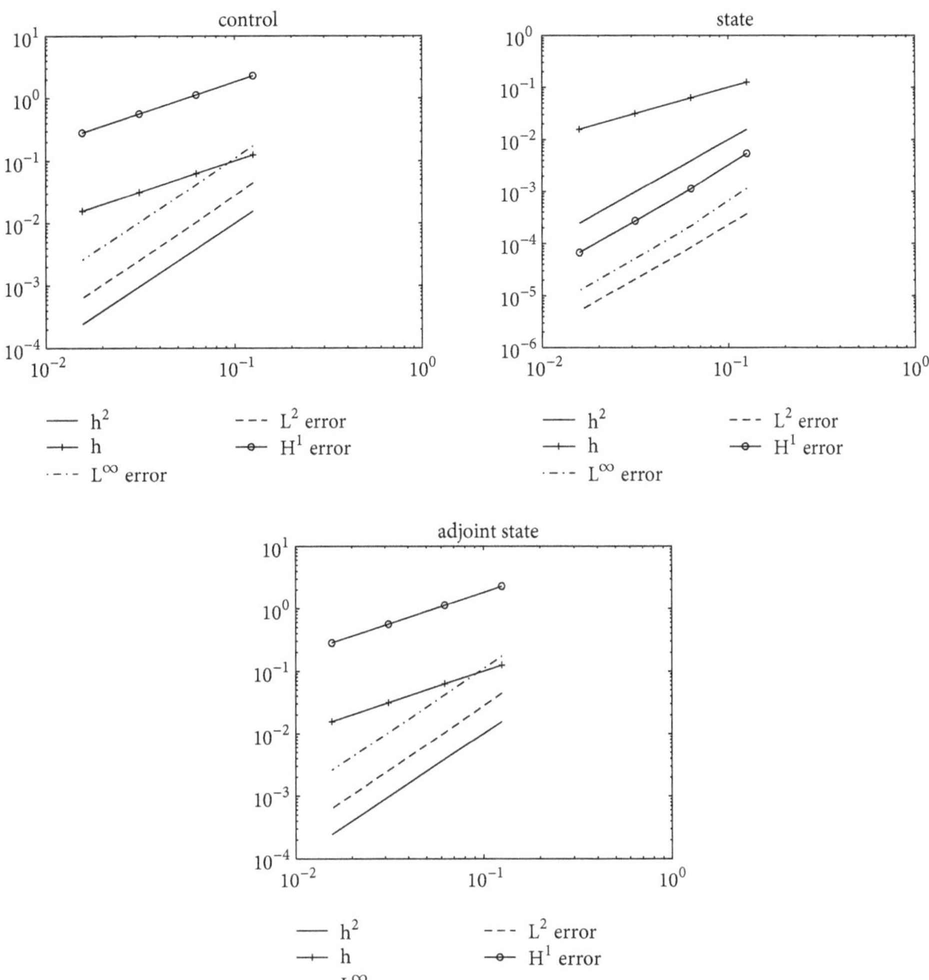

FIGURE 2: The L^∞, L^2, and H^1 error for the control, state, and adjoint state under uniform refinement of the mesh.

TABLE 3: Errors of the adjoint state for different error norms.

h	L^∞ error	r	L^2 error	r	H^1 error	r
1/8	1.7415E-02	-	4.4468E-02	-	2.3006	-
1/16	4.1867E-02	2.06	1.0373E-02	2.09	1.1313	1.02
1/32	1.0473E-02	1.99	2.5552E-03	2.02	5.6371E-01	1.00
1/64	2.6159E-03	2.00	6.3655E-04	2.01	2.8162E-01	1.00

Figure 2 depicts the development of the L^∞, L^2, and H^1 error for the control, state, and adjoint state under uniform refinement of the mesh. From the figure, the expected order $O(h^2)$ in L^∞ and L^2 norm for the control is observed, and the order $O(h)$ in H^1 norm is shown. Additionally, we observe convergence of order $O(h^2)$ in L^∞ and L^2 norm for state and adjoint state. Because of better smoothness of state, the order $O(h^2)$ in H^1 norm is also observed.

We perform a simulation with space size $h = 1/32$ for this problem. Figure 3 presents the computed state, optimal control, and adjoint state. Examination of Figure 3 shows that the approximate solutions coincide with the true solutions. At the same time, the relationship between the control and adjoint state is preserved well.

4.2. Experiment 2. Now, we consider the optimal control problem with homogeneous Dirichlet boundary condition and control constraint,

$$\min \quad J(y,u) = \frac{1}{2}\int_\Omega |y - y_\Omega|^2\, dx + \frac{\lambda}{2}\int_\Omega |u|^2\, dx, \tag{59}$$

subject to

$$\begin{aligned} -\Delta y &= u, \quad \text{in } \Omega, \\ y &= 0, \quad \text{on } \partial\Omega, \end{aligned} \tag{60}$$

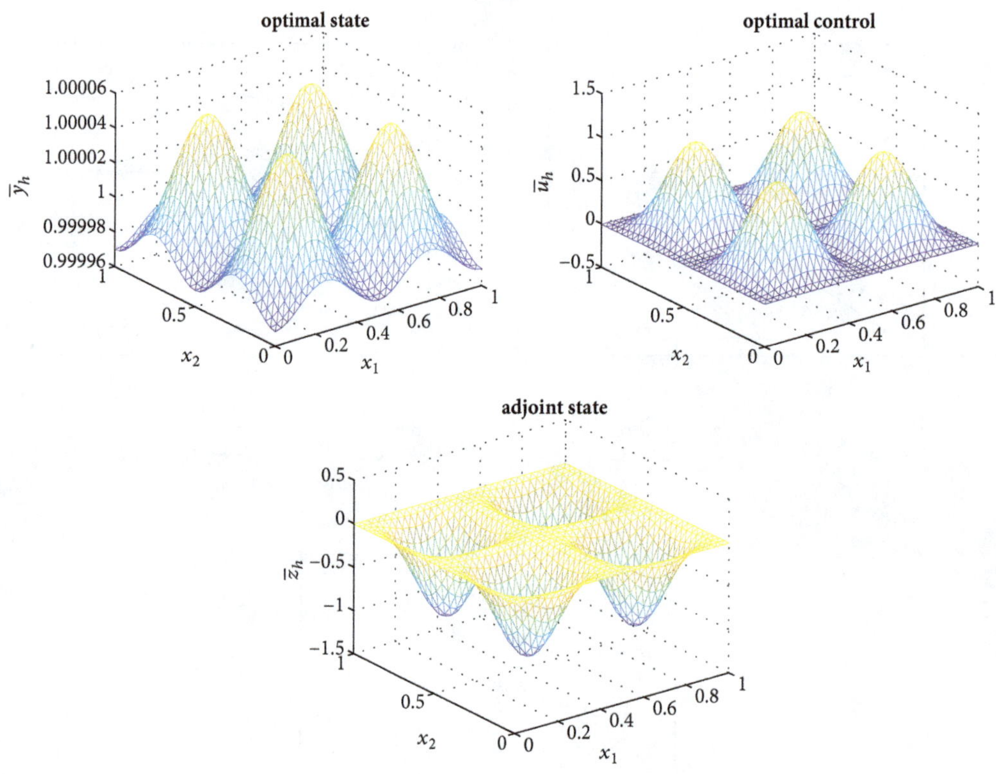

FIGURE 3: Numerical results of Experiment 1: optimal state, optimal control, and corresponding adjoint state.

TABLE 4: Errors of the control for different error norms.

h	L^∞ error	r	L^2 error	r	H^1 error	r
1/8	4.2814E-03	-	2.7687E-03	-	2.2775E-02	-
1/16	1.2186E-03	1.81	7.7173E-04	1.84	1.2642E-02	0.85
1/32	2.9931E-04	2.02	1.8098E-04	2.09	5.4512E-03	1.21
1/64	7.6218E-05	1.97	4.0938E-05	2.14	2.2949E-03	1.24

TABLE 5: Errors of the state for different error norms.

h	L^∞ error	r	L^2 error	r	H^1 error	r
1/8	5.2997E-04	-	3.8069E-04	-	2.9536E-03	-
1/16	1.3047E-04	2.02	1.0869E-04	1.81	1.7383E-03	0.76
1/32	3.2119E-05	2.02	2.7098E-05	2.00	7.6992E-04	1.17
1/64	7.9037E-06	2.02	6.2918E-06	2.10	3.3199E-04	1.21

where Ω denotes the unit circle, $U_{ad} = \{u \in L^2(\Omega) : -0.2 \leq u \leq 0.2\}$, $y_\Omega(x) = (1 - (x_1^2 + x_2^2))x_1$, and $\lambda = 0.1$.

The exact solution of the problem is not known in advance. So we use the numerical results computed on a grid with $h = 1/256$ as reference solutions. The L^∞, L^2, and H^1 errors for state, control, and adjoint state of the above problems have been computed. They are displayed in Tables 4–6 for the finite volume element schemes. Examination of the tables shows that the error measures of the schemes diminish approximately quadratically for the error in L^∞ and L^2 norm and linearly for the error in H^1 norm, which are consistent with our theoretical analysis.

In Figure 4, the development of the L^∞, L^2, and H^1 error for control, state, and adjoint state under uniform refinement of the mesh is shown. Here, the expected order $O(h^2)$ in L^∞ and L^2 norm for the control is observed. Again, we observe convergence of order $O(h^2)$ in L^∞ and L^2 norm for state and adjoint state, which is consistent with our expectation of the order of convergence. The errors in H^1 norm confirm our error estimation (11). Figure 5 displays the numerical solution computed by the finite volume element schemes with $h = 1/16$. The results are nearly the same as those in [19]. The relationship between the control and adjoint state is also preserved well.

TABLE 6: Errors of the adjoint state for different error norms.

h	L^∞ error	r	L^2 error	r	H^1 error	r
1/8	5.9019E-04	-	3.9236E-04	-	3.0287E-03	-
1/16	1.6144E-04	1.87	1.1101E-04	1.82	1.7719E-03	0.77
1/32	4.2371E-05	1.93	2.8422E-05	1.97	8.0665E-04	1.13
1/64	1.0588E-05	2.00	6.9125E-06	2.04	3.6516E-04	1.14

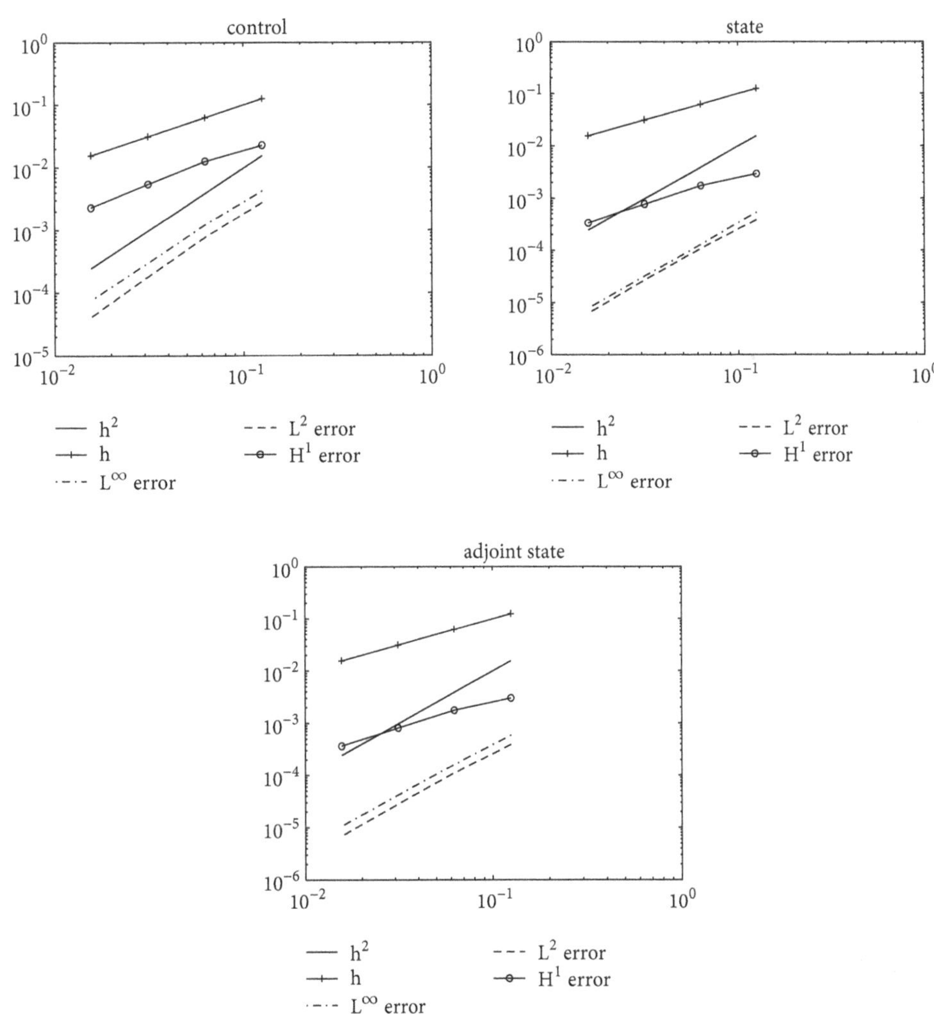

FIGURE 4: The L^∞, L^2, and H^1 error for the control, state, and adjoint state under uniform refinement of the mesh.

4.3. *Experiment 3.* Now we consider the optimal control problem (59) with $\Omega = (-1,1)^2 \setminus ([-1,0] \times [0,1])$ denoting an L-shaped domain, $U_{ad} = \{u \in L^2(\Omega) : -0.2 \le u \le 0.2\}$. Further, we set $y_\Omega = 1 - (x_1^2 + x_2^2)$ and $\lambda = 0.1$.

In this situation, the solution does not admit integrable second derivatives. The desired convergence results of finite volume element schemes cannot be expected. So we only present the numerical solutions of the finite volume element schemes in Figure 6, which are nearly the same as those in [19]. On one hand, the desired convergence results may be obtained by using graded meshes and postprocessing [20], which will need more computational cost. On the other hand, we can modify finite volume element schemes near the corner to obtain the second-order accuracy. The related results will be reported in the future.

5. Conclusions

In this article, we have investigated the finite volume element discretizations of optimal control problems governed by linear elliptic partial differential equations and subject to pointwise control constraints. Optimal order L^2, H^1, and L^∞ error estimates for the considered problems are obtained and numerical experiments validate the theoretical results. In addition, we discuss the optimal control problems in polygonal domains with corner singularities. Two effective

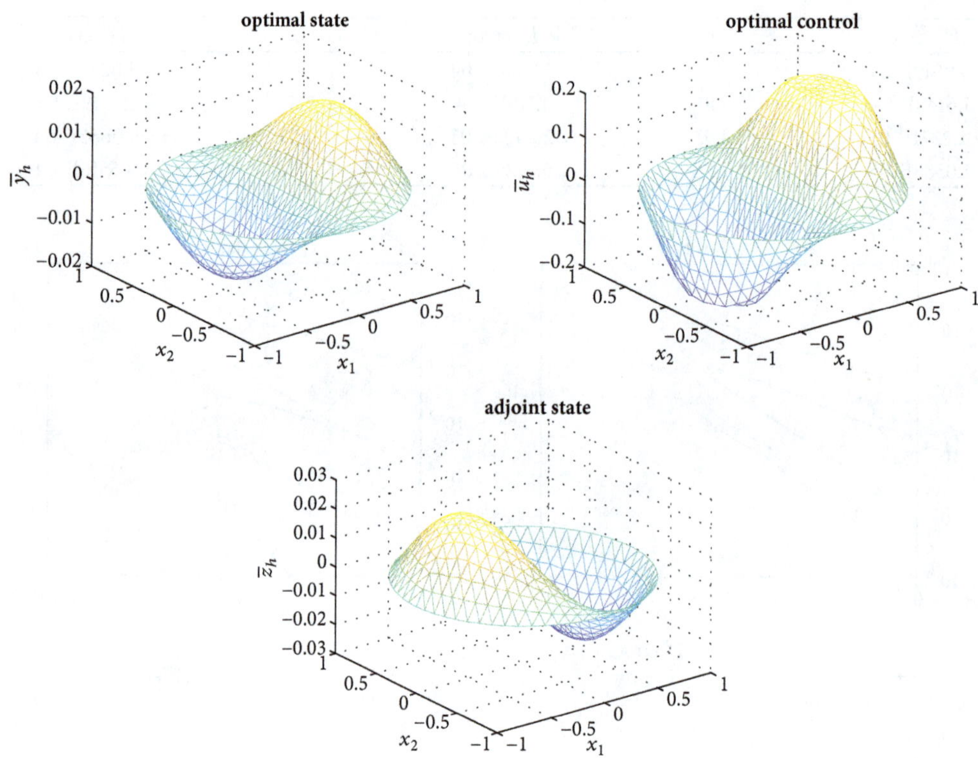

FIGURE 5: Numerical results of Experiment 2: optimal state, optimal control, and corresponding adjoint state.

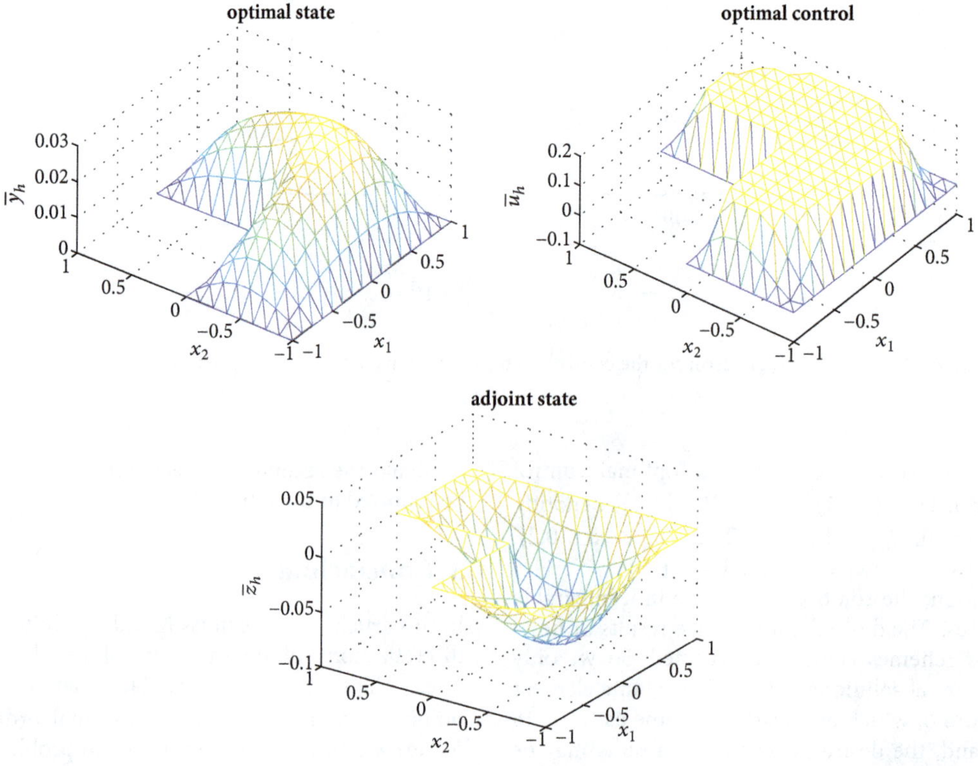

FIGURE 6: Numerical results of Experiment 3: optimal state, optimal control, and corresponding adjoint state.

methods are proposed to compensate the negative effects of the corner singularities. The corresponding results will be reported in the future.

Conflicts of Interest

The authors declare that they have no conflicts of interest.

Acknowledgments

This project is partially supported by the Fundamental Research Funds for the Central Universities (Nos. KYZ201565 and KJQN201839) and the National Natural Science Foundation of China (Nos. 11701283, 11426134, and 11471166).

References

[1] R. Becker, H. Kapp, and R. Rannacher, "Adaptive finite element methods for optimal control of partial differential equations: basic concept," *SIAM Journal on Control and Optimization*, vol. 39, no. 1, pp. 113–132, 2000.

[2] N. Arada, E. Casas, and F. Tröltzsch, "Error estimates for the numerical approximation of a semilinear elliptic control problem," *Computational Optimization and Applications*, vol. 23, no. 2, pp. 201–229, 2002.

[3] Y. Chen and W. Liu, "Error estimates and superconvergence of mixed finite element for quadratic optimal control," *International Journal of Numerical Analysis & Modeling*, vol. 3, no. 3, pp. 311–321, 2006.

[4] A. Kröner and B. Vexler, "A priori error estimates for elliptic optimal control problems with a bilinear state equation," *Journal of Computational and Applied Mathematics*, vol. 230, no. 2, pp. 781–802, 2009.

[5] T. Hou and L. Li, "Error estimates of mixed methods for optimal control problems governed by general elliptic equations," *Advances in Applied Mathematics and Mechanics*, vol. 8, no. 6, pp. 1050–1071, 2016.

[6] Z. Zhang, "Error estimates for finite volume element method for the pollution in groundwater flow," *Numerical Methods for Partial Differential Equations*, vol. 25, no. 2, pp. 259–274, 2009.

[7] Q. Wang, Z. Zhang, and Z. Li, "A Fourier finite volume element method for solving two-dimensional quasi-geostrophic equations on a sphere," *Applied Numerical Mathematics*, vol. 71, pp. 1–13, 2013.

[8] Q. Wang, S. Lin, and Z. Zhang, "Numerical methods for a fluid mixture model," *International Journal for Numerical Methods in Fluids*, vol. 71, no. 1, pp. 1–12, 2013.

[9] Q. Wang, Z. Zhang, X. Zhang, and Q. Zhu, "Energy-preserving finite volume element method for the improved Boussinesq equation," *Journal of Computational Physics*, vol. 270, pp. 58–69, 2014.

[10] R. Ruiz-Baier and H. Torres, "Numerical solution of a multidimensional sedimentation problem using finite volume-element methods," *Applied Numerical Mathematics*, vol. 95, pp. 280–291, 2015.

[11] C. Chen and W. Liu, "A two-grid finite volume element method for a nonlinear parabolic problem," *International Journal of Numerical Analysis & Modeling*, vol. 12, no. 2, pp. 197–210, 2015.

[12] C. Bi and C. Wang, "A posteriori error estimates of finite volume element method for second-order quasilinear elliptic problems," *International Journal of Numerical Analysis & Modeling*, vol. 13, no. 1, pp. 22–40, 2016.

[13] X. Luo, Y. Chen, and Y. Huang, "Some error estimates of finite volume element approximation for elliptic optimal control problems," *International Journal of Numerical Analysis & Modeling*, vol. 10, no. 3, pp. 697–711, 2013.

[14] R. Sandilya and S. Kumar, "Convergence analysis of discontinuous finite volume methods for elliptic optimal control problems," *International Journal of Computational Methods*, vol. 13, no. 2, 1640012, 20 pages, 2016.

[15] F. Tröltzsch, *Optimal Control of Partial Differential Equations: Theory, Methods and Applications*, vol. 112, American Mathematical Society, 2010.

[16] R. Li, Z. Chen, and W. Wu, *Generalized difference methods for differential equations*, vol. 226 of *Monographs and Textbooks in Pure and Applied Mathematics*, Marcel Dekker, Inc., 2000.

[17] S. Chou and Q. Li, "Error estimates in L^2,H^1 and L^∞ in covolume methods for elliptic and parabolic problems: a unified approach," *Mathematics of Computation*, vol. 69, no. 229, pp. 103–120, 2000.

[18] R. E. Ewing, T. Lin, and Y. Lin, "On the accuracy of the finite volume element method based on piecewise linear polynomials," *SIAM Journal on Numerical Analysis*, vol. 39, no. 6, pp. 1865–1888, 2002.

[19] M. Hinze, R. Pinnau, M. Ulbrich, and S. Ulbrich, *Optimization with PDE Constraints*, Springer, 2009.

[20] T. Apel, A. Rösch, and D. Sirch, "L^∞ error estimates on graded meshes with application to optimal control," *SIAM Journal on Control and Optimization*, vol. 48, no. 3, pp. 1771–1796, 2009.

10

Numerical Method for Solving Nonhomogeneous Backward Heat Conduction Problem

LingDe Su[1] **and TongSong Jiang**[2]

[1]*Institute of Mathematics and Information Science, North-Eastern Federal University, Russia*
[2]*Department of Mathematics, Heze University, Shandong, China*

Correspondence should be addressed to TongSong Jiang; jiangtongsong@sina.com

Guest Editor: Ovidiu Bagdasar

In this paper we consider a numerical method for solving nonhomogeneous backward heat conduction problem. Coupled with the likewise Crank Nicolson scheme and an intermediate variable, the backward problem is transformed to a nonhomogeneous Helmholtz type problem; the unknown initial temperature can be obtained by solving this Helmholtz type problem. To illustrate the effectiveness and accuracy of the proposed method, we solve several problems in both two and three dimensions. The results show that this numerical method can solve nonhomogeneous backward heat conduction problem effectively and precisely, even though the final temperature is disturbed by significant noise.

1. Introduction

The heat conduction equation is a kind of very important time-dependent parabolic partial differential equation; it describes the distribution of heat or temperature in a given region over time and is widely used in diverse scientific fields, such as the study of Brownian motion [1], to solve the Black-Scholes partial differential equation [2] and the research of chemical diffusion. Many works have been done to study the heat conduction problem [3–5]. The purpose of this article is to numerically solve nonhomogeneous backward heat conduction problem. This problem is one kind of inverse problem, also called final value problem or time inverse problem.

In many engineering and application areas we need to reconstruct the unknown initial heat energy or temperature from the final measured one; it is a typical inverse problem which is related to the initial boundary value problems in heat conduction. As we all know, there are many kinds of inverse problems in mathematical physics, such as boundary inverse problems [6, 7], coefficient inverse problems [8, 9], and evolutionary inverse problems (or time inverse problem) [10, 11]. The backward heat conduction problem is a time inverse problem, in which the initial conditions are unknown; instead the final data are observable.

In the sense of Hadamard [12], the inverse problem is always considered as a class of classically ill-posed problem, which means that a small error in the input data may cause enormous error to the result. The backward heat conduction problem is a typically ill-posed problem, because of the measurement error of the input data. Most standard numerical method can not achieve a good result in solving an ill-posed problem, due to the ill-posedness of the problem and the ill-conditioning of the discretized matrix, so, it is very important to find a stable and efficient numerical algorithm to solve ill-posed problems.

The backward heat conduction problem arises in many physical and chemical fields and has attracted the attention of many researchers. The improperly posed backward heat conduction problem has been considered by L. Payne [13], and Miranker [14] has given a research of the uniqueness conditions for the backward heat conduction problem. There are also many works that have been done to numerically solve the ill-posed backward heat conduction problem. In [15], Fourier regularization method is used to solve one-dimensional backward heat problem, H. Han [16] has considered this problem using boundary element method, and also some other numerical methods for solving backward heat conduction problem have been given in many works, such as finite difference method [17, 18], iterative boundary element method [19, 20], the method of fundamental solution [21], etc. Most of these methods are used to solve one-dimensional homogeneous situations; there are a few papers on the nonhomogeneous case in higher dimensional space. Scientists M. Denche and A. Abdessemed [22] gave extensions of the quasi-boundary methods to the nonhomogeneous case. Paper [23] regularized the two-dimensional nonhomogeneous backward heat problem by perturbing the final value. M. Li, T. S. Jiang, and Y. C. Hon have solved nonhomogeneous situations by using a new meshless method based on radial basis functions [11].

In this paper we consider the nonhomogeneous backward heat conduction problem in two- or three-dimensional spatial domains. We propose a new numerical method to solve this severely ill-posed problem; coupled with the likewise Crank-Nicolson scheme and a new variable, the backward problem is transformed to a nonhomogeneous Helmholtz type problem; by solving this nonhomogeneous Helmholtz type problem and submitting back to the new variable, we can get the solution of the initial condition.

The structure of the paper is organized as follows: In Section 2, we briefly introduce the formulation of the nonhomogeneous backward heat conduction problem. In Section 3, we introduce the numerical method and apply this method on the inverse problem. The results of several two- and three-dimensional numerical experiments are presented to illustrate the stability and accuracy of the proposed method in Section 4. Section 5 is dedicated to a brief conclusion. Finally, some references are introduced at the end.

2. The Formulation of the Problem

We consider the nonhomogeneous backward heat conduction problem in a bounded and connected domain $\overline{\Omega} \subset \mathbb{R}^d$ $(d = 2, 3)$ with the boundary $\Gamma = \partial\Omega$ in the following,

$$\frac{\partial}{\partial t}u(\mathbf{x}, t) = \Delta u(\mathbf{x}, t) + f(\mathbf{x}, t), \quad \mathbf{x} \in \overline{\Omega}, \ t \in (0, T), \tag{1}$$

with the final temperature condition,

$$u(\mathbf{x}, T) = g(\mathbf{x}), \quad \mathbf{x} \in \Omega, \tag{2}$$

and the Dirichlet boundary conditions,

$$u(\mathbf{x}, t) = h(\mathbf{x}, t), \quad \mathbf{x} \in \Gamma, \ t \in (0, T), \tag{3}$$

where $f(\mathbf{x}, t)$, $g(\mathbf{x})$, and $h(\mathbf{x}, t)$ are known functions, Δ is Laplace Operator, and $u(\mathbf{x}, t)$ satisfies the nonhomogeneous heat conduction equation (1). The problem which we want to solve in this paper is the backward heat conduction problem; we will determine the temperature before some particular time T from the known data (2) at time T and the boundary conditions (3).

3. The Computational Algorithm

To solve the backward problem (1)–(3) numerically, we introduce the grid of time,

$$\omega_t = \{t^n \mid t^n = n\tau, \ n = 0, 1, \ldots, N, \ N\tau = T\}. \tag{4}$$

Using the notation $u^n(\mathbf{x}) = u(\mathbf{x}, t^n)$ and a similar format as Crank-Nicolson (C-N) scheme to approximate the heat equation (1) with second order at the time moment $t^{n+1/2} = (t^{n+1} + t^n)/2$, we have

$$\frac{u^{n+1} - u^n}{\tau} = \frac{1}{2}\left(\Delta u^{n+1} + \Delta u^n\right) + f\left(\mathbf{x}, t^{n+1/2}\right), \tag{5}$$

where $n = 0, 1, \ldots, N-1$. Here, it is worth mentioning that in (5) we did not use the C-N approximation scheme, but a scheme has a similar format as C-N scheme, which we called the likewise C-N scheme [24].

By using a new intermediate variable $\widetilde{u}$, which is given as follows,

$$\widetilde{u} = \frac{1}{2}\left(u^{n+1} + u^n\right), \tag{6}$$

(5) can be transformed in the following new form,

$$\frac{2u^{n+1} - 2\widetilde{u}}{\tau} = \Delta\widetilde{u} + f\left(\mathbf{x}, t^{n+1/2}\right). \tag{7}$$

By simplifying (7), we can get

$$\Delta\widetilde{u} + \frac{2}{\tau}\widetilde{u} = \frac{2}{\tau}u^{n+1} - f\left(\mathbf{x}, t^{n+1/2}\right). \tag{8}$$

Notice that (8) is a nonhomogeneous Helmholtz equation with $\widetilde{u}$ as solution; it can be written as

$$\Delta\widetilde{u} + k^2\widetilde{u} = F(\mathbf{x}, t), \tag{9}$$

where

$$k^2 = \frac{2}{\tau}, \qquad F(\mathbf{x}, t) = \frac{2}{\tau}u^{n+1} - f\left(\mathbf{x}, t^{n+1/2}\right), \tag{10}$$

with the boundary conditions,

$$\widetilde{u} = \frac{1}{2}\left(h^{n+1}(\mathbf{x}) + h^n(\mathbf{x})\right), \quad \mathbf{x} \in \Gamma, \tag{11}$$

where $h^n(\mathbf{x}) = h(\mathbf{x}, t^n)$ is known from the Dirichlet boundary conditions (3).

To get the solution $\widetilde{u}$ from the nonhomogeneous Helmholtz equation (9) with boundary conditions (11), the finite element method is used. We multiply by a test function v, which vanishes on the boundary, and apply integration by parts, getting the variational problem with $n = 0, 1, \ldots, N-1$ as follows,

$$-\int \nabla\widetilde{u} \cdot \nabla v \, d\mathbf{x} + k^2 \int \widetilde{u} v \, d\mathbf{x} = \int F(\mathbf{x}, t) \, v \, d\mathbf{x} \tag{12}$$

where k^2, $F(\mathbf{x}, t)$ are given in (10). We can obtain the value of $\widetilde{u}$ by solving the nonhomogeneous Helmholtz equation (8) (or (9)); then, the solution u^n can be computed from (6). After repeating the above steps N times, we can get the initial temperature at $n = 0$.

The Helmholtz equation often appears in the study of physical problems; it is caused by the propagation of time harmonics and applied in many science and engineering fields, such as acoustic, electromagnetic science, and geophysical problems. Many experts have studied solving Helmholtz equation by finite element method (FEM); F. Ihlenburg and I. Babuška studied the finite element solution of the Helmholtz equation with high wave number by using the h-version [25] and h-p version [26] of the FEM. The Least-Squares stabilization of finite element computation for the Helmholtz equation is considered by I. Harari and F. Magoulès [27]. Y. Wong and G. Li [28] consider a new finite difference scheme for solving the Helmholtz equation at any wavenumber, etc.

From [25–28] and the references it can be found that, due to the so-called pollution effect [29], to compute approximate solutions of the Helmholtz equation for high wave numbers by the standard finite element method (FEM) is unreliable, so, in (9), the wave number should be not high. Suppose the numerical discrete grid size is h; in order to ensure an accurate numerical solution, the condition $k^2h < 1$ must be enforced and it is necessary to require kh to be small. A conclusion that the stability constant does not depend on k if k^2h is bounded using FEM to solve the Helmholtz equation was given in [25, 26]. Also, a convergence theorem is stated in [30] under the assumption that k^2h is sufficiently small. As a result of the fact that the wave number k in the Helmholtz equation (9) is related to the time step τ, in order to ensure stability and accuracy, the time step τ should be selected carefully; it can not be chosen too small.

4. Numerical Examples

In this section we give several numerical examples in both 2D and 3D to demonstrate the effectiveness and stability of the new computation method. In the computation, we impose the noise to the final temperature condition. We set

$$g_\delta(\mathbf{x}) = g(\mathbf{x}) \cdot (1 + \delta \times \operatorname{rand} n), \tag{13}$$

where $g(\mathbf{x})$ denotes the exact final temperature condition, δ is the tolerated noise level, and rand n is Gaussian random number with mean 0 and variance 1.

Two kinds of errors are used to compare the accuracy of the numerical solutions with the exact solutions. The maximum error (Maxerror) and the root-mean-square error (RMSE) are defined as follows,

$$\begin{aligned} \text{MaxError} &= \max_{1\leqslant i\leqslant M} \left|\tilde{u}_i - u_i\right|, \\ \text{RMSE} &= \sqrt{\frac{1}{M}\sum_{i=1}^{M}(\tilde{u}_i - u_i)^2}, \end{aligned} \tag{14}$$

where M is the total number of testing nodes within the domain and $\tilde{u}_i$ denotes the approximate solution at the ith node; u_i is the exact one.

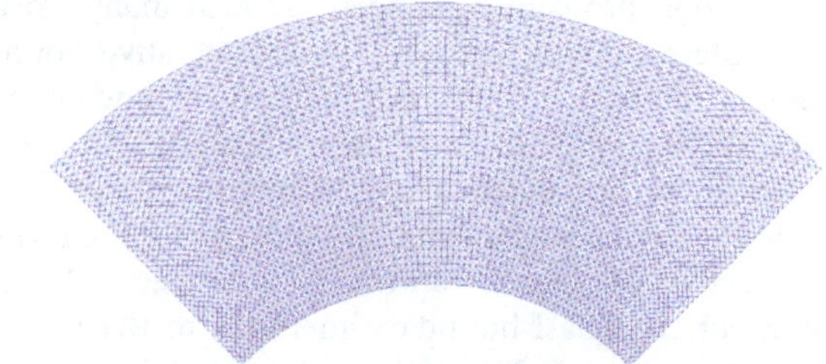

Figure 1: The domain with computational grid.

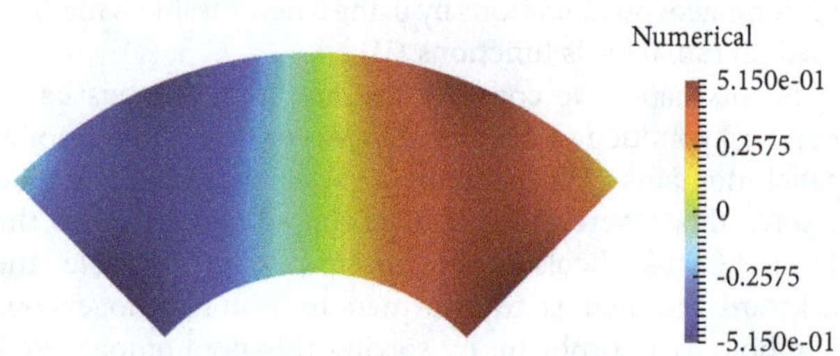

Figure 2: The numerical solutions $u(\mathbf{x}, 0)$.

Example 1. In this example we consider the two-dimensional nonhomogeneous heat conduction equation (1), with the Dirichlet boundary conditions,

$$\begin{aligned} u(\mathbf{x}, t) &= (y\sin(\pi x) + x\cos(\pi y))\cos t, \\ &\qquad \mathbf{x} = (x, y) \in \Gamma,\ t \in (0, T), \end{aligned} \tag{15}$$

the final conditions,

$$u(\mathbf{x}, T) = (y\sin(\pi x) + x\cos(\pi y))\cos T, \quad \mathbf{x} \in \Omega, \tag{16}$$

and $f(\mathbf{x}, t) = \pi^2(y\sin(\pi x) + x\cos(\pi y))\cos t - (y\sin(\pi x) + x\cos(\pi y))\sin t$.

The computational domain $\overline{\Omega}$ is a given rectangle in $(\overline{x}, \overline{y})$ space such that $0 < a \leqslant \overline{x} \leqslant b$ and $c \leqslant \overline{y} \leqslant d$, using the map

$$\begin{aligned} x &= \overline{x}\cos(\theta\overline{y}), \\ y &= \overline{x}\sin(\theta\overline{y}), \end{aligned} \tag{17}$$

to take a point in the rectangular $(\overline{x}, \overline{y})$ geometry and to map it to a point (x, y) in a big hollow cylinder.

In the computation we choose the mesh 41×81 on the rectangular and using the same map to put the mesh on the hollow cylinder, the mesh on the computational domain with $a = c = 0.5, b = 1.0, d = 1.5$, and $\theta = \pi/2$ is shown in Figure 1.

The Maxerror and RMSE for $T = 1.0$, 2.0, and 3.0 are shown in Table 1 with noise level $\delta = 10^{-3}$ and various time steps τ. It can be seen from Table 1 that the proposed method performs well for solving the backward problem.

The figure of numerical initial solutions with $T = 1.0$, $\tau = 0.25$, and $\delta = 10^{-3}$ is presented in Figure 2.

For further investigations, we give a research of the results with a disturbed δ of the final time conditions; the figure of

TABLE 1: Maxerror and RMSE for $T = 1.0, 2.0$ and 3.0 with different τ.

	$\tau = 0.2$	$\tau = 0.25$	$\tau = 0.5$	$\tau = 1.0$
$T = 1.0$				
Maxerror	4.951×10^{-4}	3.714×10^{-4}	8.721×10^{-4}	1.635×10^{-2}
RMSE	2.230×10^{-4}	2.358×10^{-4}	4.734×10^{-4}	8.604×10^{-3}
$T = 2.0$				
Maxerror	2.393×10^{-3}	2.041×10^{-3}	4.100×10^{-3}	1.569×10^{-2}
RMSE	1.172×10^{-3}	1.016×10^{-3}	2.133×10^{-3}	8.278×10^{-3}
$T = 3.0$				
Maxerror	2.336×10^{-3}	1.620×10^{-3}	6.557×10^{-3}	3.554×10^{-3}
RMSE	1.046×10^{-3}	1.766×10^{-3}	3.320×10^{-3}	8.416×10^{-4}

TABLE 2: Maxerror and RMSE for $T = 1.0, 2.0, 3.0$ with different grids.

	21×41	41×81	81×161	161×321
$T = 1.0$				
Maxerror	3.256×10^{-4}	2.205×10^{-4}	1.466×10^{-4}	3.418×10^{-4}
RMSE	6.786×10^{-5}	1.358×10^{-4}	8.334×10^{-5}	2.276×10^{-4}
$T = 2.0$				
Maxerror	1.362×10^{-3}	1.665×10^{-3}	2.013×10^{-3}	1.982×10^{-3}
RMSE	2.888×10^{-4}	8.172×10^{-4}	1.020×10^{-3}	1.009×10^{-3}
$T = 3.0$				
Maxerror	2.061×10^{-3}	5.661×10^{-3}	3.301×10^{-3}	7.441×10^{-3}
RMSE	5.832×10^{-4}	2.751×10^{-3}	1.565×10^{-3}	3.705×10^{-3}

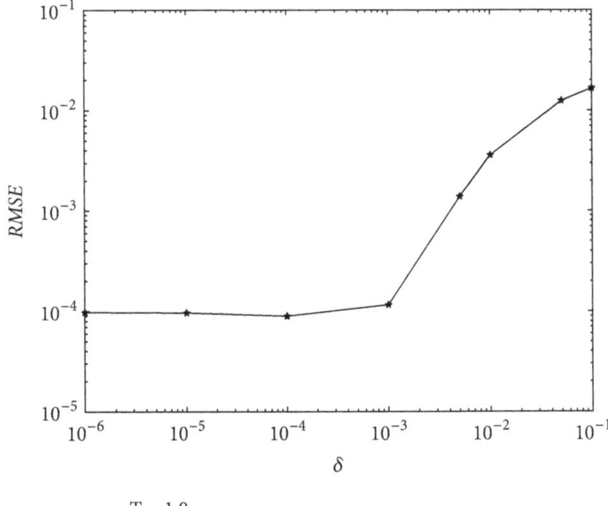

FIGURE 3: The errors of RMSE with different δ, $T = 1.0$.

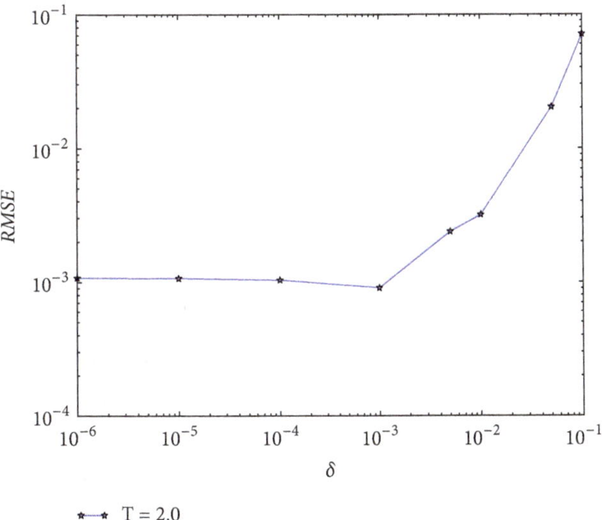

FIGURE 4: The errors of RMSE with different δ, $T = 2.0$.

RMSE with different noisy level δ with $\tau = 0.25$ is present in Figures 3 and 4 for $T = 1.0$ and $T = 2.0$, respectively.

We also discuss the results under different grids; the Maxerror and RMSE for $T = 1.0, 2.0$ and 3.0 are shown in Table 2 with various grids with noise level $\delta = 10^{-3}$ and time steps $\tau = 0.25$.

To illustrate the generality, we choose different variables for the computational domain. The figures of the exact and numerical initial solutions are given in Figures 5–7 with different variables of the computational domain. In this computation we choose the mesh 81×161, $\tau = 0.2$ and $\delta = 10^{-3}$, $T = 1$.

Example 2. We also consider the 2D nonhomogeneous backward heat conduction problem in this example; the known conditions are presented as follows,

$$h(\mathbf{x}, t) = \sin(\pi xy) \exp(-t),$$
$$g(\mathbf{x}) = \sin(\pi xy) \exp(-T), \tag{18}$$

and $f(\mathbf{x}, t) = (\pi^2 y^2 + \pi^2 x^2 - 1) \sin(\pi xy) \exp(-t)$.

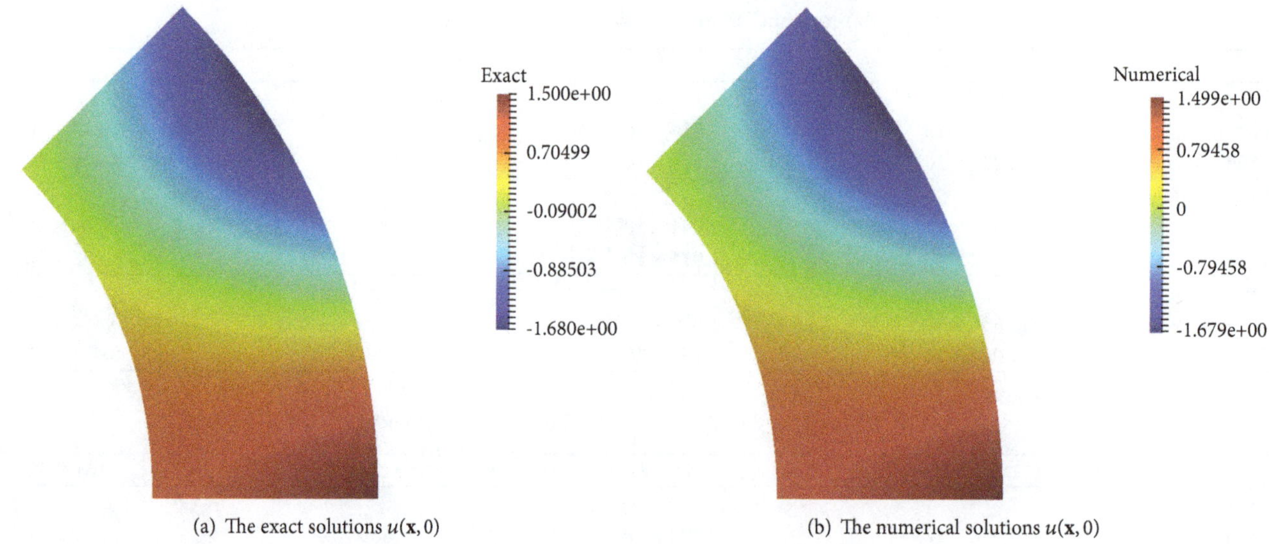

(a) The exact solutions $u(\mathbf{x}, 0)$

(b) The numerical solutions $u(\mathbf{x}, 0)$

FIGURE 5: The figure with $a = 1.0, b = 1.5, c = 0, d = 1.0$, and $\theta = \pi/4$.

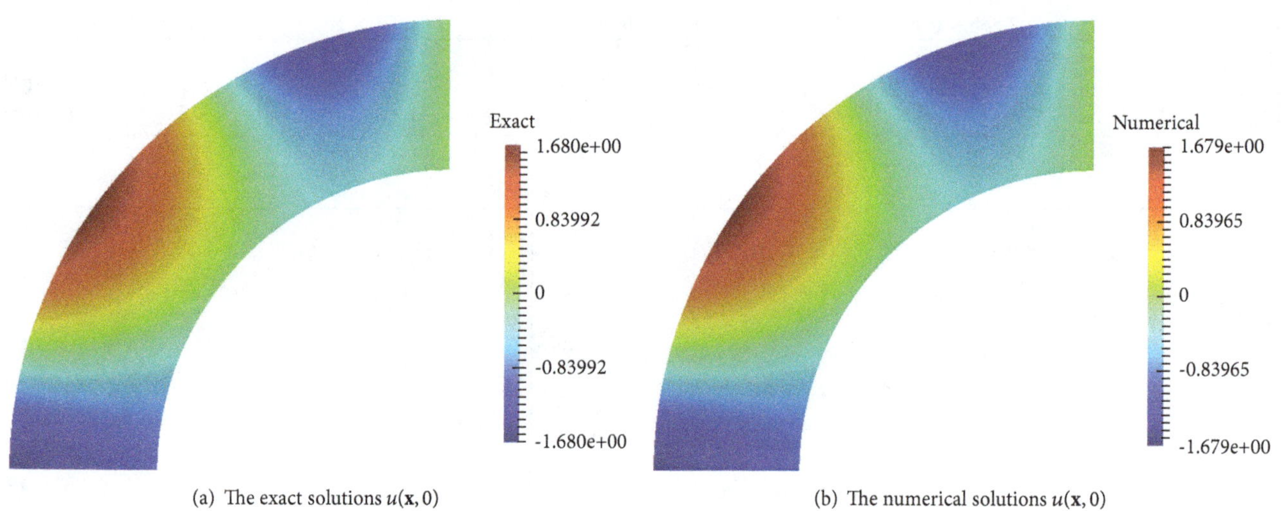

(a) The exact solutions $u(\mathbf{x}, 0)$

(b) The numerical solutions $u(\mathbf{x}, 0)$

FIGURE 6: The figure with $a = 1.0, b = 1.5, c = 1.0, d = 2.0$, and $\theta = \pi/2$.

In this example, the calculation area is a square of the center at origin; side length is 1. In the computation using the mesh 51×51, the computational grid is shown in Figure 8. The Maxerror and RMSE for $T = 1.0, 2.0, 3.0$ are shown in Table 3 with $\delta = 10^{-3}$ and various τ.

The errors of RMSE with different δ are also given in Figure 9. In this figure $\tau = 0.2$, and the mesh is the same as Figure 8.

The same with Example 1, we also consider the results with different meshes; the Maxerror and RMSE for $T = 1.0, 2.0$ and 3.0 are shown in Table 4 with various meshes with noise level $\delta = 10^{-3}$ and time steps $\tau = 0.2$.

The figure of numerical solutions $u(x, y, 0)$ with $T = 1$, $\tau = 0.2$, and $\delta = 10^{-3}$ is presented in Figure 10; the computational mesh is 51×51.

We also consider a slightly more complicated computational area, one unit square to remove a small rectangle and two small circles. In this case, the computation is a little more complicated; we define the unite square as $\overline{\Omega}_1$ and the small rectangle as Ω_2,

$$\begin{aligned} \overline{\Omega}_1 &= \{(x, y) \mid 0 \leqslant x, y \leqslant 1\}, \\ \Omega_2 &= \{(x, y) \mid 0.3 < x < 0.4,\ 0.2 < y < 0.5\}, \end{aligned} \tag{19}$$

the two small circles as

$$\begin{aligned} \Omega_3 &= \{(x, y) \mid (x - 0.1)^2 + (y - 0.9)^2 < 0.05^2\}, \\ \Omega_4 &= \{(x, y) \mid (x - 0.7)^2 + (y - 0.6)^2 < 0.1^2\}, \end{aligned} \tag{20}$$

and the computational domain is defined as

$$\overline{\Omega} = \overline{\Omega}_1 - \Omega_2 - \Omega_3 - \Omega_4. \tag{21}$$

Table 3: Maxerror and RMSE for $T = 1, 2, 3$ with different τ.

	$\tau = 1.0$	$\tau = 0.5$	$\tau = 0.25$	$\tau = 0.2$
$T = 1.0$				
Maxerror	1.301×10^{-3}	1.374×10^{-4}	2.504×10^{-4}	4.221×10^{-4}
RMSE	5.907×10^{-4}	5.727×10^{-5}	1.409×10^{-4}	2.280×10^{-4}
$T = 2.0$				
Maxerror	7.725×10^{-4}	1.513×10^{-4}	6.912×10^{-5}	2.041×10^{-4}
RMSE	3.463×10^{-4}	6.095×10^{-5}	3.407×10^{-5}	1.015×10^{-4}
$T = 3.0$				
Maxerror	9.877×10^{-4}	2.187×10^{-4}	6.443×10^{-5}	4.832×10^{-4}
RMSE	4.474×10^{-4}	9.301×10^{-5}	1.911×10^{-5}	2.457×10^{-4}

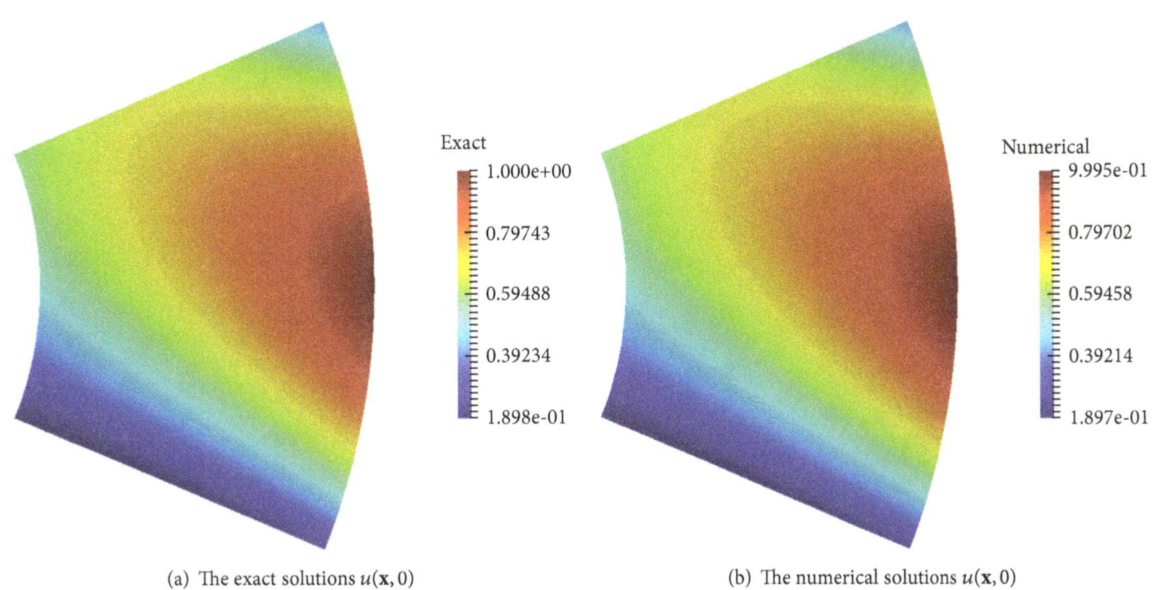

(a) The exact solutions $u(\mathbf{x}, 0)$

(b) The numerical solutions $u(\mathbf{x}, 0)$

Figure 7: The figure with $a = 0.5$, $b = 1.0$, $c = -0.5$, $d = 0.5$, and $\theta = \pi/4$.

Figure 8: The square domain with computational grid.

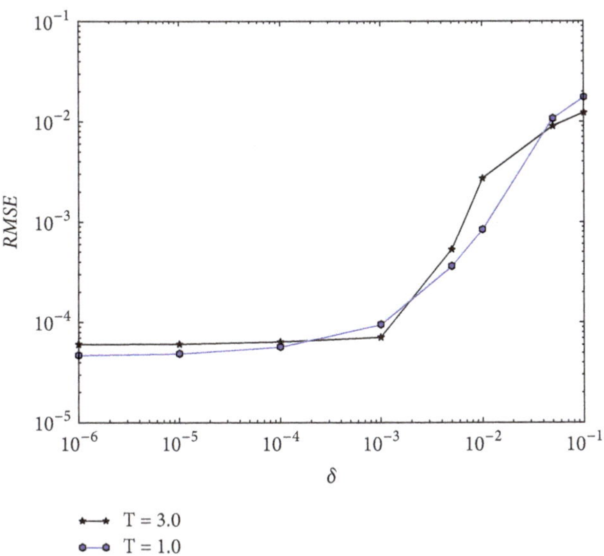

Figure 9: The errors of RMSE with different δ, $T = 1.0$ and $T = 3.0$.

TABLE 4: Maxerror and RMSE for $T = 1.0, 2.0, 3.0$ with different grids.

	21×21	51×51	101×101	201×201
$T = 1.0$				
Maxerror	4.767×10^{-4}	4.114×10^{-4}	9.978×10^{-5}	1.744×10^{-4}
RMSE	1.409×10^{-4}	2.221×10^{-4}	4.938×10^{-5}	9.225×10^{-5}
$T = 2.0$				
Maxerror	2.644×10^{-4}	2.124×10^{-4}	3.899×10^{-4}	3.138×10^{-4}
RMSE	4.982×10^{-5}	1.058×10^{-4}	1.986×10^{-4}	1.621×10^{-4}
$T = 3.0$				
Maxerror	5.485×10^{-4}	4.497×10^{-4}	5.738×10^{-4}	6.567×10^{-4}
RMSE	2.807×10^{-4}	2.288×10^{-4}	2.925×10^{-4}	3.353×10^{-4}

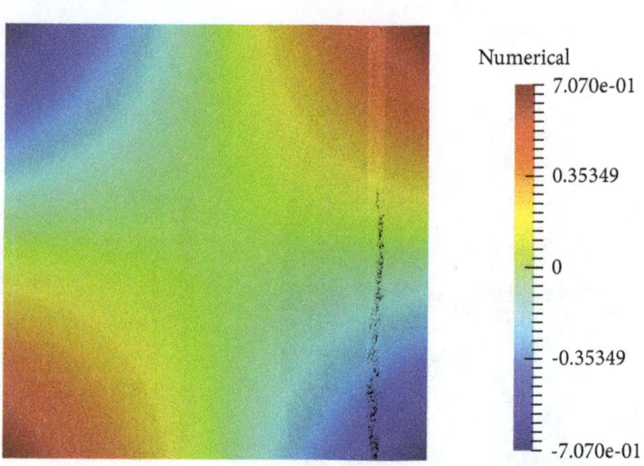

FIGURE 10: The numerical solutions of $u(x, y, 0)$.

FIGURE 11: The domain with computational grid.

The figure of the domain with the computational mesh can be seen in Figure 11 and the figure of solutions $u(x, y, 0)$ ia also given in Figure 12 with $T = 3$, $\tau = 0.2$.

Example 3. In this example we consider the inverse problem in three dimensions with the Dirichlet boundary conditions,

$$h(\mathbf{x}, t) = \sin x \sin y \sin z \cos t, \quad \mathbf{x} = (x, y, z) \in \Gamma, \; t \in (0, T), \tag{22}$$

the final conditions,

$$g(\mathbf{x}) = \sin x \sin y \sin z \cos T, \quad \mathbf{x} = (x, y, z) \in \Omega, \tag{23}$$

and $f(\mathbf{x}, t) = \sin x \sin y \sin z (3 \cos t - \sin t)$; the analytical solution is

$$u(\mathbf{x}, t) = \sin x \sin y \sin z \cos t. \tag{24}$$

In this example we choose the 3D computational domain defined as a cube without a cylinder insider. The cube is defined as

$$\Omega_1 = \{(x, y, z) \mid -0.5 \leqslant x, y \leqslant 0.5, \; -1 \leqslant z \leqslant 1\}, \tag{25}$$

the cylinder is defined as

$$\Omega_2 = \{(x, y, z) \mid (4x)^2 + (4y)^2 < 1, \; -1 \leqslant z \leqslant 1\}, \tag{26}$$

and the computational domain is $\overline{\Omega} = \Omega_1 - \Omega_2$. The domain with the computational grid is shown in Figure 13.

The Maxerror and RMSE for $T = 1.0, 3.0, 5.0$ are shown in Table 5 with noise level $\delta = 10^{-3}$ and various τ.

The figure of numerical initial solutions with $T = 1$, $\tau = 0.2$ with noisy level $\delta = 10^{-3}$ can be seen in Figure 14.

Example 4. In this example we also consider the inverse problem with the same conditions in Example 3, but the computational domain is a sphere with the center at origin and the diameter is 1. The domain with the computational grid is shown in Figure 15.

The Maxerror and RMSE for $T = 1.0, 2.0, 3.0$ are shown in Table 6 with noise level $\delta = 10^{-3}$ and various τ.

The figure of numerical initial solutions with $T = 3$, $\tau = 0.2$ with noisy level $\delta = 10^{-3}$ can be seen in Figure 16.

5. Conclusion

In this paper, we proposed a new numerical scheme to solve nonhomogeneous backward heat conduction problem

TABLE 5: Maxerror and RMSE for $T = 1, 3, 5$ with different τ.

	$\tau = 1/2$	$\tau = 1/3$	$\tau = 1/4$	$\tau = 1/5$
$T = 1.0$				
Maxerror	4.726×10^{-4}	2.014×10^{-3}	4.399×10^{-4}	1.409×10^{-3}
RMSE	5.653×10^{-5}	1.499×10^{-4}	4.827×10^{-5}	1.489×10^{-4}
$T = 3.0$				
Maxerror	1.807×10^{-3}	1.161×10^{-4}	1.717×10^{-3}	2.961×10^{-4}
RMSE	1.983×10^{-4}	3.789×10^{-5}	1.974×10^{-4}	6.815×10^{-5}
$T = 5.0$				
Maxerror	6.437×10^{-4}	1.125×10^{-3}	6.596×10^{-4}	1.206×10^{-3}
RMSE	6.681×10^{-5}	1.225×10^{-4}	7.013×10^{-5}	1.629×10^{-4}

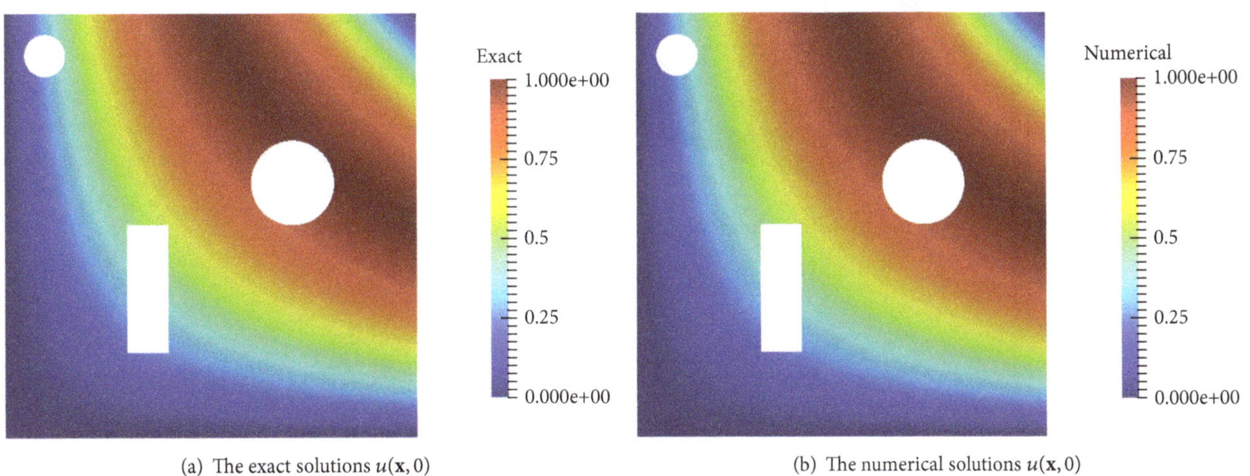

(a) The exact solutions $u(\mathbf{x}, 0)$

(b) The numerical solutions $u(\mathbf{x}, 0)$

FIGURE 12: The figures of exact and numerical solutions of $u(x, y, 0)$.

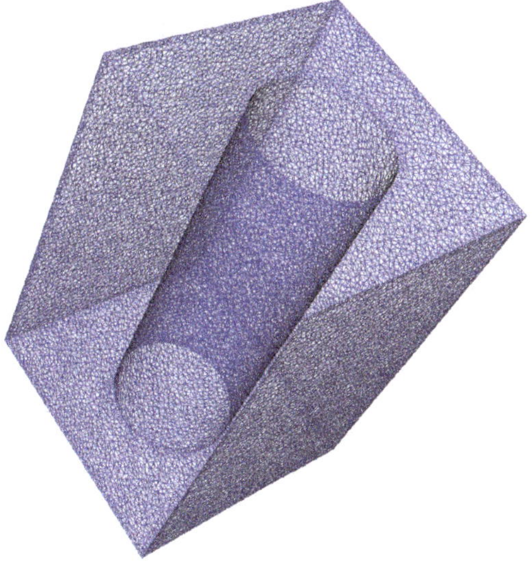

FIGURE 13: The 3D domain with computational grid.

with two- and three-dimensional computational domain. In our method, using likewise Crank Nicolson scheme and introducing a new intermediate variable, transforming the backward problem to a nonhomogeneous Helmholtz problem, and solving the Helmholtz problem using finite element method, we get the initial temperature. This new method is easy to calculate and from the numerical results presented in the previous it can be seen that this new method is effective for solving such problems.

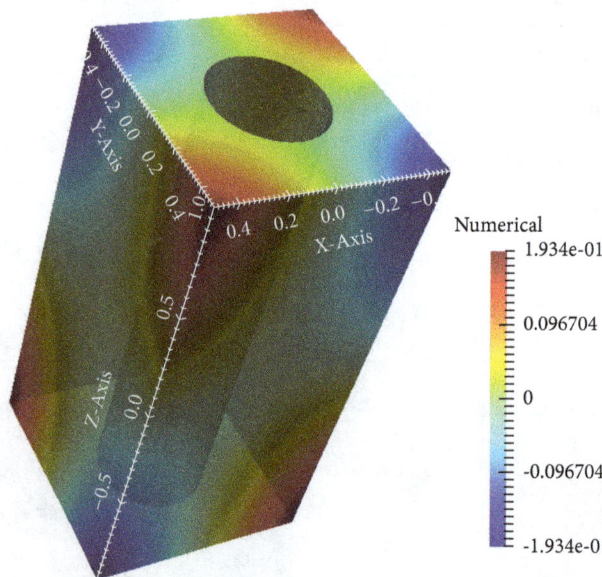

FIGURE 14: The numerical solutions $u(x, y, 0)$ with $T = 1$, $\tau = 0.2$.

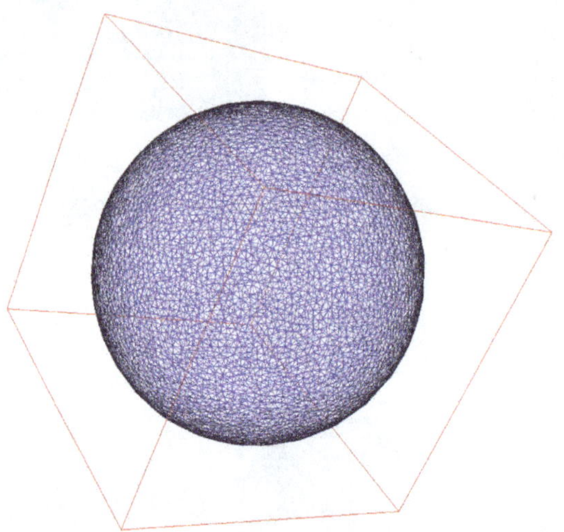

FIGURE 15: The sphere domain with computational grid.

TABLE 6: Maxerror and RMSE for $T = 1, 2, 3$ with different τ.

	$\tau = 1/2$	$\tau = 1/3$	$\tau = 1/4$	$\tau = 1/5$
$T = 1.0$				
Maxerror	3.765×10^{-5}	1.119×10^{-4}	3.521×10^{-5}	1.119×10^{-4}
RMSE	7.204×10^{-6}	1.780×10^{-5}	6.112×10^{-6}	1.766×10^{-5}
$T = 2.0$				
Maxerror	1.021×10^{-4}	1.028×10^{-4}	1.025×10^{-4}	1.031×10^{-4}
RMSE	1.742×10^{-5}	1.645×10^{-5}	1.658×10^{-5}	1.638×10^{-5}
$T = 3.0$				
Maxerror	1.431×10^{-4}	1.364×10^{-5}	1.463×10^{-4}	3.104×10^{-4}
RMSE	2.497×10^{-5}	3.921×10^{-6}	2.324×10^{-5}	1.172×10^{-5}

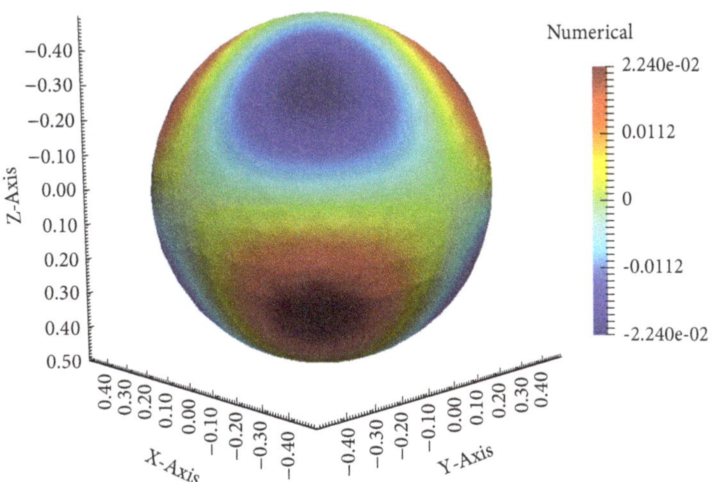

FIGURE 16: The numerical solutions $u(x, y, 0)$ with $T = 3, \tau = 0.2$.

It is worth to note that, due to the limitations of the standard finite element method in solving Helmholtz equations with high wave numbers, to solve the Helmholtz problem (9) by FEM, the time step τ should be carefully chosen, in order to ensure the wave number in (9) is not high, and for high dimension problem, dense grids need to be generated to ensure accuracy, resulting in increased computational time. Many works have been done in solving Helmholtz problem with high wave numbers [25–28]; these methods can be used instead of the standard FEM to solve the Helmholtz problem (9), so that these problems will be somewhat improved.

Conflicts of Interest

The authors declare that they have no conflicts of interest.

Acknowledgments

The authors would like to to thank professors P. N. Vabishchevish and V. I. Vasilév for their important guidance and advice in the paper writing. This work is supported by RFBR (project 17-01-00732).

References

[1] J. Fan and L. Wang, "Review of heat conduction in nanofluids," *Journal of Heat Transfer*, vol. 133, no. 4, pp. 801–815, 2011.

[2] A. Paliathanasis, R. M. Morris, and P. G. L. Leach, "Lie symmetries of (1+2) nonautonomous evolution equations in financial mathematics," *Mathematics*, vol. 4, no. 2, pp. 1–14, 2016.

[3] M. A. Bartoshevich, "A heat-conduction problem," *Journal of Engineering Physics*, vol. 28, no. 2, pp. 240–244, 1975.

[4] O. M. Alifanov, *Inverse Heat Transfer Problems*, International Series in Heat and Mass Transfer, Springer-Verlag, 1994.

[5] C.-H. Huang and S.-P. Wang, "A three-dimensional inverse heat conduction problem in estimating surface heat flux by conjugate gradient method," *International Journal of Heat and Mass Transfer*, vol. 42, no. 18, pp. 3387–3403, 1999.

[6] N. D. Aparicio and M. K. Pidcock, "The boundary inverse problem for the Laplace equation in two dimensions," *Inverse Problems*, vol. 12, no. 5, pp. 565–577, 1996.

[7] V. V. Vasilev, M. V. Vasilyeva, and A. M. Kardashevskyc, "The numerical solution of the boundary inverse problem for a parabolic equation," *American Institute of Physics*, vol. 1773, no. 1, pp. 100010(1)–100010(7), 2016.

[8] L. Su, P. N. Vabishchevish, and V. I. Vasil'ev, "The inverse problem of the simultaneous determination of the right-hand side and the lowest coefficients in parabolic equations," in *Numerical Analysis and Its Applications*, Lecture Notes in Computer Science, pp. 633–639, 2017.

[9] P. N. Vabishchevich and V. I. Vasil'ev, "Computational algorithms for solving the coefficient inverse problem for parabolic equations," *Inverse Problems in Science and Engineering*, vol. 24, no. 1, pp. 42–59, 2016.

[10] A. A. Samarskii and P. N. Vabishchevich, *Numerical Methods for Solving Inverse Problems of Mathematical Physics*, Inverse and Ill-Posed Problems Series, Walter de Gruyter, 2007.

[11] M. Li, T. Jiang, and Y. C. Hon, "A meshless method based on RBFs method for nonhomogeneous backward heat conduction problem," *Engineering Analysis with Boundary Elements*, vol. 34, no. 9, pp. 785–792, 2010.

[12] J. Hadamard, *Lectures on Cauchy Problems in Linear Partial Differential Equations*, New Haven Yale University Press, 1923.

[13] L. Payne, *Improperly Posed Problems in Partial Differential Equations*, Society for Industrial and Applied Mathematics, 1975.

[14] W. L. Miranker, "A well posed problem for the backward heat equation," *Proceedings of the American Mathematical Society*, vol. 12, no. 2, pp. 243–247, 1961.

[15] C.-L. Fu, X.-T. Xiong, and Z. Qian, "Fourier regularization for a backward heat equation," *Journal of Mathematical Analysis and Applications*, vol. 331, no. 1, pp. 472–480, 2007.

[16] H. Han, D. B. Ingham, and Y. Yuan, "The boundary element method for the solution of the backward heat conduction

equation," *Journal of Computational Physics*, vol. 116, no. 2, pp. 292–299, 1995.

[17] K. Iijima, "Numerical solution of backward heat conduction problems by a high order lattice-free finite difference method," *Journal of the Chinese Institute of Engineers*, vol. 27, no. 4, pp. 611–620, 2004.

[18] A. Shidfar and A. Zakeri, "A numerical technique for backward inverse heat conduction problems in one-dimensional space," *Applied Mathematics and Computation*, vol. 171, no. 2, pp. 1016–1024, 2005.

[19] D. Lesnic, L. Elliott, and D. B. Ingham, "An iterative boundary element method for solving the backward heat conduction problem using an elliptic approximation," *Inverse Problems in Science and Engineering*, vol. 6, no. 4, pp. 255–279, 1998.

[20] N. S. Mera, L. Elliott, D. B. Ingham, and D. Lesnic, "Iterative boundary element method for solving the one-dimensional backward heat conduction problem," *International Journal of Heat and Mass Transfer*, vol. 44, no. 10, pp. 1937–1946, 2001.

[21] Y. C. Hon and M. Li, "A discrepancy principle for the source points location in using the MFS for solving the BHCP," *International Journal of Computational Methods*, vol. 6, no. 2, pp. 181–197, 2009.

[22] M. Denche and A. Abdessemed, "On regularization and error estimates for non-homogeneous backward Cauchy problem," *Arab Journal of Mathematical Sciences*, vol. 18, no. 2, pp. 149–164, 2012.

[23] N. H. Tuan and D. D. Trong, "A new regularized method for two dimensional nonhomogeneous backward heat problem," *Applied Mathematics and Computation*, vol. 215, no. 3, pp. 873–880, 2009.

[24] S. Y. Reutskiy, C. S. Chen, and H. Y. Tian, "A boundary meshless method using Chebyshev interpolation and trigonometric basis function for solving heat conduction problems," *International Journal for Numerical Methods in Engineering*, vol. 74, no. 10, pp. 1621–1644, 2008.

[25] F. Ihlenburg and I. Babuška, "Finite element solution of the Helmholtz equation with high wave number Part I: The *h*-version of the FEM," *Computers & Mathematics with Applications*, vol. 30, no. 9, pp. 9–37, 1995.

[26] F. Ihlenburg and I. Babuka, "Finite element solution of the Helmholtz equation with high wave number part II: the h-p Version of the FEM," *SIAM Journal on Numerical Analysis*, vol. 34, no. 1, pp. 315–358, 1997.

[27] I. Harari and F. Magoulés, "Numerical investigations of stabilized finite element computations for acoustics," *Wave Motion*, vol. 39, no. 4, pp. 339–349, 2004.

[28] Y. S. Wong and G. Li, "Exact finite difference schemes for solving Helmholtz equation at any wavenumber," *International Journal of Numerical Analysis & Modeling- Series B*, vol. 2, no. 1, pp. 91–108, 2011.

[29] I. Babuka and S. A. Sauter, "Is the pollution effect of the fem avoidable for the Helmholtz equation considering high wave," *SIAM Journal on Numerical Analysis*, vol. 34, no. 1, pp. 2392–2423, 1997.

[30] A. Bayliss, C. I. Goldstein, and E. Turkel, "On accuracy conditions for the numerical computation of waves," *Journal of Computational Physics*, vol. 59, no. 3, pp. 396–404, 1985.

11

Existence of Solutions for Unbounded Elliptic Equations with Critical Natural Growth

Aziz Bouhlal,[1] **Abderrahmane El Hachimi**,[2] **Jaouad Igbida**,[3] **El Mostafa Sadek**,[4] **and Hamad Talibi Alaoui**[1]

[1] *Labo Math Appli, Faculty of Sciences, B. P. 20, El Jadida, Morocco*
[2] *Laboratory of Mathematical Analysis and Applications, Mohammed V University, Faculty of Sciences, Rabat, Morocco*
[3] *Labo DGTIC, department of Mathematics, CRMEF El Jadida, Morocco*
[4] *Laboratory LabSIPE, ENSA, d'EL Jadida, University Chouaib Doukkali, Morocco*

Correspondence should be addressed to El Mostafa Sadek; sadek.maths@gmail.com

Academic Editor: Peiguang Wang

We investigate existence and regularity of solutions to unbounded elliptic problem whose simplest model is $\{-\text{div}[(1+|u|^q)\nabla u]+u = \gamma(|\nabla u|^2/(1+|u|)^{1-q})+f$ in Ω, $u=0$ on $\partial\Omega,\}$, where $0<q<1$, $\gamma>0$ and f belongs to some appropriate Lebesgue space. We give assumptions on f with respect to q and γ to show the existence and regularity results for the solutions of previous equation.

1. Introduction

In this paper, we consider the Dirichlet problem for some nonlinear elliptic problems such as

$$-\text{div}\left(\left[a(x)+|u|^q\right]\nabla u\right)+u=H(x,u,\nabla u)+f, \quad x\in\Omega,\ u\in H_0^1(\Omega), \tag{1}$$

under the following assumptions: Ω is a bounded open subset of $\mathbb{R}^N$, where $N\geq 3$, $0<q<1$, and $f\in L^m$ with $m\geq 2$ and $a:\Omega\longrightarrow\mathbb{R}$ is a measurable function satisfying the following conditions:

$$\alpha\leq a(x)\leq\beta, \tag{2}$$

for almost every $x\in\Omega$, where α and β are positive real constants. $H(x,s,\xi)$ is a Carathéodory-type function satisfying to:

$$\left|H(x,s,\xi)\right|\leq\gamma\frac{|\xi|^2}{(1+|s|)^{1-q}} \tag{3}$$

for some $\gamma>0$.

In [1], Arcoya, Boccardo, and Leonor obtained the existence and regularity results for the following elliptic problem with degenerate coercivity:

$$-\text{div}\left(\frac{\alpha\nabla u}{(1+|u|)^2}\right)+u=\gamma\frac{|\nabla u|^2}{(1+|u|)^3}+f, \quad x\in\Omega,\ u\in H_0^1(\Omega), \tag{4}$$

where $\alpha,\gamma>0$, $f\in L^m(\Omega)$ with $m\geq 2$, and Ω is a bounded subset of $\mathbb{R}^N$, $N\geq 3$.

The purpose of the present paper is to study the same kind of lower order terms as in problem (4) in the case of an elliptic operator with unbounded coefficients such as (1).

There are several papers concerned with existence and regularity of the solution for the following problem:

$$\begin{aligned}-\text{div}(M(x,u)\nabla u)+g(x,u,\nabla u)&=f(x) \quad x\in\Omega,\\ u(x)&=0 \quad x\in\partial\Omega.\end{aligned} \tag{5}$$

We refer the intersting articles: Boccardo, Murat and Puel [2], Bensoussan, Boccardo and Murat [3], and Boccardo, Gallout [4]. In all these works g is a nonlinear lower term having natural growth with respect to ∇u, data f in suitable Lebesgue spaces, and $M(x, u)$ is a Carathéodory-type bounded function subject to certain structural inequalities.

Another motivation for studying these problem arises from the calculus of variations in the case where $0 \le f \in L^m(\Omega)$ with $m \ge N/2$ and

$$g(x, u, \nabla u) = \frac{|\nabla u|^2}{u^\theta}, \tag{6}$$

where $\theta \in (0, 1)$, which is considered by Puel in [5].

We point out that in [6] the authors considered $M(x, u)$ as a bounded function and

$$g(x, u, \nabla u) = \frac{Q(x, u)|\nabla u|^2}{u^\theta}, \tag{7}$$

where $\theta \in (0, 1]$. The function $Q(x, s) : \Omega \times \mathbb{R} \longrightarrow \mathbb{R}^{N^2}$ is symmetric, measurable with respect to x and continuous with respect to s with the following uniform ellipticity condition: for $x \in \Omega$, and $s \in \mathbb{R}$,

$$\mu|\xi|^2 \le Q(x, s)\xi\xi \le \nu|\xi|^2, \quad 0 < \mu \le \nu. \tag{8}$$

We shall prove the following main results on existence and regularity of solutions for problem (1).

Theorem 1. *Let $\tilde{\alpha} = \min\{1, \alpha\}$. Assuming that the functions a and H satisfy (2) and (3) then, if f belong to $L^m(\Omega)$, with*

$$m > 2\left(\frac{\gamma}{\tilde{\alpha}} + 1\right) + q, \tag{9}$$

there exists a distributional solution $u \in W_0^{1,1}(\Omega)$ of problem (1) such that

$$H(x, u, \nabla u) \in L^1(\Omega), \quad [a(x) + |u|^q]|\nabla u| \in L^1(\Omega),$$

$$\begin{aligned} &\int_\Omega [a(x) + |u|^q]\nabla u \nabla \psi + \int_\Omega u\psi \\ &= \int_\Omega H(x, u, \nabla u)\psi + \int_\Omega f\psi, \quad \forall \psi \in C_0^\infty(\Omega). \end{aligned} \tag{10}$$

Furthermore, any solution of the problem (1) belongs to $H_0^1(\Omega)$.

In the next result, we consider the case where f has a high summability.

Theorem 2. *Let $\tilde{\alpha} = \min\{1, \alpha\}$, and assume that (2) and (3) hold true. If u the solution given by Theorem 1 and f belongs to $L^m(\Omega)$, with*

$$m > \max\left\{2\left(\frac{\gamma}{\tilde{\alpha}} + 1\right) + q, \frac{N}{2}\left(\frac{\gamma}{\tilde{\alpha}} + 1\right)\right\}, \tag{11}$$

then u belongs to $H_0^1(\Omega) \cap L^\infty(\Omega)$.

The rest of the paper is organized as follows: Section 2 is devoted to give some a priori estimates for the approximated problem associated with problem (1); while in Section 3, we give the detailed proofs of Theorems 1 and 2.

2. The Approximated Problem

In this section, we use the hypotheses (2) and (3) and we suppose that

$$\tilde{\alpha}(m - 1) - \gamma > 0, \tag{12}$$

where $\tilde{\alpha} = \min\{1, \alpha\}$ holds true. To prove Theorem 1 and Theorem 2, we will use the following approximating problems associated with problem (1):

$$\begin{aligned} &-\operatorname{div}\left(\left[a(x) + |u_n|^q\right]\nabla u_n\right) + u_n \\ &= H_n(x, u_n, \nabla u_n) + f_n, \quad x \in \Omega, \end{aligned} \tag{13}$$

where

$$f_n(x) = \frac{f(x)}{1 + (1/n)|f(x)|}, \tag{14}$$

and

$$H_n(x, s, \xi) = \frac{H(x, s, \xi)}{1 + (1/n)|\xi|^2}. \tag{15}$$

By the results of [2, 4] there exists a weak solution u_n in $H_0^1(\Omega) \cap L^\infty(\Omega)$ of problem (13) in the sense that

$$\begin{aligned} &\int_\Omega \left[a(x) + |u_n|^q\right]\nabla u_n \nabla \varphi + \int_\Omega u_n\varphi \\ &= \int_\Omega H_n(x, u_n, \nabla u_n)\varphi + \int_\Omega f_n\varphi \end{aligned} \tag{16}$$

for every $\varphi \in H_0^1(\Omega) \cap L^\infty(\Omega)$.

The following lemma will be very useful, as it gives us an a priori estimate on the summability of the solutions to problems (13).

Lemma 3. *If u_n is a solution to problem (13), then for every $k \ge 0$,*

$$\int_\Omega |G_k(u_n)|^m \le \int_{\{|u_n| \ge k\}} |f|^m. \tag{17}$$

Moreover, there exist $R > 0$ depending on $\|f\|_{L^m(\Omega)}$, α, q, and γ, such that

$$\|u_n\|_{H_0^1(\Omega)} \le R. \tag{18}$$

Remark 4. (i) Let $\{u_n\}$ be a sequence of solutions u_n of (13). As a consequence of Lemma 3, there exists $u \in H_0^1(\Omega)$ such that, up to a subsequence, u_n converges weakly to u in $H_0^1(\Omega)$ and a.e. in Ω.

(ii) By the previous lemma we deduce from (3) that

$$\|H_n(x, u_n, \nabla u_n)\|_{L^1(\Omega)} \le \gamma \int_\Omega \frac{|\nabla u_n|^2}{(1 + |u_n|)^{1-q}} \le \gamma R^2. \tag{19}$$

Proof of Lemma 3. In order to prove (17), we claim that by assumption (2) and $q < 1$, there exist positive constant c_0 such that

$$\tilde{\alpha}(1 + |t|)^q \le a(x) + |t|^q \le c_0(1 + |t|)^q, \quad \forall t \in \mathbb{R}. \tag{20}$$

Choosing $\varphi = |G_k(u_n)|^{m-1}\operatorname{sgn}(u_n)$ in (16) and using (20), we obtain

$$\begin{aligned}&\tilde{\alpha}(m-1)\int_\Omega (1+|u_n|)^q |\nabla u_n|^2 |G_k(u_n)|^{m-2} \\ &\quad + \int_\Omega |u_n| |G_k(u_n)|^{m-1} \\ &\le \gamma \int_\Omega \frac{|\nabla u_n|^2}{(1+|u_n|)^{1-q}} |G_k(u_n)|^{m-1} \\ &\quad + \int_\Omega |f_n| |G_k(u_n)|^{m-1}.\end{aligned} \tag{21}$$

Thus, joining the terms involving the gradient, we get

$$\begin{aligned}&\int_\Omega \left[\tilde{\alpha}(m-1) - \gamma\frac{|G_k(u_n)|}{1+|u_n|}\right](1+|u_n|)^q |\nabla u_n|^2 \\ &\quad \cdot |G_k(u_n)|^{m-2} + \int_\Omega |G_k(u_n)|^m \le \int_\Omega |f_n| \\ &\quad \cdot |G_k(u_n)|^{m-1}.\end{aligned} \tag{22}$$

Using (12) we deduce that

$$\int_\Omega |G_k(u_n)|^m \le \int_\Omega |f| |G_k(u_n)|^{m-1}, \tag{23}$$

and the Hölder inequality on the right hand side yields

$$\begin{aligned}&\int_\Omega |G_k(u_n)|^m \\ &\le \left(\int_{\{|u_n|\ge k\}} |f|^m\right)^{1/m} \left(\int_\Omega |G_k(u_n)|^m\right)^{1-1/m},\end{aligned} \tag{24}$$

which implies (17).

Let us choose now $\varphi = [(1+|u_n|)^{m-1} - 1]\operatorname{sgn}(u_n)$ as a test function in (16), and we obtain

$$\begin{aligned}&(\tilde{\alpha}(m-1) - \gamma)\int_\Omega (1+|u_n|)^{m-2+q} |\nabla u_n|^2 \\ &\le \|f\|_{L^m(\Omega)} \left(\int_\Omega (1+|u_n|)^m\right)^{1-1/m}.\end{aligned} \tag{25}$$

Since $m \ge 2$, the previous calculations imply

$$\int_\Omega |\nabla u_n|^2 \le c\left(\int_\Omega (1+|u_n|)^m\right)^{1-1/m}. \tag{26}$$

Using (17) with $k = 0$, (18) follows. □

Lemma 5. *Let u_n be the sequence of solutions to problems (13) and let the function u given by Remark 4. Then u_n strongly converges to u in $L^m(\Omega)$. Moreover ∇u_n strongly converges to ∇u in $L^1(\Omega)^N$.*

Remark 6. Note that (25) implies that there exists $\delta > 0$ independent of n such that

$$\int_\Omega (1+|u_n|)^{m-2+q} |\nabla u_n|^2 \le \delta. \tag{27}$$

By using the previous lemma, we deduce that

$$(1+|u_n|)^{(m-2+q)/2} |\nabla u_n| \longrightarrow (1+|u|)^{(m-2+q)/2} |\nabla u| \quad \text{weakly in } L^2(\Omega)^N. \tag{28}$$

Proof of Lemma 5. We use (17) written for $k = 0$:

$$\int_\Omega |u_n|^m \le \int_\Omega |f|^m \le c. \tag{29}$$

Since u_n almost everywhere converges to u, we have from Fatou's lemma that

$$\int_\Omega |u|^m \le c. \tag{30}$$

Hence u belongs to $L^m(\Omega)$. Using assumption (17), for any $k > 0$ we have

$$\begin{aligned}\int_E |u_n|^m &\le \int_{E\cap\{|u_n|\le k\}} |u_n|^m + \int_{E\cap\{|u_n|\ge k\}} |u_n|^m \\ &\le k^m meas(E) + \int_{\{|u_n|\ge k\}} |f|^m.\end{aligned} \tag{31}$$

As before, we first choose k such that the second integral is small, uniformly with respect to n, and then the measure of E small enough such that the first term is small. The almost everywhere convergence of u_n to u and Vitali's theorem imply that u_n strongly converges to u in $L^m(\Omega)$.

For the second convergence, we will follow the same technique as in [1] (see also [7]). Let $h, k > 0$. In the sequel C will denote a constant independent of n, h, k. Let us consider $T_h[u_n - T_k(u)]$ as a test function in problems (16). Then,

$$\begin{aligned}&\int_\Omega [a(x) + |u_n|^2] \nabla u_n \nabla T_h[u_n - T_k(u)] \\ &\quad + \int_\Omega u_n T_h[u_n - T_k(u)] \\ &\le \left(\|f\|_{L^1(\Omega)} + \|H_n(x, u_n, \nabla u_n)\|_{L^1(\Omega)}\right) h.\end{aligned} \tag{32}$$

Moreover, thanks to the $L^m(\Omega)$ convergence of u_n, the second integral in (32) converges (as n diverges) to a positive number. Thus, it yields to

$$\begin{aligned}&\alpha\int_\Omega |\nabla T_h[u_n - T_k(u)]|^2 \\ &\le \left(\|f\|_{L^1(\Omega)} + \gamma R^2\right) h \\ &\quad - \int_\Omega [a(x) + |u_n|^q] \nabla T_k(u) \nabla T_h[u_n - T_k(u)].\end{aligned} \tag{33}$$

Let $\mathcal{K} = h+k$, observing that $\nabla T_h[u_n - T_k(u)] = 0$ if $|u_n| > \mathcal{K}$, then

$$\begin{aligned}&\int_\Omega [a(x) + |u_n|^q] \nabla T_k(u) \nabla T_h[u_n - T_k(u)] \\ &= \int_\Omega [a(x) + |T_{\mathcal{K}}(u_n)|^q] \nabla T_k(u) \nabla T_h[u_n - T_k(u)].\end{aligned} \tag{34}$$

Since $T_h[u_n - T_k(u)]$ converges to $T_h[u - T_k(u)]$ weakly in $(L^2(\Omega))^N$ and $[a(x) + |T_{\mathcal{K}}(u_n)|^q]\nabla T_k(u)$ strongly converges to $[a(x) + |T_{\mathcal{K}}(u)|^q]\nabla T_k(u)$ in $(L^2(\Omega))^N$, we have

$$\lim_{n\to+\infty}\int_\Omega \left[a(x) + |u_n|^q\right]\nabla T_k(u)\nabla T_h\left[u_n - T_k(u)\right] = 0, \tag{35}$$

thus, yielding

$$\int_\Omega \left|\nabla T_h\left[u_n - T_k(u)\right]\right|^2 \le Ch + \varepsilon(n), \tag{36}$$

where $\varepsilon(n)$ denote any quantity that vanishes as n diverges. Hence, by Hölder's inequality, we deduce that

$$\begin{aligned}\int_{\{|u_n-u|\le h, |u|\le k\}} |\nabla(u_n - u)| &= \int_\Omega \left|\nabla T_h\left[u_n - T_k(u)\right]\right| \\ &\le |\Omega|^{1/2}\sqrt{Ch + \varepsilon(n)}.\end{aligned} \tag{37}$$

Fix, now, $\epsilon > 0$ there exist h_0 such that, for $h < h_0$, we have

$$|\Omega|^{1/2}\sqrt{Ch} < \epsilon. \tag{38}$$

Thanks to the weak convergence of u_n in $H_0^1(\Omega)$ and the absolute continuity of the integral, there exists k_0 independent from n such that, for $k > k_0$, we have

$$\int_{\{|u|>k\}} |\nabla u_n| + \int_{\{|u|>k\}} |\nabla u| \le \epsilon. \tag{39}$$

In addition, by Dunford Pettis Theorem, we deduce that there exists $n(h,\epsilon)$ such that, for $n > n(h,\epsilon)$, we have

$$\int_{\{|u_n-u|>h\}} |\nabla(u_n - u)| \le \epsilon. \tag{40}$$

We can write

$$\begin{aligned}\int_\Omega |\nabla(u_n - u)| &= \int_{\{|u_n-u|\le h, |u|\le k\}} |\nabla(u_n - u)| \\ &+ \int_{\{|u_n-u|\le h, |u|>k\}} |\nabla(u_n - u)| \\ &+ \int_{\{|u_n-u|>h\}} |\nabla(u_n - u)|.\end{aligned} \tag{41}$$

Using (37), (39), and (40), for $h < h_0$ and $n > n(h,\epsilon)$, we have

$$\int_\Omega |\nabla(u_n - u)| \le 3\epsilon + \varepsilon(n). \tag{42}$$

This proves the strong convergence of ∇u_n to ∇u in $L^1(\Omega)^N$. □

The following lemma yields some a priori estimate on $\{u_n\}$.

Lemma 7. *Let u be the function given by Remark 4. Then $|u|^q|\nabla u|$ belongs to $L^r(\Omega)$, for every $r < N/(N-1)$.*

Proof. For every $\lambda > 1$, we take $[1 - 1/(1+|u_n|)^{\lambda-1}]\text{sign}(u_n)$ as a test function in (16). Droping positive terms yields

$$\begin{aligned}\tilde{\alpha}(\lambda - 1)\int_\Omega \frac{(1+|u_n|^q)|\nabla u_n|^2}{(1+|u_n|)^\lambda} \\ \le \|f\|_{L^1(\Omega)} + \|H_n(x,u_n,\nabla u_n)\|_{L^1(\Omega)}.\end{aligned} \tag{43}$$

Hence, using $q < 1$, it follows that

$$\int_\Omega \frac{|\nabla u_n|^2}{(1+|u_n|)^{\lambda-q}} \le \frac{\|f\|_{L^1(\Omega)} + \gamma R^2}{\tilde{\alpha}(\lambda - 1)}. \tag{44}$$

On the other hand, for every $\lambda > 1$; we have

$$\begin{aligned}&\int_\Omega |u_n|^{qr}|\nabla u_n|^r \\ &\le \int_\Omega \frac{|\nabla u_n|^r}{(1+|u_n|)^{r(\lambda-q)/2}}(1+|u_n|)^{r(\lambda+q)/2}, \\ &\le \left(\frac{\|f\|_{L^1(\Omega)} + \gamma R^2}{\tilde{\alpha}(\lambda - 1)}\right)^{r/2} \\ &\cdot\left(\int_\Omega (1+|u_n|)^{r(\lambda+q)/(2-r)}\right)^{(2-r)/2}.\end{aligned} \tag{45}$$

Then, we obtain

$$\left(\int_\Omega |u_n|^{(q+1)r^*}\right)^{r/r^*} \le c\left(\int_\Omega |u_n|^{r(\lambda+q)/(2-r)}\right)^{(2-r)/2}. \tag{46}$$

Let us choose r such that $(q+1)r^* = r(\lambda+q)/(2-r)$, that is

$$r = \frac{N(2+q-\lambda)}{N(q+1) - (\lambda+q)}. \tag{47}$$

Since $\lambda > 1$, we then have an estimate on $|u_n|^q|\nabla u_n|$ in $L^r(\Omega)$, for every $r < N/(N-1)$. □

The next result will be used in the proof of Theorem 2.

Lemma 8. *Suppose that (2), (3), and (11) hold true. Let $f \in L^m(\Omega)$ and $\{u_n\}$ be a solution of (13) with $f_n = f$ for every $n \in \mathbb{N}$. Then the norms of $\{u_n\}$ in $L^\infty(\Omega)$ and in $H_0^1(\Omega)$ are bounded by a constant which depends on q, m, N, α, γ, meas(Ω) and on the norm of f in $L^m(\Omega)$.*

Proof. Since $m > (N/2)(\gamma/\tilde{\alpha} + 1)$, we have $(1/2)(\gamma/\tilde{\alpha} + 1) < m/N$. Let us choose $\sigma > 0$ such that

$$\frac{1}{2}\left(\frac{\gamma}{\tilde{\alpha}} + q + 1\right) < \sigma < \frac{m}{N} + \frac{q}{2}. \tag{48}$$

The use of

$$\left[(1+|u_n|)^{2\sigma-q-1} - (1+k)^{2\sigma-q-1}\right]^+ \text{sign}(u_n), \tag{49}$$

as test function in (16), (3), and (20), implies that

$$\begin{aligned}(2\sigma - q - 1)\tilde{\alpha}\int_{A_k} |\nabla u_n|^2 (1 + |u_n|)^{2\sigma-2} \\ + \int_{A_k} |u_n| \left[(1 + |u_n|)^{2\sigma-q-1} - (1 + k)^{2\sigma-q-1}\right] \\ \leq \gamma \int_{A_k} \frac{|\nabla u_n|^2}{(1 + |u_n|)^{1-q}} (1 + |u_n|)^{2\sigma-q-1} \\ + \int_{A_k} |f_n| (1 + |u_n|)^{2\sigma-q-1},\end{aligned} \tag{50}$$

where

$$A_k = \{x \in \Omega : |u_n| > k\}. \tag{51}$$

By Young and Hölder's inequalities, we find

$$\begin{aligned}[(2\sigma - q - 1)\tilde{\alpha} - \gamma] \int_{A_k} |\nabla u_n|^2 (1 + |u_n|)^{2\sigma-2} \\ \leq C_1 \int_{A_k} (1 + |u_n|)^{2\sigma-q} + C_2 \int_{A_k} |f|^{2\sigma-q} \\ \leq C_m (\text{meas} A_k)^{1-(2\sigma-q)/m}.\end{aligned} \tag{52}$$

Then, using Sobolev's inequality gives

$$\begin{aligned}[(2\sigma - q - 1)\tilde{\alpha} - \gamma] \\ \cdot \frac{\mathcal{S}^2}{\sigma^2} \left(\int_{A_k} \left[(1 + |u_n|)^\sigma - (1 + k)^\sigma\right]^{2^*}\right)^{2/2^*} \\ \leq C_m (\text{meas} A_k)^{1-(2\sigma-q)/m},\end{aligned} \tag{53}$$

where $\mathcal{S}$ denotes the best constant in Sobolev inequality. Now, we set

$$(1 + |u_n|)^\sigma = v_n \tag{54}$$

and

$$(1 + k)^\sigma = h. \tag{55}$$

and the fact that $A_k = \{x \in \Omega : v_n > h\}.$, the last inequality gives

$$\begin{aligned}[(2\sigma - q - 1)\tilde{\alpha} - \gamma] \frac{\mathcal{S}^2}{\sigma^2} \left(\int_{A_k} (v_n - h)^{2^*}\right)^{2/2^*} \\ \leq C_m (\text{meas} A_k)^{1-(2\sigma-q)/m}.\end{aligned} \tag{56}$$

Note that $\sigma < m/N + q/2$ implies that $[1 - (2\sigma - q)/m](2^*/2) > 1$. Then Stampacchia's technique implies the following relation for some positive constant C_3,

$$\|v_n\|_{L^\infty(\Omega)} = \left\|(1 + |u_n|)^\sigma\right\|_{L^\infty(\Omega)} \leq C_3, \tag{57}$$

that is, $\|u_n\|_{L^\infty(\Omega)}$ is bounded. □

3. Proof of the Main Results

We are now ready to prove the main result of this paper. We first observe that condition (9) implies (12). Hence the results of the previous section hold true. In order to prove the result, we have to pass to the limit in (16). To this aim, let g be a function in $C^1(\mathbb{R})$ such that

$$g(s) = \begin{cases} \dfrac{1 + s}{\tilde{\alpha}\rho - \gamma} & \text{if } s \geq 0 \\ \dfrac{1}{(1 - s)(\tilde{\alpha}\rho - \gamma)} & \text{if } s < 0, \end{cases} \tag{58}$$

where

$$\rho = \frac{m - q - 2}{2}. \tag{59}$$

Observe that, by (9), g is positive, increasing, and it verifies

$$\tilde{\alpha}\rho g'(s) - \gamma \frac{g(s)}{1 + |s|} > 0, \quad \forall s \in \mathbb{R}. \tag{60}$$

We will use, for $k > 0$ and $s \in \mathbb{R}$,

$$R_k(s) = 1 - T_1(G_k(s)), \tag{61}$$

to define a test function. Remark that $R_k \geq 0, -k-1 \leq R_k(s) \leq k + 1$ and

$$R_k'(s) = \begin{cases} 1 & \text{if } -k - 1 \leq s \leq -k \\ -1 & \text{if } k \leq s \leq k + 1 \\ 0 & \text{otherwise.} \end{cases} \tag{62}$$

First of all, note that the a.e. convergence of ∇u_n (see Lemma 5), Remark 6, and (20) imply both that

$$\begin{aligned}\left[a(x) + |u_n|^q\right] g^\rho(u_n) \nabla u_n \longrightarrow \\ \left[a(x) + |u|^q\right] g^\rho(u) \nabla u \quad \text{weakly in } L^2(\Omega)^N\end{aligned} \tag{63}$$

and

$$\begin{aligned}\left[a(x) + |u_n|^q\right] \frac{1}{g^\rho(u_n)} \nabla u_n \longrightarrow \\ \left[a(x) + |u|^q\right] \frac{1}{g^\rho(u)} \nabla u \quad \text{weakly in } L^2(\Omega)^N,\end{aligned} \tag{64}$$

where ρ is defined in (59).

The proof of the result will be achieved in two steps.

Step 1 (The first inequality). We fix $\psi \in H_0^1(\Omega) \cap L^\infty(\Omega)$, with $\psi \geq 0$, and take

$$\phi = \frac{g^\rho(u_n)}{g^\rho(u)} R_k(u) \psi \tag{65}$$

As test function in (16), we have that

$$\begin{aligned}&\int_\Omega \left[a(x)+|u_n|^q\right]\nabla u_n \nabla\psi \frac{g^\rho(u_n)}{g^\rho(u)} R_k(u)\\&-\rho\int_\Omega \left[a(x)+|u_n|^q\right]\nabla u_n \nabla u \frac{g^\rho(u_n)}{g^{\rho+1}(u)} g'(u) R_k(u)\\&\cdot\psi+\int_\Omega \left[a(x)+|u_n|^q\right]\nabla u_n \nabla u \frac{g^\rho(u_n)}{g^\rho(u)} R_k'(u)\psi\\&+\rho\int_\Omega \left[a(x)+|u_n|^q\right]\nabla u_n \nabla u_n \frac{g^{\rho-1}(u_n)}{g^\rho(u)} g'(u_n)\\&\cdot R_k(u)\psi-\int_\Omega H_n(x,u_n,\nabla u_n)\frac{g^\rho(u_n)}{g^\rho(u)} R_k(u)\psi\\&+\int_\Omega u_n \frac{g^\rho(u_n)}{g^\rho(u)} R_k(u)\psi=\int_\Omega f_n \frac{g^\rho(u_n)}{g^\rho(u)} R_k(u)\psi.\end{aligned} \tag{66}$$

Remark now that, by the assumptions on a, H, relation (60) and the fact that $\psi \geq 0$, then we have

$$\begin{aligned}&\rho\left[a(x)+|u_n|^q\right]\nabla u_n \nabla u_n \frac{g^{\rho-1}(u_n)}{g^\rho(u)} g'(u_n) R_k(u)\psi\\&-H_n(x,u_n,\nabla u_n)\frac{g^\rho(u_n)}{g^\rho(u)} R_k(u)\psi \geq (1+|u_n|)^q\\&\cdot|\nabla u_n|^2 \frac{g^{\rho-1}(u_n)}{g^\rho(u)} R_k(u)\\&\cdot\psi\left[\tilde{\alpha}\rho g'(u_n)-\gamma\frac{g(u_n)}{1+|u_n|}\right]\geq 0.\end{aligned} \tag{67}$$

Therefore, using the almost everywhere convergence of both ∇u_n and u_n, and applying Fatou's lemma, we get

$$\begin{aligned}&\liminf_{n\to\infty}\rho\int_\Omega \left[a(x)+|u_n|^q\right]\nabla u_n \nabla u_n \frac{g^{\rho-1}(u_n)}{g^\rho(u)} g'(u_n)\\&\cdot R_k(u)\psi-\int_\Omega H_n(x,u_n,\nabla u_n)\frac{g^\rho(u_n)}{g^\rho(u)} R_k(u)\psi\\&\geq\rho\int_\Omega \left[a(x)+|u|^q\right]\nabla u \nabla u \frac{g'(u)}{g(u)} R_k(u)\psi\\&-\int_\Omega H(x,u,\nabla u) R_k(u)\psi.\end{aligned} \tag{68}$$

Furthermore, by using Lebesgue's theorem and (63), we obtain

$$\begin{aligned}&\lim_{n\to\infty}\int_\Omega \left[a(x)+|u_n|^q\right]\nabla u_n \nabla\psi \frac{g^\rho(u_n)}{g^\rho(u)} R_k(u)\\&=\int_\Omega \left[a(x)+|u|^q\right]\nabla u \nabla\psi R_k(u),\end{aligned} \tag{69}$$

and

$$\begin{aligned}&\lim_{n\to\infty}\rho\int_\Omega \left[a(x)+|u_n|^q\right]\nabla u_n \nabla u \frac{g^\rho(u_n)}{g^{\rho+1}(u)} g'(u) R_k(u)\\&\cdot\psi=\rho\int_\Omega \left[a(x)+|u|^q\right]\nabla u \nabla u \frac{g'(u)}{g(u)} R_k(u)\psi.\end{aligned} \tag{70}$$

Similarly, using the convergence $(u_n - f_n) \longrightarrow (u - f)$ in $L^m(\Omega)$, we have

$$\begin{aligned}&\lim_{n\to\infty}\int_\Omega (u_n-f_n)\frac{g^\rho(u_n)}{g^\rho(u)} R_k(u)\psi\\&=\int_\Omega (u-f) R_k(u)\psi.\end{aligned} \tag{71}$$

Now, from (62), we get

$$\begin{aligned}&\lim_{n\to\infty}\int_\Omega \left[a(x)+|u_n|^q\right]\nabla u_n \nabla u \frac{g^\rho(u_n)}{g^\rho(u)} R_k'(u)\psi\\&=\int_\Omega \left[a(x)+|u|^q\right]\nabla u \nabla u R_k'(u)\\&=\int_{\{k\leq|u|\leq k+1\}} \left[a(x)+|u|^q\right]\nabla u \nabla u.\end{aligned} \tag{72}$$

Passing to the limit in (66) when n tends to infinity and gathering together (68)-(72), weobtain

$$\begin{aligned}&\int_\Omega \left[a(x)+|u|^q\right]\nabla u \nabla\psi R_k(u)+\int_\Omega u R_k(u)\psi\\&+\int_{\{k\leq|u|\leq k+1\}} \left[a(x)+|u|^q\right]\nabla u \nabla u\\&-\int_\Omega H(x,u,\nabla u) R_k(u)\psi \leq \int_\Omega f R_k(u)\psi.\end{aligned} \tag{73}$$

Choosing $T_1(G_k(u_n))$ in (16), we get

$$\begin{aligned}&\int_{\{k\leq|u_n|\leq k+1\}} \left[a(x)+|u_n|^q\right]\nabla u_n \nabla u_n\\&\leq\int_{\{k\leq|u_n|\}}|f|+\int_{\{k\leq|u_n|\}}|H(x,u_n,\nabla u_n)|\\&\leq\int_{\{k\leq|u_n|\}}|f|+\frac{\gamma R^2}{(1+k)^{1-q}}.\end{aligned} \tag{74}$$

By Fatou's lemma, we have

$$\lim_{k\to\infty}\int_{\{k\leq|u|\leq k+1\}} \left[a(x)+|u|^q\right]\nabla u \nabla u = 0. \tag{75}$$

In order to pass to the limit as k tends to infinity in the inequality (73), we recall that $H(x,u,\nabla u) \in L^1(\Omega)$ and $[a(x)+|u|^q]\nabla u \in L^1(\Omega)$,. We obtain

$$\begin{aligned}&\int_\Omega \left[a(x)+|u|^q\right]\nabla u \nabla\psi+\int_\Omega u\psi\\&\leq\int_\Omega H(x,u,\nabla u)\psi+\int_\Omega f\psi,\end{aligned} \tag{76}$$

for every $\psi \in H_0^1(\Omega) \cap L^\infty(\Omega)$, with $\psi \geq 0$; that is, u is a subsolution of problem (1).

Step 2 (The second inequality). Let ψ be in $H_0^1(\Omega) \cap L^\infty(\Omega)$, with $\psi \leq 0$, and g be given by (58), and choose

$$\phi = \frac{g^\rho(u)}{g^\rho(u_n)} R_k(u)\psi \tag{77}$$

asa test function in (16). We obtain

$$\begin{aligned}
&\int_\Omega \left[a(x) + |u_n|^q\right] \nabla u_n \nabla \psi \frac{g^\rho(u)}{g^\rho(u_n)} R_k(u) \\
&+ \rho \int_\Omega \left[a(x) + |u_n|^q\right] \nabla u_n \nabla u \frac{g^{\rho-1}(u)}{g^\rho(u_n)} g'(u) R_k(u) \\
&\cdot \psi + \int_\Omega \left[a(x) + |u_n|^q\right] \nabla u_n \nabla u \frac{g^\rho(u)}{g^\rho(u_n)} R_k'(u)\psi \\
&- \rho \int_\Omega \left[a(x) + |u_n|^q\right] \nabla u_n \nabla u_n \frac{g^\rho(u)}{g^{\rho+1}(u_n)} g'(u_n) \\
&\cdot R_k(u)\psi - \int_\Omega H_n(x, u_n, \nabla u_n) \frac{g^\rho(u)}{g^\rho(u_n)} R_k(u)\psi \\
&+ \int_\Omega u_n \frac{g^\rho(u)}{g^\rho(u_n)} R_k(u)\psi = \int_\Omega f_n \frac{g^\rho(u)}{g^\rho(u_n)} R_k(u)\psi.
\end{aligned} \tag{78}$$

We observe that, by (60) and the fact that $\psi \leq 0$, we have

$$\begin{aligned}
&- \rho \left[a(x) + |u_n|^q\right] \nabla u_n \nabla u_n \frac{g^\rho(u)}{g^{\rho+1}(u_n)} g'(u_n) R_k(u)\psi \\
&- H_n(x, u_n, \nabla u_n) \frac{g^\rho(u)}{g^\rho(u_n)} R_k(u)\psi \geq -\left(1 + |u_n|\right)^q \\
&\cdot |\nabla u_n|^2 \frac{g^\rho(u)}{g^{\rho+1}(u_n)} R_k(u) \\
&\cdot \psi \left[\tilde{\alpha}\rho g'(u_n) - \gamma \frac{g(u_n)}{1 + |u_n|}\right] \geq 0.
\end{aligned} \tag{79}$$

Applying the same argument of Step 1 and using (64) instead of (63), we deduce that

$$\begin{aligned}
&\int_\Omega \left[a(x) + |u|^q\right] \nabla u \nabla \psi + \int_\Omega u\psi \\
&\leq \int_\Omega H(x, u, \nabla u)\psi + \int_\Omega f\psi,
\end{aligned} \tag{80}$$

for every $\psi \in H_0^1(\Omega) \cap L^\infty(\Omega)$, with $\psi \leq 0$.

Consequently, summarizing Steps 1 and 2, we have

$$\begin{aligned}
&\int_\Omega \left[a(x) + |u|^q\right] \nabla u \nabla \psi + \int_\Omega u\psi \\
&\leq \int_\Omega H(x, u, \nabla u)\psi + \int_\Omega f\psi,
\end{aligned} \tag{81}$$

for every $\psi \in H_0^1(\Omega) \cap L^\infty(\Omega)$.

Finally, interchanging ψ and $-\psi$ we conclude that

$$\begin{aligned}
&\int_\Omega \left[a(x) + |u|^q\right] \nabla u \nabla \psi + \int_\Omega u\psi \\
&= \int_\Omega H(x, u, \nabla u)\psi + \int_\Omega f\psi,
\end{aligned} \tag{82}$$

for every $\psi \in H_0^1(\Omega) \cap L^\infty(\Omega)$.

Disclosure

This paper has been presented at the 5th International Congress of the Moroccan Society of Applied Mathematics (SM2A-2017) and at the second edition of the "Journe doctorale de l'ENSAJ" in the "Ecole Nationale des Sciences Appliques d'El Jadida"- Morocco. We are gratefull to all people who helped in any way in this work.

Conflicts of Interest

The authors declare that they have no conflicts of interest.

Acknowledgments

The authors are grateful to all people who helped in this work.

References

[1] D. Arcoya, L. Boccardo, and T. Leonori, "$W_0^{1,1}$-solutions for elliptic problems having gradient quadratic lower order terms," *Nonlinear Differential Equations and Applications NoDEA*, vol. 20, no. 6, pp. 1741–1757, 2013.

[2] L. Boccardo, F. Murat, and J.-P. Puel, "Existence de solutions non bornées pour certaines équations quasi-linéaires," *Portugaliae Mathematica*, vol. 41, pp. 507–534, 1982.

[3] A. Bensoussan, L. Boccardo, and F. Murat, "On a nonlinear partial differential equation having natural growth terms and unbounded solution," *Ann. Inst. H. Poincaré*, vol. 5, no. 4, pp. 347–364, 1988.

[4] L. Boccardo and T. Gallouët, "Strongly nonlinear elliptic equations having natural growth terms and L^1 data," *Nonlinear Analysis: Theory, Methods & Applications*, vol. 19, no. 6, pp. 573–579, 1992.

[5] J.-P. Puel, "Existence, comportement à l'infini et stabilité dans certains problmes quasilinéaires elliptiques et paraboliques d'ordre 2," *Ann. Scuola Norm. Sup. Pisa Cl. Sci*, vol. 3, no. 1, pp. 89–119, 1976.

[6] L. Boccardo, "Dirichlet problems with singular and gradient quadratic lower order terms," *ESAIM: Control, Optimisation and Calculus of Variations*, vol. 14, no. 3, pp. 411–426, 2008.

[7] L. Boccardo, "A contribution to the theory of quasilinear elliptic equations and application to the minimization of integral functionals," *Milan Journal of Mathematics*, vol. 79, no. 1, pp. 193–206, 2011.

12

Applications of Parameterized Nonlinear Ordinary Differential Equations and Dynamic Systems: An Example of the Taiwan Stock Index

Meng-Rong Li,[1] Tsung-Jui Chiang-Lin,[2] and Yong-Shiuan Lee[3]

[1] *Department of Mathematical Sciences, National Chengchi University, Taipei 116, Taiwan*
[2] *Graduate Institute of Finance, National Taiwan University of Science and Technology, Taipei 106, Taiwan*
[3] *Department of Statistics, National Chengchi University, Taipei 116, Taiwan*

Correspondence should be addressed to Yong-Shiuan Lee; 99354501@nccu.edu.tw

Academic Editor: Tongxing Li

Considering the phenomenon of the mean reversion and the different speeds of stock prices in the bull market and in the bear market, we propose four dynamic models each of which is represented by a parameterized ordinary differential equation in this study. Based on existing studies, the models are in the form of either the logistic growth or the law of Newton's cooling. We solve the models by dynamic integration and apply them to the daily closing prices of the Taiwan stock index, Taiwan Stock Exchange Capitalization Weighted Stock Index. The empirical study shows that some of the models fit the prices well and the forecasting ability of the best model is acceptable even though the martingale forecasts the prices slightly better. To increase the forecasting ability and to broaden the scope of applications of the dynamic models, we will model the coefficients of the dynamic models in the future. Applying the models to the market without the price limit is also our future work.

1. Introduction

Stock indices draw a lot of attention in the financial field and modelling stock prices is one of the major topics. Nevertheless, in most existing studies, stock price returns are modelled instead of stock prices themselves. It results from the theory of random walks stated and published in many books and theses. Fama's 1965 article [1] is one of the most commonly known ones. Among massive researches of modelling stock prices, Chen et al. [2] have investigated and modelled the mean reversion of stock prices. They characterized the phenomenon by three dynamic models derived from the concept of Newton's law of cooling. However, the type of the stock movement other than mean reversion should be included in a more reasonable model. Also, the speed of the convergence to the implied equilibrium should be described if we try to increase the accuracy of the model.

From past experiences, the uptrend and downtrend of the stock price, usually during a bull market and a bear market, respectively, move differently. The uptrend usually goes up fast at first due to good news and then slows down gradually as a result of selling pressure, but the downtrend generally drops steeply. The stock price movement and the logistic growth perform very much alike; nevertheless, stock prices fluctuate more dynamically in the real world. Therefore, modelling this phenomenon is the main goal in the study.

We examined a series of the daily closing prices of Taiwan Stock Exchange Capitalization Weighted Stock Index (TAIEX) and treated it as a dynamic system. The only assumption in this study is that the stock prices are related to time. In order to depict the movement with respect to time, we attempt to modify the logistic growth model and bring in the technique of dynamic integration to construct some models. Furthermore, the models describing the mean reversion in the existing study only consist one time related coefficient [2]. We include two time-varying coefficients for constructing new models to increase the accuracy. Hence, we will introduce the nonlinear dynamic models in this study. From this point of view, modelling the stock prices may be more reliable.

In Section 2, we provide the literature review of the logistic growth model and dynamic integration. Section 3 is the methodology of how we build the nonlinear dynamic models. Section 4 consists of the empirical study of applying the dynamic models to TAIEX. In Section 5, we make some conclusions from the empirical study and some suggestions for future works.

2. Literature Review

The logistic growth model was named by the mathematician Verhulst. He solved the logistic equation and called the solution the logistic function [3–5] to express the population growth. The logistic model is continuous in time and can be stated by the differential equation:

$$\frac{dN}{dt} = rN\left(1 - \frac{N}{K}\right), \tag{1}$$

where t is time, r is the growth rate, K is the carrying capacity, and N is the population size. The logistic growth model was first used to describe the population growth in a restricted environment and then extensively applied to all kinds of fields such as Biology, Chemistry, Economics, Demography, and Statistics.

Concerning the topic of macroeconomics, Teräsvirta and Anderson [6] included logistic function to build a smooth transition autoregressive (STAR) model for describing business cycles. Based on their research, González-Rivera [7] applied STGARCH model on stock returns and exchange rates to explore the asymmetric response of conditional variances to positive and negative news. Other than applying logistic function, the nonlinear time series model such as GARCH is applied to explain the nonlinear phenomenon in the behaviors of the stock returns in the existing studies [8, 9]. Besides, there are neural network models applied to model the stock index [10–12]. Guresen et al. [10] list a table of existing studies of artificial neural networks and financial time series models. The exponential law of the stock movement is described and analyzed by Gkranas et al. [13] and Zarikas et al. [14]. However, applying logistic equation to stock prices directly is absent in existing studies.

As for applying dynamic system and differential equations to stock prices, Chiang-Lin et al. [15] applied "parabola approximation" and "dynamic integration" proposed in [16] to model the Taiwan stock index, TAIEX, and evaluate its derivative by considering the index as a dynamic system. Li et al. [17] also applied the same methods to model German DAX and tried to develop a risk detecting instrument of approaching financial catastrophe. Chen et al. [2] further combined Newton's law of cooling and dynamic integration to explain the phenomenon of mean reversion by modelling TAIEX. On the basis of existing researches, we try to transform the logistic model into some models by considering the stock price, the velocity of the stock price, and the stock return as the subject, respectively. We also solve the equations representing the models by dynamic integration. The details of the dynamic models are in the following section.

3. Methodology

In order to characterize the movement of the stock prices, we only assume that the stock price is a function of time t and consider it as a dynamic system. Since we suspect that the rise and fall of the stock prices may be described by the logistic growth model, we combine the concept of the logistic growth model and the dynamic integration to build dynamic models. The models are represented by differential equations which are detailed below along with their solutions.

Model A (dynamic logistic model)

$$\frac{dS(t)}{dt} = \alpha_1(t) \cdot S(t)^2 + \beta_1(t) \cdot S(t), \tag{2}$$

where $S(t)$ is the stock price at time t. We can rewrite (2) as

$$\frac{dS(t)}{dt} = \beta_1(t) \cdot S(t) \cdot \left(1 - \frac{S(t)}{(-\beta_1(t)/\alpha_1(t))}\right), \tag{3}$$

and hence model A is in the form of the logistic growth model.

Because the stock prices change dynamically in reality, we assume that the coefficients $\alpha_1(t)$ and $\beta_1(t)$ are constant during a very short time. Then we consider the parameterized differential equation:

$$\begin{aligned} &\frac{dS(t)}{dt} = \alpha_1 \cdot S^2(t) + \beta_1 \cdot S(t), \\ &S(t_0) = S_0, \\ &S(t_1) = S_1, \\ &\qquad\qquad t_0 < t_1, \end{aligned} \tag{4}$$

with given values of two points at time t_0 and t_1. Let $c = |(S(t_0) + \beta_1/\alpha_1)/S(t_0)|$ and the solution is

$$S(t) = \frac{(\beta_1/\alpha_1)(1/c)e^{\beta_1(t-t_0)}}{1-(1/c)e^{\beta_1(t-t_0)}}, \quad \text{if } S(t) > S(t_0),\ (S(t)+\beta_1/\alpha_1)\cdot S(t) > 0; \tag{5}$$

$$S(t) = \frac{-(\beta_1/\alpha_1)(1/c)e^{\beta_1(t-t_0)}}{1+(1/c)e^{\beta_1(t-t_0)}}, \quad \text{if } S(t) > S(t_0),\ (S(t)+\beta_1/\alpha_1)\cdot S(t) < 0; \tag{6}$$

$$S(t) = \frac{(\beta_1/\alpha_1)(1/c)e^{-|\beta_1|(t-t_0)}}{1-(1/c)e^{-|\beta_1|(t-t_0)}}, \quad \text{if } S(t) < S(t_0),\ (S(t)+\beta_1/\alpha_1)\cdot S(t) > 0; \tag{7}$$

$$S(t) = \frac{-(\beta_1/\alpha_1)(1/c)e^{-|\beta_1|(t-t_0)}}{1+(1/c)e^{-|\beta_1|(t-t_0)}}, \quad \text{if } S(t) < S(t_0),\ (S(t)+\beta_1/\alpha_1)\cdot S(t) < 0. \tag{8}$$

Model A describes the relationship between the stock price, $S(t)$, and its velocity, $dS(t)/dt$. From the economic point of view, there exists an implied equilibrium in the market. All other things being equal, if the stock price is below the equilibrium, it will rise towards the equilibrium price; if the stock price is above the equilibrium, it will decline. The phenomenon is also called the mean reversion. This model characterizes the mean reversion of the stock prices if α_1 is negative, β_1 is positive and when the expected future stock price is higher than the current stock price. Under the circumstances, the implied equilibrium is at $|\beta_1/\alpha_1|$ and the stock price moves as how the logistic growth model describes. The convergence of the simulated theoretical stock prices by (5), (6) is plotted in Figures 7 and 8. Otherwise, the movement of the stock prices follows a quadratic differential equation with no constant growth rate [15]. In this situation, the stock prices will not converge to the implied equilibrium.

In addition to model A, we further consider both the velocity, $dS(t)/dt$, and the acceleration, $d^2S(t)/dt^2$, of the stock price to construct a model. In this instance, the model may represent the movement more properly. Hence, we propose model B in the following.

Model B (dynamic transformed logistic model)

$$\frac{d^2S(t)}{dt^2} = \alpha_2(t)\cdot\left(\frac{dS(t)}{dt}\right)^2 + \beta_2(t)\cdot\left(\frac{dS(t)}{dt}\right), \quad (9)$$

where $S(t)$ is the stock price at time t.

Assuming the coefficients $\alpha_2(t)$ and $\beta_2(t)$ are constant during a very short time, we consider the parameterized differential equation with given values of two points at times t_0 and t_1:

$$\begin{aligned}
&\frac{d^2S(t)}{dt^2} = \alpha_2\cdot\left(\frac{dS(t)}{dt}\right)^2 + \beta_2\cdot\frac{dS(t)}{dt},\\
&S(t_0) = S_0,\\
&S(t_1) = S_1,\\
&t_0 < t_1,
\end{aligned} \quad (10)$$

and it can be solved as follows. Let $v(t) = dS(t)/dt$ and $h = |(v(t_0) + \beta_2/\alpha_2)/v(t_0)|$; then the solution is

$$\begin{aligned}
&S(t) = S(t_0) - \frac{1}{\alpha_2}\ln\left|\frac{e^{\beta_2(t-t_0)} - h}{1-h}\right|,\\
&\text{if } \left(v(t) + \frac{\beta_2}{\alpha_2}\right)\cdot v(t) > 0,\ v(t) > v(t_0),\ S(t) > S(t_0);\\
&S(t) = S(t_0) + \left|\frac{1}{\alpha_2}\right|\frac{|\beta_2|}{\beta_2}\ln\left|\frac{e^{\beta_2(t-t_0)} - h}{1-h}\right|,\\
&\text{if } \left(v(t) + \frac{\beta_2}{\alpha_2}\right)\cdot v(t) > 0,\ v(t) > v(t_0),\ S(t) < S(t_0);\\
&S(t) = S(t_0) - \frac{1}{\alpha_2}\ln\left|\frac{e^{\beta_2(t-t_0)} + h}{1+h}\right|,\\
&\text{if } \left(v(t) + \frac{\beta_2}{\alpha_2}\right)\cdot v(t) < 0,\ v(t) > v(t_0),\ S(t) > S(t_0);\\
&S(t) = S(t_0) - \left|\frac{1}{\alpha_2}\right|\cdot\frac{|\beta_2|}{\beta_2}\ln\left|\frac{e^{\beta_2(t-t_0)} + h}{1+h}\right|,\\
&\text{if } \left(v(t) + \frac{\beta_2}{\alpha_2}\right)\cdot v(t) < 0,\ v(t) > v(t_0),\ S(t) < S(t_0);\\
&S(t) = S(t_0) + \frac{1}{\alpha_2}\frac{\beta_2}{|\beta_2|}\ln\left|\frac{e^{-|\beta_2|(t-t_0)} - h}{1-h}\right|,\\
&\text{if } \left(v(t) + \frac{\beta_2}{\alpha_2}\right)\cdot v(t) > 0,\ v(t) < v(t_0),\ S(t) > S(t_0);\\
&S(t) = S(t_0) - \left|\frac{1}{\alpha_2}\right|\ln\left|\frac{e^{-|\beta_2|(t-t_0)} - h}{1-h}\right|,\\
&\text{if } \left(v(t) + \frac{\beta_2}{\alpha_2}\right)\cdot v(t) > 0,\ v(t) < v(t_0),\ S(t) < S(t_0);\\
&S(t) = S(t_0) + \frac{1}{\alpha_2}\frac{\beta_2}{|\beta_2|}\ln\left|\frac{e^{-|\beta_2|(t-t_0)} + h}{1+h}\right|,\\
&\text{if } \left(v(t) + \frac{\beta_2}{\alpha_2}\right)\cdot v(t) < 0,\ v(t) < v(t_0),\ S(t) > S(t_0);\\
&S(t) = S(t_0) + \left|\frac{1}{\alpha_2}\right|\ln\left|\frac{e^{-|\beta_2|(t-t_0)} + h}{1+h}\right|,\\
&\text{if } \left(v(t) + \frac{\beta_2}{\alpha_2}\right)\cdot v(t) < 0,\ v(t) < v(t_0),\ S(t) < S(t_0).
\end{aligned} \quad (11)$$

Model B describes the relationship between the velocity, $dS(t)/dt$, and the acceleration, $d^2S(t)/dt^2$, of the stock price. The relationship is in the form of the logistic growth model. When α_2 is negative and β_2 is positive, the velocity moves in the way that the logistic growth model depicts. That is, the velocity reverses to an implied equilibrium velocity so that the stock price goes up or down with an approximately constant speed. In the case of other combinations of the coefficients (α_2 and β_2), the velocity diverges and hence the stock prices fluctuate more dynamically.

Other than the velocity, the relative growth rate of the stock price, $(dS(t)/dt)/S(t)$, which is also known as the return is a meaningful measure in the field of finance. Therefore, we change the velocity and the acceleration in model B into the relative growth rate and its derivative to build model C. Model C provides another kind of differential equation characterizing the stock price movement.

Model C (dynamic relative growth rate transformed logistic model)

$$\frac{d\delta(t)}{dt} = \alpha_3(t)\cdot\delta^2(t) + \beta_3(t)\cdot\delta(t), \quad (12)$$

where $S(t)$ is the stock price at time t and $\delta(t) = (dS(t)/dt)/S(t)$ is the relative growth rate of the stock price at time t.

Assuming the coefficients $\alpha_3(t)$ and $\beta_3(t)$ are constant during a very short time, we consider the parameterized differential equation with given values of two points at times t_0 and t_1:

$$\frac{d\delta(t)}{dt} = \frac{d}{dt}\left(\frac{dS(t)/dt}{S(t)}\right) = \alpha_3 \cdot \delta^2(t) + \beta_3 \cdot \delta(t),$$
$$S(t_0) = S_0,$$
$$S(t_1) = S_1, \tag{13}$$
$$t_0 < t_1,$$

and it can be solved as follows. Let $k = |(\delta(t_0) + \beta_3/\alpha_3)/\delta(t_0)|$ and the solution is

$$S(t) = S(t_0) \cdot \exp\left(-\frac{1}{\alpha_3}\ln\left|\frac{e^{\beta_3(t-t_0)} - k}{1-k}\right|\right),$$
$$\text{if } \left(\delta(t) + \frac{\beta_3}{\alpha_3}\right) \cdot \delta(t) > 0,\ \delta(t) > \delta(t_0),\ S(t) > S(t_0);$$
$$S(t) = S(t_0) \cdot \exp\left(\left|\frac{1}{\alpha_3}\right| \frac{|\beta_3|}{\beta_3}\ln\left|\frac{e^{\beta_3(t-t_0)} - k}{1-k}\right|\right),$$
$$\text{if } \left(\delta(t) + \frac{\beta_3}{\alpha_3}\right) \cdot \delta(t) > 0,\ \delta(t) > \delta(t_0),\ S(t) < S(t_0);$$
$$S(t) = S(t_0) \cdot \exp\left(-\frac{1}{\alpha_3}\ln\left|\frac{e^{\beta_3(t-t_0)} + k}{1+k}\right|\right),$$
$$\text{if } \left(\delta(t) + \frac{\beta_3}{\alpha_3}\right) \cdot \delta(t) < 0,\ \delta(t) > \delta(t_0),\ S(t) > S(t_0);$$
$$S(t) = S(t_0) \cdot \exp\left(-\left|\frac{1}{\alpha_3}\right| \cdot \frac{|\beta_3|}{\beta_3}\ln\left|\frac{e^{\beta_3(t-t_0)} + k}{1+k}\right|\right),$$
$$\text{if } \left(\delta(t) + \frac{\beta_3}{\alpha_3}\right) \cdot \delta(t) < 0,\ \delta(t) > \delta(t_0),\ S(t) < S(t_0); \tag{14}$$
$$S(t) = S(t_0) \cdot \exp\left(\frac{1}{\alpha_3}\frac{\beta_3}{|\beta_3|}\ln\left|\frac{e^{-|\beta_3|(t-t_0)} - k}{1-k}\right|\right),$$
$$\text{if } \left(\delta(t) + \frac{\beta_3}{\alpha_3}\right) \cdot \delta(t) > 0,\ \delta(t) < \delta(t_0),\ S(t) > S(t_0);$$
$$S(t) = S(t_0) \cdot \exp\left(-\left|\frac{1}{\alpha_3}\right|\ln\left|\frac{e^{-|\beta_3|(t-t_0)} - k}{1-k}\right|\right),$$
$$\text{if } \left(\delta(t) + \frac{\beta_3}{\alpha_3}\right) \cdot \delta(t) > 0,\ \delta(t) < \delta(t_0),\ S(t) < S(t_0);$$
$$S(t) = S(t_0) \cdot \exp\left(\frac{1}{\alpha_3}\frac{\beta_3}{|\beta_3|}\ln\left|\frac{e^{-|\beta_3|(t-t_0)} + k}{1+k}\right|\right),$$
$$\text{if } \left(\delta(t) + \frac{\beta_3}{\alpha_3}\right) \cdot \delta(t) < 0,\ \delta(t) < \delta(t_0),\ S(t) > S(t_0);$$
$$S(t) = S(t_0) \cdot \exp\left(\left|\frac{1}{\alpha_3}\right|\ln\left|\frac{e^{-|\beta_3|(t-t_0)} + k}{1+k}\right|\right),$$
$$\text{if } \left(\delta(t) + \frac{\beta_3}{\alpha_3}\right) \cdot \delta(t) < 0,\ \delta(t) < \delta(t_0),\ S(t) < S(t_0).$$

TABLE 1: Classification of model forecasting ability by MAPE.

MAPE	<10%	10%~20%	20%~50%	>50%
Accuracy	High	Good	Reasonable	Inaccurate

Model C describes the relationship between the relative growth rate of the stock price, $\delta(t)$, and its derivative. Since the relative growth rate is the return of the stock price, when α_3 is negative and β_3 is positive, the return moves in the way that the logistic growth model depicts. That is, the return reverses to a constant implied equilibrium. In the case of other combinations of the coefficients (α_3 and β_3), the return diverges and the fluctuations of the stock prices are more dynamical but different from what model B describes.

The dynamic models above are built on the work of Chen et al. [2]. But all of them consist of two coefficients instead of one as in their study. Hence we generalize one of their models into a two-coefficient dynamic model and compare it with the three models above. This extended model is stated in the following.

Model D (dynamic general Newton model)

$$\frac{dS(t)}{dt} = \alpha_4(t) \cdot [S(t) - A(t)], \tag{15}$$

where $S(t)$ is the stock price at time t. We assume the coefficients in (15) are constant during a very short time interval $t \in [t_0, t_1]$ and consider the parameterized differential equation:

$$\frac{dS(t)}{dt} = \alpha_4 \cdot [S(t) - A],$$
$$S(t_0) = S_0,$$
$$S(t_1) = S_1, \tag{16}$$
$$t_0 < t_1,$$

between two given data points at times t_0 and t_1. The solution can be found in the article of Chen et al. [2].

As applying these dynamic models, we discretize and convert them into difference equations since the time interval between the two given points at times t_0 and t_1 is relatively short compared to the interval of the complete data. The discretization and the parameterization make the differential equations solvable. In the empirical study, we fit and forecast the data by the solutions. We also adopt the dynamic forecasting method discussed in [2] to obtain more accurate theoretical values.

The forecasting error is measured by MAPE (Mean Absolute Percentage Error) and RMSPE (Root Mean Square Percentage Error). They are defined as MAPE $= (\mathbf{1}/N)\sum_{i=1}^{N}|(\widehat{S}(i) - S(i))/S(i)| \times 100\%$ and RMSPE $= \sqrt{(\mathbf{1}/N)\sum_{i=1}^{N}((\widehat{S}(i) - S(i))/S(i))^2} \times 100\%$, where $\widehat{S}(i)$ is the theoretical value of the model at time i, $S(i)$ is the market value at time i, and N is the length of data points. Lewis [18] suggested that the forecasting ability of a model can be classified as in Table 1.

TABLE 2: MAPEs and RMSPEs, fits of the four dynamic models and the martingale.

	MAPE	RMSPE
Model A	2.0362%	3.5065%
Model D	0.3467%	0.5943%
Martingale (2015/06/02~2016/07/19)	0.7608%	1.0404%
Model B	0.6634%	1.0072%
Model C	0.6622%	1.0080%
Martingale (2015/06/02~2016/07/18)	0.7625%	1.0421%

RMSPE can also determine the forecasting ability of a model by the classification of Table 1 although RMSPE is more affected by extreme values. In brief, the smaller the values of MAPE and RMSPE are, the better the forecasting ability of a model is. Besides, MAPE and RMSPE are also calculated to measure the fitting accuracy of the models in this study.

4. Empirical Study

We collected and studied a series of Taiwan Stock Exchange Capitalization Weighted Stock Index (TAIEX) closing prices from June 1, 2015, to August 31, 2016. There are 310 trading days during this period. Since the dynamic forecasting method is applied, the length of forecasts needs not to be long [2] so that we make 30 forecasts for each model. But the forecasting periods of the four models are different because first-order difference is revolved in calculations of model A, D and second-order difference is revolved in calculations of model B, C.

We apply forward difference method to numerical analysis since the results do not differ much from applying backward difference method [16]. For fitting with the dynamic models, the future information is included in calculations. To avoid the inclusion of future information for model forecasting, the theoretical values are obtained assuming that the situation is unchanged during a longer period from three trading days before the date to be forecasted. That is, the model coefficients are assumed to be constant in the interval $t \in [t_0, t_3]$ and the theoretical values at time t_3 are obtained by solving (4), (10), (13), and (16), respectively. Therefore, the fitting periods of model A, D begin from June 2, 2015, and end at July 19, 2016, while the fitting periods of model B, C begin from June 2, 2015, and end at July 18, 2016. Forecasts of model A, D are both from July 20, 2016, to August 30, 2016; forecasts of model B, C are both from July 19, 2016, to August 29, 2016.

In numerical analysis, the theoretical values of model D sometimes blow up when the coefficient α_4 are very large because of its position in the power of exponential part of the solution. The theoretical values make sense if they are restricted in a range where market values exist. Hence, when α_4 is higher than 2 (resulting in the theoretical value out of range of possible market values on the condition) the theoretical value will be replaced with the theoretical value of the previous trading day. Moreover, a 10% daily price fluctuation limit in Taiwan stock market is set up since June 1, 2015. Accordingly, the theoretical values of the four models are restricted to the limit if they are above or below it.

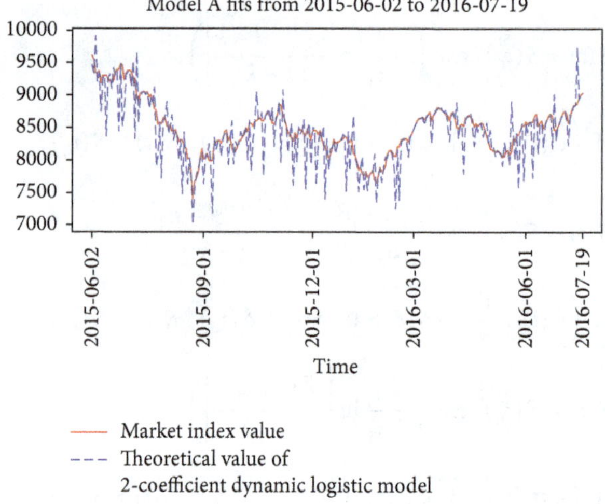

FIGURE 1: Time plot of raw data and model A fits.

The TAIEX series is also analyzed by examining its sample ACF (autocorrelation function) and PACF (partial autocorrelation function) and testing by Ljung-Box Test [19] shown in Figure 9 and Table 4. The autocorrelation of the series is not statistically significant. Under the circumstances, we suppose the stock closing prices during the period is a random walk and consequently the martingale is set as the benchmark model. Once the theoretical values of the four dynamic models are obtained, we compare them with the fits and forecasts of the martingale.

We will first present the fits of the four dynamic models and next their forecasts in the section. The fitting results are in Figures 1–4.

The graphs show that the fits of model A fluctuate much more than the other models and model D has the fewest deviations of the fits from the market values. The MAPEs and RMSPEs of the models are listed in Table 2. The order of the model performance by the values coincides with the graphs. Furthermore, the fits of the martingale are only closer to the market values than those of model A.

We then examine the model coefficients. The graphs and some statistics of the coefficients are in Table 5 and Figures 10–13. Basically the alphas and the betas move in the small

FIGURE 2: Time plot of raw data and model B fits.

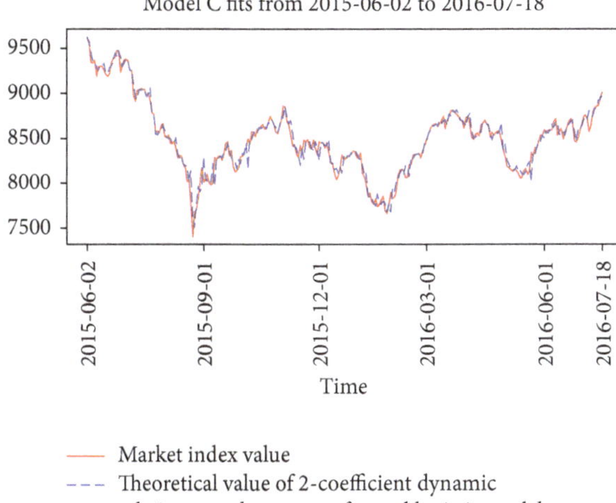

FIGURE 3: Time plot of raw data and model C fits.

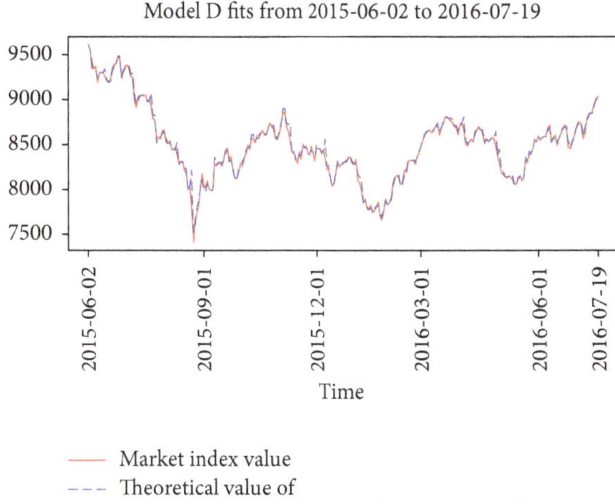

FIGURE 4: Time plot of raw data and model D fits.

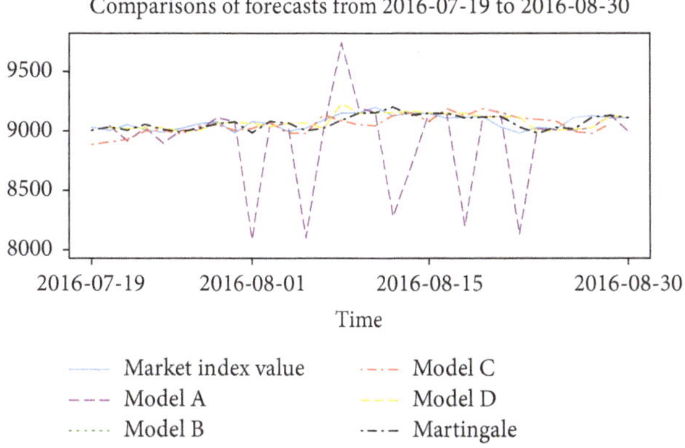

FIGURE 5: Time plot of raw data and the forecasts of the four dynamic models and the martingale.

range around their median except for some large jumps or drops. The unusual jumps and drops seem to appear on the same dates and should be studied more thoroughly.

As noted in the previous section, the phenomenon of mean reversion shows if $\alpha_1 < 0$ and $\beta_1 > 0$ for model A, $\alpha_2 < 0$ and $\beta_2 > 0$ for model B, $\alpha_3 < 0$ and $\beta_3 > 0$ for model C, and $\alpha_4 < 0$ for model D. On the condition, if the subject value is higher than the implied equilibrium, it will reverse to the equilibrium with a sharp decreasing speed; if the subject value is lower than the implied equilibrium, it will first reverse to the equilibrium with a steep rise and then gradually slow down. The phenomenon is for different subjects related to stock index: the stock index closing price for model A, D; the velocity of stock price for model B; the relative growth rate (return) of stock price for model C. Therefore, the mean reversion described in models A and D is either the bull market or the bear market for the stock prices. In models B and C, the bull market (upward trend) or the bear market (downward trend) is used to describe the velocity and the related growth rate of the stock prices, respectively. The proportion of mean reversion shown during the studied period of the data is 0.72 (199/278) for model A, D and 0.23 (64/277) for model B, C. Other than the mean reversion phenomenon, the dynamic models also characterize different types of market trends as well by different performances of the model coefficients.

Next we forecast by the four dynamic models and compare the results with the martingale in Figures 5 and 6. The details of each model are shown in Figures 14–19. The forecasts of model A fluctuate the most and have the most deviations from the market values. Models B, C, and D all forecast the trend of the stock prices well and model D seems to have the most accurate forecasts. We list the errors of the four dynamic models and the martingale in Table 3. The forecasting ability of the martingale is the best among all models; however, model D has similar results.

The sources of error include the disparity between the differential equations and the difference equations, the disagreement between the movement of the subject and the model form, and the dynamics of the model coefficients. If the model form does not represent the movement of the subject very accurately, the model will produce larger error and thus

Table 3: MAPEs and RMSPEs, forecasts of the four dynamic models and the martingale.

	MAPE	RMSPE
Model A	2.5055%	4.4041%
Model D	0.4658%	0.5973%
Martingale (2016/07/20~2016/08/30)	0.4306%	0.5377%
Model B	0.7082%	0.8786%
Model C	0.7101%	0.8799%
Martingale (2016/07/19~2016/08/29)	0.4403%	0.5404%

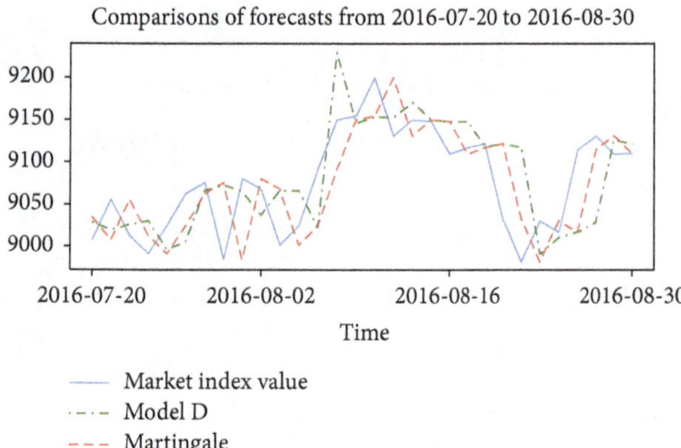

Figure 6: Time plot of raw data and the forecasts of model D and the martingale.

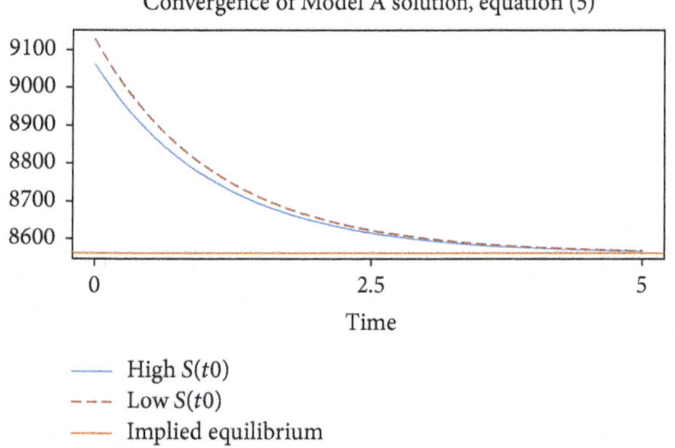

Figure 7: The convergence of the simulated theoretical stock prices by equation (5). (Implied equilibrium = $|\text{median}(\beta_1)/\text{median}(\alpha_1)|$ = 8563.101 high stock index initial value $S(t_0)$ = 8563.101 + 500 = 9063.101; low stock index initial value $S(t_0)$ = 8563.101 − 500 = 8063.101).

the forecasting ability of the model will be lower. Model A and model D both characterize the relationship between the subject and its first-order derivative. Nevertheless, model D forecasts much better than model A as a result of the concordance between model D and the data. Besides, error comes from the dynamics of the model coefficients. We assume the model coefficients are constant in a four-day interval prior to the forecast during the forecasting process. In real world,

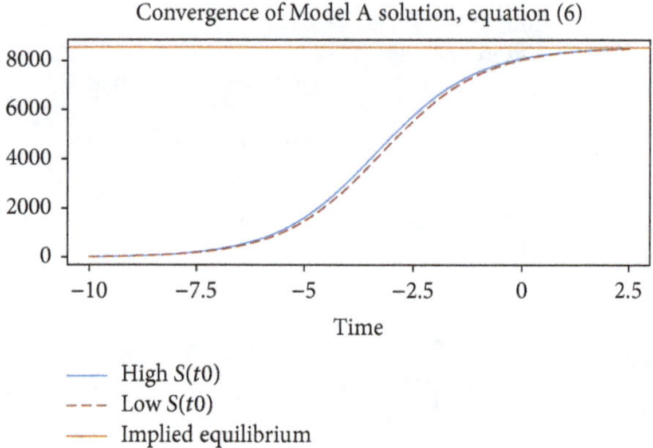

Figure 8: The convergence of the simulated theoretical stock prices by equation (6) (implied equilibrium = 8563.101; high stock index initial value $S(t_0)$ = 9063.101; low stock index initial value $S(t_0)$ = 8063.101).

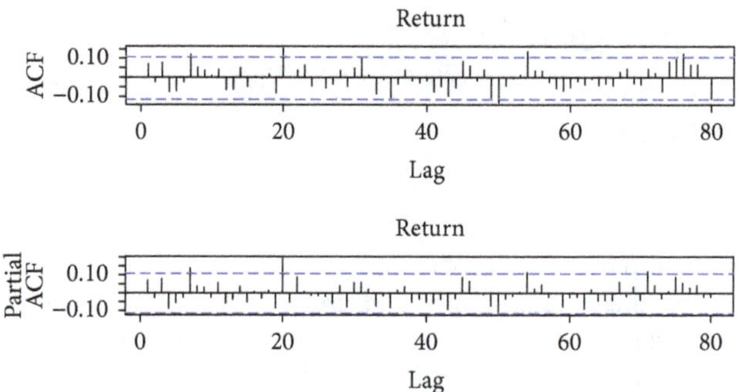

Figure 9: ACF and PACF of TAIEX daily returns from June 2, 2015, to August 31, 2016.

the coefficients are more dynamical and consequently error shows.

5. Conclusions

We propose four dynamic models to characterize the movement of the subject, the TAIEX closing prices. Some types of the movement are considered in the models: the mean reversion, different speeds of the uptrend and the downtrend, and so forth. There are two forms of the models: the logistic growth and the law of Newton's cooling. We apply the

TABLE 4: p values of Ljung-Box tests for TAIEX daily returns from June 2, 2015, to August 31, 2016.

lag = 1	lag = 2	lag = 3	lag = 4	lag = 5	lag = 6	lag = 7	lag = 8	lag = 9	lag = 10	lag = 11	lag = 12	lag = 13	lag = 14	lag = 15	lag = 16	lag = 17	lag = 18	lag = 19	lag = 20
0.2254	0.4485	0.3458	0.2858	0.2568	0.3482	0.1224	0.1421	0.1831	0.2470	0.2842	0.2744	0.2669	0.2865	0.3065	0.3719	0.4398	0.5037	0.4334	0.1304

Table 5: Centering tendency of the dynamic model coefficients.

	Mean	10% Trimmed Mean	Median
α_1	−0.000141	−0.000098	−0.000101
β_1	1.169824	0.832151	0.867041
α_2	0.031618	−0.004302	−0.000321
β_2	−1.144368	−0.044949	−0.911920
α_3	206.432875	−37.809327	−2.199180
β_3	−1.052092	−0.026376	−0.909614
α_4	−1.20276	−0.866740	−0.859210
A	8520.318324	8490.764107	8477.203518

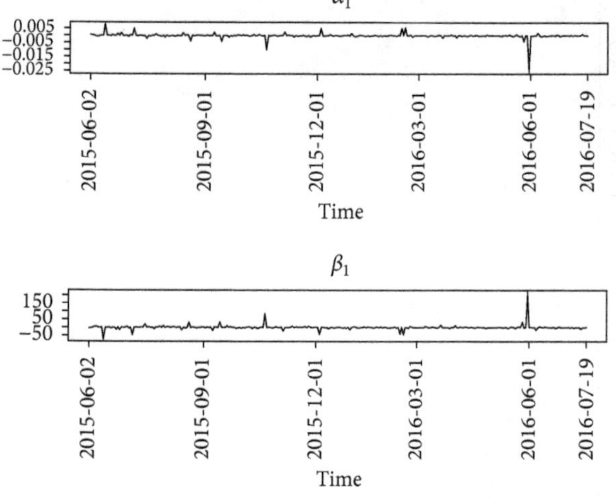

Figure 10: Time plots of model A coefficients-α_1 and β_1

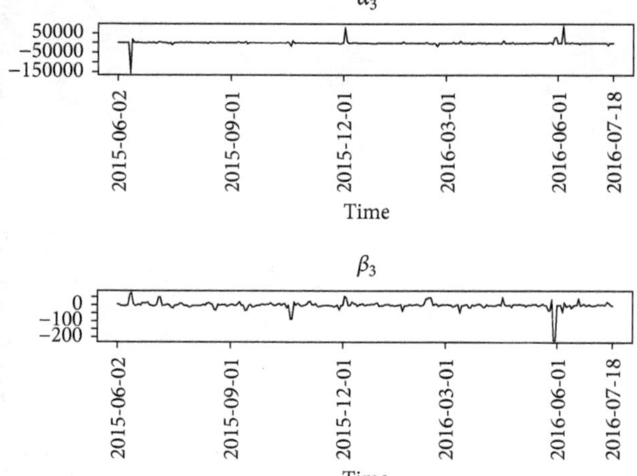

Figure 12: Time plots of model C coefficients-α_3 and β_3.

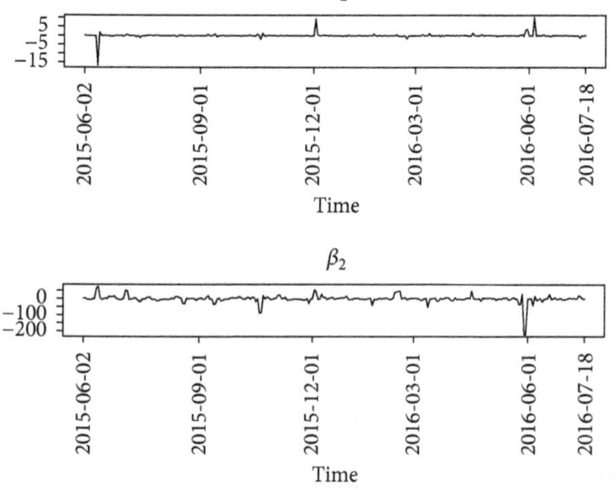

Figure 11: Time plots of model B coefficients-α_2 and β_2.

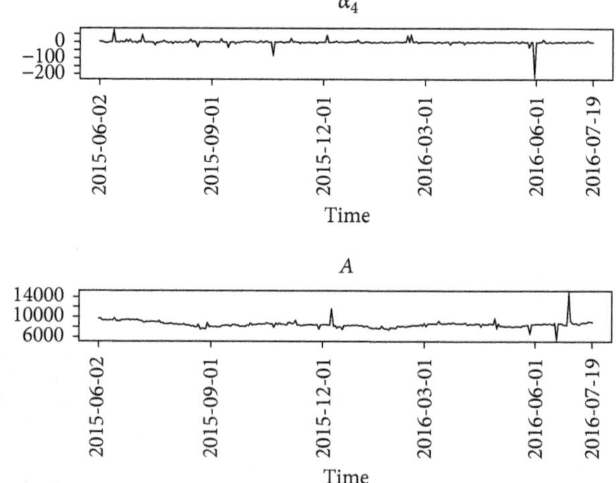

Figure 13: Time plots of model D coefficients-α_4 and A.

four dynamic models to fit and forecast the TAIEX closing prices and compare them with the martingale. The fits of models B, C, and D are closer to the market values than those of the martingale which suggests that they characterize the movement of the subject quite properly. However, the martingale outperforms the dynamic models for forecasting since the error of the martingale is the smallest. But the forecasts of model D are very similar to the martingale and has the best performance among the four dynamic models. To sum up, model D is the most proper model to characterize the TAIEX closing prices but we still need to examine and model the coefficients to increase its forecasting ability. Modelling

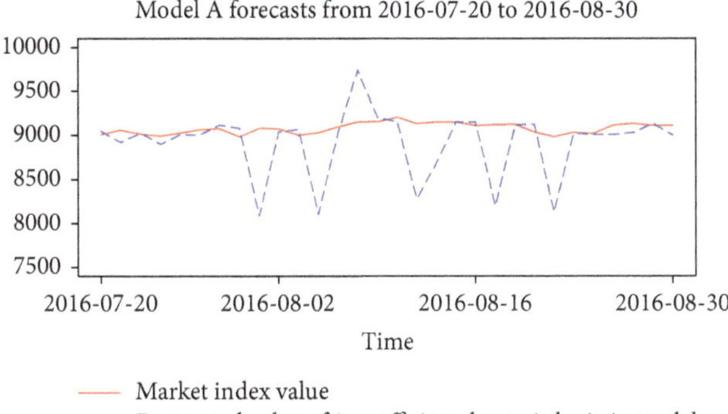

FIGURE 14: Time plots of raw data and the forecasts of model A.

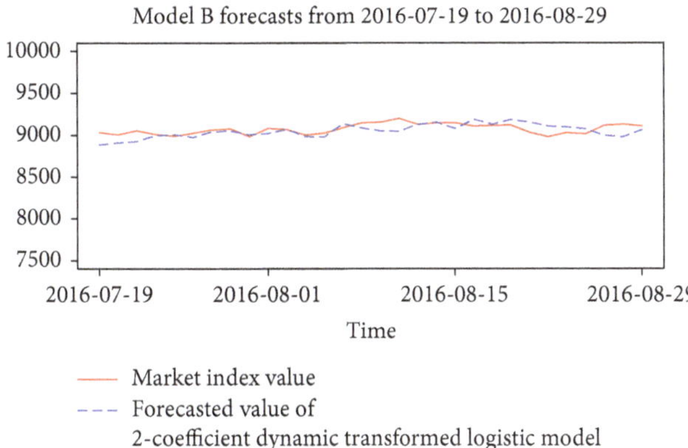

FIGURE 15: Time plots of raw data and the forecasts of model B.

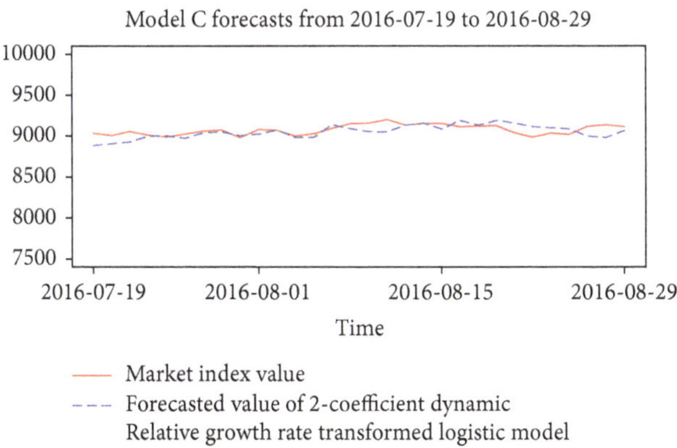

FIGURE 16: Time plots of raw data and the forecasts of model C.

the coefficients by some different methods such as time series analysis and polynomial extrapolation is our future work.

Furthermore, there are some limits to the dynamic model applications. For example, the theoretical values are limited in a range of 10% price limit because of the regulation constraint. Therefore, applying the dynamic models to the subject in the market without the price limit may not be suitable. However, the empirical study shows that the model in the form of the law of Newton's cooling (model D) is better than the models in the form of the logistic growth (models A, B, and C). This suggests that the former type of the model may be more proper for the market with price limit while the later type of the model may be more proper for the market without price limit. Confirming the conjecture and modifying the models to generalize their applications are also our future work.

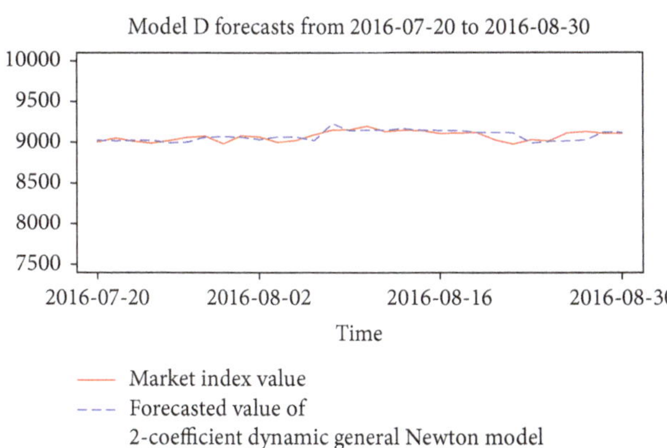

FIGURE 17: Time plots of raw data and the forecasts of model D.

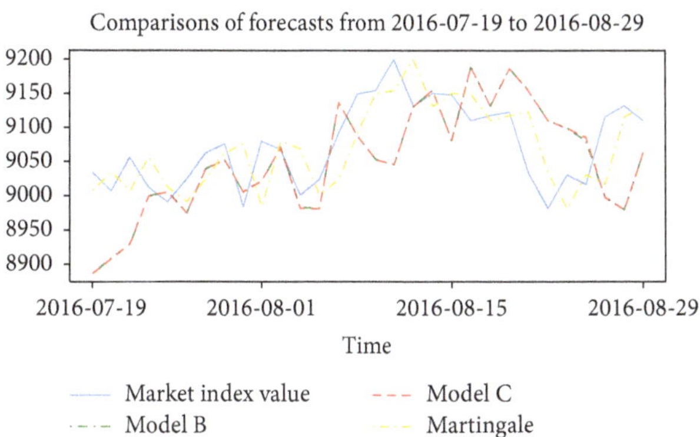

FIGURE 18: Time plots of raw data and the forecasts of model B, C and the martingale.

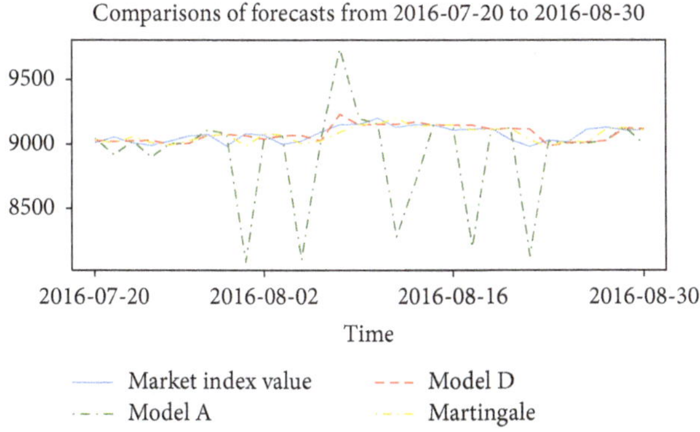

FIGURE 19: Time plots of raw data and the forecasts of model A, D and the martingale.

Conflicts of Interest

The authors declare that there are no conflicts of interest regarding the publication of this paper.

References

[1] E. F. Fama, "The behavior of stock-market prices," *The Journal of Business*, vol. 38, no. 1, pp. 34–105, 1965.

[2] N.-P. Chen, M.-R. Li, T.-J. Chiang-Lin, Y.-S. Lee, and D. W. Miao, "Applications of linear ordinary differential equations and dynamic system to economics—An example of Taiwan stock index TAIEX," *International Journal of Dynamical Systems and Differential Equations*, vol. 7, no. 2, pp. 95–111, 2017.

[3] P. F. Verhulst, "Notice sur la loi que la population poursuit dans son accroissement," *Correspondance Mathematique et Physique*, vol. 10, pp. 113–121, 1838.

[4] P. F. Verhulst, "Recherches Mathematiques sur la Loi D'accroissement de la Population (mathematical researches into the law of population growth increase)," *Nouveaux Memoires de l'Academie Royale des Sciences et Belles Lettres de Bruxelles*, vol. 18, pp. 1–42, 1845.

[5] P. F. Verhulst, "Deuxieme memoire sur la loi daccroissement de la population," *Memoires de l'Academie Royale des Sciences, des Lettres et des Beaux Arts de Belgique*, vol. 20, pp. 1–32, 1847.

[6] T. Teräsvirta and H. M. Anderson, "Characterizing nonlinearities in business cycles using smooth transition autoregressive models," *Journal of Applied Econometrics*, vol. 7, pp. S119–S136, 1992.

[7] G. González-Rivera, "Smooth-transition GARCH models," *Studies in Nonlinear Dynamics & Econometrics*, vol. 3, no. 2, pp. 61–78, 1998.

[8] A. Abhyankar, L. S. Copeland, and W. Wong, "Nonlinear dynamics in real-time equity market indices: evidence from the United Kingdom," *The Economic Journal*, vol. 105, no. 431, pp. 864–880, 1995.

[9] A. Abhyankar, L. S. Copeland, and W. Wong, "Uncovering nonlinear structure in real-time stock-market indexes: The S & P 500, the DAX, the Nikkei 225, and the FTSE-100," *Journal of Business and Economic Statistics*, vol. 15, no. 1, pp. 1–14, 1997.

[10] E. Guresen, G. Kayakutlu, and T. U. Daim, "Using artificial neural network models in stock market index prediction," *Expert Systems with Applications*, vol. 38, no. 8, pp. 10389–10397, 2011.

[11] Z. Liao and J. Wang, "Forecasting model of global stock index by stochastic time effective neural network," *Expert Systems with Applications*, vol. 37, no. 1, pp. 834–841, 2010.

[12] J. Wang, H. Pan, and F. Liu, "Forecasting crude oil price and stock price by jump stochastic time effective neural network model," *Journal of Applied Mathematics*, vol. 2012, Article ID 646475, 15 pages, 2012.

[13] A. Gkranas, V. L. Rendoumis, and H. M. Polatoglou, "Athens and Lisbon stock markets - A thermodynamic approach," *WSEAS Transactions on Business and Economics*, vol. 1, no. 1, pp. 95–100, 2004.

[14] V. Zarikas, A. G. Christopoulos, and V. L. Rendoumis, "A thermodynamic description of the time evolution of a stock market index," *European Journal of Economics, Finance and Administrative Sciences*, vol. 16, pp. 73–83, 2009.

[15] T. J. Chiang-Lin, M.-R. Li, and Y. S. Lee, "Taiex index option model by using nonlinear differential equation," *Mathematical & Computational Applications*, vol. 19, no. 1, pp. 78–92, 2014.

[16] M.-R. Li, T.-H. Shieh, C. J. Yue, P. Lee, and Y.-T. Li, "Parabola method in ordinary differential equation," *Taiwanese Journal of Mathematics*, vol. 15, no. 4, pp. 1841–1857, 2011.

[17] M. R. Li, D. W. Miao, T. J. Chiang-Lin, and Y. S. Lee, "Modelling DAX by applying parabola approximation method," *International Journal of Computing Science and Mathematics*.

[18] C. D. Lewis, *Industrial and business forecasting methods: A radical guide to exponential smoothing and curve fitting*, Butterworth Scientific, London, UK, 1982.

[19] G. M. Ljung and G. E. P. Box, "On a measure of lack of fit in time series models," *Biometrika*, vol. 65, no. 2, pp. 553–564, 1978.

13

Linearization of Fifth-Order Ordinary Differential Equations by Generalized Sundman Transformations

Supaporn Suksern[1,2] **and Kwanpaka Naboonmee**[1]

[1]*Department of Mathematics, Faculty of Science, Naresuan University, Phitsanulok 65000, Thailand*
[2]*Research Center for Academic Excellence in Mathematics, Naresuan University, Phitsanulok 65000, Thailand*

Correspondence should be addressed to Supaporn Suksern; supapornsu@nu.ac.th

Academic Editor: Patricia J. Y. Wong

In this article, the linearization problem of fifth-order ordinary differential equation is presented by using the generalized Sundman transformation. The necessary and sufficient conditions which allow the nonlinear fifth-order ordinary differential equation to be transformed to the simplest linear equation are found. There is only one case in the part of sufficient conditions which is surprisingly less than the number of cases in the same part for order 2, 3, and 4. Moreover, the derivations of the explicit forms for the linearizing transformation are exhibited. Examples for the main results are included.

1. Introduction

Nonlinear problems are of interest to engineers, physicists, mathematicians and many other scientists since most equations are inherently nonlinear in nature. Although linear ordinary differential equations can be solved by a large number of methods but this situation does not hold for nonlinear equations. One common method to solve nonlinear ordinary differential equations is to change their unknowns by suitable variables so as to get linear ordinary differential equations.

The main tools used to solve the linearization problem are transformations such as point, contact, tangent, and generalized Sundman transformations.

It was recognized that Lie [1] is the first person who solved linearization problem for ordinary differential equations in 1883. He discovered the linearization of second-order ordinary differential equations by point transformations. Later, Liouville [2] and Tresse [3] attacked the equivalence problems for second-order ordinary differential equations via group of point transformations. Moreover, Cartan [4] approached the second-order ordinary differential equations by geometric structure of a certain form.

Mahomed and Leach [5] indicated that the nth-order ($n > 3$) linear ordinary differential equation has exactly one of $n+1, n+2$, or $n+4$ point symmetries. They suggested that the necessary and sufficient conditions for the nth-order ($n \geq 3$) to be linearizable by a point transformation must admit the n dimensional Abelian algebra.

The linearization problem of third-order ordinary differential equations under point transformations was solved by Bocharov et al. [6], Grebot [7], and Ibragimov and Meleshko [8]. Fourth-order ordinary differential equation was studied by Ibragimov et al. [9]. They found the necessary and sufficient conditions for a complete linearization problem. The linearization problem of a fifth-order ordinary differential equation with respect to fiber preserving transformations was considered by Suksern and Pinyo [10].

In the series of articles [8, 11–14] the linearization problem of a third-order ordinary differential equation via the contact transformations was solved. For a fourth-order ordinary differential equation, this problem was studied in [15, 16]. The criteria of the linearization problem of fifth-order ordinary differential equations were discovered by Suksern [17].

The linearization problems of third-order and fourth-order ordinary differential equations by the tangent transformations are examined in [18, 19]. These are the first application of tangent (essentially) transformations to the linearization problems of third-order and fourth-order ordinary differential equations. Necessary and sufficient conditions for third-order and fourth-order ordinary differential equations to be linearizable are obtained there.

Sundman introduced the generalized Sundman transformations in 1992. Later on Duarte et al. [20] applied this method to transform second-order ordinary differential equations into free particle equations. In addition, Muriel and Romero [21] characterized the equations that can be linearized by means of generalized Sundman transformations in terms of first integral. A new characterization of linearizable equations in terms of the coefficients of ordinary differential equation and one auxiliary function was given by Mustafa et al. [22]. Moreover, Nakpim and Meleshko [23] pointed out that the solution given by Duarte et al. using the Laguerre form is not complete.

For the third-order ordinary differential equations, the linearization by the generalized Sundman transformation was investigated by [24] for the form $X'''(T) = 0$ and [25] for the Laguerre form. Some applications of the generalized Sundman transformation to ordinary differential equations can be found in [26]. More information of the generalized Sundman transformation are collected in the book [27].

The linearization problem of a fourth-order ordinary differential equation with respect to generalized Sundman transformations was studied in [28]. They found the necessary and sufficient conditions which allow the fourth-order ordinary differential equation to be transformed to the simplest linear equation.

In this article, we intend to use the generalized Sundman transformations to linearize the fifth-order ordinary differential equations in some particular cases. We use computer algebra system Reduce to compute the necessary and sufficient conditions of the linearization. We provide some examples to illustrate the conditions that we have found and also obtain the linearizing transformations.

2. Necessary Conditions

We now concentrate on finding the fifth-order ordinary differential equations

$$x^{(5)} = f\left(t, x, x', x'', x''', x^{(4)}\right), \tag{1}$$

which can be transformed to the linear equation

$$X^{(5)}(T) = 0, \tag{2}$$

under the generalized Sundman transformation

$$\begin{aligned} X &= F(t,x), \\ dT &= G(t,x)\,dt. \end{aligned} \tag{3}$$

It turns out that those equations must be in the form of the following theorem.

Theorem 1. *Any linearizable fifth-order ordinary differential equations that can be transformed by a generalized Sundman transformation has to be in the form*

$$\begin{aligned} & x^{(5)} + \left(A_1 x' + A_0\right) x^{(4)} \\ & + \left(B_3 x'' + B_2 x'^2 + B_1 x' + B_0\right) x''' \\ & + \left(C_1 x' + C_0\right) x''^2 \\ & + \left(D_3 x'^3 + D_2 x'^2 + D_1 x' + D_0\right) x'' + H_5 x'^5 \\ & + H_4 x'^4 + H_3 x'^3 + H_2 x'^2 + H_1 x' + H_0 = 0. \end{aligned} \tag{4}$$

Here $A_i = A_i(t,x)$, $B_i = B_i(t,x)$, $C_i = C_i(t,x)$, $D_i = D_i(t,x)$, and $H_i = H_i(t,x)$ are some functions of t and x. Expressions of these coefficients are presented in the appendix.

Proof. By a generalized Sundman transformation (3), we have

$$X'(T) = \frac{D_t F(t,x)}{D_t \int G(t,x)\,dt} = \frac{F_t + x' F_x}{G} = P\left(t, x, x'\right),$$

$$X''(T) = \frac{D_t P\left(t,x,x'\right)}{D_t \int G(t,x)\,dt} = \frac{P_t + x' P_x + x'' P_{x'}}{G}$$

$$= \frac{1}{G^3}\left[2F_{tx} G x' + F_{tt} G - F_t G_t - F_t G_x x' + F_{xx} G x'^2 - F_x G_t x' - F_x G_x x'^2 + F_x G x''\right] = Q\left(t, x, x', x''\right),$$

$$X'''(T) = \frac{D_t Q}{D_t \int G(t,x)\,dt}$$

$$= \frac{Q_t + x' Q_x + x'' Q_{x'} + x''' Q_{x''}}{G} = \frac{1}{G^5}\left[\left(F_x G^2\right) x''' + G\left(3F_{xx} G - 4F_x G_x\right) x' x'' + G\left(3F_{tx} G - F_t G_x - 3F_x G_t\right) x'' + \cdots\right]$$

$$= R\left(t, x, x', x'', x'''\right),$$

$$X^{(4)}(T) = \frac{D_t R}{D_t \int G(t,x)\,dt}$$

$$= \frac{R_t + x' R_x + x'' R_{x'} + x''' R_{x''} + x^{(4)} R_{x'''}}{G}$$

$$= \frac{1}{G^7}\left[\left(F_x G^3\right) x^{(4)} + G^2\left(4F_{xx} G - 7F_x G_x\right) x' x''' + G^2\left(4F_{tx} G - F_t G_x - 6F_x G_t\right) x''' + \cdots\right]$$

$$= S\left(t, x, x', x'', x''', x^{(4)}\right),$$

$$X^{(5)}(T) = \frac{D_t S}{D_t \int G(t,x)\,dt}$$

$$= \frac{S_t + x' S_x + x'' S_{x'} + x''' S_{x''} + x^{(4)} S_{x'''} G + x^{(5)} S_{x^{(4)}}}{G}$$

$$= \frac{1}{G^9}\left[\left(F_x G^4\right) x^{(5)} + G^3\left(5F_{xx} G - 11F_x G_x\right) x' x^{(4)}\right.$$

$$+G^3\left(5F_{tx}G - F_tF_x - 10F_xG_t\right)x^{(4)} + \cdots\Big]$$
$$= V\left(t, x, x', x'', x''', x^{(4)}, x^{(5)}\right), \tag{5}$$

where $D_t = \partial/\partial t + x'(\partial/\partial x) + x''(\partial/\partial x') + x'''(\partial/\partial x'') + x^{(4)}(\partial/\partial x''') + x^{(5)}(\partial/\partial x^{(4)}) + \cdots$ is a total derivative. Replacing $X^{(5)}(T)$ in (2), we get that

$$\begin{aligned}
x^{(5)} &+ \left(\left(\frac{5F_{xx}G - 11F_xG_x}{F_xG}\right)x' + \left(\frac{5F_{tx}G - F_tG_x - 10F_xG_t}{F_xG}\right)\right)x^{(4)} + \left(\left(\frac{5\left(2F_{xx}G - 3F_xG_x\right)}{F_xG}\right)x''\right. \\
&+ \left(\frac{10F_{xxx}G^2 - 45F_{xx}G_xG - 14F_xG_{xx}G + \cdots}{F_xG^2}\right)x'^2 \\
&+ \left.\left(\frac{20F_{txx}G^2 - 50F_{tx}G_xG - 4F_tG_{xx}G + 15F_tG_x^2 - 40F_{xx}G_tG + \cdots}{F_xG^2}\right)x' + \cdots\right)x''' \\
&+ \left(\left(\frac{15F_{xxx}G^2 - 60F_{xx}G_xG - 18F_xG_{xx}G + 70F_xG_x^2}{F_xG^2}\right)x'\right. \\
&+ \left.\left(\frac{15F_{txx}G^2 - 30F_{tx}G_xG - 3F_tG_{xx}G + 10F_tG_x^2 - 30F_{xx}G_tG + \cdots}{F_xG^2}\right)\right)x''^2 \\
&+ \left(\left(\frac{10F_{xxxx}G^3 - 70F_{xxx}G_xG^2 - 45F_{xx}G_{xx}G^2 + 195F_{xx}G_x^2G + \cdots}{F_xG^3}\right)x'^3 + \cdots\right)x'' \\
&+ \left(\frac{F_{xxxxx}G^4 - 10F_{xxxx}G_xG^3 - 10F_{xxx}G_{xx}G^3 + 45F_{xxx}G_x^2G^2 + \cdots}{F_xG^4}\right)x'^5 + \cdots = 0.
\end{aligned} \tag{6}$$

Denoting A_i, B_i, C_i, D_i, and H_i as (A.1)–(A.18), we obtain the necessary form (4). This proves the theorem. □

3. Sufficient Conditions and Linearizing Transformation

To get the sufficient conditions, we consider (A.1)–(A.18) appearing in the previous section. After using the compatibility theory to those equations, we derive the following results.

Theorem 2. *Equation (4) can be linearizable by the generalized Sundman transformation if its coefficients satisfy the following equations:*

$$S_{4x} = \frac{\left(7A_1S_4 + 49S_1 + 23S_4^2\right)}{280}, \tag{7}$$

$$S_{8t} = \left(27720S_2S_8 - 8520S_5S_8 + 62S_6S_8 + 4928S_7 - 115S_8^2\right)\left(277200S_4\right)^{-1}, \tag{8}$$

$$S_{8x} = \left(21A_1S_4S_8 + 147S_1S_8 + 73920S_2S_4^2 - 22720S_4^2S_5 - 40S_4^2S_6 + 69S_4^2S_8\right)\left(840S_4\right)^{-1}, \tag{9}$$

$$\begin{aligned}
B_{2xx} = {}& \left(-2195200B_{2x}A_1 + 823200B_{2x}S_4 - 548800S_{1xx} - 644840S_{1x}A_1 + 82320S_{1x}S_4 + 10633A_1^3S_4 - 10633A_1^2S_1\right. \\
&+ 1519A_1^2S_4^2 + 89180A_1B_2S_4 + 72716A_1S_1S_4 + 2604A_1S_4^3 - 548800B_2^2 - 260680B_2S_1 + 11760B_2S_4^2 + 54880000H_5 \\
&\left. + 43218S_1^2 + 1176S_1S_4^2 + 558S_4^4\right)\left(5488000\right)^{-1},
\end{aligned} \tag{10}$$

$$\begin{aligned}
S_{1x} = {}& \left(-78400B_{2x} - 637A_1^2S_4 - 15680A_1B_2 - 5635A_1S_1 + 105A_1S_4^2 + 2940B_2S_4 + 78400D_3 + 294S_1S_4 + 30S_4^3\right) \\
&\cdot\left(5880\right)^{-1},
\end{aligned} \tag{11}$$

$$C_{0x} = \left(-5488560A_{1t}S_4^2 - 38419920S_{1t}S_4 + 25347840S_{5x}S_4 + 75600S_{6x}S_4 - 18627840A_1C_0S_4 - 6137208A_1S_2S_4 + 6337320A_1S_4S_5 + 16929A_1S_4S_6 + 6573A_1S_4S_8 + 23950080C_0S_4^2 + 46569600D_2S_4 + 2794176S_1S_2 - 5188848S_1S_5 - 12096S_1S_6 - 6237S_1S_8 - 5207664S_2S_4^2 + 2080288S_4^2S_5 + 3682S_4^2S_6 + 5925S_4^2S_8\right)\left(931392000S_4\right)^{-1}, \tag{12}$$

$$A_{1t} = \left(-133056A_1S_2S_4 + 168240A_1S_4S_5 + 576A_1S_4S_6 - 231A_1S_4S_8 + 443520C_0S_4^2 + 25872S_1S_5 + 231S_1S_8 - 133056S_2S_4^2 + 21424S_4^2S_5 + 144S_4^2S_6 - 215S_4^2S_8\right)\left(1330560S_4^2\right)^{-1}, \tag{13}$$

$$S_{1t} = \left(135878400S_{5x}S_4 + 40320S_{6x}S_4 - 1862784A_1S_2S_4 + 2662464A_1S_4S_5 + 7056A_1S_4S_6 + 2541A_1S_4S_8 + 6209280C_0S_4^2 + 1862784S_1S_2 - 3235008S_1S_5 - 8064S_1S_6 - 2541S_1S_8 - 1862784S_2S_4^2 + 483392S_4^2S_5 + 504S_4^2S_6 + 2165S_4^2S_8\right) \cdot \left(18627840S_4\right)^{-1}, \tag{14}$$

$$S_{3x} = \Big(-\Big(\big(246400\left(144\left(3\left(308B_0 + 491S_3\right)S_4 + 5S_{6t}\right)S_4^2 - \left(337S_{5x} + S_{6x}\right)\left(112S_5 + S_8\right)\right) - \left(70209254400S_5^2 + 33795200S_5S_6 - 2623280S_5S_8 + 61440S_6^2 - 41624S_6S_8 + 2776064S_7 + 52785S_8^2\right)S_4\big)S_4 - 77\left(149022720S_3S_4^2 - 23206400S_5^2 - 62720S_5S_6 - 225680S_5S_8 - 224S_6S_8 + 21504S_7 - 765S_8^2\right)S_1 + 443520\left(32S_6 + 23S_8 + 27344S_5\right)S_2S_4^2 + 77\left(949760S_5^2 + 8960S_5S_6 - 10000S_5S_8 + 416S_6S_8 + 21504S_7 - 765S_8^2 + 149022720S_3S_4^2\right) \cdot A_1S_4\Big)\Big)\left(655699968000S_4^3\right)^{-1}, \tag{15}$$

$$S_{5x} = \left(-155232A_1S_4S_5 - 1617A_1S_4S_8 + 362208S_1S_5 + 1617S_1S_8 - 680064S_2S_4^2 - 69856S_4^2S_5 - 136S_4^2S_6 - 1449S_4^2S_8\right) \cdot \left(1034880S_4\right)^{-1}, \tag{16}$$

$$S_{7x} = \left(776160000S_{6t}S_4^3 - 64680S_{6x}S_4S_8 + 1617A_1S_4S_6S_8 + 206976A_1S_4S_7 + 20490624000B_0S_4^4 + 11319S_1S_6S_8 + 1448832S_1S_7 + 4719052800S_2S_4^2S_5 + 2069760S_2S_4^2S_6 + 17001600S_2S_4^2S_8 + 18255283200S_3S_4^4 - 3786137600S_4^2S_5^2 - 18085760S_4^2S_5S_6 - 5225600S_4^2S_5S_8 - 30520S_4^2S_6^2 - 3887S_4^2S_6S_8 - 1022336S_4^2S_7\right)\left(4139520S_4\right)^{-1}, \tag{17}$$

$$C_1 = \frac{\left(9A_1S_4 + 60B_2 + 3S_1 + S_4^2\right)}{40}, \tag{18}$$

and (A.19), (A.20), (A.21), (A.22), (A.23), (A.24), (A.25) (moved to the appendix in order to avoid the huge expressions), where

$$S_1 = -10A_{1x} - 2A_1^2 + 5B_2,$$

$$S_2 = -20A_{0x} - 4A_0A_1 + 5B_1,$$

$$S_3 = 10A_{0t} + 2A_0^2 - 5B_0,$$

$$S_4 = -2A_1 + B_3,$$

$$S_5 = -80S_{4t} + 2A_0S_4 + 7S_2,$$

$$S_6 = -462A_0S_4 + 231S_2 - 337S_5,$$

$$S_7 = -13860S_{5t}S_4 + 1386S_2S_5 + 8316S_3S_4^2 - 202S_5^2 + S_5S_6,$$

$$S_8 = -41580B_1 + 55440C_0 - 8316S_2 + 9176S_5 + 27S_6. \tag{19}$$

Proof. We start with the coefficients A_i, B_i, C_i, D_i, and H_i in Theorem 1 through the unknown functions F and G. From (A.1) and (A.2), we have the derivatives

$$F_{xx} = \frac{\left(F_x\left(11G_x + A_1G\right)\right)}{(5G)}, \tag{20}$$

$$F_{tx} = \frac{\left(F_tG_x + 10F_xG_t + F_xA_0G\right)}{(5G)}. \tag{21}$$

From (A.4), one obtains the derivative

$$G_{xx} = \frac{\left(63G_x^2 + G_xA_1G + G^2S_1\right)}{(40G)}, \tag{22}$$

where

$$S_1 = -10A_{1x} - 2A_1^2 + 5B_2. \quad (23)$$

From (A.5), one gets the derivative

$$G_{tx} = \frac{\left(-9F_tG_x^2 + 135F_xG_tG_x + F_xG\left(2G_xA_0 + GS_2\right)\right)}{\left(80F_xG\right)}, \quad (24)$$

where

$$S_2 = -20A_{0x} - 4A_0A_1 + 5B_1. \quad (25)$$

From (A.6), one finds the derivative

$$F_{tt} = \frac{\left(9F_t^2G_x^2 + 225F_tF_xG_tG_x + F_tF_xG\left(14G_xA_0 - GS_2\right) + 400F_x^2G_{tt}G - 600F_x^2G_t^2 + 8F_x^2G^2S_3\right)}{\left(120F_xG_xG\right)}, \quad (26)$$

where

$$S_3 = 10A_{0t} + 2A_0^2 - 5B_0. \quad (27)$$

From (A.3), one obtains the derivative

$$G_x = \frac{(GS_4)}{7}, \quad (28)$$

where

$$S_4 = -2A_1 + B_3. \quad (29)$$

We note that, for the case $G_x = 0$, the generalized Sundman transformations are indeed the point transformations. We then suppose $G_x \neq 0$, which also implies $S_4 \neq 0$.

The relations $(G_x)_x = G_{xx}$ and $(G_x)_t = G_{tx}$ provide condition (7) and the derivative

$$F_t = \frac{\left(385F_xG_tS_4 + 7F_xGS_5\right)}{\left(9GS_4^2\right)}, \quad (30)$$

where

$$S_5 = -80S_{4t} + 2A_0S_4 + 7S_2. \quad (31)$$

The relation $(F_t)_t = F_{tt}$ gives the derivative

$$G_{tt} = \frac{\left(-2156000G_t^2S_4^2 + 385G_tGS_4S_6 + 4G^2S_7\right)}{\left(1386000GS_4^2\right)}, \quad (32)$$

where

$$\begin{aligned} S_6 &= -462A_0S_4 + 231S_2 - 337S_5, \\ S_7 &= -13860S_{5t}S_4 + 1386S_2S_5 + 8316S_3S_4^2 - 202S_5^2 \\ &\quad + S_5S_6. \end{aligned} \quad (33)$$

Substituting A_0 into A_{0x} and A_{0t}, one obtains the conditions

$$\begin{aligned} S_{2x} &= \left(94360S_{5x}S_4 + 280S_{6x}S_4 - 11319A_1S_2S_4\right. \\ &\quad + 16513A_1S_4S_5 + 49A_1S_4S_6 + 32340B_1S_4^2 \\ &\quad + 11319S_1S_2 - 16513S_1S_5 - 49S_1S_6 - 1155S_2S_4^2 \\ &\quad \left.- 7751S_4^2S_5 - 23S_4^2S_6\right)\left(64680S_4\right)^{-1}, \\ S_{2t} &= \left(889680S_{5t}S_4 + 2640S_{6t}S_4 + 609840B_0S_4^2\right. \\ &\quad + 70224S_2S_5 + 231S_2S_6 + 121968S_3S_4^2 - 102448S_5^2 \\ &\quad \left.- 641S_5S_6 - S_6^2\right)\left(609840S_4\right)^{-1}. \end{aligned} \quad (34)$$

From (A.8), we have

$$G_t = \frac{(GS_8)}{(6160S_4)}, \quad (35)$$

where

$$\begin{aligned} S_8 &= -41580B_1 + 55440C_0 - 8316S_2 + 9176S_5 \\ &\quad + 27S_6. \end{aligned} \quad (36)$$

The relations $(G_t)_t = G_{tt}$ and $(G_t)_x = G_{tx}$ provide conditions (8) and (9). From (A.18), (A.15), (A.17), (A.13), and (A.11), we obtain conditions (A.19)–(A.21), (10), (A.22). Substituting the relation C_{0t} into C_{0tt}, one obtains condition (A.23). Equations (A.9), (A.10), and (A.12) provide conditions (11), (12), (A.24). Comparing the mixed derivatives $(F_{xx})_t = (F_{tx})_x$, $(G_{xx})_t = (G_{tx})_x$, $(F_{tt})_x = (F_{tx})_t$, $(F_t)_x = F_{tx}$, we obtain conditions (13)–(16). Substituting the relation S_{5x} into S_{5xx}, one obtains condition (A.25). Comparing the mixed derivative $(G_{tt})_x = (G_{tx})_t$, one arrives at condition (17). From (A.7), one obtains condition (18). This proves the theorem. □

Corollary 3. *Under the sufficient conditions in Theorem 2, the transformation (3) mapping equation (4) to a linear equation (2) can be solved by the compatible system of (20), (28), (30), and (35).*

Remark 4. In the part of sufficient conditions for second-order, there are 2 cases in [20] and 3 cases in [23]. For the third-order, there are 3 cases in [24] and 4 cases in [25]. For the fourth-order, there are 2 cases in [28]. But for the fifth-order there is only one case.

4. Examples

Example 1. For the fifth-order ordinary differential equation

$$\begin{aligned} & x^{(5)}x^4 - 11x'x^{(4)}x^3 - 15x''x'''x^3 + 60x'^2x'''x^2 \\ & \quad + 70x'x''^2x^2 - 210x'^3x''x + 105x'^5 = 0, \end{aligned} \tag{37}$$

we can verify that this equation cannot be linearized by a point transformation [10] or contact transformation [17]. However, (37) is in fact the form (4) in Theorem 1 with the coefficients

$$\begin{aligned} A_1 &= \frac{-11}{x}, \\ A_0 &= 0, \\ B_3 &= \frac{-15}{x}, \\ B_2 &= \frac{60}{x^2}, \\ B_1 &= 0, \\ B_0 &= 0, \\ C_1 &= \frac{70}{x^2}, \\ C_0 &= 0, \\ D_3 &= \frac{-210}{x^3}, \\ D_2 &= 0, \\ D_1 &= 0, \\ D_0 &= 0, \\ H_5 &= \frac{105}{x^4}, \\ H_4 &= 0, \\ H_3 &= 0, \\ H_2 &= 0, \\ H_1 &= 0, \\ H_0 &= 0, \\ S_1 &= \frac{-52}{x^2}, \\ S_2 &= 0, \\ S_3 &= 0, \\ S_4 &= \frac{7}{x}, \\ S_5 &= 0, \\ S_6 &= 0, \\ S_7 &= 0, \\ S_8 &= 0. \end{aligned} \tag{38}$$

Moreover, these coefficients also satisfy the conditions in Theorem 2. We now conclude that (37) is linearizable by a generalized Sundman transformation. Corollary 3 yields the linearizing transformation by solving the following equations:

$$F_{xx} = 0, \tag{39}$$

$$G_x = \frac{G}{x}, \tag{40}$$

$$G_t = 0, \tag{41}$$

$$F_t = 0. \tag{42}$$

Considering (40), one arrives at

$$G = xK_1(t). \tag{43}$$

Considering (41), one obtains

$$G = K_2(x). \tag{44}$$

From (43) and (44), one can choose $K_1(t) = 1$ and $K_2(x) = x$; then we have

$$G = x. \tag{45}$$

Considering (39), one gets

$$F = K_3(t)x + K_4(t). \tag{46}$$

Considering (42), one arrives at

$$F = K_5(x). \tag{47}$$

From (46) and (47), one can choose $K_3(t) = 1$, $K_4(t) = 0$, and $K_5(x) = x$; then we obtain

$$F = x. \tag{48}$$

So the linearizing transformation is

$$\begin{aligned} X &= x, \\ dT &= x\,dt. \end{aligned} \tag{49}$$

Hence, by (49), (37) becomes

$$X^{(5)} = 0. \tag{50}$$

The general solution of (50) is

$$X = \frac{c_1}{24}T^4 + \frac{c_2}{6}T^3 + \frac{c_3}{2}T^2 + c_4T + c_5, \tag{51}$$

where c_1, c_2, c_3, c_4, and c_5 are arbitrary constants. Substituting (49) into (51), the general solution of (37) is

$$x(t) = \frac{c_1}{24}\phi(t)^4 + \frac{c_2}{6}\phi(t)^3 + \frac{c_3}{2}\phi(t)^2 + c_4\phi(t) + c_5, \quad (52)$$

where the function $T = \phi(t)$ is a solution of the equation

$$\frac{dT}{dt} = \frac{c_1}{24}T^4 + \frac{c_2}{6}T^3 + \frac{c_3}{2}T^2 + c_4 T + c_5. \quad (53)$$

Example 2. For the fifth-order ordinary differential equation

$$\begin{aligned} &x^{(5)}tx^4 - 22x'x^{(4)}tx^3 + 3x^{(4)}x^4 - 30x''x'''tx^3 \\ &\quad + 212x'^2x'''tx^2 - 48x'x'''x^3 + 244x'x''^2tx^2 \\ &\quad - 26x''^2x^3 - 1180x'^3x''tx + 320x'^2x''x^2 \\ &\quad + 880x'^5t - 320x'^4x = 0 \end{aligned} \quad (54)$$

we can verify that this equation cannot be linearized by a point transformation [10] or contact transformation [17]. However, (54) is in fact the form (4) in Theorem 1 with the coefficients

$$\begin{aligned} A_1 &= \frac{-22}{x}, \\ A_0 &= \frac{3}{t}, \\ B_3 &= \frac{-30}{x}, \\ B_2 &= \frac{212}{x^2}, \\ B_1 &= \frac{-48}{tx}, \\ B_0 &= 0, \\ C_1 &= \frac{244}{x^2}, \\ C_0 &= \frac{-26}{tx}, \\ D_3 &= \frac{-1180}{x^3}, \\ D_2 &= \frac{320}{tx^2}, \\ D_1 &= 0, \\ D_0 &= 0, \\ H_5 &= \frac{880}{x^4}, \\ H_4 &= \frac{-320}{tx^3}, \\ H_3 &= 0, \\ H_2 &= 0, \\ H_1 &= 0, \\ H_0 &= 0, \\ S_1 &= \frac{-128}{x^2}, \\ S_2 &= \frac{24}{tx}, \\ S_3 &= \frac{-12}{t^2}, \\ S_4 &= \frac{14}{x}, \\ S_5 &= \frac{252}{tx}, \\ S_6 &= \frac{-98784}{tx}, \\ S_7 &= 0, \\ S_8 &= 0. \end{aligned} \quad (55)$$

Moreover, these coefficients also satisfy the conditions in Theorem 2. We now conclude that (54) is linearizable by a generalized Sundman transformation. Corollary 3 yields the linearizing transformation by solving the following equations:

$$F_{xx} = 0, \quad (56)$$

$$G_x = \frac{(2G)}{x}, \quad (57)$$

$$G_t = 0, \quad (58)$$

$$F_t = \frac{(F_x x)}{t}. \quad (59)$$

Considering (57), one arrives at

$$G = K_1(t)\, x^2. \quad (60)$$

Considering (58), one obtains

$$G = K_2(x). \quad (61)$$

From (60) and (61), one can choose $K_1(t) = 1$ and $K_2(x) = x^2$; then we obtain

$$G = x^2. \quad (62)$$

Equation (59) becomes

$$tF_t - xF_x = 0, \quad (63)$$

and by Cauchy method, one arrives at

$$F = tx. \quad (64)$$

This solution satisfies (56), so the linearizing transformation is

$$X = tx, \quad dT = x^2 dt. \tag{65}$$

Hence, by (65), (54) becomes

$$X^{(5)} = 0. \tag{66}$$

The general solution of (66) is

$$X = \frac{c_1}{24}T^4 + \frac{c_2}{6}T^3 + \frac{c_3}{2}T^2 + c_4T + c_5, \tag{67}$$

where c_1, c_2, c_3, c_4, and c_5 are arbitrary constants. Substituting (65) into (67), the general solution of (54) is

$$x(t) = \frac{\left((c_1/24)\,\phi(t)^4 + (c_2/6)\,\phi(t)^3 + (c_3/2)\,\phi(t)^2 + c_4\phi(t) + c_5\right)}{t}, \tag{68}$$

where the function $T = \phi(t)$ is a solution of the equation

$$\frac{dT}{dt} = \left(\frac{(c_1/24)\,T^4 + (c_2/6)\,T^3 + (c_3/2)\,T^2 + c_4T + c_5}{t}\right)^2. \tag{69}$$

Appendix

Equations for Theorem 1 in Section 2

$$A_1 = \frac{(5F_{xx}G - 11F_xG_x)}{(F_xG)}, \tag{A.1}$$

$$A_0 = \frac{(5F_{tx}G - F_tG_x - 10F_xG_t)}{(F_xG)}, \tag{A.2}$$

$$B_3 = \frac{5(2F_{xx}G - 3F_xG_x)}{(F_xG)}, \tag{A.3}$$

$$B_2 = \frac{(10F_{xxx}G^2 - 45F_{xx}G_xG - 14F_xG_{xx}G + 60F_xG_x^2)}{(F_xG^2)}, \tag{A.4}$$

$$B_1 = (20F_{txx}G^2 - 50F_{tx}G_xG - 4F_tG_{xx}G + 15F_tG_x^2 - 40F_{xx}G_tG - 24F_xG_{tx}G + 105F_xG_tG_x)(F_xG^2)^{-1}, \tag{A.5}$$

$$B_0 = (-40F_{tx}G_tG + 10F_{ttx}G^2 - 5F_{tt}G_xG - 4F_tG_{tx}G + 15F_tG_tG_x - 10F_xG_{tt}G + 45F_xG_t^2)(F_xG^2)^{-1}, \tag{A.6}$$

$$C_1 = \frac{(15F_{xxx}G^2 - 60F_{xx}G_xG - 18F_xG_{xx}G + 70F_xG_x^2)}{(F_xG^2)}, \tag{A.7}$$

$$C_0 = (15F_{txx}G^2 - 30F_{tx}G_xG - 3F_tG_{xx}G + 10F_tG_x^2 - 30F_{xx}G_tG - 15F_xG_{tx}G + 60F_xG_tG_x)(F_xG^2)^{-1}, \tag{A.8}$$

$$D_3 = (10F_{xxxx}G^3 - 70F_{xxx}G_xG^2 - 45F_{xx}G_{xx}G^2 + 195F_{xx}G_x^2G - 11F_xG_{xxx}G^2 + 125F_xG_{xx}G_xG - 210F_xG_x^3)(F_xG^3)^{-1}, \tag{A.9}$$

$$D_2 = (30F_{txxx}G^3 - 150F_{txx}G_xG^2 - 60F_{tx}G_{xx}G^2 + 255F_{tx}G_x^2G - 6F_tG_{xxx}G^2 + 65F_tG_{xx}G_xG - 105F_tG_x^3 - 60F_{xxx}G_tG^2 - 75F_{xx}G_{tx}G^2 + 330F_{xx}G_tG_xG - 27F_xG_{txx}G^2 + 205F_xG_{tx}G_xG + 105F_xG_tG_{xx}G - 525F_xG_tG_x^2)(F_xG^3)^{-1}, \tag{A.10}$$

$$D_1 = (-120F_{txx}G_tG^2 - 90F_{tx}G_{tx}G^2 + 390F_{tx}G_tG_xG + 30F_{ttxx}G^3 - 90F_{ttx}G_xG^2 - 15F_{tt}G_{xx}G^2 + 60F_{tt}G_x^2G - 12F_tG_{txx}G^2 + 85F_tG_{tx}G_xG + 45F_tG_tG_{xx}G - 210F_tG_tG_x^2 - 30F_{xx}G_{tt}G^2 + 135F_{xx}G_t^2G + 165F_xG_{tx}G_tG - 21F_xG_{ttx}G^2 + 80F_xG_{tt}G_xG - 420F_xG_t^2G_x)(F_xG^3)^{-1}, \tag{A.11}$$

$$D_0 = (-30F_{tx}G_{tt}G^2 + 135F_{tx}G_t^2G + 10F_{tttx}G^3 - 10F_{ttt}G_xG^2 - 60F_{ttx}G_tG^2 - 15F_{tt}G_{tx}G^2 + 60F_{tt}G_tG_xG + 45F_tG_{tx}G_tG - 6F_tG_{ttx}G^2 + 20F_tG_{tt}G_xG - 105F_tG_t^2G_x - 5F_xG_{ttt}G^2 + 60F_xG_{tt}G_tG - 105F_xG_t^3)(F_xG^3)^{-1}, \tag{A.12}$$

$$H_5 = (F_{xxxxx}G^4 - 10F_{xxxx}G_xG^3 - 10F_{xxx}G_{xx}G^3 + 45F_{xxx}G_x^2G^2 - 5F_{xx}G_{xxx}G^3 + 60F_{xx}G_{xx}G_xG^2 - 105F_{xx}G_x^3G - F_xG_{xxxx}G^3 + 15F_xG_{xxx}G_xG^2 + 10F_xG_{xx}^2G^2 - 105F_xG_{xx}G_x^2G + 105F_xG_x^4) \cdot (F_xG^4)^{-1}, \tag{A.13}$$

$$H_4 = (5F_{txxxx}G^4 - 40F_{txxx}G_xG^3 - 30F_{txx}G_{xx}G^3 + 135F_{txx}G_x^2G^2 - 10F_{tx}G_{xxx}G^3 + 120F_{tx}G_{xx}G_xG^2 - 210F_{tx}G_x^3G - F_tG_{xxxx}G^3 + 15F_tG_{xxx}G_xG^2$$

$$\begin{aligned}
&+ 10F_tG_{xx}^2G^2 - 105F_tG_{xx}G_x^2G + 105F_tG_x^4 \\
&- 10F_{xxxx}G_tG^3 - 20F_{xxx}G_{tx}G^3 + 90F_{xxx}G_tG_xG^2 \\
&- 15F_{xx}G_{txx}G^3 + 120F_{xx}G_{tx}G_xG^2 + 60F_{xx}G_tG_{xx}G^2 \\
&- 315F_{xx}G_tG_x^2G - 4F_xG_{txxx}G^3 + 45F_xG_{txx}G_xG^2 \\
&+ 40F_xG_{tx}G_{xx}G^2 - 210F_xG_{tx}G_x^2G + 15F_xG_tG_{xxx}G^2 \\
&- 210F_xG_tG_{xx}G_xG + 420F_xG_tG_x^3\big)\left(F_xG^4\right)^{-1},
\end{aligned} \tag{A.14}$$

$$\begin{aligned}
H_3 = {} & \big(-40F_{txxx}G_tG^3 - 60F_{txx}G_{tx}G^3 + 270F_{txx}G_tG_xG^2 \\
&- 30F_{tx}G_{txx}G^3 + 240F_{tx}G_{tx}G_xG^2 + 120F_{tx}G_tG_{xx}G^2 \\
&- 630F_{tx}G_tG_x^2G + 10F_{ttxxx}G^4 - 60F_{ttxx}G_xG^3 \\
&- 30F_{ttx}G_{xx}G^3 + 135F_{ttx}G_x^2G^2 - 5F_{tt}G_{xxx}G^3 \\
&+ 60F_{tt}G_{xx}G_xG^2 - 105F_{tt}G_x^3G - 4F_tG_{txxx}G^3 \\
&+ 45F_tG_{txx}G_xG^2 + 40F_tG_{tx}G_{xx}G^2 - 210F_tG_{tx}G_x^2G \\
&+ 15F_tG_tG_{xxx}G^2 - 210F_tG_tG_{xx}G_xG + 420F_tG_tG_x^3 \\
&- 10F_{xxx}G_{tt}G^3 + 45F_{xxx}G_t^2G^2 + 120F_{xx}G_{tx}G_tG^2 \\
&- 15F_{xx}G_{ttx}G^3 + 60F_{xx}G_{tt}G_xG^2 - 315F_{xx}G_t^2G_xG \\
&+ 45F_xG_{txx}G_tG^2 + 40F_xG_{tx}^2G^2 - 420F_xG_{tx}G_tG_xG \\
&- 6F_xG_{ttxx}G^3 + 45F_xG_{ttx}G_xG^2 + 20F_xG_{tt}G_{xx}G^2 \\
&- 105F_xG_{tt}G_x^2G - 105F_xG_t^2G_{xx}G + 630F_xG_t^2G_x^2\big) \\
&\cdot\left(F_xG^4\right)^{-1},
\end{aligned} \tag{A.15}$$

$$\begin{aligned}
H_2 = {} & \big(-30F_{txx}G_{tt}G^3 + 135F_{txx}G_t^2G^2 + 240F_{tx}G_{tx}G_tG^2 \\
&- 30F_{tx}G_{ttx}G^3 + 120F_{tx}G_{tt}G_xG^2 - 630F_{tx}G_t^2G_xG \\
&+ 10F_{tttxx}G^4 - 40F_{tttx}G_xG^3 - 10F_{ttt}G_{xx}G^3 \\
&+ 45F_{ttt}G_x^2G^2 - 60F_{ttxx}G_tG^3 - 60F_{ttx}G_{tx}G^3 \\
&+ 270F_{ttx}G_tG_xG^2 - 15F_{tt}G_{txx}G^3 + 120F_{tt}G_{tx}G_xG^2 \\
&+ 60F_{tt}G_tG_{xx}G^2 - 315F_{tt}G_tG_x^2G + 45F_tG_{txx}G_tG^2 \\
&+ 40F_tG_{tx}^2G^2 - 420F_tG_{tx}G_tG_xG - 6F_tG_{ttxx}G^3 \\
&+ 45F_tG_{ttx}G_xG^2 + 20F_tG_{tt}G_{xx}G^2 - 105F_tG_{tt}G_x^2G \\
&- 105F_tG_t^2G_{xx}G + 630F_tG_t^2G_x^2 - 5F_{xx}G_{ttt}G^3 \\
&+ 60F_{xx}G_{tt}G_tG^2 - 105F_{xx}G_t^3G + 40F_xG_{tx}G_{tt}G^2 \\
&- 210F_xG_{tx}G_t^2G - 4F_xG_{tttx}G^3 + 15F_xG_{ttt}G_xG^2 \\
&+ 45F_xG_{ttx}G_tG^2 - 210F_xG_{tt}G_tG_xG + 420F_xG_t^3G_x\big) \\
&\cdot\left(F_xG^4\right)^{-1},
\end{aligned} \tag{A.16}$$

$$\begin{aligned}
H_1 = {} & \big(-10F_{tx}G_{ttt}G^3 + 120F_{tx}G_{tt}G_tG^2 - 210F_{tx}G_t^3G \\
&+ 5F_{ttttx}G^4 - 10F_{tttt}G_xG^3 - 40F_{tttx}G_tG^3 \\
&- 20F_{ttt}G_{tx}G^3 + 90F_{ttt}G_tG_xG^2 - 30F_{ttx}G_{tt}G^3 \\
&+ 135F_{ttx}G_t^2G^2 + 120F_{tt}G_{tx}G_tG^2 - 15F_{tt}G_{ttx}G^3 \\
&+ 60F_{tt}G_{tt}G_xG^2 - 315F_{tt}G_t^2G_xG + 40F_tG_{tx}G_{tt}G^2 \\
&- 210F_tG_{tx}G_t^2G - 4F_tG_{tttx}G^3 + 15F_tG_{ttt}G_xG^2 \\
&+ 45F_tG_{ttx}G_tG^2 - 210F_tG_{tt}G_tG_xG + 420F_tG_t^3G_x \\
&- F_xG_{tttt}G^3 + 15F_xG_{ttt}G_tG^2 + 10F_xG_{tt}^2G^2 \\
&- 105F_xG_{tt}G_t^2G + 105F_xG_t^4\big)\left(F_xG^4\right)^{-1},
\end{aligned} \tag{A.17}$$

$$\begin{aligned}
H_0 = {} & \big(F_{ttttt}G^4 - 10F_{tttt}G_tG^3 - 10F_{ttt}G_{tt}G^3 \\
&+ 45F_{ttt}G_t^2G^2 - 5F_{tt}G_{ttt}G^3 + 60F_{tt}G_{tt}G_tG^2 \\
&- 105F_{tt}G_t^3G - F_tG_{tttt}G^3 + 15F_tG_{ttt}G_tG^2 \\
&+ 10F_tG_{tt}^2G^2 - 105F_tG_{tt}G_t^2G + 105F_tG_t^4\big)\left(F_xG^4\right)^{-1}.
\end{aligned} \tag{A.18}$$

Equations for Theorem 2 in Section 3

$$\begin{aligned}
S_{3ttt} = {} & \big(238560040857600000000S_{6ttt}S_4^3S_5 \\
&+ 8520001459200000000S_{6ttt}S_4^3S_8 \\
&- 4518182592000000000S_{6t}^2S_4^2S_5 \\
&- 122636384640000S_{6t}^2S_4^2S_8 \\
&- 81792014008320000000S_{7ttt}S_4^3 \\
&+ 1214611408023552000000000H_0S_4^6 \\
&+ 1288224220631040000S_2^2S_3S_4^2S_6 \\
&+ 333433269227520000S_2^2S_5^2S_6 \\
&+ 202242458880000S_2^2S_5S_6^2 \\
&+ 1247633114112000S_2^2S_5S_6S_8 \\
&- 3987862293381120000S_2^2S_5S_7 \\
&+ 2142294739200S_2^2S_6^2S_8 \\
&- 1699574317056000S_2^2S_6S_7 \\
&- 1775000304000S_2^2S_6S_8^2 \\
&- 147532492800000S_2^2S_7S_8 \\
&+ 487653901701120000S_2S_3S_4^2S_5S_6 \\
&+ 2581818624000000S_2S_3S_4^2S_6^2 \\
&+ 3947600676096000S_2S_3S_4^2S_6S_8
\end{aligned}$$

$- 1004306790187008000S_2S_3S_4^2S_7$

$- 332554631270400000S_2S_5^3S_6$

$- 1429140303360000S_2S_5^2S_6^2$

$- 1233512146944000S_2S_5^2S_6S_8$

$+ 3534379287183360000S_2S_5^2S_7$

$- 2831431680000S_2S_5S_6^3$

$- 6606314668800S_2S_5S_6^2S_8$

$+ 1176324891033600S_2S_5S_6S_7$

$+ 2489610816000S_2S_5S_6S_8^2$

$+ 149604618240000S_2S_5S_7S_8$

$- 10866683520S_2S_6^3S_8$

$+ 2607397309440S_2S_6^2S_7$

$+ 4354257600S_2S_6^2S_8^2$

$+ 130235212800S_2S_6S_7S_8$

$- 3590433000S_2S_6S_8^3$

$+ 176538275020800S_2S_7^2$

$- 512265600000S_2S_7S_8^2$

$+ 15982865547264000000S_3S_4^2S_5^3$

$+ 2184672479109120000S_3S_4^2S_5^2S_6$

$+ 672295138099200000S_3S_4^2S_5^2S_8$

$+ 10708586380800000S_3S_4^2S_5S_6^2$

$+ 1042969036032000S_3S_4^2S_5S_6S_8$

$+ 3375843865067520000S_3S_4^2S_5S_7$

$+ 5249697868800000S_3S_4^2S_5S_8^2$

$+ 14883644160000S_3S_4^2S_6^3$

$+ 4409023449600S_3S_4^2S_6^2S_8$

$- 740940963840000S_3S_4^2S_6S_7$

$+ 4756386096000S_3S_4^2S_6S_8^2$

$+ 291637463040000S_3S_4^2S_7S_8$

$+ 21679812000000S_3S_4^2S_8^3$

$- 552494117683200000S_5^5$

$+ 68289477365760000S_5^4S_6$

$- 7047118848000000S_5^4S_8$

$+ 291110400000S_5^2S_6^3$

$+ 2269611955200S_5^2S_6^2S_8$

$- 402813609984000S_5^2S_6S_7$

$- 517887216000S_5^2S_6S_8^2$

$- 57156096000000S_5^2S_7S_8$

$- 2773848000000S_5^2S_8^3$

$- 1049440000S_5S_6^4$

$+ 6279324480S_5S_6^3S_8$

$- 650224460800S_5S_6^2S_7$

$- 9707266800S_5S_6^2S_8^2$

$- 637826227200S_5S_6S_7S_8$

$+ 11823801000S_5S_6S_8^3$

$- 17672824422400S_5S_7^2$

$+ 505169280000S_5S_7S_8^2$

$- 21954900000S_5S_8^4$

$+ 4937432S_6^4S_8 + 896097408S_6^3S_7$

$- 14788620S_6^3S_8^2$

$- 1103392000S_6^2S_7S_8 - 1631850S_6^2S_8^3$

$+ 151872430080S_6S_7^2$

$- 536659200S_6S_7S_8^2$

$+ 32170875S_6S_8^4 - 12909568000S_7^2S_8$

$+ 848760000S_7S_8^3 - 62184375S_8^5$

$+ 7683984000\big(155232000S_2S_5S_6$

$+ 554400S_2S_6S_8 - 170311680S_2S_7$

$- 838252800S_3S_4^2S_6$

$- 35795200S_5^2S_6 + 112000S_5S_6^2$

$- 164800S_5S_6S_8 + 48762880S_5S_7$

$- 272S_6^2S_8 - 96000S_6S_7$

$+ 1155S_6S_8^2 + 64000S_7S_8\big)B_0S_4^2$

$+ 17740800\big(276623424000B_0S_4^2$

$- 82987027200S_2^2 + 69144768000S_2S_5$

$$
\begin{aligned}
&- 59209920S_2S_6 + 55440000S_2S_8 \\
&+ 264608467200S_3S_4^2 \\
&- 49174841600S_5^2 - 232872800S_5S_6 \\
&- 1724000S_5S_8 \\
&- 514314S_6^2 - 1160S_6S_8 - 49754880S_7 \\
&+ 144375S_8^2\big) S_{7t}S_4 \\
&+ 25818186240000\big(4158000S_{6t}S_4 \\
&- 415800S_2S_6 + 1096381440S_3S_4^2 \\
&- 129382400S_5^2 \\
&- 741880S_5S_6 - 268800S_5S_8 \\
&- 1280S_6^2 - 684S_6S_8 \\
&- 4992S_7 - 825S_8^2\big) S_{3t}S_4^3 \\
&+ 1022400175104000000(77S_6 + 20S_8 \\
&+ 23968S_5) S_{3tt}S_4^4 \\
&- 619636469760000(24213S_6 \\
&+ 4880S_8 + 4866176S_5) S_3^2S_4^4 \\
&+ 768000\big(\big(417386935S_6 \\
&+ 422682482S_8\big) S_6 \\
&- 88\big(1252844464S_7 + 2443875S_8^2\big)\big) S_5^3 \\
&- 511200087552000(S_6S_8 \\
&- 384S_7 + 280S_5S_6) S_2^3 \\
&- 8520001459200000(S_6S_8 \\
&- 192S_7 + 280S_5S_6) B_{0t}S_4^3 \\
&+ 19670999040000(327S_6 - 125S_8 \\
&+ 26080S_5 + 249480S_2) S_{7tt}S_4^2 \\
&- 76839840000\big(332640\big(\big(280S_5 \\
&+ S_8\big) S_2 - 2520S_3S_4^2\big) \\
&+ 54790400S_5^2 + 380800S_5S_6 \\
&+ 158720S_5S_8 + 688S_6S_8 \\
&- 112896S_7 + 1155S_8^2\big) S_{6tt}S_4^2 \\
&- 277200\big(6147187200\big(3\big(5B_0S_4^2 \\
&- S_2^2\big)(280S_5 + S_8) \\
&- 16S_{7t}S_4\big) - \big(5438522880000S_5^3 \\
&+ 30283456000S_5^2S_6 \\
&+ 18086822400S_5^2S_8 + 96768000S_5S_6^2 \\
&+ 991268880S_5S_6S_8 \\
&+ 6218168320S_5S_7 + 3603600S_5S_8^2 \\
&+ 211536S_6^2S_8 \\
&- 29151360S_6S_7 + 120120S_6S_8^2 \\
&+ 10810240S_7S_8 + 129525S_8^3\big) \\
&+ 93139200(3160S_6 + 1397S_8 \\
&+ 879840S_5) S_3S_4^2 \\
&+ 55440\big(81088000S_5^2 - 565600S_5S_6 \\
&+ 326560S_5S_8 - 844S_6S_8 + 467712S_7 \\
&- 1155S_8^2 + 838252800S_3S_4^2\big) S_2\big) S_{6t}S_4\big) \\
&\cdot \big(566818657077657600000000S_4^5\big)^{-1},
\end{aligned}
\tag{A.19}
$$

$$
\begin{aligned}
S_{6tx} = {} & \big(\big(2\big(40320\big(2\big(99\big(259952S_2^2 \\
&- 436299S_3S_4^2 - 19404000H_3 \\
&+ 1724800C_0^2\big) S_4 \\
&+ 24736S_{7x} - 285959520S_{3x}S_4^2 \\
&+ 672348600S_{1tt}S_4 \\
&+ 1280664000C_{0tx}S_4 \\
&- 384199200B_{0x}S_4^2 \\
&+ 96049800A_{1tt}S_4^2\big) S_4 \\
&- 1334025(112S_5 + S_8) S_{1tx}\big) \\
&- \big(48428561920S_5^2 \\
&+ 654609600S_5S_6 \\
&+ 600371360S_5S_8 \\
&+ 1929680S_6^2 + 9171338S_6S_8 \\
&+ 552802176S_7 \\
&- 10786905S_8^2\big) S_4^2
\end{aligned}
$$

$$- 1890\left(819624960S_2^2\right.$$
$$- 2209320960S_2S_5$$
$$- 7096320S_2S_6$$
$$+ 1626240S_2S_8$$
$$+ 204906240S_3S_4^2$$
$$+ 1478702080S_5^2$$
$$+ 9564160S_5S_6$$
$$- 2372480S_5S_8$$
$$+ 15360S_6^2 - 10274S_6S_8$$
$$\left.- 206976S_7 + 5775S_8^2\right) B_2)$$
$$+ 189\left(13113999360S_2^2\right.$$
$$- 357134131 20S_2S_5$$
$$- 113541120S_2S_6$$
$$+ 22767360S_2S_8$$
$$+ 1434343680S_3S_4^2$$
$$+ 24190668800S_5^2$$
$$+ 154603520S_5S_6$$
$$- 33214720S_5S_8$$
$$+ 245760S_6^2 - 95326S_6S_8$$
$$\left.+ 206976S_7 - 5775S_8^2\right) A_1^2$$
$$- 42\left(29506498560S_2^2\right.$$
$$- 79535554560S_2S_5$$
$$- 255467520S_2S_6$$
$$+ 105114240S_2S_8$$
$$- 101056032000S_3S_4^2$$
$$+ 31943598080S_5^2$$
$$+ 253961600S_5S_6$$
$$- 60480320S_5S_8 + 552960S_6^2$$
$$- 700362S_6S_8$$
$$\left.- 15044736S_7 + 976015S_8^2\right) S_1$$
$$- 663896217600\left(21S_1\right.$$
$$\left.- 11S_4^2\right) B_0S_4^2$$
$$+ 41309097984000\left(A_1\right.$$
$$\left.- S_4\right) C_{0t}S_4^2$$
$$+ 838252800\left(67A_1 + 6S_4\right) S_{6t}S_4^2$$
$$- 34927200\left(8624A_1S_5 + 77A_1S_8\right.$$
$$- 487872S_2S_4 + 265760S_4S_5$$
$$\left.+ 2112S_4S_6 - 3298S_4S_8\right) S_{1t}$$
$$- 73920\left(176576S_6 + 4499S_8\right.$$
$$\left.+ 51980272S_5\right) S_2S_4^2) S_4$$
$$- 582120\left(18627840S_3S_4^2\right.$$
$$+ 636160S_5^2 + 1120S_5S_6$$
$$+ 5680S_5S_8 + 52S_6S_8$$
$$+ 2688S_7 - 75S_8^2$$
$$\left.- 18480\left(112S_5 + S_8\right) S_2\right) S_{1x}$$
$$- 3725568000\left(48S_6 - 11S_8\right.$$
$$\left.+ 14944S_5 - 11088S_2\right) C_{0x}S_4^2$$
$$+ 698544000\left(192S_6 - 11S_8\right.$$
$$\left.+ 63472S_5 - 44352S_2\right) B_{2t}S_4^2$$
$$+ 4300800\left(17555S_6 - 2693S_8\right.$$
$$\left.+ 4417608S_5 - 5580036S_2\right) S_{5x}S_4^2$$
$$+ 13440\left(18360S_6 - 2657S_8\right.$$
$$+ 3797800S_5$$
$$\left.- 4241160S_2\right) S_{6x}S_4^2$$
$$+ 8870400\left(1617\left(\left(112S_5\right.\right.\right.$$
$$\left.\left.+ S_8\right) S_1 - 1728S_2S_4^2\right)$$
$$+ \left(8064S_6 - 4841S_8\right.$$
$$\left.\left.+ 1852816S_5\right) S_4^2\right) C_0S_4$$
$$- 3\left(4851\left(18627840S_3S_4^2\right.\right.$$
$$- 47676160S_5^2$$
$$- 142240S_5S_6 - 425680S_5S_8$$
$$- 1228S_6S_8 + 2688S_7 - 75S_8^2$$
$$\left.+ 277200\left(112S_5 + S_8\right) S_2\right) S_1$$
$$- \left(3098182348800B_0S_4^2\right.$$

$$
\begin{aligned}
&+ 1910545781760S_2^2 \\
&- 2436527681280S_2S_5 \\
&- 11120820480S_2S_6 \\
&+ 2970419760S_2S_8 \\
&+ 4109618177280S_3S_4^2 \\
&+ 394650233600S_5^2 \\
&+ 6026409760S_5S_6 \\
&- 4658100400S_5S_8 \\
&+ 12337920S_6^2 \\
&- 16712320S_6S_8 \\
&- 588303744S_7 \\
&+ 4150685S_8^2) S_4^2 \\
&- 186278400 (64S_6 - 11S_8 \\
&+ 20336S_5 - 14784S_2) C_0S_4^2) A_1 \\
&- 1663200 (1617 (192 (80C_0 - 57S_2) S_4^2 \\
&+ (112S_5 + S_8) S_1) \\
&+ (49392S_6 - 9451S_8 \\
&+ 15674288S_5) S_4^2 \\
&+ 42 (1536S_6 - 77S_8 + 509008S_5 \\
&- 354816S_2) A_1S_4) A_{1t}S_4) \\
&\cdot (167650560000S_4^3)^{-1},
\end{aligned}
\tag{A.20}
$$

$$
\begin{aligned}
S_{7tt} = {} & (852000145920000000B_{0tt}S_4^4 \\
&- 852000145920000000S_{3tt}S_4^4 \\
&+ 215151552000000S_{6tt}S_4^2S_5 \\
&+ 307359360000S_{6tt}S_4^2S_8 \\
&+ 852000145920000000B_0^2S_4^4 \\
&- 1704000291840000000H_1S_4^4 \\
&+ 5553778728960000S_2S_3S_4^2S_5 \\
&- 18646467840000S_2S_3S_4^2S_6 \\
&+ 3688312320000S_2S_3S_4^2S_8 \\
&- 1761779712 0000S_2S_5^3 \\
&- 3671754240000S_2S_5^2S_6 \\
&- 471905280000S_2S_5^2S_8 \\
&- 172480000S_2S_5S_6^2 \\
&- 11718537600S_2S_5S_6S_8 \\
&+ 4367800729600S_2S_5S_7 \\
&- 2550240000S_2S_5S_8^2 \\
&- 18098080S_2S_6^2S_8 \\
&+ 256650240S_2S_6S_7 \\
&+ 46985400S_2S_6S_8^2 \\
&+ 2848384000S_2S_7S_8 \\
&- 78540000S_2S_8^3 \\
&+ 3996655229952 0000S_3^2S_4^4 \\
&- 10295681679360000S_3S_4^2S_5^2 \\
&- 71642672640000S_3S_4^2S_5S_6 \\
&- 19314408960000S_3S_4^2S_5S_8 \\
&- 109438560000S_3S_4^2S_6^2 \\
&- 54992044800S_3S_4^2S_6S_8 \\
&+ 11248234905600S_3S_4^2S_7 \\
&- 48898080000S_3S_4^2S_8^2 \\
&+ 24367472640000S_5^4 \\
&+ 3398088960000S_5^3S_6 \\
&+ 717050880000S_5^3S_8 + 86176S_6^3S_8 \\
&- 29701056S_6^2S_7 - 56120S_6^2S_8^2 \\
&+ 4396160S_6S_7S_8 - 267075S_6S_8^3 \\
&- 1886760960S_7^2 - 15064000S_7S_8^2 \\
&+ 628125S_8^4 \\
&- 277200 (86240000S_2S_5 \\
&+ 252560S_2S_8 - 1397088000S_3S_4^2 \\
&+ 72800000S_5^2 + 672000S_5S_6 \\
&- 56160S_5S_8 \\
&+ 876S_6S_8 - 276864S_7 - 455S_8^2) S_{6t}S_4
\end{aligned}
$$

$$+ 6400\left(\left(2909900S_6 + 1186419S_8\right)S_6\right.$$
$$\left.- \left(440601056S_7 - 1100025S_8^2\right)\right)S_5^2$$
$$+ 170755200$$
$$\cdot\left(41S_6S_8 - 22848S_7 + S_6 + 14000S_5S_6\right)S_2^2$$
$$- 737662464000000\left(337S_5 - 231S_2\right)B_{0t}S_4^3$$
$$+ 18441561600000$$
$$\cdot\left(51S_6 + 4S_8 + 16312S_5 - 8008S_2\right)S_{3t}S_4^3$$
$$+ 17740800\left(2373S_6 - 500S_8\right.$$
$$\left.+ 152320S_5 + 1681680S_2\right)S_{7t}S_4$$
$$+ 40\left(2\left(2\left(224875S_6 + 349638S_8\right)S_6^2\right.\right.$$
$$\left.-75\left(520192S_7 - 29095S_8^2\right)S_8\right)$$
$$\left.- \left(304667776S_7 + 2689035S_8^2\right)S_6\right)S_5$$
$$- 30735936000\left(S_6S_8 - 672S_7\right.$$
$$\left.\left.+ 700S_5S_6 + 2217600S_3S_4^2\right)B_0S_4^2\right)$$
$$\cdot\left(103272744960000S_4^2\right)^{-1}, \tag{A.21}$$

$$C_{0t} = \left(-\left(\left(332640\left(4\left(45S_{6t}S_4\right.\right.\right.\right.\right.$$
$$- 27720B_0S_4^2 - 92400D_1S_4$$
$$\left.- 5544S_2^2 + 20110S_3S_4^2\right)S_4$$
$$\left.- 385\left(112S_5 + S_8\right)S_{1t}\right)$$
$$- \left(7120943360S_5^2\right.$$
$$+ 38913440S_5S_6$$
$$- 103088880S_5S_8 + 60480S_6^2$$
$$- 64408S_6S_8 - 769152S_7$$
$$\left.+ 57855S_8^2\right)S_4$$
$$- 9\left(819624960S_2^2\right.$$
$$- 2209320960S_2S_5$$
$$- 7096320S_2S_6$$
$$+ 1626240S_2S_8$$
$$+ 204906240S_3S_4^2$$
$$+ 1478702080S_5^2$$
$$+ 9564160S_5S_6$$
$$- 2372480S_5S_8 + 15360S_6^2$$
$$- 10274S_6S_8 - 206976S_7$$
$$\left.+ 5775S_8^2\right)A_1$$
$$+ 55440\left(828S_6 - 65S_8\right.$$
$$\left.\left.+ 330832S_5\right)S_2S_4\right)S_4$$
$$- 693\left(18627840S_3S_4^2\right.$$
$$+ 636160S_5^2$$
$$+ 1120S_5S_6 + 5680S_5S_8$$
$$+ 52S_6S_8 + 2688S_7$$
$$\left.- 75S_8^2 - 18480\left(112S_5 + S_8\right)S_2\right)S_1$$
$$- 4435200\left(48S_6 - 11S_8\right.$$
$$\left.+ 14944S_5 - 11088S_2\right)C_0S_4^2$$
$$+ 1663200\left(192S_6 - 11S_8\right.$$
$$\left.\left.+ 63472S_5 - 44352S_2\right)A_{1t}S_4^2\right)\Big)$$
$$\cdot\left(245887488000S_4^3\right)^{-1}, \tag{A.22}$$

$$S_{6tt} = \left(-\left(\left(36960\left(15\left(3049200\left(672\left(10\left(D_{1t}\right.\right.\right.\right.\right.\right.\right.$$
$$\left.\left.- 3H_2\right) -3B_{0t}S_4\right)S_4^2$$
$$\left.- \left(A_{1tt}S_4 + 7S_{1tt}\right)\left(112S_5 + S_8\right)\right)S_4$$
$$+ \left(86240A_1S_4S_5 + 2387A_1S_4S_8\right.$$
$$+ 1509200S_1S_5$$
$$+ 5390S_1S_8 + 3326400S_2S_4^2$$
$$- 3437680S_4^2S_5 - 14400S_4^2S_6$$
$$\left.+ 7604S_4^2S_8\right)S_{6t}$$
$$+ 1848\left(72S_6 + 91S_8 + 104944S_5\right)S_2^2S_4$$
$$- 24\left(330129S_6 - 20551S_8\right.$$
$$\left.+ 94386224S_5\right)S_3S_4^3$$
$$+ 231\left(2383360S_5^2 + 7840S_5S_6\right.$$
$$+ 21280S_5S_8 + 28S_6S_8 - 2688S_7$$
$$\left.\left.+ 75S_8^2 - 18627840S_3S_4^2\right)S_{1t}\right)$$

$$- \left(3544441600S_5^2 + 15097240S_5S_6\right.$$
$$+ 703120S_5S_8$$
$$+ 17280S_6^2 + 3970S_6S_8$$
$$+ 3094464S_7 + 13785S_8^2$$
$$\left.- 343790092800S_3S_4^2\right) S_2S_4$$
$$- 19958400 (2695S_1 - 7099S_4^2$$
$$+ 385A_1S_4) S_{3t}S_4^2$$
$$- 2880 \left(2695S_1 - 3183S_4^2 - 539A_1S_4\right) S_{7t}$$
$$+ 5940 \left(896S_7 - 25S_8^2\right.$$
$$\left.+ 14S_6S_8 + 2069760S_3S_4^2\right) C_0S_4$$
$$- 5544000 (24S_6 - 11S_8$$
$$\left.+ 6856S_5 - 5544S_2) D_1S_4^2\right)$$
$$+ \left(160 \left(\left(46268855S_6 - 4733737S_8\right) S_6\right.\right.$$
$$\left.+ 293259136S_7 + 5053935S_8^2\right) S_5$$
$$+ 13742136217600S_5^3$$
$$+ 150598419200S_5^2S_6$$
$$- 639894464000S_5^2S_8$$
$$+ 1382400S_6^3 - 828024S_6^2S_8$$
$$+ 190854144S_6S_7 - 2697315S_6S_8^2$$
$$\left.- 214158080S_7S_8 + 5466300S_8^3\right) S_4) S_4$$
$$- 385 \left(776160 (S_6S_8 - 192S_7\right.$$
$$+ 280S_5S_6) S_2$$
$$- \left(1341502525440S_3S_4^2S_5\right.$$
$$+ 3259872000S_3S_4^2S_6$$
$$+ 3520661760S_3S_4^2S_8$$
$$- 40040448000S_5^3 - 183456000S_5^2S_6$$
$$- 371804160S_5^2S_8 - 431200S_5S_6^2$$
$$- 274176S_5S_6S_8 + 19726336S_5S_7$$
$$- 3265920S_5S_8^2 - 364S_6^2S_8$$
$$+ 182784S_6S_7 - 441S_6S_8^2$$
$$\left.\left.+ 103936S_7S_8 - 11220S_8^3\right)\right) S_1$$
$$+ 15367968000 \left(77 \left(192 (10C_0 - 3S_2) S_4^2\right.\right.$$
$$\left.+ (112S_5 + S_8) S_1\right)$$
$$\left.+ (120S_6 + 17S_8 + 60272S_5) S_4^2\right) B_0S_4^2$$
$$- 91476000 \left(3696 \left(5040 (4B_0 + S_3) S_4^2\right.\right.$$
$$\left.+ (112S_5 + S_8) S_2\right)$$
$$- \left(603904S_5^2 + 1792S_5S_6\right.$$
$$+ 5392S_5S_8 - 26S_6S_8$$
$$\left.\left.- 2688S_7 + 75S_8^2\right)\right) A_{1t}S_4^2$$
$$+ 11 \left(11176704000 (24S_6\right.$$
$$- 11S_8 + 6856S_5$$
$$- 5544S_2) B_0S_4^2$$
$$+ 21451820544000S_3S_4^2S_5$$
$$+ 71624044800S_3S_4^2S_6$$
$$- 20676902400S_3S_4^2S_8$$
$$- 4004044800S_5^3$$
$$+ 133593600S_5^2S_6$$
$$- 107251200S_5^2S_8$$
$$- 39200S_5S_6^2 + 49007280S_5S_6S_8$$
$$+ 3140157440S_5S_7$$
$$- 81867600S_5S_8^2 + 188188S_6^2S_8$$
$$+ 12348672S_6S_7 - 476955S_6S_8^2$$
$$- 10622080S_7S_8$$
$$+ 249900S_8^3 - 55440 \left(847S_6S_8\right.$$
$$+ 59136S_7 - 1125S_8^2 + 7840S_5S_6$$
$$\left.\left.+ 279417600S_3S_4^2\right) S_2\right) A_1S_4))$$
$$\cdot \left(5532468480000S_4^4\right)^{-1}, \tag{A.23}$$

$$B_{0t} = \left(3688312320000S_{3t}S_4^3\right.$$
$$- 620928000S_{6t}S_4S_5$$
$$+ 277200S_{6t}S_4S_8$$
$$+ 372556800S_{7t}S_4$$

$$
\begin{aligned}
&- 1106493696000B_0S_2S_4^2 \\
&+ 1614235392000B_0S_4^2S_5 \\
&+ 4790016000B_0S_4^2S_6 \\
&+ 11064936960000D_0S_4^3 \\
&+ 442597478400S_2S_3S_4^2 \\
&+ 62092800S_2S_5S_6 \\
&- 27720S_2S_6S_8 - 74511360S_2S_7 \\
&- 826011648000S_3S_4^2S_5 \\
&- 2567980800S_3S_4^2S_6 \\
&- 119750400S_3S_4^2S_8 \\
&+ 572006400S_5^3 \\
&- 19084800S_5^2S_6 \\
&+ 15321600S_5^2S_8 + 5600S_5S_6^2 \\
&+ 38760S_5S_6S_8 \\
&+ 363305920S_5S_7 + 82800S_5S_8^2 \\
&+ 512S_6^2S_8 + 49728S_6S_7 \\
&- 1935S_6S_8^2 - 34880S_7S_8 \\
&+2550S_8^3\Big) \\
&\cdot \left(11064936960000S_4^3\right)^{-1},
\end{aligned}
\tag{A.24}
$$

$$
\begin{aligned}
S_{6xx} = {} & \Big(7\Big(33\Big(240\left(108662400H_4\right. \\
&- 778447S_2S_4^2 \\
&+ 2199120D_2S_4 \\
&- 18110400D_{2x} \\
&\left.- 1164240B_{2t}S_4\right)S_4^2 \\
&- 136171\left(112S_5 + S_8\right)S_1^2 \\
&- 53760\Big(4851S_1 \\
&- 2188S_4^2\Big)C_0S_4^2\Big) \\
&+ 2\Big(4303031040B_2S_2 \\
&- 2283635200B_2S_5 \\
&- 18627840B_2S_6 \\
&+ 20913200B_2S_8 \\
&- 1434343680C_0S_4^2 \\
&- 14343436800D_2S_4 \\
&+ 1290909312S_1S_2 \\
&- 1052792048S_1S_5 \\
&- 5588352S_1S_6 \\
&+ 6852307S_1S_8 \\
&+ 2189453112S_2S_4^2 \\
&- 3051480S_4^2S_5 \\
&+ 2469168S_4^2S_6 \\
&- 4097730S_4^2S_8\Big)A_1S_4 \\
&- 388080\left(48A_1 - 79S_4\right)S_{6x}S_4^2\Big) \\
&- \Big(166660080S_6 + 979694487S_8 \\
&+ 9031282208S_5\Big)S_4^4 \\
&+ 1764\Big(7528052S_5 \\
&- 27160S_6 + 99515S_8 \\
&+ 15878940S_2\Big)S_1S_4^2 \\
&+ 258720\Big(297S_6 + 14S_8 \\
&+77780S_5 - 141372S_2\Big)B_2S_4^2 \\
&- 3018400\Big(432S_6 \\
&- 485S_8 + 52960S_5 \\
&- 99792S_2\Big)\left(B_{2x} - D_3\right)S_4 \\
&- 3773\Big(6912S_6 - 13183S_8 \\
&+ 819440S_5 - 1596672S_2\Big)A_1^2S_4^2\Big) \\
&\cdot \left(162993600S_4^2\right)^{-1}.
\end{aligned}
\tag{A.25}
$$

Conflicts of Interest

The authors declare that there are no conflicts of interest regarding the publication of this paper.

Acknowledgments

The first author would like to thank Naresuan University, Thailand, for the partial financial support of this project under Grant no. R2560C104.

References

[1] S. Lie, "Klassifikation und Integration von gewöhnlichen Differentialgleichungen zwischen x,y, die eine Gruppe von Transformationen gestatten. III," *Archiv for Matematik og Naturvidenskab*, vol. 8, no. 4, pp. 371–427, 1883.

[2] R. Liouville, "Sur les Invariantes de Certaines Equations," *J. de lÉcole Polytechnique*, vol. 59, p. 88, 1889.

[3] A. M. Tresse, "Détermination des Invariants Ponctuels de lÉquation Différentielle Ordinaire du Second Ordre y"= f(x,y,y')," in *Preisschriften der fürstlichen Jablonowski schen Geselschaft XXXII*, S. Herzel, Ed., S.Herzel, Leipzig, 1896.

[4] E. Cartan, "Sur les variétés à connexion projective," *Bulletin de la Société Mathématique de France*, vol. 52, pp. 205–241, 1924.

[5] F. M. Mahomed and P. G. Leach, "Symmetry Lie algebras of *n*th order ordinary differential equations," *Journal of Mathematical Analysis and Applications*, vol. 151, no. 1, pp. 80–107, 1990.

[6] A. V. Bocharov, V. V. Sokolov, and S. I. Svinolupov, "On some equivalence problems for differential equations," Preprint ESI, Vienna 54, 12 pages, 1993.

[7] G. Grebot, "The characterization of third order ordinary differential equations admitting a transitive fiber-preserving point symmetry group," *Journal of Mathematical Analysis and Applications*, vol. 206, no. 2, pp. 364–388, 1997.

[8] N. H. Ibragimov and S. V. Meleshko, "Linearization of third-order ordinary differential equations by point and contact transformations," *Journal of Mathematical Analysis and Applications*, vol. 308, no. 1, pp. 266–289, 2005.

[9] N. H. Ibragimov, S. V. Meleshko, and S. Suksern, "Linearization of fourth-order ordinary differential equations by point transformations," *Journal of Physics A: Mathematical and Theoretical*, vol. 41, no. 23, Article ID 235206, 2008.

[10] S. Suksern and W. Pinyo, "On the fiber preserving transformations for the fifth-order ordinary differential equations," *Journal of Applied Mathematics*, vol. 2014, Article ID 735910, 2014.

[11] S.-s. Chern, "The geometry of the differential equation y'"= F(x,y,y',y")," in *Rep. Nat*, vol. 4, pp. 97–111, Tsing Hua Univ, 1940.

[12] B. Doubrov, "Contact trivialization of ordinary differential equations," in *Proceedings of the 8th International Conference on Differential Geometry and Its Applications*, O. Kowalski, D. Krupka, and., and J. Slovak, Eds., pp. 73–84, Silesian University in Opava, Opava, Czech Republic, 2001.

[13] B. Doubrov, B. Komrakov, and T. Morimoto, "Equivalence of holonomic differential equations," *Lobachevskii Journal of Mathematics*, vol. 3, pp. 39–71, 1999.

[14] V. N. Gusyatnikova and V. A. Yumaguzhin, "Contact transformations and local reducibility of ODE to the form y'"=0," *Acta Applicandae Mathematicae*, vol. 56, no. 2-3, pp. 155–179, 1999.

[15] R. Dridi and S. Neut, "On the geometry of y(4)=f(x,y,y',y",y'")," *Journal of Mathematical Analysis and Applications*, vol. 323, no. 2, pp. 1311–1317, 2006.

[16] S. Suksern, S. V. Meleshko, and N. H. Ibragimov, "Criteria for fourth-order ordinary differential equations to be linearizable by contact transformations," *Communications in Nonlinear Science and Numerical Simulation*, vol. 14, no. 6, pp. 2619–2628, 2009.

[17] S. Suksern, "Reduction of fifth-order ordinary differential equations to linearizable form by contact transformations," *Differential Equations and Dynamical Systems*, 2017.

[18] W. Nakpim, "Third-order ordinary differential equations equivalent to linear second-order ordinary differential equations via tangent transformations," *Journal of Symbolic Computation*, vol. 77, pp. 63–77, 2016.

[19] S. Suksern and S. V. Meleshko, "Applications of tangent transformations to the linearization problem of fourth-order ordinary differential equations," *Far East Journal of Applied Mathematics*, vol. 86, no. 2, pp. 145–172, 2014.

[20] L. G. S. Duarte, I. C. Moreira, and F. C. Santos, "Linearization under nonpoint transformations," *Journal of Physics A: Mathematical and General*, vol. 27, no. 19, pp. L739–L743, 1994.

[21] C. Muriel and J. L. Romero, "Nonlocal transformations and linearization of second-order ordinary differential equations," *Journal of Physics A: Mathematical and Theoretical*, vol. 43, no. 43, Article ID 434025, 2010.

[22] M. T. Mustafa, A. Y. Al-Dweik, and R. A. Mara'beh, "On the linearization of second-order ordinary differential equations to the laguerre form via generalized Sundman transformations," *Symmetry, Integrability and Geometry: Methods and Applications (SIGMA)*, vol. 9, article no. 041, 2013.

[23] W. Nakpim and S. V. Meleshko, "Linearization of second-order ordinary differential equations by generalized sundman transformations," *Symmetry, Integrability and Geometry: Methods and Applications (SIGMA)*, vol. 6, article no. 051, 2010.

[24] N. Euler, T. Wolf, P. G. Leach, and M. Euler, "Linearizable third-order ordinary differential equations and generalized Sundman transformations: The Case X'"=0," *Acta Applicandae Mathematicae*, vol. 76, no. 1, pp. 89–115, 2003.

[25] W. Nakpim and S. V. Meleshko, "Linearization of third-order ordinary differential equations by generalized Sundman transformations: The case X'"+αX=0," *Communications in Nonlinear Science and Numerical Simulation*, vol. 15, no. 7, pp. 1717–1723, 2010.

[26] L. M. Berkovich, "The integration of ordinary differential equations: factorization and transformations," *Mathematics and Computers in Simulation*, vol. 57, no. 3-5, pp. 175–195, 2001.

[27] L. M. Berkovich, "Factorization and transformations of differential equations," *Methods and applications, RC Dynamics*, 2002 (Russian).

[28] S. Suksern and S. Tammakun, "Linearizability of nonlinear fourth-order ordinary differential equations by a generalized sundman transformation," *Far East Journal of Applied Mathematics*, vol. 86, no. 3, pp. 183–210, 2014.

14

Uniqueness Results for Higher Order Elliptic Equations in Weighted Sobolev Spaces

Loredana Caso, Patrizia Di Gironimo, Sara Monsurrò, and Maria Transirico

University of Salerno, Via Giovanni Paolo II, No. 132, 84084 Fisciano, Italy

Correspondence should be addressed to Sara Monsurrò; smonsurro@unisa.it

Academic Editor: P. A. Krutitskii

We prove some uniqueness results for the solution of two kinds of Dirichlet boundary value problems for second- and fourth-order linear elliptic differential equations with discontinuous coefficients in polyhedral angles, in weighted Sobolev spaces.

1. Introduction

The Dirichlet problem for polyharmonic equations in bounded domains of $\mathbb{R}^n$ has been studied, among the first, by Sobolev in [1].

The problem was developed in various directions. For instance, Vekua in [2, 3] considers different boundary value problems in not necessarily bounded domains for harmonic, biharmonic, and metaharmonic functions. Successively, analogous problems in more general cases, for what concerns domains and operators, have been studied with different methods by many authors (see, e.g., [4–7]).

In particular, in [7], the author obtains a uniqueness result for the Dirichlet problem for polyharmonic operators of order $2m$ in polyhedral angles of $\mathbb{R}^n$. This result has been later on generalized, in [5], to the case of operators in divergence form of order $2m$ with discontinuous bounded measurable elliptic coefficients.

In [6] the authors study a boundary value problem for biharmonic functions in presence of nonregular points on the boundary of the domain. It is well known that in the neighborhood of these singular points (corners or edges) the solution of the problem presents a singularity that can be characterized by the presence of a suitable weight.

Uniqueness results for different Dirichlet problems in weighted Sobolev spaces for different classes of weights can be found in [8–12]. Studies of Dirichlet problems in the framework of weighted Sobolev spaces and in the case of unbounded domains can be found in [13–22].

In this paper, we extend the results of [5, 7] to the case of weighted Sobolev spaces. More precisely, we prove some uniqueness results for the solution of two kinds of Dirichlet boundary value problems for second- and fourth-order linear elliptic differential equations with discontinuous coefficients in the polyhedral angle $\mathbb{R}^n_l$, $0 \le l \le n-1$, $n \ge 2$, in weighted Sobolev spaces.

The first problem we consider is the following:

$$\sum_{i,j=1}^{n} \left(a_{ij} u_{x_i}\right)_{x_j} = f, \quad f \in L^2_{-s}\left(\mathbb{R}^n_l\right), \qquad u \in \overset{\circ}{W}{}^{1,2}_s\left(\mathbb{R}^n_l\right), \tag{1}$$

where, for $k \in \mathbb{N}_0$ and $s \in \mathbb{R}$, $W^{k,2}_s(\Omega)$ denotes a weighted Sobolev space where the weight is a power of the distance from the origin, $\overset{\circ}{W}{}^{k,2}_s(\Omega)$ is the closure of $C^\infty_0(\Omega)$ in $W^{k,2}_s(\Omega)$, and $W^{0,2}_s(\Omega) = L^2_s(\Omega)$; see Section 2 for details.

The second problem we study is

$$\sum_{i,j=1}^{n} \left(a_{ij} u_{x_i x_j}\right)_{x_i x_j} = f, \quad f \in L^2_{-s}\left(\mathbb{R}^n_l\right), \qquad u \in \overset{\circ}{W}{}^{2,2}_s\left(\mathbb{R}^n_l\right). \tag{2}$$

In both cases the coefficients a_{ij} belong to some weighted Sobolev spaces.

The main tool in our analysis is a generalization of the Hardy's inequality proved by Kondrat'ev and Olènik in [23].

2. Preliminary Results

Let Ω be an open subset of $\mathbb{R}^n$ with $n \geq 2$, whose boundary contains $x = 0$. For $k \in \mathbb{N}_0$ and $s \in \mathbb{R}$, $W_s^{k,2}(\Omega)$ denotes the space of all functions $u : \Omega \to \mathbb{R}$ such that $|x|^s D^\alpha u \in L^2(\Omega)$ for $|\alpha| \leq k$, normed by

$$\|u\|_{W_s^{k,2}(\Omega)} = \sum_{|\alpha|\leq k} \left\| |x|^s D^\alpha u \right\|_{L^2(\Omega)},$$

$$\overset{\circ}{W}{}_s^{k,2}(\Omega) \text{ is the closure of } C_0^\infty(\Omega) \text{ in } W_s^{k,2}(\Omega), \tag{3}$$

$$W_s^{0,2}(\Omega) = L_s^2(\Omega).$$

From [24] and Propositions 6.3 and 6.5, we get the following.

Proposition 1. *If G is a bounded open subset in $\mathbb{R}^n$ with $0 \in \partial G$, then*

$$W_s^{k,2}(G) \hookrightarrow W^{k,2}(G) \quad \text{for } s \leq 0. \tag{4}$$

Furthermore, for each $q \in [1, 2[$ there exists $\epsilon_0 = \epsilon_0(q) > 0$ such that

$$W_s^{k,2}(G) \hookrightarrow W^{k,q}(G) \quad \text{for } 0 < s \leq \epsilon_0. \tag{5}$$

In the present paper we use the following notation:

(i) $V \subset \mathbb{R}^n$ is a cone with vertex in the origin of coordinates;

(ii) B_R, $R > 0$, is the open ball of center in the origin and radius R;

(iii) $V_R = V \cap B_R$;

(iv) for every $l \in \{0, \dots, n-1\}$,

$$\mathbb{R}_l^n = \{x = (x_1, x_2, \dots, x_n) \in \mathbb{R}^n : x_i > 0,\ i = n - l, \dots, n\}, \tag{6}$$

is the "polyhedral angle" with vertex in the origin;

(v) $\mathbb{R}_+^n = \mathbb{R}_0^n$ is the half-space;

(vi) $Q_R = \mathbb{R}_l^n \cap B_R$.

To prove our main results, consisting in two uniqueness theorems, we will use the following inequality. We observe that this is a slightly modified version of a generalized Hardy's inequality that was proved by Kondrat'ev and Olènik in [23], adapted to our needs (see also [5]).

Lemma 2 (generalized Hardy's inequality). *Let $p > 1$ and $r \in \mathbb{R}$ be such that $r + n - p \neq 0$. Assume that for a sufficiently smooth function g the following condition is fulfilled:*

$$\int_{V_{R_2} \setminus V_{R_1}} |x|^r |\nabla g(x)|^p \, dx < +\infty, \tag{7}$$

where $\nabla g = (\partial g/\partial x_1, \dots, \partial g/\partial x_n)$ is the gradient of the function g and $0 < R_1 < R_2$. Then, there exist two constants $M, K > 0$ such that

$$\int_{V_{R_2} \setminus V_{R_1}} |x|^{r-p} |g(x) - M|^p \, dx < K \int_{V_{R_2} \setminus V_{R_1}} |x|^r |\nabla g(x)|^p \, dx, \tag{8}$$

where K does not depend on the function g, R_1, and R_2. If, in addition, $g(0) = 0$ then $M = 0$.

Remark 3. We remark that there are always important restrictions on the dimension n of the space, the order of "singularity" r, and the summability exponent p (see, e.g., [23, 25–29], where different variants of Hardy or Caffarelli-Kohn-Nirenberg type inequalities are proved).

3. Dirichlet Problem for Second-Order Elliptic Equations

We consider the following differential operator in divergence form in the polyhedral angle $\mathbb{R}_l^n$, $0 \leq l \leq n - 1$:

$$\sum_{i,j=1}^n \left(a_{ij} u_{x_i}\right)_{x_j}, \tag{9}$$

where the coefficients a_{ij} are measurable functions such that there exist two positive constants λ and μ such that

$$\lambda |x|^{2s} |\xi|^2 \leq \sum_{i,j=1}^n a_{ij}(x) \xi_i \xi_j \leq \mu |x|^{2s} |\xi|^2 \quad \text{a.e. in } \mathbb{R}_l^n,\ \forall \xi \in \mathbb{R}^n. \tag{10}$$

We study the Dirichlet problem

$$\sum_{i,j=1}^n \left(a_{ij} u_{x_i}\right)_{x_j} = f, \quad \text{a.e. in } \mathbb{R}_l^n, \qquad u \in \overset{\circ}{W}{}_s^{1,2}(\mathbb{R}_l^n), \tag{11}$$

where $f \in L_{-s}^2(\mathbb{R}_l^n)$.

Definition 4. We say that a function u is a generalized solution of problem (11) if it satisfies the integral identity

$$\int_{Q_R} \sum_{i,j=1}^n a_{ij} u_{x_i} v_{x_j} \, dx = -\int_{Q_R} f v \, dx, \tag{12}$$

for any $R > 0$ and any function $v \in \overset{\circ}{W}{}_s^{1,2}(Q_R)$.

Now we prove our first uniqueness result.

Theorem 5. *Let $u \in \overset{\circ}{W}{}_s^{1,2}(\mathbb{R}_l^n)$ be a generalized solution of problem (11), with $f = 0$. Then there exists $\epsilon_0 > 0$ such that if $s \leq \epsilon_0/2$ and $s \neq (2 - n)/2$ one has $u \equiv 0$ in $\mathbb{R}_l^n$.*

Proof. Let $\Theta(t)$ be an auxiliary function in $C_0^\infty([0,\infty[)$ defined by

$$\Theta(t) \equiv \begin{cases} 1 & 0 \le t \le 1, \\ \theta(t) & 1 \le t \le 2, \\ 0 & t \ge 2, \end{cases} \tag{13}$$

where $\theta(t)$ is such that $0 \le \theta(t) \le 1$. Let us also assume that there exists a positive constant K_0 such that

$$\left|\Theta'(t)\right|^2 \le K_0 \Theta(t). \tag{14}$$

Set, for any $R > 0$,

$$\Theta_R(x) = \Theta\left(\frac{|x|}{R}\right). \tag{15}$$

Note that the function Θ_R is such that, for any $j = 1, \ldots, n$, one has

$$(\Theta_R)_{x_j}(x) = \Theta'\left(\frac{|x|}{R}\right)\frac{x_j}{R|x|}. \tag{16}$$

Let $u \in \overset{\circ}{W}{}_s^{1,2}(\mathbb{R}_l^n)$ be a generalized solution of problem (11), with $f = 0$. We put

$$v_R(x) = u(x)\Theta_R(x). \tag{17}$$

Clearly, by definition of Θ_R and as a consequence of our boundary condition, one has that $v_R \in \overset{\circ}{W}{}_s^{1,2}(Q_{2R})$.

Thus, using v_R as test function in (12), we get

$$\begin{aligned} &\int_{Q_{2R}} \sum_{i,j=1}^n a_{ij} u_{x_i} u_{x_j} \Theta_R(x)\,dx \\ &\quad + \int_{Q_{2R}\setminus Q_R} \sum_{i,j=1}^n a_{ij} u_{x_i} u (\Theta_R)_{x_j}(x)\,dx = 0. \end{aligned} \tag{18}$$

From (10), (16), and (18) we deduce that there exists a positive constant $K_1 = K_1(n,\mu)$ such that

$$\begin{aligned} &\int_{Q_{2R}} \sum_{i,j=1}^n a_{ij} u_{x_i} u_{x_j} \Theta_R(x)\,dx \\ &= \left|\int_{Q_{2R}\setminus Q_R} \sum_{i,j=1}^n a_{ij} u_{x_i} u \Theta'\left(\frac{|x|}{R}\right)\frac{x_j}{R|x|}dx\right| \\ &\le K_1 \int_{Q_{2R}\setminus Q_R} |x|^{2s} u_x \Theta'\left(\frac{|x|}{R}\right)\frac{|u|}{|x|}dx \\ &= K_1 \int_{Q_{2R}\setminus Q_R} |x|^{s} u_x \Theta'\left(\frac{|x|}{R}\right)|x|^{s-1}|u|\,dx, \end{aligned} \tag{19}$$

where u_x denotes the modulus of the gradient of u.

By applying Young's inequality one gets that for any $\epsilon > 0$

$$\begin{aligned} &\int_{Q_{2R}} \sum_{i,j=1}^n a_{ij} u_{x_i} u_{x_j} \Theta_R(x)\,dx \\ &\le \frac{\varepsilon}{2} K_1 \int_{Q_{2R}\setminus Q_R} |x|^{2s} u_x^2 \Theta'^2\left(\frac{|x|}{R}\right)dx \\ &\quad + \frac{K_1}{2\varepsilon}\int_{Q_{2R}\setminus Q_R} |x|^{2s-2} u^2 dx. \end{aligned} \tag{20}$$

Thus, taking into account (14) and applying the generalized Hardy's inequality (8) (with $p = 2$ and $r = 2s$) to the second term in the right-hand side of (20), we deduce that if $s \neq (2-n)/2$,

$$\begin{aligned} &\int_{Q_{2R}} \sum_{i,j=1}^n a_{ij} u_{x_i} u_{x_j} \Theta_R(x)\,dx \\ &\le \frac{\varepsilon}{2} K_1 K_0 \int_{Q_{2R}\setminus Q_R} |x|^{2s} u_x^2 \Theta_R dx \\ &\quad + \frac{K_1}{2\varepsilon}\int_{Q_{2R}\setminus Q_R} |x|^{2s-2} u^2 dx \\ &\le \frac{\varepsilon}{2} K_1 K_0 \int_{Q_{2R}\setminus Q_R} |x|^{2s} u_x^2 \Theta_R dx \\ &\quad + \frac{K_1 K}{2\varepsilon}\int_{Q_{2R}\setminus Q_R} |x|^{2s} u_x^2 dx. \end{aligned} \tag{21}$$

From the ellipticity condition in (10) and for $\epsilon = \lambda/K_1K_0$, we have

$$\int_{Q_{2R}} |x|^{2s} u_x^2 \Theta_R dx \le K_2 \int_{Q_{2R}\setminus Q_R} |x|^{2s} u_x^2 dx, \tag{22}$$

where the constant $K_2 = K_2(n,\lambda,\mu,K_0,K)$.

Thus for any $P > 0$ and for any $R > P$ we obtain

$$\int_{Q_P} |x|^{2s} u_x^2 dx \le K_2 \int_{Q_{2R}\setminus Q_R} |x|^{2s} u_x^2 dx. \tag{23}$$

Since u is a generalized solution of problem (11), with $f = 0$, and the constant K_2 does not depend on the radius R and on the solution u, the right-hand side of (23) tends to zero when $R \to +\infty$ and then

$$\int_{Q_P} |x|^{2s} u_x^2 dx = 0 \quad \forall P > 0. \tag{24}$$

This implies that

$$|x|^{2s} u_x^2 = 0 \quad \text{a.e. in } Q_P \ \forall P > 0; \tag{25}$$

therefore

$$u_x = 0 \quad \text{a.e. in } Q_P \ \forall P > 0. \tag{26}$$

By Proposition 1 we deduce that if the solution $u \in W_s^{1,2}(Q_P)$ with $s \le 0$, then $u \in W^{1,2}(Q_P)$, for any $P > 0$. On the other hand, if $s > 0$ for any $q \in [1,2[$ there exists $\epsilon_0 = \epsilon_0(q) > 0$ such that if $0 < s \le \epsilon_0/2$, then $u \in W^{1,q}(Q_P)$ for any $P > 0$. Thus, by (26) the function $u(x)$ is a constant in $\mathbb{R}_l^n$, and since $u \in \overset{\circ}{W}{}_s^{1,2}(\mathbb{R}_l^n)$ one concludes that $u = 0$ in $\mathbb{R}_l^n$. □

4. Dirichlet Problem for 4th-Order Elliptic Equations

Let us now consider the following differential operator of 4th order in the polyhedral angle $\mathbb{R}_l^n$, $0 \le l \le n-1$,

$$\sum_{i,j=1}^{n} \left(a_{ij} u_{x_i x_j}\right)_{x_i x_j}, \tag{27}$$

where a_{ij} are measurable symmetric coefficients and there exist two positive constants λ and μ such that

$$\lambda |x|^{2s} \le a_{ij}(x) \le \mu |x|^{2s} \quad \text{a.e. in } \mathbb{R}_l^n,\ i,j = 1,\ldots n. \tag{28}$$

We want to prove a uniqueness result for the solution of the Dirichlet problem

$$\sum_{i,j=1}^{n} \left(a_{ij} u_{x_i x_j}\right)_{x_i x_j} = f, \quad \text{a.e. in } \mathbb{R}_l^n,$$
$$u \in \overset{\circ}{W}{}_s^{2,2}(\mathbb{R}_l^n), \tag{29}$$

where $f \in L^2_{-s}(\mathbb{R}_l^n)$.

Definition 6. We say that a function u is a generalized solution of problem (29) if it satisfies the integral identity

$$\int_{Q_R} \sum_{i,j=1}^{n} a_{ij} u_{x_i x_j} v_{x_i x_j} dx = \int_{Q_R} f v \, dx, \tag{30}$$

for any $R > 0$ and any function $v \in \overset{\circ}{W}{}_s^{2,2}(Q_R)$.

The result is the following.

Theorem 7. *Let $u \in \overset{\circ}{W}{}_s^{2,2}(\mathbb{R}_l^n)$ be a generalized solution of problem (29), with $f = 0$. Then there exists $\epsilon_0 > 0$ such that if $s \le \epsilon_0/2$ and $s \ne (2-n)/2, (4-n)/2$ one has $u \equiv 0$ in $\mathbb{R}_l^n$.*

Proof. We shall rely on the methods developed in [5, 7]. We consider the function $\Theta_R(x)$ defined in (13) and satisfying (14). Furthermore, we assume that there exists a positive constant K_1 such that

$$\left|\Theta''(t)\right|^2 \le K_1 \Theta(t). \tag{31}$$

Note that the function Θ_R is such that, for any $i, j = 1, \ldots, n$, one has (16) and

$$\left(\Theta_R\right)_{x_i x_j}(x) = \Theta''\left(\frac{|x|}{R}\right) \frac{x_i x_j}{R^2 |x|^2} + \Theta'\left(\frac{|x|}{R}\right) \frac{|x|^2 \delta_{ij} - x_i x_j}{R |x|^3}, \tag{32}$$

where δ_{ij} denotes the Kronecker delta.

Again we put

$$v_R(x) = u(x)\,\Theta_R(x), \tag{33}$$

where $u \in \overset{\circ}{W}{}_s^{2,2}(\mathbb{R}_l^n)$ is a generalized solution of problem (29), with $f = 0$.

Observe that the definition of Θ_R together with the boundary condition satisfied by u gives that $v_R \in \overset{\circ}{W}{}_s^{2,2}(Q_{2R})$. Hence, by the symmetry of a_{ij}, if we take v_R as test function in (30) we get

$$\begin{aligned}
&\int_{Q_{2R}} \sum_{i,j=1}^{n} a_{ij} u_{x_i x_j}^2 \Theta_R(x)\,dx \\
&\quad + 2 \int_{Q_{2R}\setminus Q_R} \sum_{i,j=1}^{n} a_{ij} u_{x_i x_j} u_{x_i} \left(\Theta_R\right)_{x_j}(x)\,dx \\
&\quad + \int_{Q_{2R}\setminus Q_R} \sum_{i,j=1}^{n} a_{ij} u_{x_i x_j} u \left(\Theta_R\right)_{x_i x_j}(x)\,dx = 0.
\end{aligned} \tag{34}$$

From (28) and (34) we deduce that

$$\begin{aligned}
&\lambda \int_{Q_{2R}} |x|^{2s} \sum_{i,j=1}^{n} u_{x_i x_j}^2 \Theta_R(x)\,dx \\
&\quad \le 2\mu \int_{Q_{2R}\setminus Q_R} |x|^{2s} \sum_{i,j=1}^{n} \left| u_{x_i x_j} u_{x_i} \left(\Theta_R\right)_{x_j}(x)\right| dx \\
&\quad + \mu \int_{Q_{2R}\setminus Q_R} |x|^{2s} \sum_{i,j=1}^{n} \left| u_{x_i x_j} u \left(\Theta_R\right)_{x_i x_j}(x)\right| dx.
\end{aligned} \tag{35}$$

By applying (16), (32), and Young's inequality one gets that there exist two positive constants $K_2 = K_2(n, \lambda, \mu, K_0)$ and $K_3 = K_3(n, \lambda, \mu, K_0, K_1)$ such that for any $\varepsilon, \varepsilon_1 > 0$

$$\begin{aligned}
&\int_{Q_{2R}} |x|^{2s} \sum_{i,j=1}^{n} u_{x_i x_j}^2 \Theta_R(x)\,dx \\
&\quad \le \varepsilon \int_{Q_{2R}\setminus Q_R} |x|^{2s} \sum_{i,j=1}^{n} u_{x_i x_j}^2 \Theta_R(x)\,dx \\
&\quad + \frac{\varepsilon_1}{2} \int_{Q_{2R}\setminus Q_R} |x|^{2s} \sum_{i,j=1}^{n} u_{x_i x_j}^2 \Theta_R(x)\,dx \\
&\quad + \frac{K_2}{\varepsilon} \int_{Q_{2R}\setminus Q_R} |x|^{2s-2} \sum_{i=1}^{n} u_{x_i}^2 dx \\
&\quad + \frac{K_3}{2\varepsilon_1} \int_{Q_{2R}\setminus Q_R} |x|^{2s-4} u^2 dx.
\end{aligned} \tag{36}$$

Thus, applying repeatedly the generalized Hardy's inequality (8) (with $p = 2$ and $r = 2s$ to the third integral on the right-hand side and with $p = 2$ and $r = 2s - 2$ to the last integral on the right-hand side and then again with $p = 2$ and $r = 2s$), we deduce that if $s \ne (2-n)/2, (4-n)/2$,

$$\begin{aligned}
&\int_{Q_{2R}} |x|^{2s} \sum_{i,j=1}^{n} u_{x_i x_j}^2 \Theta_R(x)\,dx \\
&\quad \le K_4 \int_{Q_{2R}\setminus Q_R} |x|^{2s} \sum_{i,j=1}^{n} u_{x_i x_j}^2 dx,
\end{aligned} \tag{37}$$

where the constant $K_4 = K_4(n, \lambda, \mu, K_0, K_1, K)$.

Thus for any $P > 0$ and for any $R > P$ we obtain

$$\int_{Q_P} |x|^{2s} \sum_{i,j=1}^{n} u_{x_i x_j}^2 dx \le K_4 \int_{Q_{2R}\setminus Q_R} |x|^{2s} \sum_{i,j=1}^{n} u_{x_i x_j}^2 dx. \quad (38)$$

Now, arguing as in the proof of Theorem 5, since u is a generalized solution of problem (29), with $f = 0$, the right-hand side of (38) tends to zero when $R \to +\infty$ and then

$$\int_{Q_P} |x|^{2s} \sum_{i,j=1}^{n} u_{x_i x_j}^2 dx = 0 \quad \forall P > 0. \quad (39)$$

This implies that

$$|x|^{2s} \sum_{i,j=1}^{n} u_{x_i x_j}^2 = 0 \quad \text{a.e. in } Q_P \ \forall P > 0; \quad (40)$$

therefore

$$u_{xx} = 0 \quad \text{a.e. in } Q_P \ \forall P > 0. \quad (41)$$

In view of Proposition 1 we obtain that if the solution $u \in W_s^{2,2}(Q_P)$ with $s \le 0$, then $u \in W^{2,2}(Q_P)$, for any $P > 0$, while if $s > 0$ for any $q \in [1, 2[$ there exists $\epsilon_0 = \epsilon_0(q) > 0$ such that if $0 < s \le \epsilon_0/2$, then $u \in W^{2,q}(Q_P)$ for any $P > 0$. Thus, by (41) the function u_x is constant a.e. in Q_P, and since $u \in \overset{\circ}{W}{}_s^{2,2}(\mathbb{R}_l^n)$ one concludes that $u_x = 0$ a.e. in Q_P, for any $P > 0$. The thesis follows then as the one of Theorem 5. □

Conflicts of Interest

The authors declare that they have no conflicts of interest.

References

[1] S. L. Sobolev, "Applications of functional analysis in mathematical physics," in *Translations of Mathematical Monographs*, vol. 7, American Mathematical Society, 1963, Izdat. Leningrad. Gos. Univ., Leningrad, 1950.

[2] I. N. Vekua, "On metaharmonic functions," *Lecture Notes of Tbilisi International Centre of Mathematics and Informatics Tbilisi Univ. Press*, vol. 14, 2013 (Russian).

[3] I. N. Vekua, *New Methods for Solving Elliptic Equations*, OGIZ, Moscow-Leningrad, 1948.

[4] F. Gazzola, H. Grunau, and G. Sweers, *Polyharmonic Boundary Value Problems*, vol. 1991 of *Lecture Notes in Mathematics*, Springer, Berlin, Germany, 2010.

[5] S. Monsurrò, I. Tavkhelidze, and M. Transirico, "Uniqueness results for the Dirichlet problem for higher order elliptic equations in polyhedral angles," *Boundary Value Problems*, vol. 2014, article 232, pp. 1–8, 2014.

[6] O. A. Olèinik, G. A. Iosifyan, and I. Tavkhelidze, "Asymptotic behavior of the solution of a biharmonic equation in the neighborhood of non regular points of the boundary of the domain at infinity," *Trudy Moskovskogo Matematičeskogo Obščestva*, vol. 4, pp. 166–175, 1981 (Russian).

[7] I. Tavkhelidze, "On some properties of solutions of polyharmonic equation in polyhedral angles," *Georgian Mathematical Journal*, vol. 14, no. 3, pp. 565–580, 2007.

[8] L. Caso, "Uniqueness results for elliptic problems with singular data," *Boundary Value Problems*, vol. 2006, Article ID 98923, 2006.

[9] L. Caso and R. D'Ambrosio, "Some uniqueness results for singular elliptic problems," *International Journal of Mathematics*, vol. 26, no. 3, Article ID 1550026, 2015.

[10] L. Caso, R. D'Ambrosio, and M. Transirico, "Solvability of the Dirichlet problem in $W^{2,p}$ for a class of elliptic equations with singular data," *Acta Mathematica Sinica*, vol. 30, no. 5, pp. 737–746, 2014.

[11] L. Caso, R. D'Ambrosio, and M. Transirico, "Well-posedness in weighted Sobolev spaces for elliptic equations of Cordes type," *Collectanea Mathematica*, vol. 67, no. 3, pp. 539–554, 2016.

[12] P. Di Gironimo, "Weighted system of elliptic equations: an existence and uniqueness theorem," *Far East Journal of Mathematical Sciences*, vol. 26, pp. 319–329, 2007.

[13] L. Caso, P. Cavaliere, and M. Transirico, "Existence results for elliptic equations," *Journal of Mathematical Analysis and Applications*, vol. 274, no. 2, pp. 554–563, 2002.

[14] L. Caso and M. Transirico, "A priori estimates for elliptic equations in weighted Sobolev spaces," *Mathematical Inequalities & Applications*, vol. 13, no. 3, pp. 655–666, 2010.

[15] P. Di Gironimo, "ABP inequality and weak Harnack inequality for fully nonlinear elliptic operators with coefficients in weighted spaces," *Far East Journal of Mathematical Sciences*, vol. 64, pp. 1–21, 2012.

[16] P. Di Gironimo, "Harnack inequality for fully nonlinear elliptic equations with coefficients in weighted spaces," *International Journal of Analysis and Applications*, vol. 15, no. 1, pp. 1–19, 2017.

[17] P. Di Gironimo and A. Vitolo, "Elliptic equations with discontinuous coefficients in weighted Sobolev spaces on unbounded domains," *Journal of Mathematical Analysis and Applications*, vol. 253, no. 1, pp. 297–309, 2001.

[18] P. Di Gironimo and A. Vitolo, "Existence and uniqueness results for elliptic equations with coefficients in locally Morrey spaces," *International Journal of Pure and Applied Mathematics*, vol. 4, pp. 181–194, 2003.

[19] S. Monsurrò and M. Transirico, "Dirichlet problem for divergence form elliptic equations with discontinuous coefficients," *Boundary Value Problems*, vol. 2012, article 67, 2012.

[20] S. Monsurrò and M. Transirico, "A priori bounds in L^p for solutions of elliptic equations in divergence form," *Bulletin des Sciences Mathematiques*, vol. 137, no. 7, pp. 851–866, 2013.

[21] S. Monsurrò and M. Transirico, "A $W^{2,p}$-estimate for a class of elliptic operators," *International Journal of Pure and Applied Mathematics*, vol. 83, no. 3, pp. 489–499, 2013.

[22] S. Monsurrò and M. Transirico, "Noncoercive elliptic equations with discontinuous coefficients in unbounded domains," *Nonlinear Analysis. Theory, Methods & Applications. An International Multidisciplinary Journal*, vol. 163, pp. 86–103, 2017.

[23] V. A. Kondrat'ev and O. A. Olènik, "Boundary value problems for the system of elasticity theory in unbounded domains. Korn's inequalities," *Akademiya Nauk SSSR i Moskovskoe Matematicheskoe Obshchestvo. Uspekhi Matematicheskikh Nauk*, vol. 43, article 5, pp. 55–98, 1988 (Russian).

[24] A. Kufner, *Weighted Sobolev Spaces*, vol. 31 of *Teubner-Texte zur Mathematik*, 1980.

[25] B. Bojarski and P. Hajlasz, "Pointwise inequalities for Sobolev functions and some applications," *Studia Mathematica*, vol. 106, pp. 77–92, 1993.

[26] H. Brezis and M. Marcus, "Hardy's inequalities revisited," *Annali della Scuola Normale Superiore di Pisa. Classe di Scienze*, vol. 25, pp. 217–237, 1997.

[27] F. Catrina and Z.-Q. Wang, "On the Caffarelli-Kohn-Nirenberg inequalities: sharp constants, existence (and nonexistence), and symmetry of extremal functions," *Communications on Pure and Applied Mathematics*, vol. 54, no. 2, pp. 229–258, 2001.

[28] J. Dávila and L. Dupaigne, "Hardy-type inequalities," *Journal of the European Mathematical Society*, vol. 6, no. 3, pp. 335–365, 2004.

[29] S. Filippas, A. Tertikas, and J. Tidblom, "On the structure of Hardy-Sobolev-Maz'ya inequalities," *Journal of the European Mathematical Society*, vol. 11, no. 6, pp. 1165–1185, 2009.

15

Asymptotics for the Ostrovsky-Hunter Equation in the Critical Case

Fernando Bernal-Vílchis,[1] Nakao Hayashi,[2] and Pavel I. Naumkin[3]

[1] *Instituto de Matemáticas, UNAM, Campus Morelia, AP 61-3 Xangari, 58089 Morelia, MICH, Mexico*
[2] *Department of Mathematics, Graduate School of Science, Osaka University, Osaka, Toyonaka 560-0043, Japan*
[3] *Centro de Ciencias Matemáticas, UNAM, Campus Morelia, AP 61-3 Xangari, 58089 Morelia, MICH, Mexico*

Correspondence should be addressed to Nakao Hayashi; nhayashi@math.sci.osaka-u.ac.jp

Academic Editor: Baoxiang Wang

We consider the Cauchy problem for the Ostrovsky-Hunter equation $\partial_x(\partial_t u-(b/3)\partial_x^3 u-\partial_x\mathscr{K}u^3)=au$, $(t,x)\in\mathbb{R}^2$, $u(0,x)=u_0(x)$, $x\in\mathbb{R}$, where $ab>0$. Define $\xi_0=(27a/b)^{1/4}$. Suppose that $\mathscr{K}$ is a pseudodifferential operator with a symbol $\widehat{K}(\xi)$ such that $\widehat{K}(\pm\xi_0)=0$, $\operatorname{Im}\widehat{K}(\xi)=0$, and $|\widehat{K}(\xi)|\le C$. For example, we can take $\widehat{K}(\xi)=(\xi^2-\xi_0^2)/(\xi^2+1)$. We prove the global in time existence and the large time asymptotic behavior of solutions.

1. Introduction

We consider the Cauchy problem for the generalized Ostrovsky-Hunter equation

$$\partial_x\left(\partial_t u-\frac{b}{3}\partial_x^3 u-\partial_x f(u)\right)=au,\quad (t,x)\in\mathbb{R}^2,$$
$$u(0,x)=u_0(x),\quad x\in\mathbb{R}, \tag{1}$$

where $ab>0$, $f(u)=\mathscr{K}u^3$. We assume that $\mathscr{K}$ is a pseudodifferential operator with a symbol $\widehat{K}(\xi)$ such that $\widehat{K}(\pm\xi_0)=0$ with $\xi_0=(27a/b)^{1/4}$. Also we suppose that $\operatorname{Im}\widehat{K}(\xi)=0$ and $|\widehat{K}(\xi)|\le C$. For example, we can choose $\widehat{K}(\xi)=(\xi^2-\xi_0^2)/(\xi^2+1)$. Denote by $\Lambda(\xi)=a/\xi+(b/3)\xi^3$ the symbol of the linear part of (1). The constant $\xi_0=(27a/b)^{1/4}$ is a positive root of $\Omega(\xi)=\Lambda(\xi)-3\Lambda(\xi/3)=(8b/27)\xi^{-1}(\xi^4-27a/b)=0$. Our strategy of the proof of the main result is similar to the one used in [1]. We translate (1) into the ordinary differential equation by using the evolution operator related to the linear problem; then we divide the nonlinear term into resonance and nonresonance parts. Nonresonance part has an oscillating term $e^{it\Omega(\xi)}$ which yields better time decay through the integration by parts; however the factor $1/\Omega(\xi)$ gives us a singularity at ξ_0; see (37) for details. This is the reason why we assume the additional condition $\widehat{K}(\pm\xi_0)=0$ on the symbol $\widehat{K}(\xi)$.

We define the evolution group $\mathscr{U}(t)=\mathscr{F}^{-1}E\mathscr{F}$, where the multiplication factor $E=e^{-it\Lambda(\xi)}$, $\Lambda(\xi)=a/\xi+(b/3)\xi^3$. It is well known that the operator $\mathscr{J}=\mathscr{U}(t)x\mathscr{U}(-t)$ is a useful tool for obtaining the $\mathbf{L}^\infty$-time decay estimates of solutions and has been used widely for studying the asymptotic behavior of solutions to various nonlinear dispersive equations. We have

$$\begin{aligned}\mathscr{J}&=\mathscr{U}(t)x\mathscr{U}(-t)=\mathscr{F}^{-1}e^{-it\Lambda(\xi)}i\partial_\xi e^{it\Lambda(\xi)}\mathscr{F}\\&=\mathscr{F}^{-1}\left(i\partial_\xi-t\Lambda'(\xi)\right)\mathscr{F}=x-t\Lambda'(-i\partial_x)\\&=x-ta\partial_x^{-2}+tb\partial_x^2,\end{aligned} \tag{2}$$

where $\Lambda'(-i\partial_x)=a\partial_x^{-2}-b\partial_x^2$, and the antiderivative ∂_x^{-1} is defined by the Fourier transform such that

$$\widehat{\partial_x^{-1}\phi}(\xi)=(i\xi)^{-1}\widehat{\phi}(\xi). \tag{3}$$

Note that the commutators are true $[\mathscr{J},\mathscr{L}]=0$, $[\partial_x,\mathscr{L}]=0$, $[\mathscr{J},\partial_x]=-1$, $[\partial_x^{-1},x\partial_x]=-\partial_x^{-1}$, where $\mathscr{L}=\partial_t+\Lambda(-i\partial_x)=\partial_t-a\partial_x^{-1}-(b/3)\partial_x^3$. However, it seems that $\mathscr{J}$ does not work well on the nonlinear terms. In order to avoid the derivative loss, when estimating the norm $\|\partial_x\mathscr{J}u\|_{\mathbf{L}^2}$ instead of the operators $\mathcal{J}$ we apply the modified dilation operator defined by

$$\mathcal{P} = t\partial_t + \frac{1}{3}x\partial_x - \frac{4}{3}a\partial_a. \quad (4)$$

Note that $\mathcal{P}$ acts well on the nonlinear terms as the first-order differential operator and it almost commutes with $\mathcal{L}$: $[\mathcal{P}, \mathcal{L}] = -\mathcal{L}$. Also $\mathcal{J}$ and $\mathcal{P}$ are related via the identity

$$\mathcal{P} = t\mathcal{L} + \frac{1}{3}\mathcal{J}\partial_x - \frac{4}{3}a\mathcal{J}, \quad (5)$$

where

$$\mathcal{I} = \partial_a - t\partial_x^{-1}. \quad (6)$$

Note that $[\mathcal{I}, \mathcal{L}] = 0$. In order to get the estimate of $\partial_x \mathcal{J}u$, we will show the a priori estimates of , $t\mathcal{L}u$, and $\mathcal{I}u$. Different point compared to the previous works is to consider the estimate of $\mathcal{I}u$ since $\mathcal{I}u$ contains the term $t\partial_x^{-1}$ with an additional explicit time growth.

When $f(u) = u^2$, then (1) was introduced in [2] for modelling the small-amplitude long waves in a rotating fluid of finite depth. Therefore (1) with $f(u) = u^2$ is called the Ostrovsky equation. It was studied by many authors (see, e.g., [3–5] and references cited therein). When $b = 0$, (1) is called the reduced Ostrovsky equation. Equation (1) has some conservation quantities, when $f(u) = \lambda|u|^{p-1}u$, $\lambda \in \mathbb{R}$. One of them is the zero mass conservation law which is obtained by integrating in space

$$a \int u(t,x)\, dx = 0 \quad (7)$$

under the restriction $\int u_0(x)dx = 0$. Rewrite (1) as

$$\partial_t u - \frac{b}{3}\partial_x^3 u - \partial_x f(u) = aD_x^{-1}u. \quad (8)$$

Multiplying both sides of (8) by u, integrating in space, using (7), we obtain

$$\frac{d}{dt}\int |u(t,x)|^2\, dx + \frac{2\lambda}{p+1}\int |u(t,x)|^{p+1}\, dx = 0 \quad (9)$$

which is the conservation of the momentum. The same approach as in deriving (9) will be used for the high frequency part in order to avoid the derivative loss, when proving the existence of solutions of (1).

Local well-posedness for the Ostrovsky equation was shown in [5] in the case of the initial data

$$u_0 \in \mathbf{H}^s \cap \dot{\mathbf{H}}^{-1}, \quad s > \frac{3}{2} \quad (10)$$

by using the parabolic regularization technique and limiting arguments. Their method works also for the case of the generalized nonlinearity $f(u) = |u|^{p-1}u$ and also generalized reduced Ostrovsky equation (1), since the dispersive effects were not used in the proof. Thanks to the high frequency part u_{xxx}, the solutions to the linear equation $(u_t - \beta u_{xxx})_x = \gamma u$ obtain a smoothing property. By using this property, in [3], the local well-posedness for the Ostrovsky equation was shown under the condition

$$u_0 \in \mathbf{H}^s \cap \dot{\mathbf{H}}^{-1}, \quad s > \frac{3}{4}. \quad (11)$$

The method of [3] depends on the linear part of the equation and also works for the nonlinearities of a general order. In [4, 6–8] the local well-posedness for the Ostrovsky equation was treated by the Fourier restriction norm method of [9] and in [4] the $\mathbf{H}^{-3/4+}$ local well-posedness was shown. We note here that the Sobolev space $\mathbf{H}^{-3/4}$ is considered as critical regularity concerning the Korteweg-de Vries equation.

Global well-posedness in the energy class was obtained for the Ostrovsky equation in [3] through the energy conservation law, when the initial data

$$u_0 \in \mathbf{H}^1 \cap \dot{\mathbf{H}}^{-1}, \quad (12)$$

and $ab > 0$. After their work, the global well-posedness in

$$\mathbf{L}^2 \cap \dot{\mathbf{H}}^{-s}, \quad 0 \le s \le 1, \quad (13)$$

was proved in [4, 6] due to the $\mathbf{L}^2$-conservation law. The global well-posedness, in the negative order Sobolev space $\mathbf{H}^{-3/10+}$, was shown in [8] by using the I method of [10].

We now turn to the case of the reduced Ostrovsky equation. The local well-posedness was shown in the space $\mathbf{H}^2$ in paper [11] and after that in $\mathbf{H}^{3/2+}$ in [12]. Their methods work also in the case of the general nonlinear dispersive equations with different nonlinearities. We also refer to [13, 14] for the local well-posedness in the class

$$u_0 \in \mathbf{H}^m \cap \dot{\mathbf{H}}^{-1} \quad m \ge 2. \quad (14)$$

However there are few works on the global well-posedness for the reduced Ostrovsky equation due to the lack of the smoothing property. The global well-posedness for reduced Ostrovsky equation (1) with $b = 0$ and cubic nonlinearity $f(u) = u^3$ (which is called the short pulse equation) was obtained in [15], when the initial data

$$\|\partial_x u_0\|_{\mathbf{H}^1} < 1, \quad u_0 \in \mathbf{H}^2, \quad (15)$$

whereas for the quadratic nonlinearity $f(u) = u^2$ (which is called the reduced Ostrovsky equation or the Ostrovsky-Hunter equation; see [16, 17]), it was shown in [18] when the initial data

$$\left(1 - 3\partial_x^2\right) u_0(x) < 0, \quad u_0 \in \mathbf{H}^3, \quad (16)$$

for all $x \in \mathbf{R}$. The time decay properties of solutions to the corresponding linear problem can be studied if we assume that the initial data decay rapidly at infinity. So the global existence was shown in [12], for the nonlinearity $f(u) = u^p$ with an integer $p \ge 4$, when the initial data are small and sufficiently regular:

$$u_0 \in \mathbf{H}^5 \cap \mathbf{H}_1^3. \quad (17)$$

In [1, 19, 20], we considered the large time asymptotics for reduced Ostrovsky equation (1) with $b = 0$ and some conditions on the order of nonlinearity.

To state our results precisely we introduce *Notation and Function Spaces.* We denote the Lebesgue space by $\mathbf{L}^p = \{\phi \in \mathbf{S}'; \|\phi\|_{\mathbf{L}^p} < \infty\}$, where the norm $\|\phi\|_{\mathbf{L}^p} = (\int |\phi(x)|^p dx)^{1/p}$ for $1 \le p < \infty$ and $\|\phi\|_{\mathbf{L}^\infty} = \text{ess.sup}_{x\in\mathbf{R}}|\phi(x)|$ for $p = \infty$. The weighted Sobolev space is

$$\mathbf{H}^{k,s}_p = \left\{\varphi \in \mathbf{S}'; \|\phi\|_{\mathbf{H}^{k,s}_p} = \left\| \langle x\rangle^s \langle i\partial_x\rangle^k \phi \right\|_{\mathbf{L}^p} < \infty \right\}. \quad (18)$$

$k, s \in \mathbf{R}$, $1 \le p \le \infty$, $\langle x\rangle = \sqrt{1+x^2}$, and $\langle i\partial_x\rangle = \sqrt{1-\partial_x^2}$. We also use the notations $\mathbf{H}^{k,s} = \mathbf{H}^{k,s}_2$, $\mathbf{H}^k = \mathbf{H}^{k,0}$ shortly, if they do not cause any confusion. Let $\mathbf{C}(\mathbf{I};\mathbf{B})$ be the space of continuous functions from an interval $\mathbf{I}$ to a Banach space $\mathbf{B}$. Different positive constants might be denoted by the same letter C. We define the free evolution group $\mathscr{U}(t) = e^{-it\Lambda(-i\partial_x)} = \mathscr{F}^{-1}E\mathscr{F}$, where the multiplication factor $E(t,\xi) = e^{-it\Lambda(\xi)}$.

We are now in a position to state our main result.

Theorem 1. *Assume that the initial data $u_0 \in \mathbf{H}^2 \cap \mathbf{H}^{1,1}$ are real-valued with a sufficiently small norm $\|u_0\|_{\mathbf{H}^2\cap\mathbf{H}^{1,1}} \le \varepsilon$. Then there exists a unique global solution $u \in \mathbf{C}([0,\infty);\mathbf{H}^2)$ of Cauchy problem (1) satisfying the time decay estimate*

$$\|u(t)\|_{\mathbf{L}^\infty} \le C\varepsilon t^{-1/2}. \quad (19)$$

Moreover there exists a unique modified final state $W_+ \in \mathbf{L}^\infty$ such that the asymptotics

$$\begin{aligned} u(t) &= 2\text{Re}\, t^{-1/2} e^{-2it(a/\eta(x/t)-(b/3)\eta(x/t)^3)} \frac{W_+(\eta(x/t))}{\sqrt{\Lambda''(\eta(x/t))}} \\ &\cdot \exp\left(\frac{3i\eta(x/t)\widehat{K}(\eta(x/t))}{\langle\eta(x/t)\rangle\Lambda''(\eta(x/t))} \left|W_+\left(\eta\left(\frac{x}{t}\right)\right)\right|^2 \right. \\ &\left. \cdot \log t \right) + O\left(\varepsilon t^{-1/2-\delta}\right) \end{aligned} \quad (20)$$

is valid for $t \to \infty$ uniformly with respect to $x \in \mathbf{R}$, where $\delta > 0$ is a small constant and

$$\eta(x) = \sqrt{\frac{1}{2b}\left(x + \sqrt{4ab+x^2}\right)}. \quad (21)$$

2. Factorization Technique

We now introduce the factorization formulas for (1). We have for the free evolution group $\mathscr{U}(t) = \mathscr{F}^{-1}E\mathscr{F}$, where the multiplication factor $E = e^{-it\Lambda(\xi)}$, $\Lambda(\xi) = a/\xi + (b/3)\xi^3$. Denote the Heaviside function $\theta(\xi) = 1$ for $\xi > 0$ and $\theta(\xi) = 0$ for $\xi \le 0$. Then for the real-valued function $u(t) = \mathscr{U}(t)\mathscr{F}^{-1}\widehat{\varphi}$ we find

$$\begin{aligned} \mathscr{U}(t)\mathscr{F}^{-1}\widehat{\varphi} &= 2\text{Re}\mathscr{F}^{-1}\theta E\widehat{\varphi} \\ &= 2\text{Re}\frac{1}{\sqrt{2\pi}}\int_0^\infty e^{it((x/t)\xi-\Lambda(\xi))}\widehat{\varphi}(\xi)\,d\xi \\ &= 2\text{Re}\mathscr{D}_t\frac{|t|^{1/2}}{\sqrt{2\pi}}\int_0^\infty e^{it(x\xi-\Lambda(\xi))}\widehat{\varphi}(\xi)\,d\xi, \end{aligned} \quad (22)$$

where the dilation operator $\mathscr{D}_t\phi = |t|^{-1/2}\phi(xt^{-1})$. Note that there is a unique stationary point in the integral $\int_0^\infty e^{it(x\xi-\Lambda(\xi))}\phi(\xi)d\xi$, which is defined by the root $\xi = \eta(x) = \sqrt{(1/2b)(x+\sqrt{4ab+x^2})} > 0$ of the equation $\Lambda'(\xi) = -a/\xi^2 + b\xi^2 = x$ for all $x \in \mathbb{R}$. Thus $\Lambda'(\eta(x)) = x$ and we introduce the so-called scaling operator

$$\left(\mathscr{B}^{-1}\phi\right)(x) = \frac{1}{\sqrt{\Lambda''(\eta(x))}}\phi(\eta(x)) \quad (23)$$

and the multiplication factor

$$M(t,\eta) = e^{it(\eta\Lambda'(\eta)-\Lambda(\eta))} = e^{-2it(a/\eta-(b/3)\eta^3)}. \quad (24)$$

Note that, in the case of $b = 0$, then $\eta(x)$ is defined by $\Lambda'(\xi) = -a/\xi^2 = x$; namely, $\eta(x) = \sqrt{a/|x|}$ for $x < 0$. Hence

$$\left(\mathscr{B}^{-1}\phi\right)(x) = \frac{1}{\sqrt{3a}}\left(\frac{a}{|x|}\right)^{3/4}\phi\left(\sqrt{\frac{a}{|x|}}\right), \quad (25)$$

for $b = 0$, $x < 0$; see [1]. Therefore $\mathscr{B}^{-1}$ is the scaling operator if the symbol $\Lambda(\xi)$ is homogeneous.

By the definition of $\mathscr{B}^{-1}$, its inverse operator is defined by

$$(\mathscr{B}\phi)(\eta) = \sqrt{\Lambda''(\eta)}\phi\left(\Lambda'(\eta)\right). \quad (26)$$

Then we have

$$\begin{aligned} \mathscr{U}(t)\mathscr{F}^{-1}\phi &= 2\text{Re}\mathscr{D}_t\frac{|t|^{1/2}}{\sqrt{2\pi}}\int_0^\infty e^{it(x\xi-\Lambda(\xi))}\phi(\xi)\,d\xi \\ &= 2\text{Re}\mathscr{D}_t\mathscr{B}^{-1}\mathscr{B}\frac{|t|^{1/2}}{\sqrt{2\pi}}\int_0^\infty e^{it(\Lambda'(\eta(x))\xi-\Lambda(\xi))}\phi(\xi)\,d\xi \\ &= 2\text{Re}\mathscr{D}_t\mathscr{B}^{-1}M\sqrt{\frac{|t|\Lambda''(\eta)}{2\pi}}\int_0^\infty e^{-itS(\eta,\xi)}\phi(\xi)\,d\xi \\ &= 2\text{Re}\mathscr{D}_t\mathscr{B}^{-1}M\mathscr{V}\phi, \end{aligned} \quad (27)$$

where the phase function $S(\eta,\xi) = \Lambda(\xi) - \Lambda(\eta) - \Lambda'(\eta)(\xi-\eta)$ and the operator

$$\begin{aligned} \mathscr{V}\phi &= \overline{M}\mathscr{B}^{-1}\mathscr{D}_t^{-1}\mathscr{F}^{-1}E\theta\phi \\ &= \sqrt{\frac{|t|\Lambda''(\eta)}{2\pi}}\int_0^\infty e^{-itS(\eta,\xi)}\phi(\xi)\,d\xi. \end{aligned} \quad (28)$$

We have $\|\mathscr{D}_t^{-1}\phi\|_{\mathbf{L}^2} = \|\phi\|_{\mathbf{L}^2}$, $\|\mathscr{F}^{-1}\phi\|_{\mathbf{L}^2} = \|\phi\|_{\mathbf{L}^2}$, and $\|\mathscr{B}^{-1}\phi\|_{\mathbf{L}^2} = \|\phi\|_{\mathbf{L}^2}$. Hence

$$\|\mathscr{V}\phi\|_{\mathbf{L}^2} = \left\|\overline{M}\mathscr{B}^{-1}\mathscr{D}_t^{-1}\mathscr{F}^{-1}E\theta\phi\right\|_{\mathbf{L}^2} = \|\theta\phi\|_{\mathbf{L}^2} \leq \|\phi\|_{\mathbf{L}^2}. \quad (29)$$

Also we decompose the inverse operator

$$\begin{aligned}\mathscr{F}\mathscr{U}(-t)\phi &= \overline{E}\mathscr{F}\phi = \frac{1}{\sqrt{2\pi}}\int_{\mathbb{R}} e^{-it((x/t)\xi-\Lambda(\xi))}\phi(x)\,dx \\ &= \frac{|t|^{1/2}}{\sqrt{2\pi}}\int_{\mathbb{R}} e^{-it(\xi x-\Lambda(\xi))}\mathscr{D}_t^{-1}\phi(x)\,dx \\ &= \frac{|t|^{1/2}}{\sqrt{2\pi}}\int_0^\infty e^{-it(\xi\Lambda'(\eta)-\Lambda(\xi))}\sqrt{\Lambda''(\eta)}\left(\mathscr{B}\mathscr{D}_t^{-1}\phi\right)d\eta \\ &= \frac{|t|^{1/2}}{\sqrt{2\pi}}\int_0^\infty e^{itS(\eta,\xi)}\overline{M}\left(\mathscr{B}\mathscr{D}_t^{-1}\phi\right)\sqrt{\Lambda''(\eta)}d\eta \\ &= \mathscr{V}^*\overline{M}\mathscr{B}\mathscr{D}_t^{-1}\phi.\end{aligned} \quad (30)$$

Since $x = \Lambda'(\eta)$, then

$$\begin{aligned}&\int_{\mathbb{R}} e^{-it(\xi x-\Lambda(\xi))}\mathscr{D}_t^{-1}\phi(x)\,dx \\ &= \int_{\mathbb{R}} e^{-it(\xi\Lambda'(\eta)-\Lambda(\xi))}\mathscr{D}_t^{-1}\phi\left(\Lambda'(\eta)\right)\Lambda''(\eta)\,d\eta \\ &= \int_0^\infty e^{-it(\xi\Lambda'(\eta)-\Lambda(\xi))}\sqrt{\Lambda''(\eta)}\left(\mathscr{B}\mathscr{D}_t^{-1}\phi\right)d\eta \\ &= \int_0^\infty e^{itS(\eta,\xi)}\overline{M}\left(\mathscr{B}\mathscr{D}_t^{-1}\phi\right)\sqrt{\Lambda''(\eta)}d\eta \\ &= \frac{\sqrt{2\pi}}{|t|^{1/2}}\mathscr{V}^*\overline{M}\mathscr{B}\mathscr{D}_t^{-1}\phi,\end{aligned} \quad (31)$$

where $\Lambda''(\xi) = 2\xi^{-3}(a + b\xi^4) > 0$ and the operator

$$\begin{aligned}\mathscr{V}^*\phi &= \mathscr{F}\mathscr{U}(-t)\mathscr{D}_t\mathscr{B}^{-1}M\phi \\ &= \frac{|t|^{1/2}}{\sqrt{2\pi}}\int_0^\infty e^{itS(\eta,\xi)}\phi(\eta)\sqrt{\Lambda''(\eta)}d\eta.\end{aligned} \quad (32)$$

Define the new dependent variable $\widehat{\varphi} = \mathscr{F}\mathscr{U}(-t)u(t)$. Since $\mathscr{F}\mathscr{U}(-t)\mathscr{L} = \partial_t\mathscr{F}\mathscr{U}(-t)$, where $\mathscr{L} = \partial_t - \partial_x^{-1} - (1/3)\partial_x^3$, applying the operator $\mathscr{F}\mathscr{U}(-t)$ to (1) we get

$$\begin{aligned}\partial_t\widehat{\varphi} &= \partial_t\mathscr{F}\mathscr{U}(-t)u = \mathscr{F}\mathscr{U}(-t)\mathscr{L}u \\ &= \mathscr{F}\mathscr{U}(-t)\partial_x\mathscr{K}u^3 = i\xi\widehat{K}(\xi)\mathscr{F}\mathscr{U}(-t)u^3 \\ &= i\xi\widehat{K}(\xi)\mathscr{V}^*\overline{M}\mathscr{B}^{-1}\mathscr{D}_t^{-1}\left(u^3\right).\end{aligned} \quad (33)$$

Then since

$$\begin{aligned}&\mathscr{U}(t)\mathscr{F}^{-1}\phi = 2\mathrm{Re}\mathscr{D}_t\mathscr{B}^{-1}M\mathscr{V}\phi, \\ &u = \mathscr{U}(t)\mathscr{F}^{-1}\widehat{\varphi} = 2\mathrm{Re}\mathscr{D}_t\mathscr{B}^{-1}M\mathscr{V}\widehat{\varphi} \\ &\quad = \mathscr{D}_t\mathscr{B}^{-1}\left(M\mathscr{V}\widehat{\varphi} + \overline{M\mathscr{V}\widehat{\varphi}}\right),\end{aligned} \quad (34)$$

we find the following representation:

$$\begin{aligned}\partial_t\widehat{\varphi} &= \mathscr{V}^*\overline{M}\mathscr{B}\mathscr{D}_t^{-1}\left(\partial_x\mathscr{K}u^3\right) = i\xi\widehat{K}(\xi) \\ &\cdot\mathscr{V}^*\overline{M}\mathscr{B}\mathscr{D}_t^{-1}\left(\left(\mathscr{D}_t\mathscr{B}^{-1}\left(M\mathscr{V}\widehat{\varphi} + \overline{M\mathscr{V}\widehat{\varphi}}\right)\right)^3\right) \\ &= i\xi\widehat{K}(\xi)t^{-1}\mathscr{V}^*\overline{M}\mathscr{B}\left(\left(\mathscr{B}^{-1}\left(M\mathscr{V}\widehat{\varphi} + \overline{M\mathscr{V}\widehat{\varphi}}\right)\right)^3\right) \\ &= i\xi\widehat{K}(\xi)t^{-1}\mathscr{V}^*\frac{1}{\Lambda''}\overline{M}\left(M\mathscr{V}\widehat{\varphi} + \overline{M\mathscr{V}\widehat{\varphi}}\right)^3 \\ &= i\xi\widehat{K}(\xi)t^{-1}\mathscr{V}^*M^2\frac{1}{\Lambda''}\left(\mathscr{V}\widehat{\varphi}\right)^3 + 3i\xi\widehat{K}(\xi)t^{-1}\mathscr{V}^* \\ &\cdot\frac{1}{\Lambda''}\left(\mathscr{V}\widehat{\varphi}\right)^2\left(\overline{\mathscr{V}\widehat{\varphi}}\right) + 3i\xi\widehat{K}(\xi)t^{-1}\mathscr{V}^*\overline{M}^2\frac{1}{\Lambda''}\left(\mathscr{V}\widehat{\varphi}\right) \\ &\cdot\left(\overline{\mathscr{V}\widehat{\varphi}}\right)^2 + i\xi\widehat{K}(\xi)t^{-1}\mathscr{V}^*\overline{M}^4\frac{1}{\Lambda''}\left(\overline{\mathscr{V}\widehat{\varphi}}\right)^3.\end{aligned} \quad (35)$$

Note that for $\alpha \neq -1$

$$\begin{aligned}\mathscr{V}^*(t)M^\alpha\phi &= \frac{|t|^{1/2}}{\sqrt{2\pi}}\int_0^\infty e^{itS(\eta,\xi)}M^\alpha\phi(\eta)\Lambda''(\eta)\,d\eta \\ &= \frac{|t|^{1/2}}{\sqrt{2\pi}} \\ &\cdot e^{it(\Lambda(\xi)-(1+\alpha)\Lambda(\xi/(1+\alpha)))}\int_0^\infty e^{it(1+\alpha)S(\eta,\xi/(1+\alpha))}\phi(\eta) \\ &\cdot\Lambda''(\eta)\,d\eta = e^{it(\Lambda(\xi)-(1+\alpha)\Lambda(\xi/(1+\alpha)))}|1+\alpha|^{1/2} \\ &\cdot\mathscr{D}_{1+\alpha}\mathscr{V}^*((1+\alpha)t)\phi.\end{aligned} \quad (36)$$

Thus we obtain the following equation for the new dependent variable $\widehat{\varphi}(t,\xi) = \mathscr{F}\mathscr{U}(-t)u(t)$:

$$\begin{aligned}&\partial_t\widehat{\varphi}(t,\xi) \\ &\quad = \sqrt{3}i\xi\widehat{K}(\xi)t^{-1}e^{it\Omega(\xi)}\mathscr{D}_3\mathscr{V}^*(3t)\frac{1}{\Lambda''}\left(\mathscr{V}\widehat{\varphi}\right)^3 \\ &\quad + 3i\xi\widehat{K}(\xi)t^{-1}\mathscr{V}^*(t)\frac{1}{\Lambda''}\left(\mathscr{V}\widehat{\varphi}\right)^2\left(\overline{\mathscr{V}\widehat{\varphi}}\right) \\ &\quad + 3i\xi\widehat{K}(\xi)t^{-1}\mathscr{D}_{-1}\mathscr{V}^*(-t)\frac{1}{\Lambda''}\left(\mathscr{V}\widehat{\varphi}\right)\left(\overline{\mathscr{V}\widehat{\varphi}}\right)^2 \\ &\quad + \sqrt{3}i\xi\widehat{K}(\xi)t^{-1}e^{it\Omega(\xi)}\mathscr{D}_{-3}\mathscr{V}^*(-3t)\frac{1}{\Lambda''}\left(\overline{\mathscr{V}\widehat{\varphi}}\right)^3,\end{aligned} \quad (37)$$

where $\Omega(\xi) = \Lambda(\xi) - 3\Lambda(\xi/3)$.

Now we explain how to use (37) for estimating $|\widehat{\varphi}(t,\xi)|$ uniformly with respect to ξ. For the real-valued solution u, we have $\overline{\widehat{\varphi}(t,\xi)} = \widehat{\varphi}(t,-\xi)$; hence it is sufficient to consider the case $\xi > 0$ only. From Lemmas 2 and 3, we find that the last two terms of the right-hand side of (37) are the remainders. We need to consider the first and the second terms of the right-hand side of (37). Due to the oscillating factor $\widehat{K}(\xi)e^{it\Omega(\xi)}$, integrating by parts with respect to time, we will show that the first term of (37) is also a remainder, since $\widehat{K}(\xi)/\Omega(\xi)$ is bounded in view of the conditions for the symbol $\widehat{K}(\xi)$.

We organize the rest of our paper as follows. In Section 3, we state main estimates for the decomposition operators $\mathcal{V}(t)$ and $\mathcal{V}^*(t)$ related to the evolution group $\mathcal{U}(t)$. We prove a priori estimates of solutions in Section 4. Section 5 is devoted to the proof of Theorem 1.

3. Preliminaries

3.1. Two Kernels. Define the kernel

$$A_j(t,\eta)=\theta(\eta)\sqrt{\frac{t\Lambda''(\eta)}{2\pi}}\int_0^\infty e^{-itS(\eta,\xi)}\xi^j\chi(\xi\eta^{-1})\,d\xi, \quad (38)$$

where $j=-1,0,1,2$, the phase function $S(\eta,\xi)=\Lambda(\xi)-\Lambda(\eta)-\Lambda'(\eta)(\xi-\eta)=(1/3)\xi^{-1}\eta^{-2}(3a+2b\eta^3\xi+b\eta^2\xi^2)(\xi-\eta)^2$, $\Lambda''(\eta)=2\eta^{-3}(a+b\eta^4)$, the cut-off function $\chi(z)\in\mathbf{C}^2(\mathbb{R})$ is such that $\chi(z)=0$ for $z\le 1/3$ or $z\ge 3$ and $\chi(z)=1$ for $2/3\le z\le 3/2$, and the Heaviside function $\theta(\eta)=1$ for $\eta>0$ and $\theta(\eta)=0$ for $\eta\le 0$. We change $\xi=\eta y$; then we get

$$A_j(t,\eta)=\eta^j\theta(\eta)\cdot\sqrt{\frac{t\eta^2\Lambda''(\eta)}{2\pi}}\int_0^\infty e^{-it(ay^{-1}\eta^{-1}+(2/3)b\eta^3+(b/3)\eta^3y)(y-1)^2}y^j\chi(y)\,dy. \quad (39)$$

To compute the asymptotics of the kernel $A_j(t,\eta)$ for large t we apply the stationary phase method (see [21, 22], p. 163):

$$\int e^{itg(y)}f(y)\,dy=e^{itg(y_0)}f(y_0)\sqrt{\frac{2\pi}{t|g''(y_0)|}}e^{i(\pi/4)\operatorname{sgn}g''(y_0)}+O(t^{-3/2}) \quad (40)$$

for $t\to+\infty$, where the stationary point y_0 is defined by $g'(y_0)=0$. By virtue of formula (40) with $g(y)=-(ay^{-1}\eta^{-1}+(2/3)b\eta^3+(b/3)\eta^3y)(y-1)^2$, $f(y)=y^j\chi(y)$, $y_0=1$, we get

$$A_j(t,\eta)=\eta^j\theta(\eta)\left(e^{-i(\pi/4)}+O(t^{-1})\right). \quad (41)$$

In particular we have the estimate $\|\eta^{-j}A_j(t)\|_{\mathbf{L}^\infty}\le C$. Also we define the kernel

$$A^*(t,\xi)=\theta(\xi)\frac{t^{1/2}}{\sqrt{2\pi}}\int_0^\infty e^{itS(\eta,\xi)}\chi(\eta\xi^{-1})\sqrt{\Lambda''(\eta)}d\eta. \quad (42)$$

We change $\eta=\xi y$; then we get

$$A^*(t,\xi)=\theta(\xi)\frac{t^{1/2}}{\sqrt{\pi\xi}}\cdot\int_0^\infty e^{it(a\xi^{-1}y^{-2}+(2b/3)\xi^3y+(b/3)\xi^3)(y-1)^2}\chi(y)\cdot\sqrt{y^{-3}(a+b\xi^4y^4)}dy. \quad (43)$$

By virtue of formula (40) with $g(y)=(a\xi^{-1}y^{-2}+(2b/3)\xi^3y+(b/3)\xi^3)(y-1)^2$, $f(y)=\chi(y)\sqrt{y^{-3}(a+b\xi^4y^4)}$, $y_0=1$, we obtain

$$A^*(t,\xi)=\theta(\xi)e^{i(\pi/4)}+O(t^{-1}). \quad (44)$$

In particular we have the estimate $\|A^*(t)\|_{\mathbf{L}^\infty}\le C$.

3.2. Estimates in the Uniform Norm. In the next lemma we estimate the operator $\mathcal{V}$ in the uniform norm. Denote $\mu_{-1}=5/4$, $\mu_0=1/4$, $\mu_1=0$, $\nu_{-1}=-5/4$, $\nu_0=1$, and $\nu_1=1/4$.

Lemma 2. *Let $j=-1,0,1$. Then the estimates*

$$\begin{aligned}&\left\||\eta|^{\mu_j}\langle\eta\rangle^{\nu_j}\left(\mathcal{V}\xi^j\phi-A_j(t)\phi\right)\right\|_{\mathbf{L}^\infty(0,\infty)}\\&\quad\le Ct^{-1/2}\left\||\xi|^{1/2}\phi\right\|_{\mathbf{L}^\infty}+Ct^{-1/4}\left\|\xi\phi_\xi\right\|_{\mathbf{L}^2},\\&\left\||\eta|^{\nu_j}\langle\eta\rangle^{-2\nu_j+1}\mathcal{V}\xi^j\phi\right\|_{\mathbf{L}^\infty(-\infty,0)}\\&\quad\le Ct^{-1/2}\left(\left\||\xi|^{1/2}\phi\right\|_{\mathbf{L}^\infty}+\left\|\xi\phi_\xi\right\|_{\mathbf{L}^2}\right)\end{aligned} \quad (45)$$

are valid for all $t\ge 1$.

Proof. We write

$$\begin{aligned}&\mathcal{V}\xi^j\phi-A_j(t,\eta)\phi(\eta)\\&=\sqrt{\frac{t\Lambda''(\eta)}{2\pi}}\int_0^\infty e^{-itS(\eta,\xi)}(\phi(\xi)-\phi(\eta))\\&\cdot\xi^j\chi(\xi\eta^{-1})\,d\xi+\sqrt{\frac{t\Lambda''(\eta)}{2\pi}}\int_0^\infty e^{-itS(\eta,\xi)}\phi(\xi)\\&\cdot\left(1-\chi(\xi\eta^{-1})\right)\xi^jd\xi=I_1+I_2\end{aligned} \quad (46)$$

for $\eta>0$. For the first summand I_1 we integrate by parts via the identity

$$e^{-itS(\eta,\xi)}=H_1\partial_\xi\left((\xi-\eta)e^{-itS(\eta,\xi)}\right) \quad (47)$$

with $H_1=(1-it(\xi-\eta)\partial_\xi S(\eta,\xi))^{-1}$, $\partial_\xi S(\eta,\xi)=\xi^{-2}\eta^{-2}(a+b\eta^2\xi^2)(\xi^2-\eta^2)$, to get

$$\begin{aligned}I_1&=C\sqrt{t\Lambda''(\eta)}\int_0^\infty e^{-itS(\eta,\xi)}(\phi(\xi)-\phi(\eta))(\xi-\eta)\\&\cdot\partial_\xi\left(\xi^jH_1\chi(\xi\eta^{-1})\right)+e^{-itS(\eta,\xi)}(\xi-\eta)\\&\cdot\xi^jH_1\chi(\xi\eta^{-1})\phi_\xi(\xi)\,d\xi.\end{aligned} \quad (48)$$

Using the estimates

$$|\phi(\xi)-\phi(\eta)| \le C\eta^{-1}\int_{\xi}^{\eta}\xi\phi_{\xi}(\xi)\,d\xi \le C\eta^{-1}|\xi-\eta|^{1/2}\|\xi\phi_{\xi}\|_{\mathbf{L}^2},$$
$$\left|(\xi-\eta)\xi^{j}H_1\chi(\xi\eta^{-1})\right| \le \frac{C|\xi-\eta|\eta^{j}}{1+t\eta^{-3}\langle\eta\rangle^4(\xi-\eta)^2},$$
$$\left|(\xi-\eta)\partial_{\xi}\left(\xi^{j}H_1\chi(\xi\eta^{-1})\right)\right| \le \frac{C\eta^{j}}{1+t\eta^{-3}\langle\eta\rangle^4(\xi-\eta)^2} \tag{49}$$

in the domain $0<(1/3)\eta\le\xi\le 3\eta$, we find

$$\begin{aligned}|I_1| &\le Ct^{1/2}\eta^{j-5/2}\langle\eta\rangle^2\|\xi\phi_{\xi}\|_{\mathbf{L}^2}\cdot\int_{(1/3)\eta}^{3\eta}\frac{|\xi-\eta|^{1/2}d\xi}{1+t\eta^{-3}\langle\eta\rangle^4(\xi-\eta)^2}+Ct^{1/2}\eta^{j-5/2}\langle\eta\rangle^2\\&\quad\cdot\int_{(1/3)\eta}^{3\eta}\frac{|\xi-\eta|\,|\xi\phi_{\xi}(\xi)|\,d\xi}{1+t\eta^{-3}\langle\eta\rangle^4(\xi-\eta)^2}\le Ct^{1/2}\eta^{j-5/2}\langle\eta\rangle^2\\&\quad\cdot\|\xi\phi_{\xi}\|_{\mathbf{L}^2}\left(\int_{(1/3)\eta}^{3\eta}\frac{|\xi-\eta|^{1/2}d\xi}{1+t\eta^{-3}\langle\eta\rangle^4(\xi-\eta)^2}\right.\\&\quad\left.+\left(\int_{(1/3)\eta}^{3\eta}\frac{(\xi-\eta)^2d\xi}{\left(1+t\eta^{-3}\langle\eta\rangle^4(\xi-\eta)^2\right)^2}\right)^{1/2}\right).\end{aligned} \tag{50}$$

Changing $\xi=\eta y$ we have

$$\begin{aligned}&\int_{(1/3)\eta}^{3\eta}\frac{|\xi-\eta|^{1/2}d\xi}{1+t\eta^{-3}\langle\eta\rangle^4(\xi-\eta)^2}\le C\eta^{3/2}\int_{1/3}^{3}\frac{|y-1|^{1/2}dy}{1+t\eta^{-1}\langle\eta\rangle^4(y-1)^2}\\&\quad\le C\eta^{3/2}\left(t\eta^{-1}\langle\eta\rangle^4\right)^{-3/4}\le Ct^{-3/4}\eta^{9/4}\langle\eta\rangle^{-3},\\&\int_{(1/3)\eta}^{3\eta}\frac{(\xi-\eta)^2d\xi}{\left(1+t\eta^{-3}\langle\eta\rangle^4(\xi-\eta)^2\right)^2}\le C\eta^3\int_{1/3}^{3}\frac{(y-1)^2dy}{\left(1+t\eta^{-1}\langle\eta\rangle^4(y-1)^2\right)^2}\\&\quad\le C\eta^3\left(t\eta^{-1}\langle\eta\rangle^4\right)^{-3/2}\le Ct^{-3/2}\eta^{9/2}\langle\eta\rangle^{-6}.\end{aligned} \tag{51}$$

Thus we obtain

$$|I_1|\le Ct^{-1/4}\eta^{j-1/4}\langle\eta\rangle^{-1}\|\xi\phi_{\xi}\|_{\mathbf{L}^2} \tag{52}$$

for all $t\ge1$, $\eta>0$, and $j=-1,0,1$.

To estimate the second integral I_2 we integrate by parts via the identity

$$e^{-itS(x,\xi)}=H_2\partial_{\xi}\left(\xi e^{-itS(x,\xi)}\right) \tag{53}$$

with $H_2=(1-it\xi\partial_{\xi}S(\eta,\xi))^{-1}$, $\partial_{\xi}S(\eta,\xi)=\xi^{-2}\eta^{-2}(a+b\eta^2\xi^2)(\xi^2-\eta^2)$, to get

$$\begin{aligned}I_2&=C\sqrt{t\Lambda''(\eta)}\int_0^{\infty}e^{-itS(\eta,\xi)}\phi(\xi)\\&\quad\cdot\xi\partial_{\xi}\left(\left(1-\chi(\xi\eta^{-1})\right)H_2\xi^{j}\right)\\&\quad+e^{-itS(x,\xi)}H_2\left(1-\chi(\xi\eta^{-1})\right)\xi^{j+1}\phi_{\xi}(\xi)\,d\xi.\end{aligned} \tag{54}$$

Using the estimate

$$\begin{aligned}&\left|H_2\left(1-\chi(\xi\eta^{-1})\right)\xi^{j}\right|+\left|\xi\partial_{\xi}\left(\left(1-\chi(\xi\eta^{-1})\right)H_2\xi^{j}\right)\right|\\&\quad\le\frac{C\xi^{j}}{1+t\xi^{-1}\eta^{-2}(1+\xi^2\eta^2)(\xi^2+\eta^2)}\end{aligned} \tag{55}$$

in the domain $0<\xi\le(2/3)\eta$, or $\xi\ge(3/2)\eta>0$, we obtain

$$\begin{aligned}|I_2|&\le Ct^{1/2}\eta^{-3/2}\langle\eta\rangle^2\cdot\int_0^{\infty}\frac{\xi^{j}\left(|\phi(\xi)|+|\xi\phi_{\xi}(\xi)|\right)d\xi}{1+t\xi^{-1}\eta^{-2}(1+\xi^2\eta^2)(\xi^2+\eta^2)}\\&\le Ct^{1/2}\eta^{-3/2}\langle\eta\rangle^2\|\xi^{1/2}\phi\|_{\mathbf{L}^{\infty}}\cdot\int_0^{\infty}\frac{\xi^{j-1/2}d\xi}{1+t\xi^{-1}\eta^{-2}(1+\xi^2\eta^2)(\xi^2+\eta^2)}\\&\quad+Ct^{1/2}\eta^{-3/2}\langle\eta\rangle^2\|\xi\phi_{\xi}\|_{\mathbf{L}^2}\\&\quad\cdot\left(\int_0^{\infty}\frac{\xi^{2j}d\xi}{\left(1+t\xi^{-1}\eta^{-2}(1+\xi^2\eta^2)(\xi^2+\eta^2)\right)^2}\right)^{1/2}.\end{aligned} \tag{56}$$

Changing $\xi=\eta y$ we obtain

$$\begin{aligned}&\int_0^{\infty}\frac{\xi^{j-1/2}d\xi}{1+t\xi^{-1}\eta^{-2}(1+\xi^2\eta^2)(\xi^2+\eta^2)}\\&\quad\le\int_0^{\infty}\frac{\eta^{j+1/2}y^{j-1/2}dy}{1+t\eta^{-1}y^{-1}(1+\eta^4y^2)\langle y\rangle^2}.\end{aligned} \tag{57}$$

For $\eta\le1$

$$\begin{aligned}&\int_0^{1}\frac{\eta^{j+1/2}y^{j-1/2}dy}{1+t\eta^{-1}y^{-1}(1+\eta^4y^2)}\le Ct^{-1}\eta^{j+3/2}\int_0^1 y^{j+1/2}dy\\&\quad\le Ct^{-1}\eta^{j+3/2},\\&\int_1^{\infty}\frac{\eta^{j+1/2}y^{j-1/2}dy}{1+t\eta^{-1}y(1+\eta^4y^2)}\le Ct^{-1}\eta^{j+3/2}\int_1^{\infty}\frac{y^{j-3/2}dy}{1+\eta^4y^2}\\&\quad\le Ct^{-1}\eta^{3/2-\gamma_j}\end{aligned} \tag{58}$$

$\nu_{-1} = 1$, and $\nu_0 = \nu_1 = 0$. For $\eta > 1$

$$\begin{aligned}
&\int_0^1 \frac{\eta^{j+1/2} y^{j-1/2} dy}{1 + t\eta^{-1} y^{-1} (1 + \eta^4 y^2)} \\
&\quad \le Ct^{-1}\eta^{j+3/2} \int_0^{1/\eta^2} y^{j+1/2} dy \\
&\quad + Ct^{-1}\eta^{j-5/2} \int_{1/\eta^2}^1 y^{j-3/2} dy \\
&\quad \le Ct^{-1} \langle\eta\rangle^{-j-3/2} + Ct^{-1} \langle\eta\rangle^{j-5/2} \\
&\int_1^\infty \frac{\eta^{j+1/2} y^{j-1/2} dy}{1 + t\eta^{-1} y (1 + \eta^4 y^2)} \le Ct^{-1}\eta^{j-5/2} \int_1^\infty y^{j-7/2} dy \\
&\quad \le Ct^{-1} \langle\eta\rangle^{j-5/2}.
\end{aligned} \tag{59}$$

Hence

$$\begin{aligned}
&\eta^{-3/2} \langle\eta\rangle^2 \int_0^\infty \frac{\xi^{j-1/2} d\xi}{1 + t\xi^{-1}\eta^{-2} (1 + \xi^2\eta^2)(\xi^2 + \eta^2)} \\
&\quad \le Ct^{-1} \left(\{\eta\}^j + \{\eta\}^{-\nu_j}\right)\left(\langle\eta\rangle^{-j-1} + \langle\eta\rangle^{j-2}\right) \\
&\quad \le Ct^{-1} \{\eta\}^{-\nu_j} \langle\eta\rangle^{\nu_j - 1} \le Ct^{-1}\eta^{-\nu_j} \langle\eta\rangle^{2\nu_j - 1}.
\end{aligned} \tag{60}$$

In the same manner changing $\xi = \eta y$ we get

$$\begin{aligned}
&\int_0^\infty \frac{\xi^{2j} d\xi}{\left(1 + t\xi^{-1}\eta^{-2} (1 + \xi^2\eta^2)(\xi^2 + \eta^2)\right)^2} \\
&\quad \le C \int_0^\infty \frac{\eta^{2j+1} y^{2j} dy}{\left(1 + t\eta^{-1} y^{-1} (1 + \eta^4 y^2) \langle y\rangle^2\right)^2}.
\end{aligned} \tag{61}$$

For $\eta \le 1$

$$\begin{aligned}
&\int_0^1 \frac{\eta^{2j+1} y^{2j} dy}{\left(1 + t\eta^{-1} y^{-1} (1 + \eta^4 y^2)\right)^2} \\
&\quad \le Ct^{-2}\eta^{2j+3} \int_0^1 y^{2j+2} dy \le Ct^{-2}\eta^{2j+3} \le Ct^{-2}\eta^{3-2\nu_j}, \\
&\int_1^\infty \frac{\eta^{2j+1} y^{2j} dy}{\left(1 + t\eta^{-1} y (1 + \eta^4 y^2)\right)^2} \\
&\quad \le Ct^{-2}\eta^{2j+3} \int_1^\infty \frac{y^{2j-2} dy}{\left(1 + \eta^4 y^2\right)^2} \le Ct^{-2}\eta^{2j+3-2(2j-1)} \\
&\quad \le Ct^{-2}\eta^{3-2\nu_j}
\end{aligned} \tag{62}$$

$\nu_{-1} = 1$, and $\nu_0 = \nu_1 = 0$. For $\eta > 1$

$$\begin{aligned}
&\int_0^1 \frac{\eta^{2j+1} y^{2j} dy}{\left(1 + t\eta^{-1} y^{-1} (1 + \eta^4 y^2)\right)^2} \\
&\quad \le Ct^{-2}\eta^{2j+3} \int_0^{1/\eta^2} y^{2j+2} dy \\
&\quad + Ct^{-2}\eta^{2j-5} \int_{1/\eta^2}^1 y^{2j-2} dy \\
&\quad \le Ct^{-2} \langle\eta\rangle^{-2j-3} + Ct^{-2} \langle\eta\rangle^{2j-5} \\
&\int_1^\infty \frac{\eta^{2j+1} y^{2j} dy}{\left(1 + t\eta^{-1} y (1 + \eta^4 y^2)\right)^2} \le Ct^{-2}\eta^{2j-5} \int_1^\infty y^{2j-6} dy \\
&\quad \le Ct^{-2} \langle\eta\rangle^{2j-5}.
\end{aligned} \tag{63}$$

Hence

$$\begin{aligned}
&\eta^{-3} \langle\eta\rangle^4 \int_0^\infty \frac{\xi^{2j} d\xi}{\left(1 + t\xi^{-1}\eta^{-2} (1 + \xi^2\eta^2)(\xi^2 + \eta^2)\right)^2} \\
&\quad \le Ct^{-2} \{\eta\}^{-2\nu_j} \left(\langle\eta\rangle^{-2j-2} + \langle\eta\rangle^{2j-4}\right) \\
&\quad \le Ct^{-2} \{\eta\}^{-2\nu_j} \langle\eta\rangle^{2\nu_j - 2} \le Ct^{-2}\eta^{-2\nu_j} \langle\eta\rangle^{4\nu_j - 2}
\end{aligned} \tag{64}$$

for $j = -1, 0, 1$. Thus we have

$$\left|I_2\right| \le Ct^{-1/2}\eta^{-\nu_j} \langle\eta\rangle^{2\nu_j - 1} \left(\left\|\xi\phi_\xi\right\|_{\mathbf{L}^2} + \left\|\xi^{1/2}\phi\right\|_{\mathbf{L}^\infty}\right) \tag{65}$$

for all $t \ge 1$, $\eta > 0$, and $j = -1, 0, 1$.

For the case of $\eta < 0$ we integrate by parts using identity (53):

$$\begin{aligned}
\mathcal{V}\xi^j\phi &= C\sqrt{t\Lambda''(\eta)} \int_0^\infty e^{-itS(\eta,\xi)} \phi(\xi)\, \xi\partial_\xi \left(H_2\xi^j\right) \\
&\quad + e^{-itS(x,\xi)} H_2\xi^j \xi\phi_\xi(\xi)\, d\xi.
\end{aligned} \tag{66}$$

Using the estimate

$$\begin{aligned}
&\left|H_2\xi^j\right| + \left|\xi\partial_\xi \left(H_2\xi^j\right)\right| \\
&\quad \le \frac{C\xi^j}{1 + t\xi^{-1} |\eta|^{-2} (1 + \xi^2\eta^2)(\xi^2 + \eta^2)}
\end{aligned} \tag{67}$$

in the domain $\xi > 0$, $\eta < 0$, we obtain

$$\begin{aligned}
&\left|\mathcal{V}\xi^j\phi\right| \le Ct^{1/2} |\eta|^{-3/2} \langle\eta\rangle^2 \\
&\quad \cdot \int_0^\infty \frac{\xi^j \left(|\phi(\xi)| + |\xi\phi_\xi(\xi)|\right) d\xi}{1 + t\xi^{-1} |\eta|^{-2} (1 + \xi^2\eta^2)(\xi^2 + \eta^2)} \\
&\quad \le Ct^{1/2} |\eta|^{-3/2} \langle\eta\rangle^2 \left\|\xi^{1/2}\phi\right\|_{\mathbf{L}^\infty}
\end{aligned}$$

$$\cdot \int_0^\infty \frac{\xi^{j-1/2} d\xi}{1 + t\xi^{-1} |\eta|^{-2} (1+\xi^2\eta^2)(\xi^2+\eta^2)} + Ct^{1/2} |\eta|^{-3/2} \langle\eta\rangle^2 \|\xi\phi_\xi\|_{\mathbf{L}^2} \cdot \left(\int_0^\infty \frac{\xi^{2j} d\xi}{\left(1 + t\xi^{-1} |\eta|^{-2} (1+\xi^2\eta^2)(\xi^2+\eta^2)\right)^2} \right)^{1/2}. \quad (68)$$

Then as above we get

$$\left|\mathcal{V}\xi^j\phi\right| \le Ct^{-1/2} |\eta|^{-\nu_j} \langle\eta\rangle^{2\nu_j-1} \left(\|\xi\phi_\xi\|_{\mathbf{L}^2} + \left\||\xi|^{1/2}\phi\right\|_{\mathbf{L}^\infty} \right) \quad (69)$$

for all $t \ge 1, \eta < 0$, and $j = -1, 0, 1$. Lemma 2 is proved. □

By Lemma 2, we have the estimate

$$\begin{aligned} \left|\mathcal{V}\xi^j\phi\right| &\le C|\eta|^j |\phi(\eta)| + Ct^{-1/4} |\eta|^{-\mu_j} \langle\eta\rangle^{-\nu_j} \left(\left\||\xi|^{1/2}\phi\right\|_{\mathbf{L}^\infty} + \|\xi\phi_\xi\|_{\mathbf{L}^2} \right) \\ &\le C|\eta|^{j-1/2} \left\||\xi|^{1/2}\phi\right\|_{\mathbf{L}^\infty} + Ct^{-1/4} |\eta|^{-\mu_j} \langle\eta\rangle^{-\nu_j} \left(\left\||\xi|^{1/2}\phi\right\|_{\mathbf{L}^\infty} + \|\xi\phi_\xi\|_{\mathbf{L}^2} \right). \end{aligned} \quad (70)$$

We next consider the operator $\mathcal{V}^*$.

Lemma 3. *The estimates*

$$\begin{aligned} &\|\langle\xi\rangle (\mathcal{V}^*\phi - A^*(t,\xi)\phi(\xi))\|_{\mathbf{L}^\infty(0,\infty)} \le Ct^{-1/4} \left(\|\phi\|_{\mathbf{L}^\infty} + \left\||\eta|^{-9/4} \langle\eta\rangle^4 t\mathcal{A}_0\phi\right\|_{\mathbf{L}^2} \right), \\ &\|\langle\xi\rangle \mathcal{V}^*\phi\|_{\mathbf{L}^\infty(-\infty,0)} \le Ct^{-1/2} \left(\|\phi\|_{\mathbf{L}^\infty} + \left\||\eta|^{-9/4} \langle\eta\rangle^4 t\mathcal{A}_0\phi\right\|_{\mathbf{L}^2} \right) \end{aligned} \quad (71)$$

are valid for all $t \ge 1$, where $\mathcal{A}_0 = (1/t\sqrt{\Lambda''})\partial_\eta(1/\sqrt{\Lambda''})$.

Proof. We find

$$\begin{aligned} \mathcal{V}^*\phi - A^*(t,\xi)\phi(\xi) &= \frac{t^{1/2}}{\sqrt{2\pi}} \cdot \int_0^\infty e^{itS(\eta,\xi)} (\phi(\eta) - \phi(\xi)) \chi(\eta\xi^{-1}) \sqrt{\Lambda''(\eta)} d\eta \\ &+ \frac{t^{1/2}}{\sqrt{2\pi}} \int_0^\infty e^{itS(\eta,\xi)} \phi(\eta) \left(1 - \chi(\eta\xi^{-1})\right) \sqrt{\Lambda''(\eta)} d\eta \\ &= I_3 + I_4 \end{aligned} \quad (72)$$

for $\xi > 0$. In the first integral I_3 using the identity

$$e^{itS(x,\xi)} = H_3\partial_\eta \left((\eta - \xi) e^{itS(\eta,\xi)} \right) \quad (73)$$

with $H_3 = (1+it(\eta-\xi)S_\eta(\eta,\xi))^{-1}$, $\partial_\eta S(\eta,\xi) = 2\eta^{-3}(a+b\eta^4)(\eta-\xi)$, we integrate by parts

$$\begin{aligned} I_3 = Ct^{1/2} \int_0^\infty &e^{itS(\eta,\xi)} (\phi(\eta) - \phi(\xi)) (\eta - \xi) \\ &\cdot \partial_\eta \left(H_3\chi(\eta\xi^{-1}) \sqrt{\Lambda''(\eta)} \right) + e^{itS(\eta,\xi)} \phi_\eta(\eta) (\eta - \xi) \\ &\cdot H_3\chi(\eta\xi^{-1}) \sqrt{\Lambda''(\eta)} d\eta. \end{aligned} \quad (74)$$

Then using the identity

$$\phi_\eta(\eta) = \Lambda''(\eta) t\mathcal{A}_0\phi(\eta) + \frac{\Lambda'''(\eta)}{2\Lambda''(\eta)} \phi(\eta) \quad (75)$$

we get

$$\begin{aligned} I_3 = Ct^{1/2} \int_0^\infty &e^{itS(\eta,\xi)} (\phi(\eta) - \phi(\xi)) (\eta - \xi) \\ &\cdot \partial_\eta \left(H_3\chi(\eta\xi^{-1}) \sqrt{\Lambda''(\eta)} \right) + e^{itS(\eta,\xi)} (\eta - \xi) \\ &\cdot H_3\chi(\eta\xi^{-1}) \left(\Lambda''(\eta)\right)^{3/2} t\mathcal{A}_0\phi(\eta) + e^{itS(\eta,\xi)} \\ &\cdot \frac{\Lambda'''(\eta)}{\sqrt{\Lambda''(\eta)}} \phi(\eta) (\eta - \xi) H_3\chi(\eta\xi^{-1}) d\eta. \end{aligned} \quad (76)$$

Applying the estimates

$$\begin{aligned} |\phi(\eta) - \phi(\xi)| &= \left| \int_\eta^\xi \partial_\eta\phi d\eta \right| \\ &= \left| \int_\eta^\xi \Lambda''(\eta) t\mathcal{A}_0\phi(\eta) d\eta + \int_\eta^\xi \frac{\Lambda'''(\eta)}{2\Lambda''(\eta)} \phi(\eta) d\eta \right| \\ &\le C|\eta - \xi|^{1/2} |\xi|^{9/4} \langle\xi\rangle^{-3} \Lambda''(\xi) \\ &\cdot \left\||\eta|^{-9/4} \langle\eta\rangle^3 t\mathcal{A}_0\phi\right\|_{\mathbf{L}^2} + C|\xi|^{-3/2} \langle\xi\rangle^2 |\eta - \xi| \\ &\cdot \left\||\eta|^{1/2} \langle\eta\rangle^{-2} \phi\right\|_{\mathbf{L}^\infty}, \\ |H_3| &\le C\left(1 + t\xi^{-3}\langle\xi\rangle^4 (\xi - \eta)^2\right)^{-1}, \\ &\left|(\eta - \xi) H_3\chi(\eta\xi^{-1}) \left(\Lambda''(\eta)\right)^{3/2}\right| \\ &\le \frac{C\xi^{-9/2} \langle\xi\rangle^6 |\eta - \xi|}{1 + t\xi^{-3}\langle\xi\rangle^4 (\xi - \eta)^2} \end{aligned} \quad (77)$$

and $\Lambda''(\xi) = O(\xi^{-3}\langle\xi\rangle^4)$, $\Lambda'''(\xi) = O(\xi^{-4}\langle\xi\rangle^4)$, and

$$\left|(\eta - \xi)\partial_\eta \left(H_3\chi(\eta\xi^{-1}) \sqrt{\Lambda''(\eta)} \right)\right| \le \frac{C\xi^{-3/2} \langle\xi\rangle^2}{1 + t\xi^{-3}\langle\xi\rangle^4 (\xi - \eta)^2} \quad (78)$$

for $(1/3)\xi \le \eta \le 3\xi$, we find

$$\begin{aligned} |I_3| &\le Ct^{1/2}|\xi|^{-9/4}\langle\xi\rangle^3\left\||\eta|^{-9/4}\langle\eta\rangle^3 t\mathscr{A}_0\phi\right\|_{\mathbf{L}^2} \\ &\cdot\int_{(1/3)\xi}^{3\xi}\frac{|\eta-\xi|^{1/2}d\eta}{1+t\xi^{-3}\langle\xi\rangle^4(\xi-\eta)^2}+Ct^{1/2}\xi^{-3}\langle\xi\rangle^4 \\ &\cdot\left\||\eta|^{1/2}\langle\eta\rangle^{-2}\phi\right\|_{\mathbf{L}^\infty} \\ &\cdot\int_{(1/3)\xi}^{3\xi}\frac{|\eta-\xi|\,d\eta}{1+t\xi^{-3}\langle\xi\rangle^4(\xi-\eta)^2}+Ct^{1/2}|\xi|^{-9/4} \\ &\cdot\langle\xi\rangle^3\left\||\eta|^{-9/4}\langle\eta\rangle^3 t\mathscr{A}_0\phi\right\|_{\mathbf{L}^2} \\ &\cdot\left(\int_{(1/3)\xi}^{3\xi}\frac{(\eta-\xi)^2d\eta}{\left(1+t\xi^{-3}\langle\xi\rangle^4(\xi-\eta)^2\right)^2}\right)^{1/2}. \end{aligned} \tag{79}$$

Hence

$$\begin{aligned} |I_3| &\le Ct^{1/2}|\xi|^{-9/4}\langle\xi\rangle^2\left\langle t\xi^{-3}\langle\xi\rangle^4\right\rangle^{-3/4} \\ &\cdot\left\||\eta|^{-9/4}\langle\eta\rangle^4 t\mathscr{A}_0\phi\right\|_{\mathbf{L}^2}+Ct^{1/2}\xi^{-5/2}\langle\xi\rangle^3 \\ &\cdot\left\langle t\xi^{-3}\langle\xi\rangle^4\right\rangle^{\gamma-1}\|\phi\|_{\mathbf{L}^\infty}\le Ct^{-1/4}\langle\xi\rangle^{-1} \\ &\cdot\left\||\eta|^{-9/4}\langle\eta\rangle^3 t\mathscr{A}_0\phi\right\|_{\mathbf{L}^2}+Ct^{\gamma-1/2}\langle\xi\rangle^{-1}\|\phi\|_{\mathbf{L}^\infty}. \end{aligned} \tag{80}$$

In the second integral I_4, using the identity $e^{itS(\eta,\xi)} = H_4\partial_\eta(\eta e^{itS(\eta,\xi)})$ with $H_4 = (1+it\eta S_\eta(\eta,\xi))^{-1}$ we integrate by parts

$$\begin{aligned} I_4 &= Ct^{1/2}\int_0^\infty e^{itS(\eta,\xi)}\phi(\eta) \\ &\cdot\eta\partial_\eta\left(H_4\left(1-\chi\left(\eta\xi^{-1}\right)\right)\sqrt{\Lambda''(\eta)}\right) \\ &+e^{itS(\eta,\xi)}\phi_\eta(\eta)\,\eta H_4\left(1-\chi\left(\eta\xi^{-1}\right)\right)\sqrt{\Lambda''(\eta)}d\eta. \end{aligned} \tag{81}$$

Then using

$$\phi_\eta(\eta)=\Lambda''(\eta)\,t\mathscr{A}_0\phi(\eta)+\frac{\Lambda'''(\eta)}{2\Lambda''(\eta)}\phi(\eta) \tag{82}$$

we get

$$\begin{aligned} |I_4| &\le Ct^{1/2}\int_0^\infty|\phi(\eta)| \\ &\cdot\left|\eta\partial_\eta\left(H_4\left(1-\chi\left(\eta\xi^{-1}\right)\right)\sqrt{\Lambda''(\eta)}\right)\right| \\ &+\left|\frac{\Lambda'''(\eta)}{\sqrt{\Lambda''(\eta)}}\eta H_4\left(1-\chi\left(\eta\xi^{-1}\right)\right)\right||\phi(\eta)| \\ &+\left|\left(\Lambda''(\eta)\right)^{3/2}\eta H_4\left(1-\chi\left(\eta\xi^{-1}\right)\right)\right| \\ &\cdot\left|t\mathscr{A}_0\phi(\eta)\right|d\eta. \end{aligned} \tag{83}$$

Then using the estimates

$$\begin{aligned} &\left|\eta\partial_\eta\left(H_4\left(1-\chi\left(\eta\xi^{-1}\right)\right)\sqrt{\Lambda''(\eta)}\right)\right| \\ &+\left|\frac{\Lambda'''(\eta)}{\sqrt{\Lambda''(\eta)}}\eta H_4\left(1-\chi\left(\eta\xi^{-1}\right)\right)\right| \\ &\le\frac{C\eta^{-3/2}\langle\eta\rangle^2}{1+t\eta^{-2}\langle\eta\rangle^4(\xi+\eta)}, \\ &\left|\left(\Lambda''(\eta)\right)^{3/2}\eta H_4\left(1-\chi\left(\eta\xi^{-1}\right)\right)\right| \\ &\le\frac{C\eta^{-7/2}\langle\eta\rangle^6}{1+t\eta^{-2}\langle\eta\rangle^4(\xi+\eta)} \end{aligned} \tag{84}$$

in the domain $\eta \ge 3\xi > 0$ or $0 < \eta < (1/3)\xi$, we get

$$\begin{aligned} |I_4| &\le Ct^{1/2}\|\phi\|_{\mathbf{L}^\infty}\int_0^\infty\frac{\eta^{-3/2}\langle\eta\rangle^2d\eta}{1+t\eta^{-2}\langle\eta\rangle^4(\xi+\eta)} \\ &+Ct^{1/2}\left\||\eta|^{-9/4}\langle\eta\rangle^4 t\mathscr{A}_0\phi\right\|_{\mathbf{L}^2} \\ &\cdot\left(\int_0^\infty\frac{\eta^{-5/2}\langle\eta\rangle^5d\eta}{\left(1+t\eta^{-2}\langle\eta\rangle^4(\xi+\eta)\right)^2}\right)^{1/2}. \end{aligned} \tag{85}$$

Therefore

$$\begin{aligned} |I_4| &\le Ct^{-1/2}\|\phi\|_{\mathbf{L}^\infty}\int_0^\infty\frac{\eta^{1/2}\langle\eta\rangle^{-2}d\eta}{\xi+\eta} \\ &+Ct^{-1/2}\left\||\eta|^{-9/4}\langle\eta\rangle^4 t\mathscr{A}_0\phi\right\|_{\mathbf{L}^2} \\ &\cdot\left(\int_0^\infty\frac{\eta^{3/2}\langle\eta\rangle^{-3}d\eta}{(\xi+\eta)^2}\right)^{1/2}\le Ct^{-1/2}\langle\xi\rangle^{-1}\|\phi\|_{\mathbf{L}^\infty} \\ &+Ct^{-1/2}\langle\xi\rangle^{-1}\left\||\eta|^{-9/4}\langle\eta\rangle^4 t\mathscr{A}_0\phi\right\|_{\mathbf{L}^2}. \end{aligned} \tag{86}$$

Next we consider $\xi < 0$. Using the identity $e^{itS(\eta,\xi)} = H_4\partial_\eta(\eta e^{itS(\eta,\xi)})$ with $H_4 = (1+it\eta S_\eta(\eta,\xi))^{-1}$ we integrate by parts

$$\begin{aligned} \mathscr{V}^*\phi &= Ct^{1/2}\int_0^\infty e^{itS(\eta,\xi)}\phi(\eta)\,\eta\partial_\eta\left(H_4\sqrt{\Lambda''(\eta)}\right) \\ &+e^{itS(\eta,\xi)}\phi_\eta(\eta)\,\eta H_4\sqrt{\Lambda''(\eta)}d\eta. \end{aligned} \tag{87}$$

Then using formula (82), we get

$$\begin{aligned} |\mathscr{V}^*\phi| &\le Ct^{1/2}\int_0^\infty|\phi(\eta)|\left|\eta\partial_\eta\left(H_4\sqrt{\Lambda''(\eta)}\right)\right| \\ &+\left|\frac{\Lambda'''(\eta)}{\sqrt{\Lambda''(\eta)}}\eta H_4\right||\phi(\eta)| \\ &+\left|\left(\Lambda''(\eta)\right)^{3/2}\eta H_4\right|\left|t\mathscr{A}_0\phi(\eta)\right|d\eta. \end{aligned} \tag{88}$$

Then using the estimates

$$\left|\eta\partial_\eta\left(H_4\sqrt{\Lambda''(\eta)}\right)\right| + \left|\frac{\Lambda'''(\eta)}{\sqrt{\Lambda''(\eta)}}\eta H_4\right| \le \frac{C\eta^{-3/2}\langle\eta\rangle^2}{1+t\eta^{-2}\langle\eta\rangle^4(|\xi|+\eta)}, \tag{89}$$

$$\left|(\Lambda''(\eta))^{3/2}\eta H_4\right| \le \frac{C\eta^{-7/2}\langle\eta\rangle^6}{1+t\eta^{-2}\langle\eta\rangle^4(|\xi|+\eta)}$$

in the domain $\xi < 0$ and $\eta > 0$, we get

$$|\mathscr{V}^*\phi| \le Ct^{1/2}\|\phi\|_{\mathbf{L}^\infty}\int_0^\infty \frac{\eta^{-3/2}\langle\eta\rangle^2 d\eta}{1+t\eta^{-2}\langle\eta\rangle^4(|\xi|+\eta)} + Ct^{1/2}\left\||\eta|^{-9/4}\langle\eta\rangle^4 t\mathscr{A}_0\phi\right\|_{\mathbf{L}^2} \cdot\left(\int_0^\infty \frac{\eta^{-5/2}\langle\eta\rangle^6 d\eta}{\left(1+t\eta^{-2}\langle\eta\rangle^4(|\xi|+\eta)\right)^2}\right)^{1/2}. \tag{90}$$

Therefore

$$|\mathscr{V}^*\phi| \le Ct^{-1/2}\|\phi\|_{\mathbf{L}^\infty}\int_0^\infty \frac{\eta^{1/2}\langle\eta\rangle^{-2} d\eta}{|\xi|+\eta} + Ct^{-1/2}\left\||\eta|^{-9/4}\langle\eta\rangle^4 t\mathscr{A}_0\phi\right\|_{\mathbf{L}^2} \cdot\left(\int_0^\infty \frac{\eta^{3/2}\langle\eta\rangle^{-3} d\eta}{(|\xi|+\eta)^2}\right)^{1/2} \le Ct^{-1/2}\langle\xi\rangle^{-1}\|\phi\|_{\mathbf{L}^\infty} + Ct^{-1/2}\langle\xi\rangle^{-1}\left\||\eta|^{-9/4}\langle\eta\rangle^4 t\mathscr{A}_0\phi\right\|_{\mathbf{L}^2}. \tag{91}$$

Lemma 3 is proved. □

3.3. Estimates for Derivatives. Denote $\mathscr{A}(t) = \overline{M}(1/t\sqrt{\Lambda''})\partial_\eta(1/\sqrt{\Lambda''})M = \mathscr{A}_0(t)+i\eta$, $\mathscr{A}_0(t) = (1/t\sqrt{\Lambda''})\partial_\eta(1/\sqrt{\Lambda''})$ such that

$$i\xi\mathscr{V}^*(t)\phi = \frac{|t|^{1/2}}{\sqrt{2\pi}}e^{it\Lambda(\xi)}\int_0^\infty i\xi e^{-it\Lambda'(\eta)\xi}M(t,\eta)\phi(\eta) \cdot\sqrt{\Lambda''(\eta)}d\eta = \frac{|t|^{1/2}}{\sqrt{2\pi}} \cdot e^{it\Lambda(\xi)}\int_0^\infty e^{-it\Lambda'(\eta)\xi}\partial_\eta\left(\frac{1}{t\sqrt{\Lambda''(\eta)}}M(t,\eta) \cdot\phi(\eta)\right)d\eta = \mathscr{V}^*(t)\mathscr{A}(t)\phi. \tag{92}$$

Since $\|\mathscr{V}^*(t)\phi\|_{\mathbf{L}^\infty} \le C|t|^{1/2}\|\sqrt{\Lambda''}\phi\|_{\mathbf{L}^1(0,\infty)}$ and

$$\|\mathscr{V}^*(t)\phi\|_{\mathbf{L}^2} = \|\mathscr{F}\mathscr{U}(-t)\mathscr{D}_t\mathscr{B}M\phi\|_{\mathbf{L}^2} = \|\mathscr{D}_t\mathscr{B}M\phi\|_{\mathbf{L}^2} = \|\mathscr{B}M\phi\|_{\mathbf{L}^2} = \|\phi\|_{\mathbf{L}^2}, \tag{93}$$

then by the Riesz interpolation theorem (see [23], p. 52) we have

$$\|\mathscr{V}^*(t)\phi\|_{\mathbf{L}^p} \le C|t|^{1/2-1/p}\left\|(\Lambda'')^{1/2-1/p}\phi\right\|_{\mathbf{L}^{p/(p-1)}} \tag{94}$$

for $2 \le p \le \infty$. We now estimate the derivative $\partial_\eta\mathscr{V}\phi$.

Lemma 4. *The estimate*

$$\left\|\eta^{\gamma+j}\langle\eta\rangle^{-2\gamma}t\mathscr{A}_0\mathscr{V}\xi^j\phi\right\|_{\mathbf{L}^2} \le C\|\xi\phi_\xi\|_{\mathbf{L}^2} + \left\||\xi|^{1/2}\phi\right\|_{\mathbf{L}^\infty} \tag{95}$$

is true for all $t \ge 1$, $j = -1, 0$, where $\gamma > 0$.

Proof. Since $\mathscr{A}_0 = \mathscr{A} - i\eta$ and $\mathscr{A}\mathscr{V} = \mathscr{V}i\xi$, we have

$$t\mathscr{A}_0\mathscr{V}\xi^j\phi = it\left(\mathscr{V}\xi^{j+1}\phi - \eta\mathscr{V}\xi^j\phi\right) = \sqrt{\frac{t\Lambda''(\eta)}{2\pi}}\int_0^\infty e^{-itS(\eta,\xi)}it(\xi-\eta)\xi^j\phi(\xi)d\xi \tag{96}$$

for $\eta > 0$. So we need to estimate

$$\eta^{\gamma+j}\langle\eta\rangle^{-2\gamma}t\mathscr{A}_0\mathscr{V}\xi^j\phi = \eta^\gamma\langle\eta\rangle^{-2\gamma}\mathscr{V}\psi_1(\xi,\xi_1)\xi\phi_\xi + \eta^\gamma\langle\eta\rangle^{-2\gamma} \cdot\sqrt{\frac{t\Lambda''(\eta)}{2\pi}}\int_0^\infty e^{-itS(\eta,\xi)}(\psi_1(\xi,\eta)-\psi_1(\xi,\xi_1)) \cdot\xi\phi_\xi(\xi)d\xi + \eta^\gamma\langle\eta\rangle^{-2\gamma} \cdot\sqrt{\frac{t\Lambda''(\eta)}{2\pi}}\int_0^\infty e^{-itS(\eta,\xi)}\phi(\xi)\psi_2(\xi,\eta)d\xi = I_1 + I_2 + I_3, \tag{97}$$

where $\xi_1 = (a/b)^{1/4}$, and

$$\psi_1(\xi,\eta) = \frac{\eta^{2+j}\xi^{j+1}}{(a+b\eta^2\xi^2)(\xi+\eta)}, \qquad \psi_2(\xi,\eta) = \partial_\xi\frac{\eta^{2+j}\xi^{j+2}}{(a+b\eta^2\xi^2)(\xi+\eta)}. \tag{98}$$

For the first summand using $\|\mathscr{V}\phi\|_{\mathbf{L}^2} \le \|\phi\|_{\mathbf{L}^2}$ we have for $j = -1, 0$

$$\|I_1\|_{\mathbf{L}^2} = \|\mathscr{V}\psi_1(\xi,\xi_1)\xi\phi_\xi\|_{\mathbf{L}^2} \le \left\|\langle\xi\rangle^{-3}\xi^{j+2}\phi_\xi\right\|_{\mathbf{L}^2} \le \|\xi\phi_\xi\|_{\mathbf{L}^2}. \tag{99}$$

Consider the second summand

$$\|I_2\|^2_{\mathbf{L}^2(0,\infty)} = Ct\int_0^\infty d\eta\,\Lambda''(\eta)\,\eta^{2\gamma}\langle\eta\rangle^{-4\gamma}$$
$$\cdot\int_0^\infty e^{-itS(\eta,\xi)}\left(\psi_1(\xi,\eta)-\psi_1(\xi,\xi_1)\right)\xi\phi_\xi(\xi)\,d\xi$$
$$\cdot\int_0^\infty e^{itS(\eta,\zeta)}\left(\psi_1(\zeta,\eta)-\psi_1(\zeta,\xi_1)\right)\overline{\zeta\phi_\zeta(\zeta)}d\zeta \quad (100)$$
$$= C\int_0^\infty d\xi\,e^{-it\Lambda(\xi)}\xi\phi_\xi(\xi)\int_0^\infty d\zeta$$
$$\cdot\,e^{it\Lambda(\zeta)}\overline{\zeta\phi_\zeta(\zeta)}K(t,\xi,\zeta),$$

where

$$K(t,\xi,\zeta) = t\int_0^\infty d\eta\,\Lambda''(\eta)\,\eta^{2\gamma}\langle\eta\rangle^{-4\gamma}$$
$$\cdot\,e^{it\Lambda'(\eta)(\xi-\zeta)}\left(\psi_1(\xi,\eta)-\psi_1(\xi,\xi_1)\right) \quad (101)$$
$$\cdot\left(\psi_1(\zeta,\eta)-\psi_1(\zeta,\xi_1)\right).$$

Changing $x = \Lambda'(\eta)$ we get

$$K(t,\xi,\zeta) = t\int_{-\infty}^\infty dx e^{itx(\xi-\zeta)}\eta^{2\gamma}\langle\eta\rangle^{-4\gamma}$$
$$\cdot\left(\psi_1(\xi,\eta(x))-\psi_1(\xi,\xi_1)\right) \quad (102)$$
$$\cdot\left(\psi_1(\zeta,\eta(x))-\psi_1(\zeta,\xi_1)\right).$$

We can rotate the contour of integration $x = re^{i(\pi/8)\mathrm{sgn}\,(\xi-\zeta)}$, since we see that $\eta(x) = Cx^{1/2}$ for $x \to +\infty$, $\eta(x) = \eta(0) = (a/b)^{1/4}$ for $x \to 0$, and $\eta(x) = C|x|^{-1/2}$ for $x \to -\infty$, and hence

$$|K(t,\xi,\zeta)|$$
$$\le Ct\int_{-\infty}^\infty e^{-Ct|\xi-\zeta||r|}\left|\psi_1\left(\xi,\eta\left(re^{i(\pi/8)\mathrm{sgn}\,(\xi-\zeta)}\right)\right)\right.$$
$$\left.-\psi_1(\xi,\eta(0))\right|\left|\psi_1\left(\zeta,\eta\left(re^{i(\pi/8)\mathrm{sgn}\,(\xi-\zeta)}\right)\right)\right.$$
$$\left.-\psi_1(\zeta,\eta(0))\right|\eta^{2\gamma}\langle\eta\rangle^{-4\gamma}dr$$
$$\le Ct\int_1^\infty e^{-Ct|\xi-\zeta||r|}\langle r\rangle^{-2\gamma}dr$$
$$+Ct\int_{-1}^1 e^{-Ct|\xi-\zeta||r|}|r|\,dr$$
$$+Ct\int_{-\infty}^{-1} e^{-Ct|\xi-\zeta||r|}|r|^{-\gamma}dr \le Ct\left(|\xi-\zeta|t\right)^{\gamma-1}$$
$$\cdot\left\langle(\xi-\zeta)t\right\rangle^{-2\gamma}. \quad (103)$$

Then by the Young inequality we obtain

$$\|I_2\|^2_{\mathbf{L}^2(0,\infty)} \le Ct\left\|\xi\phi_\xi\right\|_{\mathbf{L}^2}$$
$$\cdot\left\|\int_{\mathbf{R}}\left(|\xi-\zeta|t\right)^{\gamma-1}\left\langle(\xi-\zeta)t\right\rangle^{-2\gamma}\left|\zeta\phi_\zeta(\zeta)\right|d\zeta\right\|_{\mathbf{L}^2} \quad (104)$$
$$\le Ct\left\|\xi\phi_\xi\right\|^2_{\mathbf{L}^2}\left\|(\xi t)^{\gamma-1}\langle\xi t\rangle^{-2\gamma}\right\|_{\mathbf{L}^1} \le C\left\|\xi\phi_\xi\right\|^2_{\mathbf{L}^2}.$$

To estimate I_3 we integrate by parts via identity (47)

$$I_3 = Ct^{1/2}\eta^{\gamma-3/2}\langle\eta\rangle^{2-2\gamma}$$
$$\cdot\int_0^\infty e^{-itS(\eta,\xi)}(\xi-\eta)H_1\psi_2(\xi,\eta)\,\phi_\xi(\xi)\,d\xi$$
$$+Ct^{1/2}\eta^{\gamma-3/2}\langle\eta\rangle^{2-2\gamma} \quad (105)$$
$$\cdot\int_0^\infty e^{-itS(\eta,\xi)}\phi(\xi)(\xi-\eta)\,\partial_\xi\left(H_1\psi_2(\xi,\eta)\right)d\xi.$$

Using the estimates

$$|(\xi-\eta)H_1\psi_2(\xi,\eta)|$$
$$\le \frac{C|\xi-\eta|\,\eta^{2+j}\xi^{j+1}}{\left(1+t(\xi-\eta)^2\xi^{-2}\eta^{-2}(1+\xi^2\eta^2)(\xi+\eta)\right)(1+\xi^2\eta^2)(\xi+\eta)},$$
$$|(\xi-\eta)\,\partial_\xi\left(H_1\psi_2(\xi,\eta)\right)| \quad (106)$$
$$\le \frac{C\eta^{2+j}\left(|\xi-\eta|\,\xi^j+\xi^{j+1}\right)}{\left(1+t(\xi-\eta)^2\xi^{-2}\eta^{-2}(1+\xi^2\eta^2)(\xi+\eta)\right)(1+\xi^2\eta^2)(\xi+\eta)}$$

we obtain

$$|I_3| \le Ct^{1/2}\int_0^\infty \frac{\eta^{\gamma-3/2}\langle\eta\rangle^{2-2\gamma}|\xi-\eta|\,\eta^{2+j}\xi^j\left|\xi\phi_\xi(\xi)\right|d\xi}{\left(1+t(\xi-\eta)^2\xi^{-2}\eta^{-2}(1+\xi^2\eta^2)(\xi+\eta)\right)(1+\xi^2\eta^2)(\xi+\eta)}$$
$$+Ct^{1/2}\int_0^\infty \frac{\eta^{\gamma-3/2}\langle\eta\rangle^{2-2\gamma}\eta^{2+j}\left(|\xi-\eta|\,\xi^j+\xi^{j+1}\right)|\phi(\xi)|\,d\xi}{\left(1+t(\xi-\eta)^2\xi^{-2}\eta^{-2}(1+\xi^2\eta^2)(\xi+\eta)\right)(1+\xi^2\eta^2)(\xi+\eta)}$$

$$\leq Ct^{1/2}\left\|\xi\phi_{\xi}\right\|_{\mathbf{L}^2}\left(\int_0^{\infty}\frac{\eta^{2\gamma+2j+1}\langle\eta\rangle^{4-4\gamma}|\xi-\eta|^2\xi^{2j}d\xi}{\left(1+t(\xi-\eta)^2\xi^{-2}\eta^{-2}(1+\xi^2\eta^2)(\xi+\eta)\right)^2(1+\xi^2\eta^2)^2(\xi+\eta)^2}\right)^{1/2}$$
$$+Ct^{1/2}\left\||\xi|^{1/2}\phi\right\|_{\mathbf{L}^\infty}\int_0^{\infty}\frac{\eta^{\gamma-3/2}\langle\eta\rangle^{2-2\gamma}\eta^{2+j}\left(|\xi-\eta|\xi^j+\xi^{j+1}\right)\xi^{-1/2}d\xi}{\left(1+t(\xi-\eta)^2\xi^{-2}\eta^{-2}(1+\xi^2\eta^2)(\xi+\eta)\right)(1+\xi^2\eta^2)(\xi+\eta)}. \tag{107}$$

Since changing $\xi=\eta y$

$$\int_0^{\infty}\frac{t\eta^{2\gamma+2j+1}\langle\eta\rangle^{4-4\gamma}|\xi-\eta|^2\xi^{2j}d\xi}{\left(1+t(\xi-\eta)^2\xi^{-2}\eta^{-2}(1+\xi^2\eta^2)(\xi+\eta)\right)^2(1+\xi^2\eta^2)^2(\xi+\eta)^2}$$
$$=\int_0^{\infty}\frac{t\eta^{2\gamma+4j+2}\langle\eta\rangle^{4-4\gamma}|y-1|^2y^{2j}dy}{\left(1+t(y-1)^2y^{-2}\eta^{-1}(1+y^2\eta^4)(y+1)\right)^2(1+y^2\eta^4)^2(y+1)^2}\leq Ct^{-1}\eta^{2\gamma+4j+4}\langle\eta\rangle^{4-4\gamma}\int_0^{1/2}\frac{y^{2j+4}dy}{(1+y\eta^2)^8}$$
$$+Ct\eta^{2\gamma+4j+2}\langle\eta\rangle^{-4-4\gamma}\int_{1/2}^{3/2}\frac{|y-1|^2dy}{\left(1+t\eta^{-1}\langle\eta\rangle^4(y-1)^2\right)^2}+Ct^{-1}\eta^{2\gamma+4j+4}\langle\eta\rangle^{4-4\gamma}\int_{3/2}^{\infty}\frac{y^{2j-2}dy}{(1+y\eta^2)^8}\leq C\eta^{2\gamma-1}\langle\eta\rangle^{-4\gamma},$$
$$t^{1/2}\int_0^{\infty}\frac{\eta^{\gamma-3/2}\langle\eta\rangle^{2-2\gamma}\eta^{2+j}\left(|\xi-\eta|\xi^j+\xi^{j+1}\right)\xi^{-1/2}d\xi}{\left(1+t(\xi-\eta)^2\xi^{-2}\eta^{-2}(1+\xi^2\eta^2)(\xi+\eta)\right)(1+\xi^2\eta^2)(\xi+\eta)}$$
$$=t^{1/2}\int_0^{\infty}\frac{\eta^{2j+1+\gamma}\langle\eta\rangle^{2-2\gamma}(|y-1|+y)y^{j-1/2}dy}{\left(1+t(y-1)^2y^{-2}\eta^{-1}(1+y^2\eta^4)(y+1)\right)(1+y^2\eta^4)(y+1)}\leq Ct^{-1/2}\eta^{2j+2+\gamma}\langle\eta\rangle^{2-2\gamma}\int_0^{1/2}\frac{y^{j+3/2}dy}{(1+y\eta^2)^4}$$
$$+Ct^{1/2}\eta^{2j+1+\gamma}\langle\eta\rangle^{-2-2\gamma}\int_{1/2}^{3/2}\frac{dy}{1+t\eta^{-1}\langle\eta\rangle^4(y-1)^2}+Ct^{-1/2}\eta^{2j+2+\gamma}\langle\eta\rangle^{2-2\gamma}\int_{3/2}^{\infty}\frac{y^{j-3/2}dy}{(1+y\eta^2)^4}\leq C\eta^{\gamma-1/2}\langle\eta\rangle^{-2\gamma}, \tag{108}$$

we get

$$\left\|I_3\right\|_{\mathbf{L}^2(0,\infty)}\leq C\left(\left\|\xi\phi_\xi\right\|_{\mathbf{L}^2}+\left\||\xi|^{1/2}\phi\right\|_{\mathbf{L}^\infty}\right)\left\|\eta^{\gamma-1/2}\langle\eta\rangle^{-2\gamma}\right\|_{\mathbf{L}^2(0,\infty)}\leq C\left\|\xi\phi_\xi\right\|_{\mathbf{L}^2}+C\left\||\xi|^{1/2}\phi\right\|_{\mathbf{L}^\infty}. \tag{109}$$

In the case of $\eta<0$, the same estimate is obtained easier than the case of the positive line. Lemma 4 is proved. □

We need estimate of $\mathcal{V}_t$.

Lemma 5. *The estimate*

$$\left\|\eta^{\gamma}\langle\eta\rangle^{-2\gamma-2}t\partial_t\mathcal{V}\phi\right\|_{\mathbf{L}^2}\leq C\left\|\xi\phi_\xi\right\|_{\mathbf{L}^2}+C\left\||\xi|^{1/2}\phi\right\|_{\mathbf{L}^\infty} \tag{110}$$

is true for all $t\geq1$, where $\gamma>0$.

Proof. Since

$$\frac{S(\eta,\xi)}{\partial_\xi S(\eta,\xi)}=\frac{\left(3a+2b\eta^3\xi+b\eta^2\xi^2\right)(\xi-\eta)\xi}{3\left(a+b\eta^2\xi^2\right)(\xi+\eta)}, \tag{111}$$

integrating by parts we get with $\xi_1=(a/b)^{1/4}$

$$\eta^{\gamma}\langle\eta\rangle^{-2\gamma-2}t\mathcal{V}_t\phi=\eta^{\gamma}\langle\eta\rangle^{-2\gamma-2}\frac{1}{2}\mathcal{V}\phi-it\eta^{\gamma}\langle\eta\rangle^{-2\gamma-2}\sqrt{\frac{t\Lambda''(\eta)}{2\pi}}\int_0^{\infty}e^{-itS(\eta,\xi)}S(\eta,\xi)$$
$$\cdot\phi(\xi)\,d\xi=\frac{1}{2}\eta^{\gamma}\langle\eta\rangle^{-2\gamma-2}\mathcal{V}\phi-\eta^{\gamma}\langle\eta\rangle^{-2\gamma-1}$$
$$\cdot\mathcal{V}\left(\psi_1(\xi,\xi_1)\xi\phi_\xi\right)-\eta^{\gamma}\langle\eta\rangle^{-2\gamma-2}$$
$$\cdot\sqrt{\frac{t\Lambda''(\eta)}{2\pi}}\int_0^{\infty}e^{-itS(\eta,\xi)}\left(\psi_1(\xi,\eta)-\psi_1(\xi,\xi_1)\right)$$
$$\cdot\xi\phi_\xi(\xi)\,d\xi-\eta^{\gamma}\langle\eta\rangle^{-2\gamma-2}$$
$$\cdot\sqrt{\frac{t\Lambda''(\eta)}{2\pi}}\int_0^{\infty}e^{-itS(\eta,\xi)}\psi_2(\xi,\eta)\phi(\xi)\,d\xi$$

$$- \eta^{\gamma} \langle \eta \rangle^{-2\gamma-2} \sqrt{\frac{t\Lambda''(\eta)}{2\pi}} \int_0^\infty e^{-itS(\eta,\xi)} \psi_3(\xi,\eta) \cdot \phi(\xi)\, d\xi = I_1 + I_2 + I_3 + I_4 + I_5, \tag{112}$$

where

$$\psi_1(\xi,\eta) = \frac{\left(3a + 2b\eta^3\xi + b\eta^2\xi^2\right)(\xi-\eta)}{3\left(a + b\eta^2\xi^2\right)(\xi+\eta)},$$
$$\psi_2(\xi,\eta) = (\xi-\eta)\,\partial_\xi \frac{\left(3a + 2b\eta^3\xi + b\eta^2\xi^2\right)\xi}{3\left(a + b\eta^2\xi^2\right)(\xi+\eta)} \tag{113}$$
$$\psi_3(\xi,\eta) = \frac{\left(3a + 2b\eta^3\xi + b\eta^2\xi^2\right)\xi}{3\left(a + b\eta^2\xi^2\right)(\xi+\eta)}.$$

Using the estimate $\|\mathcal{V}\phi\|_{\mathbf{L}^2} \le \|\phi\|_{\mathbf{L}^2}$ we find for the first summand

$$\|I_1\|_{\mathbf{L}^2} \le \|\mathcal{V}\phi\|_{\mathbf{L}^2} \le \|\phi\|_{\mathbf{L}^2} \tag{114}$$

and for the second summand

$$\begin{aligned} \|I_2\|_{\mathbf{L}^2(0,\infty)} &\le \left\|\eta^{\gamma}\langle\eta\rangle^{-2\gamma-2}\mathcal{V}\left(\psi_1(\xi,\xi_1)\,\xi\phi_\xi\right)\right\|_{\mathbf{L}^2(0,\infty)} \\ &\le \left\|\psi_1(\xi,\xi_1)\,\xi\phi_\xi\right\|_{\mathbf{L}^2(0,\infty)} \le C\left\|\xi\phi_\xi\right\|_{\mathbf{L}^2}. \end{aligned} \tag{115}$$

Consider the third summand

$$\begin{aligned} \left\|\eta^{\gamma}\langle\eta\rangle^{-2\gamma-2} I_3\right\|^2_{\mathbf{L}^2(0,\infty)} &= Ct\int_0^\infty d\eta\Lambda''(\eta) \cdot \eta^{2\gamma}\langle\eta\rangle^{-4\gamma-4} \\ &\cdot \int_0^\infty e^{-itS(\eta,\xi)}\left(\psi_1(\xi,\eta)-\psi_1(\xi,\xi_1)\right)\xi\phi_\xi(\xi)\,d\xi \\ &\cdot \int_0^\infty e^{itS(\eta,\zeta)}\left(\psi_1(\xi,\eta)-\psi_1(\xi,\xi_1)\right)\overline{\zeta\phi_\zeta(\zeta)}d\zeta \\ &= C\int_0^\infty d\xi\, e^{-it\Lambda(\xi)}\xi\phi_\xi(\xi)\int_0^\infty d\zeta \cdot e^{it\Lambda(\zeta)}\overline{\zeta\phi_\zeta(\zeta)}K_1(t,\xi,\zeta), \end{aligned} \tag{116}$$

where

$$\begin{aligned} K_1(t,\xi,\zeta) &= t\int_0^\infty d\eta\,\Lambda''(\eta)\,\eta^{2\gamma}\langle\eta\rangle^{-4\gamma-4} \cdot e^{it\Lambda'(\eta)(\xi-\zeta)}\left(\psi_1(\xi,\eta)-\psi_1(\xi,\xi_1)\right) \\ &\cdot \left(\psi_1(\zeta,\eta)-\psi_1(\zeta,\xi_1)\right). \end{aligned} \tag{117}$$

Changing $x = \Lambda'(\eta)$, $\eta(x) = \sqrt{(1/2b)(x + \sqrt{4ab + x^2})}$, we get

$$\begin{aligned} K_1(t,\xi,\zeta) &= t\int_{-\infty}^\infty dx\,\eta^{2\gamma}(x)\langle\eta(x)\rangle^{-4\gamma-4} \cdot e^{itx(\xi-\zeta)}\left(\psi_1(\xi,\eta(x))-\psi_1(\xi,\eta(0))\right) \\ &\cdot \left(\psi_1(\zeta,\eta(x))-\psi_1(\zeta,\eta(0))\right). \end{aligned} \tag{118}$$

We can rotate the contour of integration $x = re^{i(\pi/8)\operatorname{sgn}(\xi-\zeta)}$, since we see that $\eta(x) = Cx^{1/2}$ for $x \to +\infty$, $\eta(x) = \eta(0) = \xi_1$ for $x \to 0$, and $\eta(x) = C|x|^{-1/2}$ for $x \to -\infty$, and hence

$$\begin{aligned} |K_1(t,\xi,\zeta)| &\le Ct\int_{-\infty}^\infty e^{-Ct|\xi-\zeta||r|}\psi_1\left(\psi_1\left(\xi,\eta\left(re^{i(\pi/8)\operatorname{sgn}(\xi-\zeta)}\right)\right)\right. \\ &\left.-\psi_1(\xi,\eta(0))\right)\left(\psi_1\left(\zeta,\eta\left(re^{i(\pi/8)\operatorname{sgn}(\xi-\zeta)}\right)\right)\right. \\ &\left.-\psi_1(\zeta,\eta(0))\right)\eta^{2\gamma}\langle\eta\rangle^{-4\gamma-4}dr \\ &\le Ct\int_1^\infty e^{-Ct|\xi-\zeta||r|}\langle r\rangle^{-\gamma}dr \\ &+ Ct\int_{-1}^1 e^{-Ct|\xi-\zeta||r|}|r|\,dr \\ &+ Ct\int_{-\infty}^{-1} e^{-Ct|\xi-\zeta||r|}|r|^{-2\gamma}dr \le Ct\left(|\xi-\zeta|\,t\right)^{\gamma-1} \\ &\cdot \langle(\xi-\zeta)\,t\rangle^{-2\gamma}. \end{aligned} \tag{119}$$

Then by the Young inequality we obtain

$$\begin{aligned} \|I_3\|^2_{\mathbf{L}^2(0,\infty)} &\le Ct\left\|\xi\phi_\xi\right\|_{\mathbf{L}^2} \\ &\cdot \left\|\int_{\mathbf{R}}\left(|\xi-\zeta|\,t\right)^{\gamma-1}\langle(\xi-\zeta)\,t\rangle^{-2\gamma}\left|\zeta\phi_\zeta(\zeta)\right|d\zeta\right\|_{\mathbf{L}^2} \\ &\le Ct\left\|\xi\phi_\xi\right\|^2_{\mathbf{L}^2}\left\|(\xi t)^{\gamma-1}\langle\xi t\rangle^{-2\gamma}\right\|_{\mathbf{L}^1} \le C\left\|\xi\phi_\xi\right\|^2_{\mathbf{L}^2}. \end{aligned} \tag{120}$$

Next we estimate I_4. We integrate by parts via identity (53) to get

$$\begin{aligned} I_4 &= Ct^{1/2}\eta^{\gamma-3/2}\langle\eta\rangle^{-2\gamma} \cdot \int_0^\infty e^{-itS(\eta,\xi)}\phi(\xi)\,\xi\partial_\xi\left(H_2\psi_2(\xi,\eta)\right) \\ &+ e^{-itS(x,\xi)}H_2\psi_2(\xi,\eta)\,\xi\phi_\xi(\xi)\,d\xi. \end{aligned} \tag{121}$$

Using the estimates $|\psi_2(\xi,\eta)| \le C(|\xi-\eta|\langle\eta\rangle^2/(\xi+\eta))$,

$$\begin{aligned} |H_2\psi_2(\xi,\eta)| &\le \frac{C|\xi-\eta|\langle\eta\rangle^2}{\left(1+t\xi^{-1}\eta^{-2}\left(1+\eta^2\xi^2\right)\left|\xi^2-\eta^2\right|\right)(\xi+\eta)}, \\ \left|\xi\partial_\xi\left(H_2\psi_2\right)\right| &\le \frac{C\langle\eta\rangle^2}{1+t\xi^{-1}\eta^{-2}\left(1+\eta^2\xi^2\right)\left|\xi^2-\eta^2\right|} \end{aligned} \tag{122}$$

we obtain

$$|I_4| \le Ct^{1/2}\eta^{\gamma-3/2}\langle\eta\rangle^{2-2\gamma} \cdot \int_0^\infty \frac{|\phi(\xi)|\,d\xi}{1+t\xi^{-1}\eta^{-2}(1+\eta^2\xi^2)|\xi^2-\eta^2|} + Ct^{1/2}\eta^{\gamma-3/2}\langle\eta\rangle^{2-2\gamma} \cdot \int_0^\infty \frac{|\xi-\eta|\,|\xi\phi_\xi(\xi)|\,d\xi}{(1+t\xi^{-1}\eta^{-2}(1+\eta^2\xi^2)|\xi^2-\eta^2|)(\xi+\eta)} \le C\left\||\xi|^{1/2}\phi\right\|_{\mathbf{L}^\infty}\int_0^\infty \frac{t^{1/2}\eta^{\gamma-3/2}\langle\eta\rangle^{2-2\gamma}\xi^{-1/2}d\xi}{1+t\xi^{-1}\eta^{-2}(1+\eta^2\xi^2)|\xi^2-\eta^2|} + C\|\xi\phi_\xi\|_{\mathbf{L}^2} \cdot \left(\int_0^\infty \frac{t\eta^{2\gamma-3}\langle\eta\rangle^{4-4\gamma}(\xi-\eta)^2\,d\xi}{(1+t\xi^{-1}\eta^{-2}(1+\eta^2\xi^2)|\xi^2-\eta^2|)^2(\xi+\eta)^2}\right)^{1/2}. \quad (123)$$

Changing $\xi=\eta y$ we find

$$\int_0^\infty \frac{t^{1/2}\eta^{\gamma-3/2}\langle\eta\rangle^{2-2\gamma}\xi^{-1/2}d\xi}{1+t\xi^{-1}\eta^{-2}(1+\eta^2\xi^2)|\xi^2-\eta^2|} = \int_0^\infty \frac{t^{1/2}\eta^{\gamma-1}\langle\eta\rangle^{2-2\gamma}y^{-1/2}dy}{1+t\eta^{-1}y^{-1}(1+\eta^4y^2)(y+1)|y-1|} \le Ct^{-1/2}\eta^{\gamma}\langle\eta\rangle^{2-2\gamma}\int_0^{1/2}\frac{y^{1/2}dy}{(1+\eta^2y)^2} + Ct^{1/2}\eta^{\gamma-1}\langle\eta\rangle^{2-2\gamma}\int_{1/2}^{3/2}\frac{dy}{1+t\eta^{-1}\langle\eta\rangle^4|y-1|} + Ct^{-1/2}\eta^{\gamma}\langle\eta\rangle^{2-2\gamma}\int_{3/2}^\infty\frac{y^{-3/2}dy}{1+\eta^4y^2} \le C\eta^{\gamma-1/2}\langle\eta\rangle^{-2\gamma},$$

$$\int_0^\infty \frac{t\eta^{2\gamma-3}\langle\eta\rangle^{4-4\gamma}(\xi-\eta)^2\,d\xi}{(1+t\xi^{-1}\eta^{-2}(1+\eta^2\xi^2)|\xi^2-\eta^2|)^2(\xi+\eta)^2} = \int_0^\infty \frac{t\eta^{2\gamma-2}\langle\eta\rangle^{4-4\gamma}(y-1)^2\,dy}{(1+t\eta^{-1}y^{-1}(1+\eta^4y^2)(y+1)|y-1|)^2(y+1)^2} \le Ct^{-1}\eta^{2\gamma}\langle\eta\rangle^{4-4\gamma}\int_0^{1/2}\frac{y^2dy}{(1+\eta^2y)^4} + Ct^{-1}\eta^{2\gamma}\langle\eta\rangle^{-4-4\gamma}\int_{1/2}^{3/2}dy + Ct^{-1}\eta^{2\gamma}\langle\eta\rangle^{4-4\gamma}\int_{3/2}^\infty\frac{dy}{y^2(1+\eta^2y)^4} \le C\eta^{2\gamma-1}\langle\eta\rangle^{-4\gamma}. \quad (124)$$

Hence we get

$$\|I_4\|_{\mathbf{L}^2(0,\infty)} \le C\left(\|\xi\phi_\xi\|_{\mathbf{L}^2} + \left\||\xi|^{1/2}\phi\right\|_{\mathbf{L}^\infty}\right)\left\|\eta^{\gamma-1/2}\langle\eta\rangle^{-2\gamma}\right\|_{\mathbf{L}^2(0,\infty)} \le C\|\xi\phi_\xi\|_{\mathbf{L}^2(0,\infty)} + C\left\||\xi|^{1/2}\phi\right\|_{\mathbf{L}^\infty}. \quad (125)$$

In the last integral I_5 we integrate by parts via identity (47)

$$I_5 = Ct^{1/2}\eta^{\gamma-3/2}\langle\eta\rangle^{-2\gamma} \cdot \int_0^\infty e^{-itS(\eta,\xi)}(\xi-\eta)H_1\psi_3(\xi,\eta)\phi_\xi(\xi) + e^{-itS(\eta,\xi)}\phi(\xi)(\xi-\eta)\partial_\xi(H_1\psi_3(\xi,\eta))\,d\xi. \quad (126)$$

Using the estimates $|\psi_3(\xi,\eta)| \le C(\xi\langle\eta\rangle^2/(\xi+\eta))$,

$$|(\xi-\eta)H_1\psi_3(\xi,\eta)| \le \frac{C\xi|\xi-\eta|\langle\eta\rangle^2}{(1+t(\xi-\eta)^2\xi^{-2}\eta^{-2}(1+\xi^2\eta^2)(\xi+\eta))(\xi+\eta)},$$
$$|(\xi-\eta)\partial_\xi(H_1\psi_3(\xi,\eta))| \le \frac{C\langle\eta\rangle^2}{1+t(\xi-\eta)^2\xi^{-2}\eta^{-2}(1+\xi^2\eta^2)(\xi+\eta)} \quad (127)$$

we obtain

$$|I_5| \le C\int_0^\infty \frac{t^{1/2}\eta^{\gamma-3/2}\langle\eta\rangle^{2-2\gamma}|\xi-\eta|\,|\xi\phi_\xi(\xi)|\,d\xi}{(1+t(\xi-\eta)^2\xi^{-2}\eta^{-2}(1+\xi^2\eta^2)(\xi+\eta))(\xi+\eta)} + C\int_0^\infty \frac{t^{1/2}\eta^{\gamma-3/2}\langle\eta\rangle^{2-2\gamma}|\phi(\xi)|\,d\xi}{1+t(\xi-\eta)^2\xi^{-2}\eta^{-2}(1+\xi^2\eta^2)(\xi+\eta)}$$
$$\le C\|\xi\phi_\xi\|_{\mathbf{L}^2}\left(\int_0^\infty \frac{t\eta^{2\gamma-3}\langle\eta\rangle^{4-4\gamma}(\xi-\eta)^2\,d\xi}{(1+t(\xi-\eta)^2\xi^{-2}\eta^{-2}(1+\xi^2\eta^2)(\xi+\eta))^2(\xi+\eta)^2}\right)^{1/2} + C\left\||\xi|^{1/2}\phi\right\|_{\mathbf{L}^\infty}\int_0^\infty \frac{t^{1/2}\eta^{\gamma-3/2}\langle\eta\rangle^{2-2\gamma}\xi^{-1/2}d\xi}{1+t(\xi-\eta)^2\xi^{-2}\eta^{-2}(1+\xi^2\eta^2)(\xi+\eta)}. \quad (128)$$

Since changing $\xi = \eta y$

$$\int_0^{\infty} \frac{t\eta^{2\gamma-3} \langle \eta \rangle^{4-4\gamma} (\xi-\eta)^2 d\xi}{\left(1+t(\xi-\eta)^2 \xi^{-2}\eta^{-2}(1+\xi^2\eta^2)(\xi+\eta)\right)^2 (\xi+\eta)^2}$$
$$= \int_0^{\infty} \frac{t\eta^{2\gamma-2} \langle \eta \rangle^{4-4\gamma} (y-1)^2 dy}{\left(1+t(y-1)^2 y^{-2}\eta^{-1}(1+\eta^4 y^2)(y+1)\right)^2 (y+1)^2}$$
$$\leq Ct^{-1}\eta^{2\gamma} \langle \eta \rangle^{4-4\gamma} \int_0^{1/2} \frac{y^4 dy}{(1+\eta^2 y)^4}$$
$$+ Ct\eta^{2\gamma-2} \langle \eta \rangle^{4-4\gamma} \int_{1/2}^{3/2} \frac{(y-1)^2 dy}{\left(1+t\eta^{-1}\langle \eta \rangle^4 (y-1)^2\right)^2}$$
$$+ Ct^{-1}\eta^{2\gamma} \langle \eta \rangle^{4-4\gamma} \int_{3/2}^{\infty} \frac{dy}{(1+\eta^2 y)^4 y^2} \leq C\eta^{2\gamma-1} \langle \eta \rangle^{-4\gamma}, \tag{129}$$
$$\int_0^{\infty} \frac{t^{1/2}\eta^{\gamma-3/2} \langle \eta \rangle^{2-2\gamma} \xi^{-1/2} d\xi}{1+t(\xi-\eta)^2 \xi^{-2}\eta^{-2}(1+\xi^2\eta^2)(\xi+\eta)}$$
$$= \int_0^{\infty} \frac{t^{1/2}\eta^{\gamma-1} \langle \eta \rangle^{2-2\gamma} y^{-1/2} dy}{1+t(y-1)^2 y^{-2}\eta^{-1}(1+y^2\eta^4)(y+1)}$$
$$\leq Ct^{-1/2}\eta^{\gamma} \langle \eta \rangle^{2-2\gamma} \int_0^{1/2} \frac{y^{3/2} dy}{(1+\eta^2 y)^2}$$
$$+ Ct^{1/2}\eta^{\gamma-1} \langle \eta \rangle^{2-2\gamma} \int_{1/2}^{3/2} \frac{dy}{1+t\eta^{-1}\langle \eta \rangle^4 (y-1)^2}$$
$$+ Ct^{-1/2}\eta^{\gamma} \langle \eta \rangle^{2-2\gamma} \int_{3/2}^{\infty} \frac{y^{-3/2} dy}{(1+\eta^2 y)^2} \leq C\eta^{\gamma-1/2} \langle \eta \rangle^{-2\gamma},$$

we get

$$\|I_5\|_{\mathbf{L}^2(0,\infty)} \leq C\left(\|\xi\phi_\xi\|_{\mathbf{L}^2} + \left\||\xi|^{1/2}\phi\right\|_{\mathbf{L}^\infty}\right) \left\|\eta^{\gamma-1/2} \langle \eta \rangle^{-2\gamma}\right\|_{\mathbf{L}^2(0,\infty)} \leq C\|\xi\phi_\xi\|_{\mathbf{L}^2(0,\infty)} + C\left\||\xi|^{1/2}\phi\right\|_{\mathbf{L}^\infty}. \tag{130}$$

In the case of $\eta < 0$, the same estimate is obtained easier than the case of the positive line. Lemma 5 is proved. □

3.4. Asymptotics for the Nonlinearity. We obtain the asymptotic representation for the nonlinear term. Define the norm

$$\|\widehat{\varphi}\|_{\mathbf{Z}} = \left\|\langle \xi \rangle^{1/2} \widehat{\varphi}\right\|_{\mathbf{L}^\infty} + \|\xi\widehat{\varphi}_\xi\|_{\mathbf{L}^2}. \tag{131}$$

Lemma 6. *The asymptotics*

$$\mathcal{F}\mathcal{U}(-t)\partial_x \mathcal{K} u^3 = \frac{\sqrt{3}i\xi\widehat{K}}{it} e^{it\Omega(\xi)} \mathcal{D}_3 \frac{1}{\Lambda''} \widehat{\varphi}^3 + \frac{3i\xi\widehat{K}}{t\Lambda''} |\widehat{\varphi}|^2 \widehat{\varphi} + O\left(t^{-5/4}\|\widehat{\varphi}\|_{\mathbf{Z}}^3\right) \tag{132}$$

is true for all $t \geq 1$, $\xi \geq 0$, where $\widehat{\varphi}(t) = \mathcal{F}\mathcal{U}(-t)u(t)$.

Proof. In view of (37) we find for the new dependent variable $\widehat{\varphi} = \mathcal{F}\mathcal{U}(-t)u(t)$

$$\mathcal{F}\mathcal{U}(-t)\partial_x \mathcal{K} u^3 = \sqrt{3}i\xi\widehat{K}t^{-1} e^{it\Omega(\xi)} \mathcal{D}_3 \mathcal{V}^*(3t) \frac{1}{\Lambda''} \psi_0^3 + 3i\xi\widehat{K}t^{-1} \mathcal{V}^*(t) \frac{1}{\Lambda''} \psi_0^2 \overline{\psi_0} + 3i\xi\widehat{K}t^{-1} \mathcal{D}_{-1} \mathcal{V}^*(-t) \frac{1}{\Lambda''} \psi_0 \overline{\psi_0}^2 + \sqrt{3}i\xi\widehat{K}t^{-1} e^{it\Omega(\xi)} \mathcal{D}_{-3} \mathcal{V}^*(-3t) \frac{1}{\Lambda''} \overline{\psi_0}^3, \tag{133}$$

where $\psi_j = \mathcal{V}(i\xi)^j \widehat{\varphi}$. By Lemma 3 we have

$$3i\xi\widehat{K}t^{-1}\mathcal{V}^*(t) \frac{1}{\Lambda''} \psi_0^2 \overline{\psi_0} = 3i\xi\widehat{K}t^{-1} A^*(t) \frac{1}{\Lambda''} \psi_0^2 \overline{\psi_0} + Ct^{-5/4} \left\||\eta|^{-9/4} \langle \eta \rangle^4 t\mathcal{A}_0 \frac{1}{\Lambda''} \psi_0^2 \overline{\psi_0}\right\|_{\mathbf{L}^2} + Ct^{-5/4} \left\|\frac{1}{\Lambda''} \psi_0^2 \overline{\psi_0}\right\|_{\mathbf{L}^\infty}. \tag{134}$$

By Lemma 2

$$\psi_0 = A_0(t)\widehat{\varphi} + O\left(t^{-1/4}|\eta|^{-1/4} \langle \eta \rangle^{-1} \left(\left\|\xi^{1/2}\widehat{\varphi}\right\|_{\mathbf{L}^\infty} + \|\xi\widehat{\varphi}_\xi\|_{\mathbf{L}^2}\right)\right), \tag{135}$$
$$|\psi_0| \leq C|\eta|^{-1/4} \langle \eta \rangle^{-1/4} \|\widehat{\varphi}\|_{\mathbf{Z}}.$$

Therefore

$$\left\|\frac{1}{\Lambda''} \psi_0^2 \overline{\psi_0}\right\|_{\mathbf{L}^\infty} \leq C\|\widehat{\varphi}\|_{\mathbf{Z}}^3 \left\|\frac{1}{\Lambda''} |\eta|^{-3/4} \langle \eta \rangle^{-3/4}\right\|_{\mathbf{L}^\infty} \leq C\|\widehat{\varphi}\|_{\mathbf{Z}}^3. \tag{136}$$

Also by the Leibnitz rule

$$\mathcal{A}_0 \frac{1}{\Lambda''} \psi_0^2 \overline{\psi_0} = \frac{1}{t\sqrt{\Lambda''}} \partial_\eta \left(\left(\frac{\psi_0}{\sqrt{\Lambda''}}\right)^2 \overline{\frac{\psi_0}{\sqrt{\Lambda''}}}\right) = \frac{1}{\Lambda''} \psi_0^2 \overline{\mathcal{A}_0 \psi_0} + \frac{2}{\Lambda''} \psi_0 \overline{\psi_0} \mathcal{A}_0 \psi_0. \tag{137}$$

Then by Lemma 4 we get

$$\left\||\eta|^{-9/4} \langle \eta \rangle^4 t\mathcal{A}_0 \frac{1}{\Lambda''} \psi_0^2 \overline{\psi_0}\right\|_{\mathbf{L}^2} \leq C\left\||\eta|^{3/4} |\psi_0|^2 t\mathcal{A}_0 \psi_0\right\|_{\mathbf{L}^2} \leq C\|\widehat{\varphi}\|_{\mathbf{Z}}^2 \left\||\eta|^{1/4} \langle \eta \rangle^{-1/2} t\mathcal{A}_0 \psi_0\right\|_{\mathbf{L}^2} \leq C\|\widehat{\varphi}\|_{\mathbf{Z}}^3. \tag{138}$$

Hence we obtain

$$\begin{aligned} &3i\xi\widehat{K}t^{-1}\mathcal{V}^*(t)\frac{1}{\Lambda''}\psi_0^2\overline{\psi_0} \\ &= 3i\xi\widehat{K}t^{-1}A^*(t)\frac{1}{\Lambda''}(A_0(t)\widehat{\varphi})^2\overline{A_0(t)\widehat{\varphi}} \\ &+ O\left(t^{-5/4}\|\widehat{\varphi}\|_{\mathbf{Z}}^3\right) \\ &= \frac{3i\xi\widehat{K}}{t\Lambda''}|\widehat{\varphi}|^2\widehat{\varphi} + O\left(t^{-5/4}\|\widehat{\varphi}\|_{\mathbf{Z}}^3\right). \end{aligned} \tag{139}$$

In the same manner

$$\begin{aligned} &\sqrt{3}i\xi\widehat{K}t^{-1}e^{it\Omega(\xi)}\mathcal{D}_3\mathcal{V}^*(3t)\frac{1}{\Lambda''}\psi_0^3 \\ &= \frac{\sqrt{3}i\xi\widehat{K}}{it}e^{it\Omega(\xi)}\mathcal{D}_3\frac{1}{\Lambda''}\widehat{\varphi}^3 + O\left(t^{-5/4}\|\widehat{\varphi}\|_{\mathbf{Z}}^3\right), \\ &3i\xi\widehat{K}t^{-1}\mathcal{D}_{-1}\mathcal{V}^*(-t)\frac{1}{\Lambda''}\psi_0\overline{\psi_0}^2 = O\left(t^{-5/4}\|\widehat{\varphi}\|_{\mathbf{Z}}^3\right), \\ &\sqrt{3}i\xi\widehat{K}t^{-1}e^{it\Omega(\xi)}\mathcal{D}_{-3}\mathcal{V}^*(-3t)\frac{1}{\Lambda''}\overline{\psi_0}^3 \\ &= O\left(t^{-5/4}\|\widehat{\varphi}\|_{\mathbf{Z}}^3\right). \end{aligned} \tag{140}$$

Hence the result of the lemma follows. □

4. A Priori Estimates

We define

$$\begin{aligned} \mathbf{X}_T &= \left\{\mathcal{U}(-t)u \in \mathbf{C}\left([0,T];\mathbf{H}^2\right); \|u\|_{\mathbf{X}_T} < \infty\right\}, \\ \|u\|_{\mathbf{X}_T} &= \sup_{t\in[0,T]}\left(\left\|\langle\xi\rangle^{1/2}\widehat{\varphi}\right\|_{\mathbf{L}^\infty} + \langle t\rangle^{-\gamma}\|u(t)\|_{\mathbf{H}^2}\right. \\ &\left. + \langle t\rangle^{-\gamma}\|\partial_x\mathcal{J}u(t)\|_{\mathbf{L}^2}\right), \end{aligned} \tag{141}$$

where $\mathcal{J} = \mathcal{U}(t)x\mathcal{U}(-t)$, $\widehat{\varphi} = \mathcal{F}\mathcal{U}(-t)u(t)$, and $\gamma > 0$ is small. We have the local in time existence of solutions.

Theorem 7. *Let the initial data $u_0 \in \mathbf{H}^2 \cap \mathbf{H}^{1,1}$. Then there exists a time $T_0 > 0$ such that (1) has a unique solution u in $\mathbf{X}_{T_0}$.*

To get the desired results, we prove a priori estimates of solutions uniformly in time.

Lemma 8. *Assume $u_0 \in \mathbf{H}^2 \cap \mathbf{H}^{1,1}$ and the norm $\|u_0\|_{\mathbf{H}^{1,1}} + \|u_0\|_{\mathbf{H}^2} \le \varepsilon$. Then the estimate*

$$\|u\|_{\mathbf{X}_T} < C\varepsilon \tag{142}$$

is true for all $T \ge 1$.

Proof. By continuity of the norm $\|u\|_{\mathbf{X}_T}$ with respect to T, arguing by the contradiction we can find the first time $T \ge 1$ such that $\|u\|_{\mathbf{X}_T} = C\varepsilon$. Consider a priori estimates of $\|\partial_x\mathcal{J}u(t)\|_{\mathbf{L}^2} = \|\xi\partial_\xi\widehat{\varphi}\|_{\mathbf{L}^2}$. To avoid the derivative loss in (1) we apply the operator

$$\mathcal{P} = t\partial_t + \frac{1}{3}x\partial_x - \frac{4}{3}a\partial_a \tag{143}$$

and use the commutators $[\mathcal{P},\mathcal{L}] = -\mathcal{L}$, $[\mathcal{P},\partial_x\mathcal{K}] = \mathcal{F}^{-1}\widehat{K}_1(\xi)\mathcal{F} = \mathcal{K}_1$, where $\widehat{K}_1(\xi) = (i/3)\xi\partial_\xi(i\xi\widehat{K}(\xi)) - (4/3)i\xi a\partial_a\widehat{K}(\xi) = O(\xi)$. Also we represent $\widehat{K}(\xi) = \lambda + \widehat{K}_2(\xi)$ with $\widehat{K}_2(\xi) = O(\langle\xi\rangle^{-1})$; that is, $\mathcal{K} = \mathcal{K}_2 + \lambda$. Then we get

$$\begin{aligned} \mathcal{L}\mathcal{P}u &= (\mathcal{P}+1)\mathcal{L}u = \mathcal{P}\partial_x\mathcal{K}u^3 + \partial_x\mathcal{K}u^3 \\ &= 3\partial_x\mathcal{K}\left(u^2\mathcal{P}u\right) + \partial_x\mathcal{K}u^3 + [\mathcal{P},\partial_x\mathcal{K}]u^3 \\ &= 3\lambda\partial_x\left(u^2\mathcal{P}u\right) + 3\partial_x\mathcal{K}_2\left(u^2\mathcal{P}u\right) + \partial_x\mathcal{K}u^3 \\ &+ \mathcal{K}_1u^3. \end{aligned} \tag{144}$$

Define the high and short frequency projectors $\mathcal{Q}_j\phi = \mathcal{F}^{-1}\chi_j\widehat{\phi}$, where $\chi_1(\xi) = 1$ for $|\xi| \ge 1$ and $\chi_1(\xi) = 0$ for $|\xi| \le 1$, and also $\chi_2(\xi) = 1 - \chi_1(\xi)$. Then we get

$$\begin{aligned} \mathcal{L}\mathcal{Q}_1\mathcal{P}u &= 3\lambda\partial_x\mathcal{Q}_1\left(u^2\mathcal{Q}_1\mathcal{P}u\right) \\ &+ 3\lambda\partial_x\mathcal{Q}_1\left(u^2\mathcal{Q}_2\mathcal{P}u\right) \\ &+ 3\mathcal{Q}_1\partial_x\mathcal{K}_2\left(u^2\mathcal{P}u\right) + \mathcal{Q}_1\partial_x\mathcal{K}u^3 \\ &+ \mathcal{Q}_1\mathcal{K}_1u^3 \end{aligned} \tag{145}$$

and the integral equation

$$\begin{aligned} \mathcal{Q}_2\mathcal{P}u &= \mathcal{U}(t)\mathcal{Q}_2\mathcal{P}u_0 + \int_0^t d\tau\,\mathcal{U}(t-\tau) \\ &\cdot\left(3\lambda\partial_x\mathcal{Q}_2\left(u^2\mathcal{Q}_1\mathcal{P}u\right) + 3\lambda\partial_x\mathcal{Q}_2\left(u^2\mathcal{Q}_2\mathcal{P}u\right)\right. \\ &\left.+ 3\mathcal{Q}_2\partial_x\mathcal{K}_2\left(u^2\mathcal{P}u\right) + \mathcal{Q}_2\partial_x\mathcal{K}u^3 + \mathcal{Q}_2\mathcal{K}_1u^3\right). \end{aligned} \tag{146}$$

Hence applying the energy method to the first equation we find

$$\begin{aligned} \frac{d}{dt}\|\mathcal{Q}_1\mathcal{P}u\|_{\mathbf{L}^2}^2 &\le C\left(\|uu_x\|_{\mathbf{L}^\infty} + \|u\|_{\mathbf{L}^\infty}^2\right) \\ &\cdot\left(\|\mathcal{P}u\|_{\mathbf{L}^2} + \|u\|_{\mathbf{L}^2}\right)\|\mathcal{Q}_1\mathcal{P}u\|_{\mathbf{L}^2} \end{aligned} \tag{147}$$

and by the integral equation

$$\begin{aligned} &\|\mathcal{Q}_2\mathcal{P}u\|_{\mathbf{L}^2} \\ &\le \|\mathcal{Q}_2\mathcal{P}u_0\|_{\mathbf{L}^2} \\ &+ C\int_0^t\left(\|uu_x\|_{\mathbf{L}^\infty} + \|u\|_{\mathbf{L}^\infty}^2\right)\left(\|\mathcal{P}u\|_{\mathbf{L}^2} + \|u\|_{\mathbf{L}^2}\right)d\tau. \end{aligned} \tag{148}$$

Applying the estimate of Lemma 2 we have

$$\begin{aligned} &\left|\mathcal{V}\xi^j\phi\right| \\ &\le C\langle\eta\rangle^{1/2}|\eta|^{-1/4}\left(\left\|\langle\xi\rangle^{1/2}\widehat{\varphi}\right\|_{\mathbf{L}^\infty} + t^{-1/4}\|\xi\phi_\xi\|_{\mathbf{L}^2}\right). \end{aligned} \tag{149}$$

Hence

$$\begin{aligned}\left\|\partial_x^j u\right\|_{\mathbf{L}^\infty} &\le \left\|2\operatorname{Re}\mathscr{D}_t\mathscr{B}M\mathscr{V}(i\xi)^j\widehat{\varphi}\right\|_{\mathbf{L}^\infty}\\ &\le Ct^{-1/2}\left\|\eta^{3/2}\langle\eta\rangle^{-2}\mathscr{V}\xi^j\widehat{\varphi}\right\|_{\mathbf{L}^\infty}\\ &\le Ct^{-1/2}\left(\left\|\langle\xi\rangle^{1/2}\widehat{\varphi}\right\|_{\mathbf{L}^\infty}+t^{-1/4}\left\|\xi\phi_\xi\right\|_{\mathbf{L}^2}\right).\end{aligned} \tag{150}$$

Therefore

$$\begin{aligned}&\frac{d}{dt}\left\|\mathscr{Q}_1\mathscr{P}u\right\|_{\mathbf{L}^2} \le C\langle t\rangle^{-1}\left(\left\|\langle\xi\rangle^{1/2}\widehat{\varphi}\right\|_{\mathbf{L}^\infty}\right.\\ &\quad\left.+t^{-1/4}\left\|\xi\phi_\xi\right\|_{\mathbf{L}^2}\right)\left(\|\mathscr{P}u\|_{\mathbf{L}^2}+\|u\|_{\mathbf{L}^2}\right),\\ &\left\|\mathscr{Q}_2\mathscr{P}u\right\|_{\mathbf{L}^2} \le \left\|\mathscr{Q}_2\mathscr{P}u_0\right\|_{\mathbf{L}^2}+C\int_0^t\langle t\rangle^{-1}\\ &\quad\cdot\left(\left\|\langle\xi\rangle^{1/2}\widehat{\varphi}\right\|_{\mathbf{L}^\infty}+t^{-1/4}\left\|\xi\phi_\xi\right\|_{\mathbf{L}^2}\right)\\ &\quad\cdot\left(\|\mathscr{P}u\|_{\mathbf{L}^2}+\|u\|_{\mathbf{L}^2}\right)d\tau.\end{aligned} \tag{151}$$

And similarly

$$\begin{aligned}&\frac{d}{dt}\|u\|_{\mathbf{H}^2}\\ &\quad\le C\langle t\rangle^{-1}\left(\left\|\langle\xi\rangle^{1/2}\widehat{\varphi}\right\|_{\mathbf{L}^\infty}+t^{-1/4}\left\|\xi\phi_\xi\right\|_{\mathbf{L}^2}\right)\|u\|_{\mathbf{H}^2},\end{aligned} \tag{152}$$

from which it follows that

$$\|\mathscr{P}u\|_{\mathbf{L}^2}+\|u\|_{\mathbf{H}^2} \le \varepsilon + C\varepsilon^3\langle t\rangle^\gamma. \tag{153}$$

By the identity $\mathscr{P}=t\mathscr{L}+(1/3)\mathscr{J}\partial_x-(4/3)a\mathscr{I}$, we obtain

$$\begin{aligned}\left\|\xi\partial_\xi\widehat{\varphi}\right\|_{\mathbf{L}^2} &= \left\|\partial_x\mathscr{J}u\right\|_{\mathbf{L}^2}\\ &\le C\|\mathscr{P}u\|_{\mathbf{L}^2}+t\|\mathscr{L}u\|_{\mathbf{L}^2}+C\|\mathscr{I}u\|_{\mathbf{L}^2}\\ &\le C\|\mathscr{P}u\|_{\mathbf{L}^2}+Ct\left\|uu_x\right\|_{\mathbf{L}^\infty}\|u\|_{\mathbf{L}^2}\\ &\quad+C\|\mathscr{I}u\|_{\mathbf{L}^2} \le C\varepsilon+C\varepsilon^3t^\gamma+C\|\mathscr{I}u\|_{\mathbf{L}^2}.\end{aligned} \tag{154}$$

Next we estimate the norm $\|\mathscr{I}u\|_{\mathbf{L}^2}$. Denote $\widehat{K}_3(\xi) = i\xi\partial_a\widehat{K}(\xi) = O(1)$. Applying the operator $\mathscr{I}=\partial_a-t\partial_x^{-1}$ to (1) via the commutator $[\mathscr{I},\mathscr{L}]=0$, we get

$$\begin{aligned}\mathscr{L}\mathscr{I}u &= \mathscr{I}\mathscr{L}u = \mathscr{I}\partial_x\mathscr{K}u^3\\ &= 3\lambda\partial_x\left(u^2\mathscr{I}u\right)+\mathscr{K}_3u^3+3i\partial_x\mathscr{K}_2\left(u^2\mathscr{I}u\right)\\ &\quad+3t\mathscr{K}\partial_x\left(u^2\partial_x^{-1}u\right)-t\mathscr{K}u^3\\ &= 3\lambda\partial_x\left(u^2\mathscr{I}u\right)+\mathscr{K}_3u^3+3\mathscr{K}_2\left(u^2\mathscr{I}u\right)+N,\end{aligned} \tag{155}$$

where $N = 2t\mathscr{K}(3uu_x\partial_x^{-1}u+u^3)$. Using the factorization formulas as in the derivation of (37) we find

$$\begin{aligned}&\mathscr{F}\mathscr{U}(-t)N = 6t\widehat{K}(\xi)\mathscr{F}\mathscr{U}(-t)\left(uu_x\partial_x^{-1}u\right)\\ &\quad+2t\widehat{K}_1(\xi)\mathscr{F}\mathscr{U}(-t)\left(u^3\right) = 6t\widehat{K}(\xi)\\ &\quad\cdot\mathscr{V}^*\overline{M}\mathscr{B}^{-1}\mathscr{D}_t^{-1}\left(\mathscr{D}_t\mathscr{B}\left(M\psi_0+\overline{M\psi_0}\right)\right)\\ &\quad\cdot\left(\mathscr{D}_t\mathscr{B}\left(M\psi_1+\overline{M\psi_1}\right)\right)\\ &\quad\cdot\left(\mathscr{D}_t\mathscr{B}\left(M\psi_{-1}+\overline{M\psi_{-1}}\right)\right)+2t\widehat{K}(\xi)\\ &\quad\cdot\mathscr{V}^*\overline{M}\mathscr{B}^{-1}\mathscr{D}_t^{-1}\left(\mathscr{D}_t\mathscr{B}\left(M\psi_0+\overline{M\psi_0}\right)^3\right)\\ &\quad= 6\widehat{K}(\xi)\mathscr{V}^*\frac{1}{\Lambda''}\overline{M}\left(M\psi_0+\overline{M\psi_0}\right)\left(M\psi_1+\overline{M\psi_1}\right)\\ &\quad\cdot\left(M\psi_{-1}+\overline{M\psi_{-1}}\right)+2\widehat{K}(\xi)\mathscr{V}^*\frac{1}{\Lambda''}\\ &\quad\cdot\overline{M}\left(M\psi_0+\overline{M\psi_0}\right)^3,\end{aligned} \tag{156}$$

where we denote $\psi_j=\mathscr{V}(i\xi)^j\widehat{\varphi}$. Then we get

$$\begin{aligned}&\mathscr{F}\mathscr{U}(-t)N = 2\widehat{K}(\xi)\mathscr{V}^*M^2\frac{1}{\Lambda''}\left(3\psi_0\psi_1\psi_{-1}+\psi_0^3\right)\\ &\quad+6\widehat{K}(\xi)\mathscr{V}^*\\ &\quad\cdot\frac{1}{\Lambda''}\left(\overline{\psi_0}\psi_1\psi_{-1}+\psi_0\overline{\psi_1}\psi_{-1}+\psi_0\psi_1\overline{\psi_{-1}}+\psi_0^2\overline{\psi_0}\right)\\ &\quad+6\widehat{K}(\xi)\mathscr{V}^*\overline{M}^2\\ &\quad\cdot\frac{1}{\Lambda''}\left(\overline{\psi_0\psi_1}\psi_{-1}+\overline{\psi_0}\psi_1\overline{\psi_{-1}}+\psi_0\overline{\psi_1\psi_{-1}}+\psi_0\overline{\psi_0}^2\right)\\ &\quad+2\widehat{K}(\xi)\mathscr{V}^*\overline{M}^4\frac{1}{\Lambda''}\left(3\overline{\psi_0\psi_1\psi_{-1}}+\overline{\psi_0}^3\right).\end{aligned} \tag{157}$$

Next using identity (36) we find

$$\begin{aligned}&\mathscr{F}\mathscr{U}(-t)N = 2\sqrt{3}\widehat{K}(\xi)e^{it\Omega(\xi)}\mathscr{D}_3\mathscr{V}^*(3t)\\ &\quad\cdot\frac{1}{\Lambda''}\left(3\psi_0\psi_1\psi_{-1}+\psi_0^3\right)+6\widehat{K}(\xi)\mathscr{V}^*\\ &\quad\cdot\frac{1}{\Lambda''}\left(\overline{\psi_0}\psi_1\psi_{-1}+\psi_0\overline{\psi_1}\psi_{-1}+\psi_0\psi_1\overline{\psi_{-1}}+\psi_0^2\overline{\psi_0}\right)\\ &\quad+6\widehat{K}(\xi)\mathscr{D}_{-1}\mathscr{V}^*(-t)\\ &\quad\cdot\frac{1}{\Lambda''}\left(\overline{\psi_0\psi_1}\psi_{-1}+\overline{\psi_0}\psi_1\overline{\psi_{-1}}+\psi_0\overline{\psi_1\psi_{-1}}+\psi_0\overline{\psi_0}^2\right)\\ &\quad+2\sqrt{3}\widehat{K}(\xi)e^{it\Omega(\xi)}\mathscr{D}_{-3}\mathscr{V}^*(-3t)\\ &\quad\cdot\frac{1}{\Lambda''}\left(3\overline{\psi_0\psi_1\psi_{-1}}+\overline{\psi_0}^3\right)\end{aligned} \tag{158}$$

with $\Omega(\xi)=\Lambda(\xi)-3\Lambda(\xi/3)$. Next using the relations $\psi_j = \mathscr{V}(i\xi)^j\widehat{\varphi} = i\eta\psi_{j-1}+\mathscr{A}_0\psi_{j-1}$ and $i\eta\psi_j=\psi_{j+1}-\mathscr{A}_0\psi_j$, we get

$\psi_1\psi_{-1} = \psi_0^2 + R_1$, $\overline{\psi_1}\psi_{-1} = -\overline{\psi_0}\psi_0 + R_2$, and $\overline{\psi_1\psi_{-1}} = \overline{\psi_0}^2 + \overline{R_1}$, where $R_1 = -\psi_0\mathscr{A}_0\psi_{-1} + \psi_{-1}\mathscr{A}_0\psi_0$, and $R_2 = \overline{\psi_0}\mathscr{A}_0\psi_{-1} + \psi_{-1}\overline{\mathscr{A}_0\psi_0}$. Therefore we obtain

$$\begin{aligned}\mathscr{F}\mathscr{U}(-t)N &= 8\sqrt{3}\widehat{K}(\xi)e^{it\Omega(\xi)}\mathscr{D}_3\mathscr{V}^*(3t)\frac{1}{\Lambda''}\psi_0^3 \\ &+ 8\sqrt{3}\widehat{K}(\xi)e^{it\Omega(\xi)}\mathscr{D}_{-3}\mathscr{V}^*(-3t)\frac{1}{\Lambda''}\overline{\psi_0}^3 + R_3,\end{aligned} \tag{159}$$

where

$$\begin{aligned}R_3 &= 6\sqrt{3}\widehat{K}(\xi)e^{it\Omega(\xi)}\mathscr{D}_3\mathscr{V}^*(3t)\frac{1}{\Lambda''}\psi_0R_1 + 6\widehat{K}(\xi) \\ &\cdot\mathscr{V}^*\frac{1}{\Lambda''}\left(\overline{\psi_0}R_1 + \psi_0\left(R_2 + \overline{R_2}\right)\right) + 6\widehat{K}(\xi) \\ &\cdot\mathscr{D}_{-1}\mathscr{V}^*(-t)\frac{1}{\Lambda''}\left(\psi_0\overline{R_1} + \overline{\psi_0}\left(R_2 + \overline{R_2}\right)\right) \\ &+ 6\sqrt{3}\widehat{K}(\xi)e^{it\Omega(\xi)}\mathscr{D}_{-3}\mathscr{V}^*(-3t)\frac{1}{\Lambda''}\overline{\psi_0}\overline{R_1}.\end{aligned} \tag{160}$$

By Lemma 2 we have

$$\begin{aligned}&\left|\psi_j\right| \\ &\le C|\eta|^{j-1/4}\langle\eta\rangle^{1/4-j}\left(\left\|\langle\xi\rangle^{1/2}\widehat{\varphi}\right\|_{\mathbf{L}^\infty} + t^{-1/4}\left\|\xi\widehat{\varphi}_\xi\right\|_{\mathbf{L}^2}\right)\end{aligned} \tag{161}$$

for $j = -1, 0$, and then by Lemma 4 we obtain

$$\begin{aligned}&\left\|\mathscr{U}(t)\mathscr{F}^{-1}R_3\right\|_{\mathbf{L}^2} \le \left\|R_3\right\|_{\mathbf{L}^2} \le C\left\|\frac{1}{\Lambda''}\left|\psi_0\right|^2\mathscr{A}_0\psi_{-1}\right\|_{\mathbf{L}^2} \\ &+ C\left\|\frac{1}{\Lambda''}\left|\psi_0\psi_{-1}\right|\mathscr{A}_0\psi_0\right\|_{\mathbf{L}^2} \\ &\le C\left\|\eta^{4-\gamma}\langle\eta\rangle^{2\gamma-4}\left|\psi_0\right|^2\right\|_{\mathbf{L}^\infty}\left\|\eta^{\gamma-1}\langle\eta\rangle^{-2\gamma}\mathscr{A}_0\psi_{-1}\right\|_{\mathbf{L}^2} \\ &+ C\left\|\eta^{3-\gamma}\langle\eta\rangle^{2\gamma-4}\left|\psi_0\psi_{-1}\right|\right\|_{\mathbf{L}^\infty}\left\|\eta^{\gamma}\langle\eta\rangle^{-2\gamma}\mathscr{A}_0\psi_0\right\|_{\mathbf{L}^2} \\ &\le Ct^{-1}\left(\left\|\langle\xi\rangle^{1/2}\widehat{\varphi}\right\|_{\mathbf{L}^\infty} + t^{-1/4}\left\|\xi\widehat{\varphi}_\xi\right\|_{\mathbf{L}^2}\right)^2 \\ &\cdot\left(\left\|\xi\widehat{\varphi}_\xi\right\|_{\mathbf{L}^2} + \left\|\xi^{1/2}\widehat{\varphi}\right\|_{\mathbf{L}^\infty}\right).\end{aligned} \tag{162}$$

Then we represent

$$\begin{aligned}\mathscr{F}\mathscr{U}(-t)N &= 8\sqrt{3}\widehat{K}(\xi)e^{it\Omega(\xi)}\mathscr{D}_3\mathscr{V}^*(3t)\frac{1}{\Lambda''}\psi_0^3 \\ &+ 8\sqrt{3}\widehat{K}(\xi)e^{it\Omega(\xi)}\mathscr{D}_{-3}\mathscr{V}^*(-3t)\frac{1}{\Lambda''}\overline{\psi_0}^3 + R_3 \\ &= \partial_t\Psi + R_3 + R_4,\end{aligned} \tag{163}$$

where

$$\begin{aligned}\Psi &= 8\sqrt{3}\frac{\widehat{K}(\xi)}{i\Omega(\xi)}e^{it\Omega(\xi)}\left(\mathscr{D}_3\mathscr{V}^*(3t)\frac{1}{\Lambda''}\psi_0^3\right. \\ &\left.+ \mathscr{D}_{-3}\mathscr{V}^*(-3t)\frac{1}{\Lambda''}\overline{\psi_0}^3\right), \\ R_4 &= 8\sqrt{3}\frac{\widehat{K}(\xi)}{i\Omega(\xi)}e^{it\Omega(\xi)}\partial_t\left(\mathscr{D}_3\mathscr{V}^*(3t)\frac{1}{\Lambda''}\psi_0^3\right. \\ &\left.+ \mathscr{D}_{-3}\mathscr{V}^*(-3t)\frac{1}{\Lambda''}\overline{\psi_0}^3\right).\end{aligned} \tag{164}$$

We need to estimate the derivative $\mathscr{V}_t^*$. We have

$$\begin{aligned}&\mathscr{V}_t^*\phi \\ &= \frac{1}{2t}\mathscr{V}^*\phi \\ &+ \frac{|t|^{1/2}}{\sqrt{2\pi}}\int_0^\infty e^{itS(\eta,\xi)}iS(\eta,\xi)\phi(\eta)\sqrt{\Lambda''(\eta)}d\eta.\end{aligned} \tag{165}$$

Since $S(\eta,\xi) = \Lambda(\xi) - \Lambda(\eta) - \Lambda'(\eta)(\xi-\eta) = (1/3)\xi^{-1}\eta^{-2}(3a + 2b\eta^3\xi + b\eta^2\xi^2)(\xi-\eta)^2$ and also $i\xi\mathscr{V}^*(t)\phi - \mathscr{V}^*(t)i\eta\phi = \mathscr{V}^*(t)\mathscr{A}_0(t)\phi$ we find for the second summand

$$\begin{aligned}&\frac{|t|^{1/2}}{\sqrt{2\pi}}\int_0^\infty e^{itS(\eta,\xi)}iS(\eta,\xi)\phi(\eta)\sqrt{\Lambda''(\eta)}d\eta = \frac{i}{3}\frac{|t|^{1/2}}{\sqrt{2\pi}} \\ &\cdot\int_0^\infty e^{itS(\eta,\xi)}\left(b\eta\xi + 3\frac{a}{\eta^2} - 2b\eta^2 + b\xi^2 - 3\frac{a}{\eta\xi}\right) \\ &\cdot(\xi-\eta)\phi(\eta)\sqrt{\Lambda''(\eta)}d\eta = \frac{ib}{3}\xi\mathscr{V}^*(t)\eta\mathscr{A}_0(t)\phi \\ &+ ia\mathscr{V}^*(t)\eta^{-2}\mathscr{A}_0(t)\phi - \frac{2i}{3}b\mathscr{V}^*(t)\eta^2\mathscr{A}_0(t)\phi \\ &+ \frac{i}{3}b\xi^2\mathscr{V}^*(t)\mathscr{A}_0(t)\phi - ia\xi^{-1}\mathscr{V}^*(t)\eta^{-1}\mathscr{A}_0(t)\phi.\end{aligned} \tag{166}$$

Since $\partial_t\psi_0 = \mathscr{V}_t\widehat{\varphi} + \mathscr{V}\widehat{\varphi_t}$, we obtain

$$\begin{aligned}&\left\|\mathscr{U}(t)\mathscr{F}^{-1}R_4\right\|_{\mathbf{L}^2} \le \left\|R_4\right\|_{\mathbf{L}^2} \\ &\le \left\|\xi\langle\xi\rangle^{-4}\mathscr{V}_t^*(3t)\frac{1}{\Lambda''}\psi_0^3\right\|_{\mathbf{L}^2} \\ &+ \left\|\mathscr{V}^*(3t)\frac{1}{\Lambda''}\psi_0^2\partial_t\psi_0\right\|_{\mathbf{L}^2} \\ &\le Ct^{-1}\left\|\frac{1}{\Lambda''}\psi_0^3\right\|_{\mathbf{L}^2} \\ &+ C\left\|\left(\eta^2 + \eta^{-2}\right)\mathscr{A}_0(3t)\frac{1}{\Lambda''}\psi_0^3\right\|_{\mathbf{L}^2} \\ &+ C\left\|\frac{1}{\Lambda''}\left|\psi_0\right|^2\partial_t\psi_0\right\|_{\mathbf{L}^2}.\end{aligned} \tag{167}$$

By Lemmas 4 and 5

$$\begin{aligned}
&\left\|\frac{1}{\Lambda''}\psi_0^3\right\|_{\mathbf{L}^2} \le C\|\widehat{\varphi}\|_{\mathbf{Z}}^3\left\|\frac{1}{\Lambda''}|\eta|^{-3/4}\langle\eta\rangle^{-3/4}\right\|_{\mathbf{L}^2} \\
&\quad\le C\|\widehat{\varphi}\|_{\mathbf{Z}}^3, \\
&\left\|\left(\eta^2+\eta^{-2}\right)\mathscr{A}_0(3t)\frac{1}{\Lambda''}\psi_0^3\right\|_{\mathbf{L}^2} \le C\left\|\eta|\psi_0|^2\mathscr{A}_0\psi_0\right\|_{\mathbf{L}^2} \\
&\quad\le Ct^{-1}\|\widehat{\varphi}\|_{\mathbf{Z}}^2\left\||\eta|^{1/4}\langle\eta\rangle^{-1/2}t\mathscr{A}_0\psi_0\right\|_{\mathbf{L}^2} \\
&\quad\le Ct^{-1}\|\widehat{\varphi}\|_{\mathbf{Z}}^3, \\
&\left\|\frac{1}{\Lambda''}|\psi_0|^2\partial_t\psi_0\right\|_{\mathbf{L}^2} \le Ct^{-1}\left\|\frac{1}{\Lambda''}|\psi_0|^2\eta^{-\gamma}\langle\eta\rangle^{2\gamma+2}\right\|_{\mathbf{L}^\infty} \\
&\quad\cdot\left\|\eta^{\gamma}\langle\eta\rangle^{-2\gamma-2}t\partial_t\psi_0\right\|_{\mathbf{L}^2} \le Ct^{-1}\|\widehat{\varphi}\|_{\mathbf{Z}}^3.
\end{aligned} \tag{168}$$

Therefore

$$\left\|\mathscr{U}(t)\mathscr{F}^{-1}R_4\right\|_{\mathbf{L}^2} \le Ct^{-1}\|\widehat{\varphi}\|_{\mathbf{Z}}^3. \tag{169}$$

Thus we get

$$\begin{aligned}
\mathscr{L}\mathscr{I}u &= 3\lambda\partial_x\left(u^2\mathscr{I}u\right)+\mathscr{K}_3u^3+3\mathscr{K}_2\left(u^2\mathscr{I}u\right) \\
&\quad+\mathscr{U}(t)\mathscr{F}^{-1}\partial_t\Psi+\mathscr{U}(t)\mathscr{F}^{-1}\left(R_3+R_4\right) \\
&= 3\lambda\partial_x\left(u^2\mathscr{I}u\right)+\mathscr{K}_3u^3+3\mathscr{K}_2\left(u^2\mathscr{I}u\right) \\
&\quad+\mathscr{L}\mathscr{U}(t)\mathscr{F}^{-1}\Psi+\mathscr{U}(t)\mathscr{F}^{-1}\left(R_3+R_4\right).
\end{aligned} \tag{170}$$

Hence

$$\begin{aligned}
&\mathscr{L}\left(\mathscr{I}u-\mathscr{U}(t)\mathscr{F}^{-1}\Psi\right) \\
&\quad= 3\lambda\partial_x\left(u^2\left(\mathscr{I}u-\mathscr{U}(t)\mathscr{F}^{-1}\Psi\right)\right) \\
&\quad+3\lambda\partial_x\left(u^2\left(\mathscr{U}(t)\mathscr{F}^{-1}\Psi\right)\right)+\mathscr{K}_3u^3 \\
&\quad+3\mathscr{K}_2\left(u^2\mathscr{I}u\right)+\mathscr{U}(t)\mathscr{F}^{-1}\left(R_3+R_4\right).
\end{aligned} \tag{171}$$

Then as the above using the projectors $\mathscr{Q}_1$ and $\mathscr{Q}_2$ we find

$$\left\|\mathscr{I}u-\mathscr{U}(t)\mathscr{F}^{-1}\Psi\right\|_{\mathbf{L}^2} \le C\varepsilon+C\varepsilon^3t^{\gamma-1}. \tag{172}$$

We have

$$\begin{aligned}
\left\|\mathscr{U}(t)\mathscr{F}^{-1}\Psi\right\|_{\mathbf{L}^2} &\le \|\Psi\|_{\mathbf{L}^2} \le C\left\|\eta^3\langle\eta\rangle^{-7}\psi_0^3\right\|_{\mathbf{L}^2} \\
&\le C\varepsilon^3t^{\gamma-1},
\end{aligned} \tag{173}$$

and then we get

$$\|\mathscr{I}u\|_{\mathbf{L}^2} \le C\varepsilon+C\varepsilon^3t^{\gamma-1}. \tag{174}$$

Therefore

$$\left\|\xi\partial_\xi\widehat{\varphi}\right\|_{\mathbf{L}^2} \le 5\varepsilon+C\varepsilon^3t^{\gamma}. \tag{175}$$

Next we estimate $\|\langle\xi\rangle^{1/2}\widehat{\varphi}\|_{\mathbf{L}^\infty}$. In the domain $|\xi| \ge \langle t\rangle^{\nu}$ we get by the Sobolev imbedding theorem

$$\begin{aligned}
&\left\|\langle\xi\rangle^{1/2}\widehat{\varphi}(t,\xi)\right\|_{\mathbf{L}^\infty(|\xi|\ge\langle t\rangle^{\nu})} \\
&\quad\le C\langle t\rangle^{-\nu}\left\|\langle\xi\rangle^{3/2}\widehat{\varphi}(t,\xi)\right\|_{\mathbf{L}^\infty(|\xi|\ge\langle t\rangle^{\nu})} \\
&\quad\le C\langle t\rangle^{-\nu}\left(\left\|\xi\partial_\xi\widehat{\varphi}\right\|_{\mathbf{L}^2}+\left\|\langle\xi\rangle^2\widehat{\varphi}\right\|_{\mathbf{L}^2}\right) \le C\varepsilon\langle t\rangle^{-\nu+\gamma},
\end{aligned} \tag{176}$$

if $\nu > \gamma$, so we need to estimate the function $\langle\xi\rangle^{(1/2)}\widehat{\varphi}(t,\xi)$ in the domain $|\xi| \le \langle t\rangle^{\nu}$. Next by (37) for $\widehat{\varphi} = \mathscr{F}\mathscr{U}(-t)u(t)$, using Lemma 6, we get

$$\begin{aligned}
\partial_t\widehat{\varphi}(t,\xi) &= \mathscr{F}\mathscr{U}(-t)\partial_x\mathscr{K}u^3 \\
&= \frac{\sqrt{3}i\xi\widehat{K}}{it}e^{it\Omega(\xi)}\mathscr{D}_3\frac{1}{\Lambda''}\widehat{\varphi}^3+\frac{3i\xi\widehat{K}}{t\Lambda''}|\widehat{\varphi}|^2\widehat{\varphi} \\
&\quad+O\left(t^{-5/4}\|\widehat{\varphi}\|_{\mathbf{Z}}^3\right)
\end{aligned} \tag{177}$$

for all $t \ge 1$, $|\xi| \le \langle t\rangle^{\nu}$. Multiplying this formula by $\langle\xi\rangle^{1/2}$ we get

$$\begin{aligned}
\partial_t\langle\xi\rangle^{1/2}\widehat{\varphi}(t,\xi) &= \frac{\sqrt{3}i\xi\widehat{K}}{it}\langle\xi\rangle^{1/2}e^{it\Omega(\xi)}\mathscr{D}_3\frac{1}{\Lambda''}\widehat{\varphi}^3 \\
&\quad+\frac{3i\xi\widehat{K}}{t\Lambda''}\langle\xi\rangle^{1/2}|\widehat{\varphi}|^2\widehat{\varphi} \\
&\quad+O\left(t^{\nu-5/4}\|\widehat{\varphi}\|_{\mathbf{Z}}^3\right)
\end{aligned} \tag{178}$$

in the domain $|\xi| \le \langle t\rangle^{\nu}$. Define the cut-off function $\chi \in \mathbf{C}^1(\mathbf{R})$, such that $\chi(x) = 1$ for $|x| < 1$ and $\chi(x) = 0$ for $|x| > 2$, and define $\widehat{\varphi}_1(t,\xi) = \chi(\xi\langle t\rangle^{-\nu})\langle\xi\rangle^{1/2}\widehat{\varphi}(t,\xi)$. Thus we get

$$\begin{aligned}
\partial_t\widehat{\varphi}_1 &= \frac{\sqrt{3}i\xi\widehat{K}}{it}\chi\left(\xi\langle t\rangle^{-\nu}\right)\langle\xi\rangle^{1/2}e^{it\Omega(\xi)}\mathscr{D}_3\frac{1}{\Lambda''}\widehat{\varphi}^3 \\
&\quad+\frac{3i\xi\widehat{K}}{t\Lambda''}|\widehat{\varphi}|^2\widehat{\varphi}_1 \\
&\quad+\langle\xi\rangle^{1/2}\xi\langle t\rangle^{-1-\nu}\chi'\left(\xi\langle t\rangle^{-\nu}\right)\widehat{\varphi}(t,\xi) \\
&\quad+O\left(t^{\nu-5/4}\|\widehat{\varphi}\|_{\mathbf{Z}}^3\right)
\end{aligned} \tag{179}$$

for all $t \ge 1$. The third term is estimated by $C\varepsilon\langle t\rangle^{-1-\nu+\gamma}$. To exclude the resonant term we make a change $\widehat{\varphi}_1(t,\xi) = y(t,\xi)\Theta(t,\xi)$, where

$$\Theta(t,\xi) = \exp\left(\frac{3i\xi\widehat{K}}{\Lambda''}\int_1^t|\widehat{\varphi}(\tau,\xi)|^2\frac{d\tau}{\tau}\right). \tag{180}$$

Then we get

$$\begin{aligned}
&y_t(t,\xi) \\
&\quad= \frac{\sqrt{3}i\xi\widehat{K}}{it}\chi\left(\xi\langle t\rangle^{-\nu}\right)\langle\xi\rangle^{1/2}e^{it\Omega(\xi)}\overline{\Theta(t,\xi)}\mathscr{D}_3\frac{1}{\Lambda''}\widehat{\varphi}^3 \\
&\quad+O\left(\varepsilon^3t^{-1-\delta}\right)
\end{aligned} \tag{181}$$

with $\delta > 0$. Integrating by parts we obtain

$$y(t,\xi) - y(1,\xi) = \int_1^t \sqrt{3} i\xi\widehat{K}\chi(\xi\langle\tau\rangle^{-\nu})\langle\xi\rangle^{1/2} \cdot e^{i\tau\Omega(\xi)}\overline{\Theta(\tau,\xi)\mathscr{D}_3\frac{1}{\Lambda''}\widehat{\varphi}^3}\frac{d\tau}{i\tau} + O(\varepsilon^3) = \sqrt{3} \cdot \frac{i\xi\widehat{K}}{i\Omega(\xi)} e^{i\tau\Omega(\xi)}\chi(\xi\langle\tau\rangle^{-\nu})\langle\xi\rangle^{1/2}\overline{\Theta(\tau,\xi)\mathscr{D}_3\frac{1}{\tau\Lambda''} \cdot \widehat{\varphi}^3}\Bigg|_{\tau=1}^{\tau=t} + \int_1^t \sqrt{3}\frac{i\xi\widehat{K}}{i\Omega(\xi)} e^{i\tau\Omega(\xi)}\partial_\tau\Big(\chi(\xi\langle\tau\rangle^{-\nu}) \cdot \langle\xi\rangle^{1/2}\overline{\Theta(\tau,\xi)\mathscr{D}_3\frac{1}{\tau\Lambda''}\widehat{\varphi}^3}\Big)d\tau + O(\varepsilon^3) = O(\varepsilon^3). \tag{182}$$

Thus we get the estimate $|\widehat{\varphi}_1(\xi)| = |y(t,\xi)| \le |\widehat{\varphi}_1(1,\xi)| + O(\varepsilon^3)$ in the domain $|\xi| \le \langle t\rangle^{\nu}$. Therefore we find the desired estimate $\|\langle\xi\rangle^{1/2}\widehat{\varphi}\|_{\mathbf{L}^\infty} \le C\varepsilon$. This is the desired contradiction. Lemma 8 is proved. □

5. Proof of Theorem 1

The global existence of solution $u \in \mathbf{C}([0,T];\mathbf{H}^2)$ to Cauchy problem (1) satisfying a priori estimate

$$\left\|\langle\xi\rangle^{1/2}\widehat{\varphi}\right\|_{\mathbf{L}^\infty} + \langle t\rangle^{-\gamma}\|u(t)\|_{\mathbf{H}^2} + \langle t\rangle^{-\gamma}\left\|\partial_x\mathcal{J}u(t)\right\|_{\mathbf{L}^2} \le C\varepsilon \tag{183}$$

follows by a standard continuation argument from Lemma 8 and local existence Theorem 7. We need only to prove asymptotic formula (20).

We need to compute the asymptotics of the function $\widehat{\varphi}(t,\xi)$. As in the proof of Lemma 8 we get

$$y(t,\xi) - y(s,\xi) = \int_s^t \sqrt{3} i\xi\widehat{K}\chi(\xi\langle\tau\rangle^{-\nu})\langle\xi\rangle^{1/2} \cdot e^{i\tau\Omega(\xi)}\overline{\Theta(\tau,\xi)\mathscr{D}_3\frac{1}{\Lambda''}\widehat{\varphi}^3}\frac{d\tau}{i\tau} + O(\varepsilon^3 s^{-\delta}) = \sqrt{3} \cdot \frac{i\xi\widehat{K}}{i\Omega(\xi)} e^{i\tau\Omega(\xi)}\chi(\xi\langle\tau\rangle^{-\nu})\langle\xi\rangle^{1/2}\overline{\Theta(\tau,\xi)\mathscr{D}_3\frac{1}{\tau\Lambda''} \cdot \widehat{\varphi}^3}\Bigg|_{\tau=s}^{\tau=t} + \int_s^t \sqrt{3}\frac{i\xi\widehat{K}}{i\Omega(\xi)} e^{i\tau\Omega(\xi)}\partial_\tau\Big(\chi(\xi\langle\tau\rangle^{-\nu}) \cdot \langle\xi\rangle^{1/2}\overline{\Theta(\tau,\xi)\mathscr{D}_3\frac{1}{\tau\Lambda''}\widehat{\varphi}^3}\Big)d\tau + O(\varepsilon^3 s^{-\delta}) = O(\varepsilon^3 s^{-\delta}) \tag{184}$$

for any $t > s > 0$. Therefore there exists a unique final state $y_+ \in \mathbf{L}^\infty$ such that

$$\|y(t) - y_+\|_{\mathbf{L}^\infty} \le C\varepsilon^3 t^{-\delta} \tag{185}$$

for all $t > 0$. We write

$$\frac{3\widehat{K}}{\Lambda''}\int_1^t |\widehat{\varphi}(\tau,\xi)|^2\frac{d\tau}{\tau} = \frac{3i\xi\widehat{K}}{\langle\xi\rangle\Lambda''}\int_1^t |y(\tau,\xi)|^2\frac{d\tau}{\tau} = \frac{3i\xi\widehat{K}}{\langle\xi\rangle\Lambda''}|y_+|^2\log t + \Phi_2(t). \tag{186}$$

We study the asymptotics in time of the remainder term $\Phi_2(t)$. We have

$$\Phi_2(t) - \Phi_2(s) = \frac{3i\xi\widehat{K}}{\langle\xi\rangle\Lambda''}\int_s^t \left(|y(\tau)|^2 - |y(t)|^2\right)\frac{d\tau}{\tau} + \frac{3i\xi\widehat{K}}{\langle\xi\rangle\Lambda''}\left(|y(t)|^2 - |y_+|^2\right)\log\frac{t}{s}. \tag{187}$$

By (185) we obtain $\|\Phi_2(t)-\Phi_2(s)\|_{\mathbf{L}^\infty} \le C\varepsilon^3 s^{-\delta}$ for any $t > s > 0$. Hence there exists a unique real-valued function $\Phi_+ \in \mathbf{L}^\infty$ such that

$$\|\Phi_2(t) - \Phi_+\|_{\mathbf{L}^\infty} \le C\varepsilon^3 t^{-\delta} \tag{188}$$

for all $t > 0$. Representation (186) and estimate (188) yield

$$\left\|\Theta(\tau,\xi) - \exp\left(\frac{3i\xi\widehat{K}}{\langle\xi\rangle\Lambda''}|y_+|^2\log t + \Phi_+\right)\right\|_{\mathbf{L}^\infty} \le Ct^{-\delta} \tag{189}$$

for all $t > 0$. Thus we get the large time asymptotics

$$\left\|\chi(\xi\langle t\rangle^{-\nu})\langle\xi\rangle^{1/2}\widehat{\varphi}(t,\xi) - y_+\Theta(\tau,\xi)\right\|_{\mathbf{L}^\infty} = \|y(t) - y_+\|_{\mathbf{L}^\infty} \le Ct^{-\delta}, \tag{190}$$

$$\left\|y_+\Theta(\tau,\xi) - y_+\exp\left(\frac{3i\xi\widehat{K}}{\langle\xi\rangle\Lambda''}|y_+|^2\log t + \Phi_+\right)\right\|_{\mathbf{L}^\infty} \le Ct^{-\delta}. \tag{191}$$

Therefore we obtain the estimate

$$\left\|\chi(\xi\langle t\rangle^{-\nu})\widehat{\varphi}(t,\xi) - W_+\exp\left(\frac{3i\xi\widehat{K}}{\langle\xi\rangle\Lambda''}|W_+|^2\log t\right)\right\|_{\mathbf{L}^\infty} \le Ct^{-\delta} \tag{192}$$

with $W_+ = y_+ e^{\Phi_+}$. Using the factorization of $\mathscr{U}(t)$ we have

$$\begin{aligned} &\Bigg\| u(t) \\ &\quad - 2\mathrm{Re}\mathscr{D}_t\mathscr{B}^{-1}MW_+\exp\left(\frac{3i\xi\widehat{K}}{\langle\xi\rangle\Lambda''}|W_+|^2\log t\right)\Bigg\|_{\mathbf{L}^\infty} \\ &\le C\Bigg\|\mathscr{D}_t\mathscr{B}^{-1}M\Bigg(\chi(\xi\langle t\rangle^{-\nu})\widehat{\varphi}(t,\xi) \\ &\quad - W_+\exp\left(\frac{3i\xi\widehat{K}}{\langle\xi\rangle\Lambda''}|W_+|^2\log t\right)\Bigg)\Bigg\|_{\mathbf{L}^\infty} + Ct^{-1/2-\delta} \\ &\le Ct^{-1/2-\delta}. \end{aligned} \tag{193}$$

This completes the proof of asymptotics (20). Theorem 1 is proved.

Competing Interests

The authors declare that they have no competing interests.

Acknowledgments

The work of Nakao Hayashi is partially supported by JSPS KAKENHI Grant nos. JP25220702 and JP15H03630. The work of Pavel I. Naumkin is partially supported by CONACYT and PAPIIT Project IN100616.

References

[1] N. Hayashi and P. I. Naumkin, "Large time asymptotics for the reduced Ostrovsky equation," *Communications in Mathematical Physics*, vol. 335, no. 2, pp. 713–738, 2015.

[2] L. A. Ostrovsky and Y. A. Stepanyants, "Nonlinear surface and internal waves in rotating fluids," in *Nonlinear Waves 3*, Research Reports in Physics, pp. 106–128, Springer, Berlin, Germany, 1990.

[3] F. Linares and A. Milanes, "Local and global well-posedness for the Ostrovsky equation," *Journal of Differential Equations*, vol. 222, no. 2, pp. 325–340, 2006.

[4] K. Tsugawa, "Well-posedness and weak rotation limit for the Ostrovsky equation," *Journal of Differential Equations*, vol. 247, no. 12, pp. 3163–3180, 2009.

[5] V. Varlamov and Y. Liu, "Cauchy problem for the Ostrovsky equation," *Discrete and Continuous Dynamical Systems Series A*, vol. 10, no. 3, pp. 731–753, 2004.

[6] G. Gui and Y. Liu, "On the Cauchy problem for the Ostrovsky equation with positive dispersion," *Communications in Partial Differential Equations*, vol. 32, no. 10–12, pp. 1895–1916, 2007.

[7] Z. Huo and Y. Jia, "Low-regularity solutions for the Ostrovsky equation," *Proceedings of the Edinburgh Mathematical Society*, vol. 49, no. 1, pp. 87–100, 2006.

[8] P. Isaza and J. Mejía, "Cauchy problem for the Ostrovsky equation in spaces of low regularity," *Journal of Differential Equations*, vol. 230, no. 2, pp. 661–681, 2006.

[9] J. Bourgain, "Fourier transform restriction phenomena for certain lattice subsets and applications to nonlinear evolution equations. II. The KdV-equation," *Geometric and Functional Analysis*, vol. 3, no. 3, pp. 209–262, 1993.

[10] J. Colliander, M. Keel, G. Staffilani, H. Takaoka, and T. Tao, "Sharp global well-posedness for KdV and modified KdV on $\mathbb{R}$ and $\mathbb{T}$," *Journal of the American Mathematical Society*, vol. 16, no. 3, pp. 705–749, 2003.

[11] T. Schäfer and C. E. Wayne, "Propagation of ultra-short optical pulses in cubic nonlinear media," *Physica D. Nonlinear Phenomena*, vol. 196, no. 1-2, pp. 90–105, 2004.

[12] A. Stefanov, Y. Shen, and P. G. Kevrekidis, "Well-posedness and small data scattering for the generalized Ostrovsky equation," *Journal of Differential Equations*, vol. 249, no. 10, pp. 2600–2617, 2010.

[13] Y. Liu, D. Pelinovsky, and A. Sakovich, "Wave breaking in the short-pulse equation," *Dynamics of Partial Differential Equations*, vol. 6, no. 4, pp. 291–310, 2009.

[14] Y. Liu, D. Pelinovsky, and A. Sakovich, "Wave breaking in the Ostrovsky-Hunter equation," *SIAM Journal on Mathematical Analysis*, vol. 42, no. 5, pp. 1967–1985, 2010.

[15] D. Pelinovsky and A. Sakovich, "Global well-posedness of the short-pulse and sine-Gordon equations in energy space," *Communications in Partial Differential Equations*, vol. 35, no. 4, pp. 613–629, 2010.

[16] J. P. Boyd, "Ostrovsky and Hunter's generic wave equation for weakly dispersive waves: matched asymptotic and pseudospectral study of the paraboloidal travelling waves (corner and near-corner waves)," *European Journal of Applied Mathematics*, vol. 16, no. 1, pp. 65–81, 2005.

[17] J. Hunter, "Numerical solutions of some nonlinear dispersive wave equations," *Lectures in Applied Mathematics*, vol. 26, pp. 301–316, 1990.

[18] R. Grimshaw and D. Pelinovsky, "Global existence of small-norm solutions in the reduced Ostrovsky equation," *Discrete and Continuous Dynamical Systems Series A*, vol. 34, no. 2, pp. 557–566, 2014.

[19] N. Hayashi, P. I. Naumkin, and T. Niizato, "Asymptotics of solutions to the generalized Ostrovsky equation," *Journal of Differential Equations*, vol. 255, no. 8, pp. 2505–2520, 2013.

[20] N. Hayashi, P. I. Naumkin, and T. Niizato, "Nonexistence of the usual scattering states for the generalized Ostrovsky-Hunter equation," *Journal of Mathematical Physics*, vol. 55, no. 5, Article ID 053502, 2014.

[21] M. V. Fedoryuk, *Asymptotics: Integrals and Series, Mathematical Reference Library*, Nauka, Moscow, Russia, 1987.

[22] M. V. Fedoryuk, "Asymptotic methods in analysis," in *Analysis I*, vol. 13 of *Encyclopaedia of Mathematical Sciences*, pp. 83–191, Springer, New York, NY, USA, 1989.

[23] E. M. Stein and R. Shakarchi, *Functional Analysis. Introduction to Further Topics in Analysis*, vol. 4 of *Princeton Lectures in Analysis*, Princeton University Press, Princeton, NJ, USA, 2011.

16

On FTCS Approach for Box Model of Three-Dimension Advection-Diffusion Equation

Jeffry Kusuma, Agustinus Ribal, and Andi Galsan Mahie

Department of Mathematics, Hasanuddin University, Makassar, Indonesia

Correspondence should be addressed to Jeffry Kusuma; jeffry.kusuma@gmail.com

Guest Editor: Manoj Kumar

This paper describes a numerical solution for mathematical model of the transport equation in a simple rectangular box domain. The model of street tunnel pollution distribution using two-dimension advection and three-dimension diffusion is solved numerically. Because of the nature of the problem, the model is extended to become three-dimension advection and three-dimension diffusion to study the sea-sand mining pollution distribution. This model with various advection and diffusion parameters and the boundaries conditions is also solved numerically using a finite difference (FTCS) method.

1. Introduction

The solution of partial differential equation and their associated boundary and initial condition play an important role in modelling of phenomena in fields as diverse as physics, chemistry, geology, biology, engineering, and economics. The transport of pollutants occurs in a large variety of environmental, agricultural, and industrial processes. The phenomenon is usually modelled into partial differential equations with boundary and/or initial conditions. The models, however, in most cases have no analytical solution. Numerical solution becomes an alternative solution to models such as partial differential equation models in order to investigate, predict, and conclude the models.

Numerical solution for an advection-diffusion equation or transport equation had become an interesting subject for many authors recently. Several improvements in finite difference approach had been noted. A stability limit for a finite difference scheme such as the forward time and space-centered numerical scheme applied the convection-diffusion equation is discussed in [1]. Three-dimension solution of advection-diffusion base on the two-level fully explicit and fully implicit finite difference approximation is discussed in [2]. The comparison of the two standard finite difference schemes such as FTCS and Crank-Nicolson methods is carried out by [3]. The method of time splitting to divide complicated time dependent partial differential equation into sets of simpler equations which could then be solved separately by numerical means over fraction time step had been done as in [4]. Numerical solution of a three-dimension advection-diffusion for models in a street tunnel is discussed in [5]. More recently, an application of the generalized finite difference method to solve the advection-diffusion equation using the explicit method is discussed in [6]. Also, practical use of finite difference methods in order to study the pollution distribution on Unhas lake had been carried out as in [7].

The studying of a box model had been carried out numerically as in [5] using two-dimension advection and three-dimension diffusion. In this paper we extend the box model into three-dimension advection and three-dimension diffusion to determine the pollutants spread in the water or in the air. This is because of the nature of the pollutant particles. For certain dimension of particle, gravitational force due to the particle mass should be taken into account. We solve the three-dimension advection-diffusion equation by using the forward in time, center in space (FTCS) finite difference method. The domain of this study model is a rectangular box with length L, width W, and height H. Numerical results for several different pollutant source configurations are presented and discussed.

The paper is organized as follows. In Section 2, we introduce the basic equation and the problems. In Section 3,

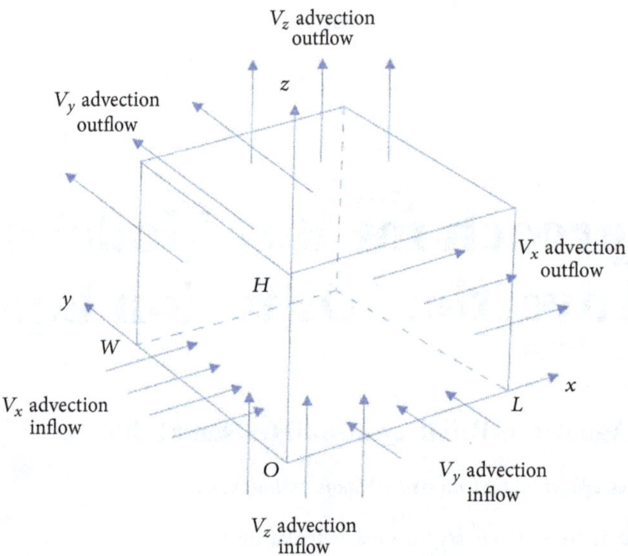

FIGURE 1: Domain of advection-diffusion.

the finite difference schemes for the computation approach are introduced. The stability condition for FTCS scheme is discussed in Section 4. In Section 5, some numerical models, results, and discussions are presented. Finally, conclusions are found in Section 6.

2. Basic Equation and Problems

In this paper we consider the three-dimension advection-diffusion equation [2, 5, 6]

$$\frac{\partial C}{\partial t} + V_x \frac{\partial C}{\partial x} + V_y \frac{\partial C}{\partial y} + V_z \frac{\partial C}{\partial z} = D_x \frac{\partial^2 C}{\partial x^2} + D_y \frac{\partial^2 C}{\partial y^2} + D_z \frac{\partial^2 C}{\partial z^2}, \quad 0 < t \le T, \tag{1}$$

in the domains $0 \le x \le L$, $0 \le y \le W$, and $0 \le z \le H$, with initial condition

$$C(x, y, z, 0) = f(x, y, z), \tag{2}$$

and the boundary conditions

$$C(0, y, z, t) = g_0(y, z, t), \quad 0 < t \le T, \tag{3}$$

$$C(L, y, z, t) = g_L(y, z, t), \quad 0 < t \le T, \tag{4}$$

$$C(x, 0, z, t) = h_0(x, z, t), \quad 0 < t \le T, \tag{5}$$

$$C(x, W, z, t) = h_W(x, z, t), \quad 0 < t \le T, \tag{6}$$

$$C(x, y, 0, t) = k_0(x, y, t), \quad 0 < t \le T, \tag{7}$$

$$C(x, y, H, t) = k_H(x, y, t), \quad 0 < t \le T, \tag{8}$$

where $f, g_0, g_L, h_0, h_W, k_0$, and k_H are known functions, while the function C is unknown. Here $C(x, y, z, t)$ denote materials concentration which is transported. The constants V_x, V_y, V_z represent speeds of advection with respect to $x-$axis, $y-$axis, and $z-$axis, respectively. Also, the constants D_x, D_y, D_z, represent the speeds of diffusivities with respect to $x-$axis, $y-$axis, and $z-$axis, respectively (see illustration in Figure 1).

3. Finite Difference Schemes

The main idea behind the finite difference schemes for obtaining the solution of a given partial differential equation is to approximate the derivatives appearing in the equation by a set of values of the function at selected number of points. The most usual way to generate these approximations is through the use of Taylor series.

The solution domain of the problem over a time $0 \le t \le T$ is covered by a mesh of uniformly spaced grid-lines parallel to the space and time coordinates axes, respectively.

$$x_i = i\Delta x, \quad i = 0, 1, 2, \ldots, M; \tag{9}$$

$$y_j = j\Delta y, \quad j = 0, 1, 2, \ldots, N; \tag{10}$$

$$z_k = k\Delta z, \quad k = 0, 1, 2, \ldots, O; \tag{11}$$

$$t_n = n\Delta t, \quad n = 0, 1, 2, \ldots, R; \tag{12}$$

Approximations $C^n_{i,j,k}$ to $C(i\Delta x, j\Delta y, k\Delta z, n\Delta t)$ are calculated at the point of intersection of these lines according to the (i, j, k, n) grid points. The uniform spatial and temporal grid spacings are $\Delta x = L/M$, $\Delta y = W/N$, $\Delta z = H/O$, and $\Delta t = T/R$, where L is the length, W is the width, and H is the height of the domain of the interested rectangular box.

Since the grid points are in three dimensions of the box form, several numerical approximations are needed for the grid points. These depend on the position of the grid points. The approximation for interior points is forward time center space (FTCS). The points obtained from intersection of two planes of the box are approximated using forward time center space forward or backward space while the edge points with intersection of three planes are approximated using forward time and combination of forward and backward approximation.

Forward time centered space (FTCS) approximation of (1) for the interior points as in [2, 5, 7] is

$$\begin{aligned}\frac{C^{n+1}_{i,j,k}-C^n_{i,j,k}}{\Delta t} &+ V_x\left(\frac{C^n_{i+1,j,k}-C^n_{i-1,j,k}}{2\Delta x}\right) + V_y\left(\frac{C^n_{i,j+1,k}-C^n_{i,j-1,k}}{2\Delta y}\right) + V_z\left(\frac{C^n_{i,j,k+1}-C^n_{i,j,k-1}}{2\Delta z}\right)\\ &= D_x\left(\frac{C^n_{i+1,j,k}-2C^n_{i,j,k}+C^n_{i-1,j,k}}{\Delta x^2}\right) + D_y\left(\frac{C^n_{i,j+1,k}-2C^n_{i,j,k}+C^n_{i,j-1,k}}{\Delta y^2}\right) + D_z\left(\frac{C^n_{i,j,k+1}-2C^n_{i,j,k}+C^n_{i,j,k-1}}{\Delta z^2}\right).\end{aligned} \tag{13}$$

Another finite difference approximation such as forward time backward space x center space y center space z is used for the boundary points on $C(L,y,z,t)=g_L(y,z,t)$

$$\begin{aligned}\frac{C^{n+1}_{i,j,k}-C^n_{i,j,k}}{\Delta t} &+ V_x\left(\frac{3C^n_{i,j,k}-4C^n_{i-1,j,k}+3C^n_{i-2,j,k}}{2\Delta x}\right) + V_y\left(\frac{C^n_{i,j+1,k}-C^n_{i,j-1,k}}{2\Delta y}\right) + V_z\left(\frac{C^n_{i,j,k+1}-C^n_{i,j,k-1}}{2\Delta z}\right)\\ &= D_x\left(\frac{C^n_{i,j,k}-2C^n_{i-1,j,k}+C^n_{i-2,j,k}}{\Delta x^2}\right) + D_y\left(\frac{C^n_{i,j+1,k}-2C^n_{i,j,k}+C^n_{i,j-1,k}}{\Delta y^2}\right) + D_z\left(\frac{C^n_{i,j,k+1}-2C^n_{i,j,k}+C^n_{i,j,k-1}}{\Delta z^2}\right).\end{aligned} \tag{14}$$

Forward time backward space x backward space y center space z is used for the boundary points on $C(L,W,z,t)$ which is

$$\begin{aligned}\frac{C^{n+1}_{i,j,k}-C^n_{i,j,k}}{\Delta t} &+ V_x\left(\frac{3C^n_{i,j,k}-4C^n_{i-1,j,k}+3C^n_{i-2,j,k}}{2\Delta x}\right) + V_y\left(\frac{3C^n_{i,j,k}-4C^n_{i,j-1,k}+3C^n_{i,j-2,k}}{2\Delta y}\right) + V_z\left(\frac{C^n_{i,j,k+1}-C^n_{i,j,k-1}}{2\Delta z}\right)\\ &= D_x\left(\frac{C^n_{i,j,k}-2C^n_{i-1,j,k}+C^n_{i-2,j,k}}{\Delta x^2}\right) + D_y\left(\frac{C^n_{i,j,k}-2C^n_{i,j-1,k}+C^n_{i,j-2,k}}{\Delta y^2}\right) + D_z\left(\frac{C^n_{i,j,k+1}-2C^n_{i,j,k}+C^n_{i,j,k-1}}{\Delta z^2}\right).\end{aligned} \tag{15}$$

Also forward time backward space x forward space y center space z is used for the boundary points on $C(L,0,z,t)$ which is

$$\begin{aligned}\frac{C^{n+1}_{i,j,k}-C^n_{i,j,k}}{\Delta t} &+ V_x\left(\frac{3C^n_{i,j,k}-4C^n_{i-1,j,k}+3C^n_{i-2,j,k}}{2\Delta x}\right) + V_y\left(\frac{C^n_{i,j,k}-4C^n_{i,j+1,k}+3C^n_{i,j+2,k}}{2\Delta y}\right) + V_z\left(\frac{C^n_{i,j,k+1}-C^n_{i,j,k-1}}{2\Delta z}\right)\\ &= D_x\left(\frac{C^n_{i,j,k}-2C^n_{i-1,j,k}+C^n_{i-2,j,k}}{\Delta x^2}\right) + D_y\left(\frac{C^n_{i,j,k}-2C^n_{i,j+1,k}+C^n_{i,j+2,k}}{\Delta y^2}\right) + D_z\left(\frac{C^n_{i,j,k+1}-2C^n_{i,j,k}+C^n_{i,j,k-1}}{\Delta z^2}\right).\end{aligned} \tag{16}$$

All approximations have error of first order in time interval and second order is spatial coordinate grid spacing.

4. Stability

To ensure obtained solutions have a nonpropagate error, the approximation or schemes should meet certain conditions. The approximation schemes on the boundary usually have a nonpropagate error since boundary exact conditions are supplied on the boundary. However, for the interior points the error might propagate. For the FTCS approximations, (13) might be rewritten as

$$\begin{aligned}C^{n+1}_{i,j,k} &= \left(\frac{D_x\Delta t}{\Delta x^2}+\frac{V_x\Delta t}{2\Delta x}\right)C^n_{i-1,j,k}\\ &+ \left(\frac{D_y\Delta t}{\Delta y^2}+\frac{V_y\Delta t}{2\Delta y}\right)C^n_{i,j-1,k}\\ &+ \left(\frac{D_z\Delta t}{\Delta z^2}+\frac{V_z\Delta t}{2\Delta z}\right)C^n_{i,j,k-1}\end{aligned}$$

$$+\left(\frac{D_x\Delta t}{\Delta x^2}-\frac{V_x\Delta t}{2\Delta x}\right)C^n_{i+1,j,k}$$
$$+\left(\frac{D_y\Delta t}{\Delta y^2}-\frac{V_y\Delta t}{2\Delta y}\right)C^n_{i,j+1,k}$$
$$+\left(\frac{D_z\Delta t}{\Delta z^2}-\frac{V_z\Delta t}{2\Delta z}\right)C^n_{i,j,k-1}$$
$$+\left(1-2\left[\frac{D_x\Delta t}{\Delta x^2}+\frac{D_y\Delta t}{\Delta y^2}+\frac{D_z\Delta t}{\Delta z^2}\right]\right)C^n_{i,j,k} \quad (17)$$

or simply

$$\begin{aligned}C^{n+1}_{i,j,k} &= \left(s_x+\frac{c_x}{2}\right)C^n_{i-1,j,k}+\left(s_y+\frac{c_y}{2}\right)C^n_{i,j-1,k}\\ &+\left(s_z+\frac{c_z}{2}\right)C^n_{i,j,k-1}+\left(s_x-\frac{c_x}{2}\right)C^n_{i+1,j,k}\\ &+\left(s_y-\frac{c_y}{2}\right)C^n_{i,j+1,k}+\left(s_z-\frac{c_z}{2}\right)C^n_{i,j,k+1}\\ &+\left(1-2\left[s_x+s_y+s_z\right]\right)C^n_{i,j,k}\end{aligned} \quad (18)$$

in which

$$s_x=\frac{D_x\Delta t}{\Delta x^2},\quad s_y=\frac{D_y\Delta t}{\Delta y^2},\quad s_z=\frac{D_z\Delta t}{\Delta z^2}, \quad (19)$$

$$c_x=\frac{V_x\Delta t}{\Delta x},\quad c_y=\frac{V_y\Delta t}{\Delta y},\quad c_z=\frac{V_z\Delta t}{\Delta z}. \quad (20)$$

The stability of this three-dimensional difference scheme may be investigated using the von Neumann method. As in [2, 5] application of this method of stability analysis shows that (18) will be stable if it satisfies both equations

$$s_x+s_y+s_z\le\frac{1}{2} \quad (21)$$

and

$$\frac{c_x^2}{s_x}+\frac{c_y^2}{s_y}+\frac{c_z^2}{s_z}\le 3. \quad (22)$$

For one-dimension version of the forward time centered space (FTCS) type formula for the case that $s_x = s_y = s_z = s$ and $c_x = c_x = c_x = c$, it should also satisfy

$$c^2\le s\le\frac{1}{6}. \quad (23)$$

This clearly is much more strict than one-dimension advection-diffusion equation of the forward time centered space (FTCS) approximation which is $s \le 1/2$.

5. Numerical Results and Discussions

Problem 1. In order to verify the accuracy of the procedure that has been built, we consider the three-dimension advection-diffusion equation (1), with initial condition (2) and boundary conditions (3)-(8). By taking $V_y = V_z = 0$, $L = 1, W = 1, H = 1$, this procedure simply reduced to three-dimension advection-diffusion equation for transport of pollutants in street tunnel as discussed in [5]

$$\frac{\partial C}{\partial t}+V_x\frac{\partial C}{\partial x}=D_x\frac{\partial^2 C}{\partial x^2}+D_y\frac{\partial^2 C}{\partial y^2}+D_z\frac{\partial^2 C}{\partial z^2},\quad 0<t<T. \quad (24)$$

Furthermore, taking appropriate function for $f, g_0, g_L, h_0, h_W, k_0, k_H$, the initial condition and boundary conditions yield

$$\begin{aligned}&C(x,y,z,0)=0 && 0\le x\le 1;\ 0\le y\le 1;\ 0\le z\le 1\\ &C(0,y,z,t)=0 && 0\le y<0.5;\ 0\le z\le 1\\ &C(0,y,z,t)=1 && 0.5\le y\le 1;\ 0\le z\le 1\\ &C(1,y,z,t)=0 && 0\le y\le 1;\ 0\le z\le 1\\ &\frac{\partial C}{\partial y}(x,0,z,t)=0 && t>0\\ &\frac{\partial C}{\partial y}(x,1,z,t)=0 && t>0\\ &\frac{\partial C}{\partial z}(x,y,0,t)=0 && t>0\\ &\frac{\partial C}{\partial z}(x,y,1,t)=0 && t>0\end{aligned} \quad (25)$$

This partial differential equation and its initial and boundary conditions come from pollution distribution on a street tunnel, where the wind flows steadily only in the x direction. There is no disperse flux of the pollutant through the solid side-walls nor through the base and the roof. By using nondimensionalised parameters $\Delta x = \Delta y = \Delta z = 0.1$, $\Delta t = 0.01$, $D_x = D_y = D_z = 0.1$, $V_x = 0.02$, and for the time $T = 20$. These given values will certainly satisfy both conditions $s_x+s_y+s_z = 3/10 \le 1/2$ and $c_x^2/s_x+c_y^2/s_y+c_z^2/s_z = 3/25000 \le 3$. Thus numerical results are stable and can be found as in the Figure 2.

There is no significant difference from the results obtained from [5], as from Figure 2 left one can see the mesh plot on $z = 0$. This is reasonable, with the boundary conditions satisfied, the concentration decreases away from the source and is less than one-half of the source value over more than three-quarters of the tunnel. Also, the solutions are

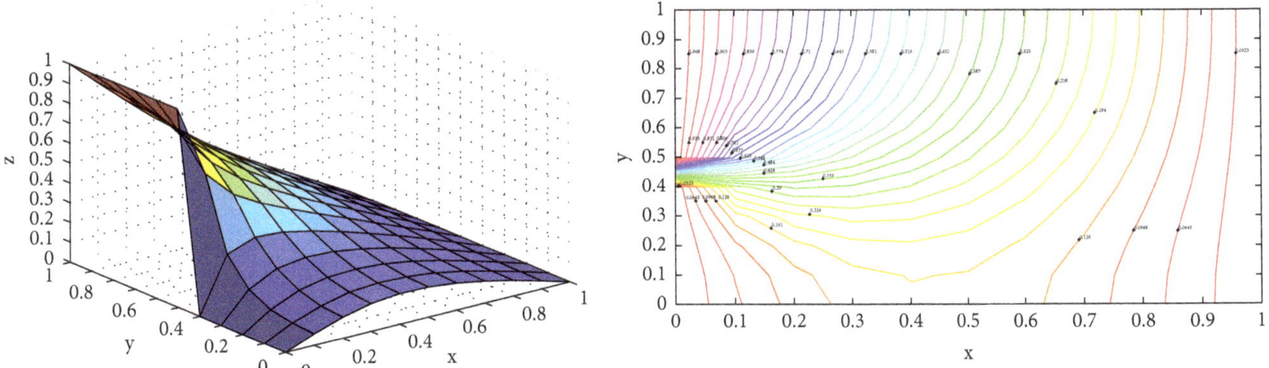

FIGURE 2: Pollutant distribution in a street tunnel with advection only in x direction.

independent of height since all contour plots on $z = 0, z = 0.2, \ldots, z = 1$ are the same as Figure 2 right. The difference may be noted from the previous work as in [5] lies only on $x = 0.1$.

Problem 2. Here, we consider the three-dimension advection-diffusion equation (1), with initial condition (2) and boundary conditions (3)-(8). By taking $V_z = 0$, $L = 1, W = 1, H = 1$, this simply reduces to two-dimension advection three-dimension diffusion equation for transport of pollutants in street tunnel problem as discussed in [5]

$$\frac{\partial C}{\partial t} + V_x \frac{\partial C}{\partial x} + V_y \frac{\partial C}{\partial y} = D_x \frac{\partial^2 C}{\partial x^2} + D_y \frac{\partial^2 C}{\partial y^2} + D_z \frac{\partial^2 C}{\partial z^2}, \quad 0 < t < T. \tag{26}$$

Furthermore, taking appropriate function for $f, g_0, g_L, h_0, h_W, k_0, k_H$, the initial condition and boundary conditions yield

$$\begin{aligned}
&C(x, y, z, 0) = 0 \quad 0 \le x \le 1;\ 0 \le y \le 1;\ 0 \le z \le 1\\
&C(0, y, z, t) = 0 \quad 0 \le y < 0.5;\ 0 \le z \le 1\\
&C(0, y, z, t) = 1 \quad 0.5 \le y \le 1;\ 0 \le z \le 1\\
&C(x, 0, z, t) = 0 \quad 0 \le y < 0.3;\ 0 \le z \le 1\\
&C(x, 0, z, t) = 1 \quad 0.3 \le y \le 0.6;\ 0 \le z \le 1\\
&C(x, 0, z, t) = 0 \quad 0.6 < y \le 1;\ 0 \le z \le 1\\
&\frac{\partial C}{\partial y}(x, 0, z, t) = 0 \quad t > 0\\
&\frac{\partial C}{\partial y}(x, 1, z, t) = 0 \quad t > 0\\
&\frac{\partial C}{\partial z}(x, y, 0, t) = 0 \quad t > 0\\
&\frac{\partial C}{\partial z}(x, y, 1, t) = 0 \quad t > 0
\end{aligned} \tag{27}$$

This model of partial differential equation and its initial and boundary conditions comes from pollution distribution on a street tunnel, where the wind flows steadily only in the x and y directions. There is no disperse flux of the pollutant through the solid side-walls nor through the base and the roof. By using nondimensionalised parameters $\Delta x = \Delta y = \Delta z = 0.1$, $\Delta t = 0.005$, $D_x = D_y = D_z = 0.2$, $V_x = 0.6$, $V_y = 0.4$ and for the time $T = 20$. These values satisfy both conditions $s_x + s_y + s_z = 3/10 \le 1/2$ and $c_x^2/s_x + c_y^2/s_y + c_z^2/s_z = 13/1000 \le 3$. The numerical results are found as in Figure 3.

Figure 3 shows a mesh plot and a contour plot on $z = 0.2$. The other mesh plot and contour plot on the other z show no difference in $z = 0.2$. This is reasonable, with the boundary conditions satisfied, the concentration decreases away from the source faster in the x direction rather than in the y direction.

Problem 3. We consider a three-dimensional advection-diffusion equation for transport of pollutants in the water such as a sea-sand mining or in the air such as limestone rock mining to produce cement. Such activity usually spread pollutants to the environment nearby. By assuming that the domain of the physical problem had been nondimensionalized, and taking the box model, it will satisfy three-dimension advection and three-dimension diffusion equation. There is no disperse flux of the pollutant through the base and the roof. The boundary condition on top of the box will be sea surface and the bottom of the box will be sea floor and satisfy $(\partial C/\partial z)(x, y, H, t) = 0$ and $(\partial C/\partial z)(x, y, 0, t) = 0$. The advection constant $V_z \ne 0$ due to the gravitational force acting on the particle. The constants V_x and V_y are the sea current speed in x and y directions. Another initial condition and boundary conditions yield

$$\begin{aligned}
&C(x, y, z, 0) = 0 \quad 0 \le x \le 1;\ 0 \le y \le 1;\ 0 \le z \le 1\\
&C(0, y, z, t) = 0\\
&\qquad 0 \le y < 0.4;\ 0.6 < y \le 1;\ 0 \le z \le 1\\
&C(0, y, z, t) = 1 \quad 0.4 \le y \le 0.6;\ 0 \le z \le 1\\
&C(x, 0, z, t) = 0\\
&\qquad 0 \le x < 0.4;\ 0.6 < x \le 1;\ 0 \le z \le 1
\end{aligned}$$

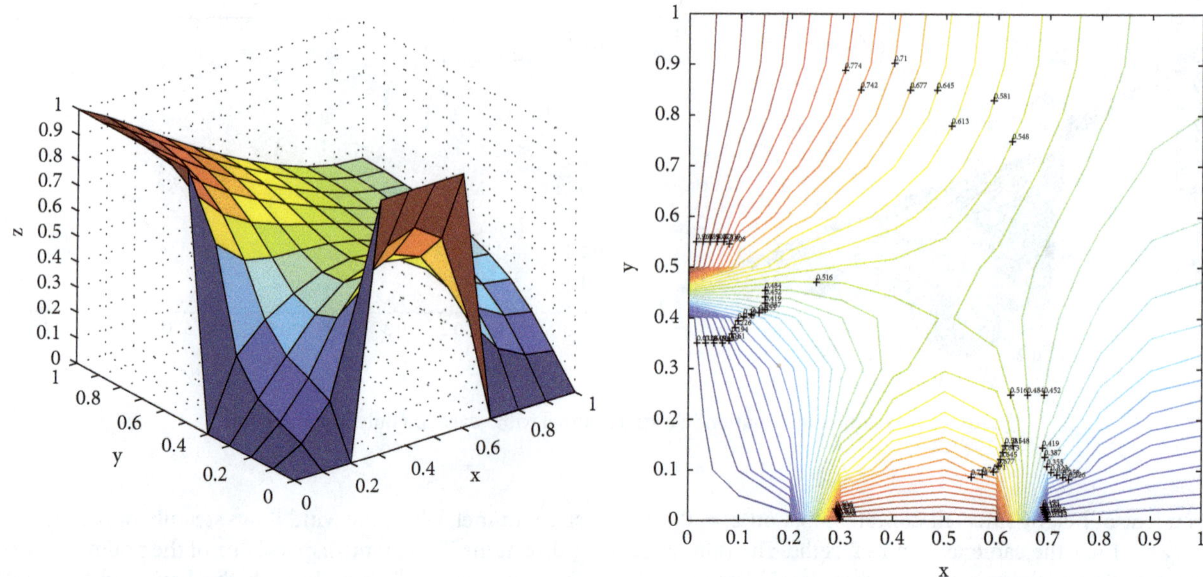

FIGURE 3: Pollutant distribution in a street tunnel with advection in x and y directions.

$$
\begin{aligned}
&C(x,0,z,t)=1 \quad 0.4\le x\le 0.6;\ 0\le z\le 1\\
&C(1,y,z,t)=0 \quad 0\le y\le 1;\ 0\le z\le 1\\
&\frac{\partial C}{\partial y}(x,0,z,t)=0 \quad t>0\\
&\frac{\partial C}{\partial y}(x,1,z,t)=0 \quad t>0\\
&\frac{\partial C}{\partial z}(x,y,0,t)=0 \quad t>0\\
&\frac{\partial C}{\partial z}(x,y,1,t)=0 \quad t>0
\end{aligned} \tag{28}
$$

Using nondimensionalised parameters $\Delta x = \Delta y = \Delta z = 0.1$, $\Delta t = 0.005$; $D_x = D_y = D_z = 0.1$, $V_x = 1.0$, $V_y = 0.3$, $V_z = -0.3$ and for the time $T = 2$. Here, the values satisfy also both stability conditions $s_x + s_y + s_z = 3/20 \le 1/2$ and $c_x^2/s_x + c_y^2/s_y + c_z^2/s_z = 59/1000 \le 3$. The numerical results for surface plot and contour plot on $z = 0.5$ and several values of time t are found as in Figure 4. Surface plot and contour plot for other values of z are similar as in Figure 4.

Problem 4. We consider a three-dimensional advection-diffusion equation for transport of pollutants in the water such as a sea-sand mining again. Such activity usually lasts for certain period of time and spreads pollutants to the environment nearby. Of our interest is to find out the pollutant distribution of this activity. By assuming that the domain of the physical problem had been nondimensionalized and taking the box model, it will satisfy three-dimension advection and three-dimension diffusion equation. There is no disperse flux of the pollutant through the base and the roof. The boundary condition on top of the box will be sea surface and the bottom of the box will be sea floor and satisfy $(\partial C/\partial z)(x,y,H,t) = 0$ and $(\partial C/\partial z)(x,y,0,t) = 0$. Other initial condition and boundary conditions yield

$$
\begin{aligned}
&C(x,y,z,0)=0 \quad 0\le x\le 1;\ 0\le y\le 1;\ 0\le z\le 1\\
&C(0,y,z,t)=0\\
&\qquad 0\le y<0.4;\ 0.6<y\le 1;\ 0\le z\le 1\\
&C(0,y,z,t)=1\\
&\qquad 0.4\le y\le 0.6;\ 0\le z\le 1;\ 0\le t\le 0.5\\
&C(x,0,z,t)=0\\
&\qquad 0\le x<0.4;\ 0.6<x\le 1;\ 0\le z\le 1\\
&C(x,0,z,t)=1\\
&\qquad 0.4\le x\le 0.6;\ 0\le z\le 1;\ 0\le t\le 0.5\\
&C(1,y,z,t)=0 \quad 0\le y\le 1;\ 0\le z\le 1\\
&C(0,y,z,t)=0\\
&\qquad 0.4\le y\le 0.6;\ 0\le z\le 1;\ t>0.5\\
&C(x,0,z,t)=0\\
&\qquad 0.4\le x\le 0.6;\ 0\le z\le 1;\ t>0.5\\
&\frac{\partial C}{\partial y}(x,0,z,t)=0 \quad t>0\\
&\frac{\partial C}{\partial y}(x,1,z,t)=0 \quad t>0
\end{aligned}
$$

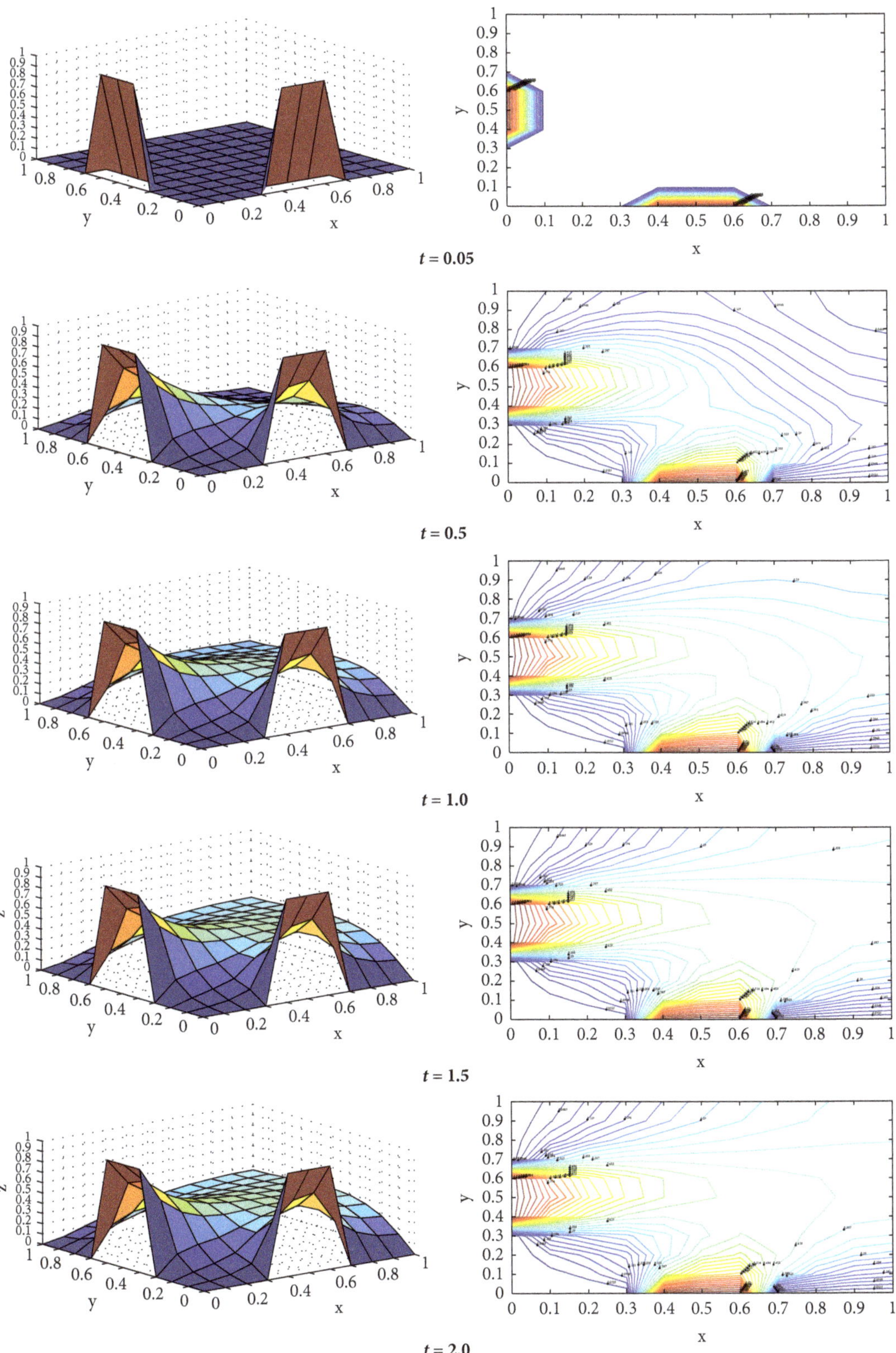

FIGURE 4: Pollutant distribution in a box model under the sea with advection in x, y, and z directions.

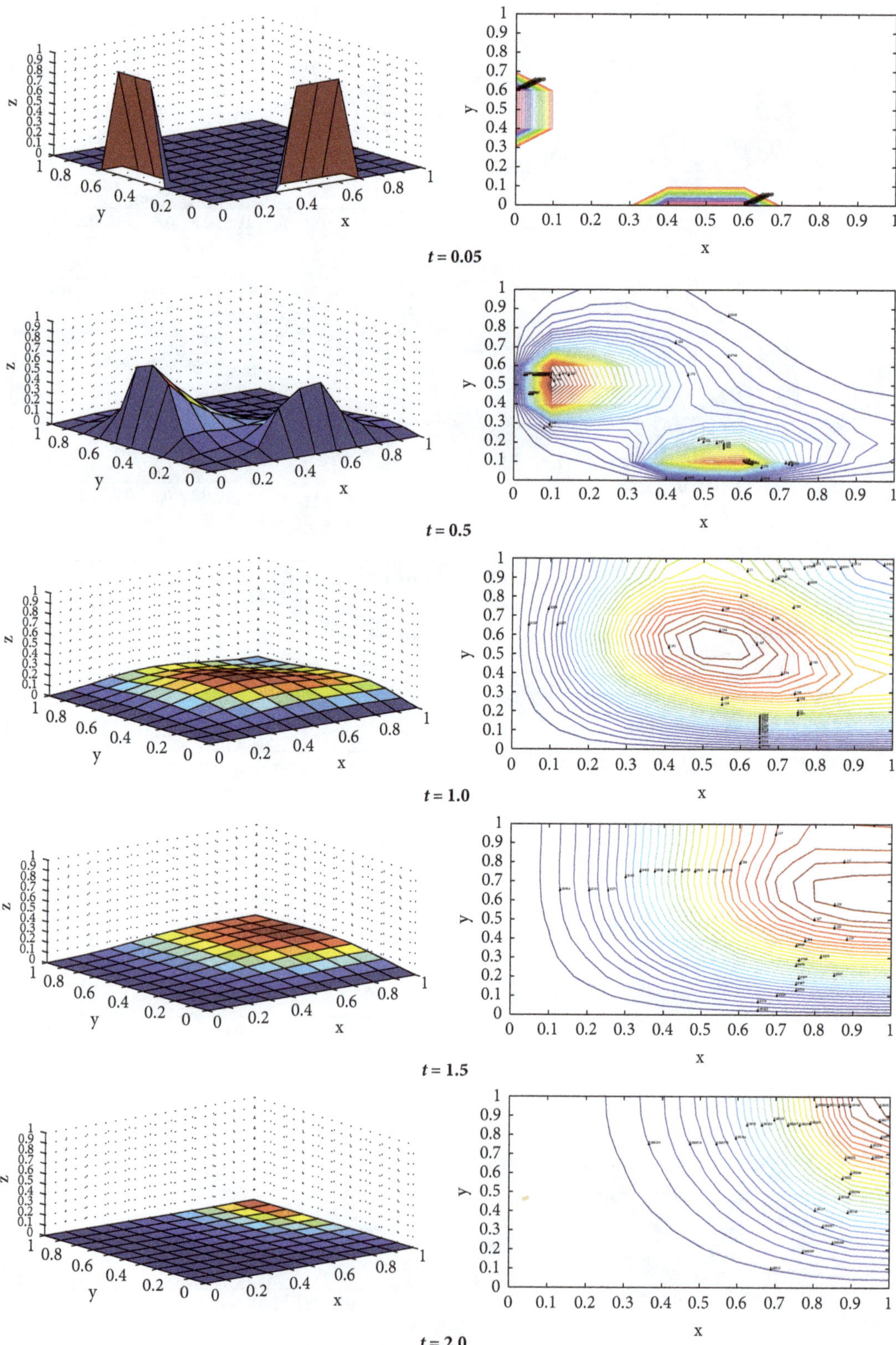

FIGURE 5: Pollutant distribution in a box model under the sea with advection in x, y, and z directions.

$$\frac{\partial C}{\partial z}(x, y, 0, t) = 0 \quad t > 0$$
$$\frac{\partial C}{\partial z}(x, y, 1, t) = 0 \quad t > 0 \tag{29}$$

Using nondimensionalised parameters $\Delta x = \Delta y = \Delta z = 0.1$, $\Delta t = 0.005$; $D_x = D_y = D_z = 0.2$, $V_x = 1.0$, $V_y = 0.3$, $V_z = -0.3$ and for the time $T = 2$. These values also satisfy both stability conditions $s_x + s_y + s_z = 3/10 \leq 1/2$ and $c_x^2/s_x + c_y^2/s_y + c_z^2/s_z = 295/10000 \leq 3$. There are two pollutant sources on the x-axis and y-axis and they last for certain period of time $0 \leq t \leq 0.5$. The numerical results for surface plot and contour plot on $z = 0.4$ for several different times are found as in Figure 5. Unlike the previous Problem 3, surface and contour plot for other values of z show that pollutant materials move to x, y, z directions as expected.

6. Conclusions

We have employed a standard finite difference scheme to study the pollution distribution for two-dimension advection and three-dimension diffusion equation and extend into three-dimension advection and three-dimension diffusion equation. The stability conditions ensure that the solution of all the interior points can be obtained. The scheme works well as one can see numerical results obtained in solutions of the problems above.

Conflicts of Interest

The authors declare that they have no conflicts of interest.

Acknowledgments

The authors acknowledge that this research work is done under the grant of Penelitian Dasar Unggulan Perguruan Tinggi (PDUPT) scheme of the Ministry of Higher Education and Research of the Republic of Indonesia.

References

[1] E. Sousa, "The controversial stability analysis," *Applied Mathematics and Computation*, vol. 145, no. 2-3, pp. 777–794, 2003.

[2] M. Dehghan, "Numerical solution of the three-dimensional advection-diffusion equation," *Applied Mathematics and Computation*, vol. 150, no. 1, pp. 5–19, 2004.

[3] M. Thongmoon and R. McKibbin, "A comparison of some numerical methods for advection-diffusion equation," *Research Letters in the Information and Mathematical Sciences*, vol. 10, pp. 49–62, 2006.

[4] M. Dehghan, "Time-splitting procedures for the solution of the two-dimensional transport equation," *Kybernetes*, vol. 36, no. 5-6, pp. 791–805, 2007.

[5] M. Thongmoon, R. McKibbin, and S. Tangmanee, "Numerical solution of a 3-D advection-dispersion model for pollutant transport," *Thai Journal of Mathematics*, vol. 5, no. 1, pp. 91–108, 2007.

[6] F. U. Prieto, J. J. Benito Muñoz, and L. G. Corvinos, "Application of the generalized finite difference method to solve the advection-diffusion equation," *Journal of Computational and Applied Mathematics*, vol. 235, no. 7, pp. 1849–1855, 2011.

[7] J. Kusuma, A. Ribal, A. G. Mahie, and N. Aris, "On pollution distribution on unhas lake using two dimension advection-diffusion equation," *Far East Journal of Mathematical Sciences*, vol. 101, no. 8, pp. 1721–1729, 2017.

Well-Posedness and Numerical Study for Solutions of a Parabolic Equation with Variable-Exponent Nonlinearities

Jamal H. Al-Smail, Salim A. Messaoudi, and Ala A. Talahmeh

Department of Mathematics and Statistics, King Fahd University of Petroleum and Minerals, P.O. Box 546, Dhahran 31261, Saudi Arabia

Correspondence should be addressed to Salim A. Messaoudi; messaoud@kfupm.edu.sa

Academic Editor: Dongfang Li

We consider the following nonlinear parabolic equation: $u_t - \operatorname{div}(|\nabla u|^{p(x)-2}\nabla u) = f(x,t)$, where $f : \Omega \times (0,T) \to \mathbb{R}$ and the exponent of nonlinearity $p(\cdot)$ are given functions. By using a nonlinear operator theory, we prove the existence and uniqueness of weak solutions under suitable assumptions. We also give a two-dimensional numerical example to illustrate the decay of solutions.

1. Introduction

Let Ω be a bounded domain in $\mathbb{R}^n$ with a smooth boundary $\partial\Omega$. We consider the following initial and boundary value problem:

$$\begin{aligned} u_t - \operatorname{div}\left(|\nabla u|^{p(x)-2}\nabla u\right) &= f(x,t), \quad \text{in } \Omega \times (0,T) \\ u(x,t) &= 0, \quad \text{on } \partial\Omega \times (0,T) \\ u(x,0) &= u_0(x), \quad \text{in } \Omega, \end{aligned} \tag{P}$$

where $f : \Omega \times (0,T) \to \mathbb{R}$ and $u_0 : \Omega \to \mathbb{R}$ are given functions. The exponent $p(\cdot)$ is a given measurable function on $\overline{\Omega}$ such that

$$2 < p_1 \le p(x) \le p_2 < +\infty, \tag{1}$$

with

$$\begin{aligned} p_1 &:= \operatorname*{essinf}_{x\in\Omega} p(x), \\ p_2 &:= \operatorname*{esssup}_{x\in\Omega} p(x). \end{aligned} \tag{2}$$

We also assume that $p(\cdot)$ satisfies the log-Hölder continuity condition:

$$|p(x) - p(y)| \le -\frac{A}{\log|x-y|}, \quad \forall x, y \in \Omega, \text{ with } |x-y| < \delta, \tag{3}$$

where $A > 0$ and $0 < \delta < 1$ are constants. The term $\operatorname{div}(|\nabla u|^{p(\cdot)-2}\nabla u)$ is called the $p(\cdot)$-Laplacian and denoted by $\Delta_{p(\cdot)}u$.

The study of partial differential equations involving variable-exponent nonlinearities has attracted the attention of researchers in recent years. The interest in studying such problems is stimulated and motivated by their applications in elastic mechanics, fluid dynamics, nonlinear elasticity, electrorheological fluids, and so forth. In particular, parabolic equations involving the $p(\cdot)$-Laplacian are related to the field of image restoration and electrorheological fluids which are characterized by their ability to change the mechanical properties under the influence of the exterior electromagnetic field. The rigorous study of these physical problems has been facilitated by the development of the Lebesgue and Sobolev spaces with variable exponents.

Regarding parabolic problems with nonlinearities of variable-exponent type, many works have appeared. We note here that most of the results deal with blow-up and global nonexistence. Let us mention some of these works. For instance, Alaoui et al. [1] considered the following nonlinear heat equation:

$$u_t(x,t) - \operatorname{div}\left(|\nabla u|^{m(x)-2}\nabla u\right) = u|u|^{p(x)-2} + f, \quad (4)$$

in a bounded domain in $\Omega \subset \mathbb{R}^n$ $(n \geq 1)$ with a smooth boundary $\partial\Omega$. Under appropriate conditions on the exponent functions m, p and for $f = 0$, they showed that any solution with nontrivial initial datum blows up in finite time. They also gave a two-dimensional numerical example to illustrate their result. Pinasco [2] studied the following problem:

$$\begin{aligned} u_t - \Delta u &= f(u), \quad \text{in } \Omega \times [0,T) \\ u(x,t) &= 0, \quad \text{on } \partial\Omega \times [0,T) \\ u(x,0) &= u_0(x), \quad \text{in } \Omega, \end{aligned} \quad (5)$$

where $\Omega \subset \mathbb{R}^n$ is a bounded domain with a smooth boundary $\partial\Omega$, and the source term is of the following form:

$$\begin{aligned} f(u) &= a(x)\,u^{p(x)} \\ \text{or } f(u) &= a(x)\int_\Omega u^{q(y)}(y,t)\,dy, \end{aligned} \quad (6)$$

with p,q: $\Omega \rightarrow (1,\infty)$ and the continuous function a: $\Omega \rightarrow \mathbb{R}$ being given functions satisfying specific conditions. They established the local existence of positive solutions and proved that solutions with initial data sufficiently large blow up in finite time. Parabolic problems with sources like the ones in (5) appear in several branches of applied mathematics and they have been used to model chemical reactions, heat transfer, or population dynamics.

Recently, Shangerganesh et al. [3] studied the following fourth-order degenerate parabolic equation:

$$u_t + \operatorname{div}\left(|\nabla\Delta u|^{p(x)-2}\nabla\Delta u\right) = f - \operatorname{div} g, \quad (7)$$

in a bounded domain $\Omega \subset \mathbb{R}^n$ $(n \geq 1)$ with a smooth boundary $\partial\Omega$, and proved the existence and uniqueness of weak solutions of (7) by using the difference and variation methods under suitable assumptions on f, g and the exponents p.

Equation (P) is a nonlinear diffusion equation which has been used to study image restoration and electrorheological fluids (see [4–11]). In particular, Bendahmane et al. [12] proved the well-posedness of a solution, for L^1-data. Akagi and Matsuura [13] gave the well-posedness for L^2 initial datum and discussed the long-time behaviour of the solution using the subdifferential calculus approach. In our paper, we give an alternative proof of the well-posedness of (P) which is simpler than that in [13] using a theory of nonlinear evolution equations. In addition, we give a numerical example in 2D to illustrate the decay result obtained in [13].

This paper consists of three sections in addition to the introduction. In Section 2, we recall the definitions of the variable-exponent Lebesgue spaces, $L^{p(\cdot)}(\Omega)$, the Sobolev spaces, $W^{1,p(\cdot)}(\Omega)$, as well as some of their properties. We also state, without proof, a proposition to be used in the proof of our main result. In Section 3, we state and prove the well-posedness of solution to our problem. In Section 4, we give a numerical verification of the decay result.

2. Preliminaries

We present some preliminary facts about the Lebesgue and Sobolev spaces with variable exponents (see [1, 14–16]). Let p : $\Omega \rightarrow [1,\infty]$ be a measurable function, where Ω is a domain of $\mathbb{R}^n$. We define the Lebesgue space with a variable-exponent $p(\cdot)$ by

$$\begin{aligned} L^{p(\cdot)}(\Omega) := \{u : \Omega &\longrightarrow \mathbb{R};\ \text{measurable in } \Omega : \varrho_{L^{p(\cdot)}(\Omega)}(\lambda u) \\ &< \infty, \text{ for some } \lambda > 0\}, \end{aligned} \quad (8)$$

where

$$\varrho_{L^{p(\cdot)}(\Omega)}(u) = \int_\Omega |u(x)|^{p(x)}\,dx \quad (9)$$

is called a modular. Equipped with the Luxembourg-type norm,

$$\|u\|_{L^{p(\cdot)}(\Omega)} := \inf\left\{\lambda > 0 : \varrho_{L^{p(\cdot)}(\Omega)}\left(\frac{u}{\lambda}\right) \leq 1\right\}, \quad (10)$$

$L^{p(\cdot)}(\Omega)$ is a Banach space (see [10]).

Lemma 1 (Hölder's inequality [10]). *Let $p, q, s \geq 1$ be measurable functions defined on Ω such that*

$$\frac{1}{s(y)} = \frac{1}{p(y)} + \frac{1}{q(y)}, \quad (11)$$

for a.e. $y \in \Omega$. If $f \in L^{p(\cdot)}(\Omega)$ and $g \in L^{q(\cdot)}(\Omega)$, then $fg \in L^{s(\cdot)}(\Omega)$ and

$$\|fg\|_{s(\cdot)} \leq 2\|f\|_{p(\cdot)}\|g\|_{q(\cdot)}. \quad (12)$$

Lemma 2 (see [10]). *Let p be a measurable function on Ω. Then,*

(a) $\|f\|_{p(\cdot)} \leq 1$ *if and only if* $\varrho_{p(\cdot)}(f) \leq 1$;

(b) *for* $f \in L^{p(\cdot)}(\Omega)$, *if* $\|f\|_{p(\cdot)} \leq 1$, *then* $\varrho_{p(\cdot)}(f) \leq \|f\|_{p(\cdot)}$; *and if* $\|f\|_{p(\cdot)} \geq 1$, *then* $\|f\|_{p(\cdot)} \leq \varrho_{p(\cdot)}(f)$;

(c) $\|f\|_{p(\cdot)} \leq 1 + \varrho_{p(\cdot)}(f)$.

Lemma 3 (see [10]). *If (1) holds, then*

$$\begin{aligned} \min\left\{\|u\|_{p(\cdot)}^{p_1}, \|u\|_{p(\cdot)}^{p_2}\right\} &\leq \varrho_{p(\cdot)}(u) \\ &\leq \max\left\{\|u\|_{p(\cdot)}^{p_1}, \|u\|_{p(\cdot)}^{p_2}\right\}, \end{aligned} \quad (13)$$

for any $u \in L^{p(\cdot)}(\Omega)$.

We next define the variable-exponent Sobolev space $W^{1,p(\cdot)}(\Omega)$ as follows:

$$W^{1,p(\cdot)}(\Omega) = \{u \in L^{p(\cdot)}(\Omega) \text{ such that } \nabla u \text{ exists, } |\nabla u| \in L^{p(\cdot)}(\Omega)\}. \quad (14)$$

This space is a Banach space with respect to the norm $\|u\|_{W^{1,p(\cdot)}(\Omega)} = \|u\|_{p(\cdot)} + \|\nabla u\|_{p(\cdot)}$. Furthermore, $W_0^{1,p(\cdot)}(\Omega)$ is the closure of $C_0^\infty(\Omega)$ in $W^{1,p(\cdot)}(\Omega)$. The dual of $W_0^{1,p(\cdot)}(\Omega)$ is defined as $W^{-1,p'(\cdot)}(\Omega)$, by the same way as the usual Sobolev spaces where $1/p(\cdot) + 1/p'(\cdot) = 1$.

Lemma 4 (see [10]). *Let Ω be a bounded domain of $\mathbb{R}^n$ and $p(\cdot)$ satisfies (1) and (3), and then*

$$\|u\|_{p(\cdot)} \leq C\|\nabla u\|_{p(\cdot)}, \quad \forall u \in W_0^{1,p(\cdot)}(\Omega), \quad (15)$$

where the positive constant C depends on $p(\cdot)$ and Ω. In particular, the space $W_0^{1,p(\cdot)}(\Omega)$ has an equivalent norm given by $\|u\|_{W^{1,p(\cdot)}(\Omega)} = \|\nabla u\|_{p(\cdot)}$.

Lemma 5 (see [10]). *If $p(\cdot) \in C(\overline{\Omega})$, $q : \Omega \to [1,\infty)$ is a continuous function and*

$$\operatorname*{essinf}_{x\in\Omega}(p^*(x) - q(x)) > 0$$

$$\text{with } p_*(x) = \begin{cases} \dfrac{np(x)}{\operatorname*{esssup}_{x\in\Omega}(n-p(x))} & \text{if } p_2 < n \\ \infty & \text{if } p_2 \geq n. \end{cases} \quad (16)$$

Then the embedding $W_0^{1,p(\cdot)}(\Omega) \hookrightarrow L^{q(\cdot)}(\Omega)$ is continuous and compact.

Definition 6 (see [17]). Let V be a separable Banach space and H be a Hilbert space such that $V \subset H \subset V'$ with continuous embedding and V is dense in H. Let $A : V \to V'$ be a nonlinear operator.

(1) A is said to be monotone if $\langle A(u) - A(v), u - v\rangle_{V'\times V} \geq 0$. If, in addition, we have

$$\langle A(u) - A(v), u - v\rangle_{V'\times V} \neq 0, \quad \forall u \neq v, \quad (17)$$

then A is said to be strictly monotone.

(2) A is said to be bounded, if $A(S)$ is bounded in V', whenever S is bounded in V.

(3) A is said to be hemicontinuous, if the real function

$$\lambda \longmapsto \langle A(u + \lambda v), w\rangle \quad (18)$$

is continuous from $\mathbb{R}$ to $\mathbb{R}$, for any fixed $u, v, w \in V$.

We end this section with a proposition which is exactly like Theorem 7.1 [17].

Proposition 7. *Let $u_0 \in H$ and $f \in L^{p'}((0,T), V')$. Suppose that $A: V \to V'$ is a bounded monotone and hemicontinuous (nonlinear) operator satisfying, for some $\alpha, \beta > 0$ and for some $p > 1$,*

$$\langle A(v), v\rangle \geq \alpha\|v\|_V^p - \beta, \quad \forall v \in V. \quad (19)$$

Then the following problem

$$\begin{aligned} u_t + A(u) &= f \\ u(\cdot, 0) &= u_0, \end{aligned} \quad (20)$$

has a unique weak solution:

$$u \in L^p((0,T), V) \quad \text{with } u_t \in L^{p'}((0,T), V'), \quad (21)$$

where $1/p + 1/p' = 1$.

3. Well-Posedness

In this section, we state and prove the well-posedness of our problem.

Theorem 8. *Let $u_0 \in L^2(\Omega)$, $f \in L^{p'(\cdot)}((0,T), W^{-1,p'(\cdot)}(\Omega))$. Assume that (1) and (3) hold. Then* (P) *has a unique weak solution:*

$$\begin{aligned} &u \in L^{p_1}((0,T), W_0^{1,p(\cdot)}(\Omega)) \cap L^\infty((0,T), L^2(\Omega)), \\ &u_t \in L^{p_2'}((0,T), W^{-1,p'(\cdot)}(\Omega)), \end{aligned} \quad (22)$$

where $1/p(\cdot) + 1/p'(\cdot) = 1$.

Proof. We verify the conditions of Proposition 7. Let $V = W_0^{1,p(\cdot)}(\Omega)$, and equip it with the norm,

$$\|u\|_{W^{1,p(\cdot)}(\Omega)} = \|\nabla u\|_{p(\cdot)}. \quad (23)$$

So, $V' = W^{-1,p'(\cdot)}(\Omega)$. Define $A : V \to V'$ by

$$A(u) = -\operatorname{div}(|\nabla u|^{p(x)-2}\nabla u). \quad (24)$$

Boundedness of A. For all $u, v \in V$,

$$\begin{aligned} |\langle A(u), v\rangle| &= \left|\int_\Omega |\nabla u|^{p(x)-2}\nabla u \cdot \nabla v\right| \\ &\leq \int_\Omega |\nabla u|^{p(x)-1}|\nabla v|. \end{aligned} \quad (25)$$

Hölder inequality gives

$$\int_\Omega |\nabla u|^{p(x)-1}|\nabla v| \leq 2\left\||\nabla u|^{p(\cdot)-1}\right\|_{p'(\cdot)}\|v\|_V. \quad (26)$$

Combining (25) and (26), we obtain

$$\|A(u)\|_{V'} \leq 2\left\||\nabla u|^{p(\cdot)-1}\right\|_{p'(\cdot)}. \quad (27)$$

Then, Lemma 2 implies

$$\left\||\nabla u|^{p(\cdot)-1}\right\|_{p'(\cdot)} \leq (1 + \varrho_{p(\cdot)}(\nabla u)). \quad (28)$$

Combining (27) and (28), we arrive at

$$\|A(u)\|_{V'} \leq 2\left(1 + \varrho_{p(\cdot)}(\nabla u)\right). \quad (29)$$

Let $S \subset V$ such that $\|u\|_V \leq M$, for all $u \in S$. That is, $\|\nabla u\|_{p(\cdot)} \leq M$.

If $M \leq 1$, then Lemma 2 implies $\varrho_{p(\cdot)}(\nabla u) \leq 1$ and (29) gives $\|A(u)\|_{V'} \leq 4 < +\infty$.

If $M > 1$, then $\varrho_{p(\cdot)}(\nabla u) \leq M^{p_2}$. Thus, (29) implies $\|A(u)\|_{V'} \leq 2(1 + M^{p_2}) < +\infty$.

Hence A is bounded.

Monotonicity of A. Let $u, v \in V$. Then

$$\begin{aligned}&\langle A(u) - A(v), u - v\rangle_{V'\times V}\\ &= -\int \operatorname{div}\left(|\nabla u|^{p(x)-2}\nabla u - |\nabla v|^{p(x)-2}\nabla v\right)(u - v)\,dx\\ &= \int \left(|\nabla u|^{p(x)-2}\nabla u - |\nabla v|^{p(x)-2}\nabla v\right)(\nabla u - \nabla v)\,dx.\end{aligned} \quad (30)$$

By using the inequality,

$$\left(|a|^{p(x)-2}a - |b|^{p(x)-2}b\right)\cdot(a - b) \geq 0, \quad (31)$$

for all $a, b \in \mathbb{R}^n$ and a.e. $x \in \Omega$. Thus we obtain $\langle A(u) - A(v), u - v\rangle \geq 0$.

Hence, A is monotone.

To verify (19), we note that, for all $u \in V$, we have

$$\langle A(u), u\rangle = \int_\Omega |\nabla u|^{p(x)}\,dx = \varrho_{p(\cdot)}(\nabla u). \quad (32)$$

If $\|u\|_V = \|\nabla u\|_{p(\cdot)} > 1$, then by Lemma 3, we get

$$\varrho_{p(\cdot)}(\nabla u) \geq \|\nabla u\|_{p(\cdot)}^{p_1}. \quad (33)$$

Combining (32) and (33), we easily see that

$$\langle A(u), u\rangle \geq \|u\|_V^{p_1}. \quad (34)$$

If $\|u\|_V = \|\nabla u\|_{p(\cdot)} \leq 1$, then by Lemma 2, we obtain

$$\langle A(u), u\rangle = \varrho_{p(\cdot)}(\nabla u) \geq \|u\|_V - 1 \geq \|u\|_V^{p_1} - 1. \quad (35)$$

Therefore, we have

$$\langle A(u), u\rangle \geq \|u\|_V^{p_1} - 1, \quad \forall u \in V. \quad (36)$$

Hemicontinuity of A. Let $u, v, w \in V$ be fixed. Let

$$\begin{aligned}g(\lambda) &= \langle A(u + \lambda v), w\rangle\\ &= \int_\Omega |\nabla u + \lambda\nabla v|^{p(x)-2}(\nabla u + \lambda\nabla v)\cdot\nabla w\,dx.\end{aligned} \quad (37)$$

Let $\lambda_k \to \lambda$ (real) and consider

$$g(\lambda_k) = \int_\Omega |\nabla u + \lambda_k\nabla v|^{p(x)-2}(\nabla u + \lambda_k\nabla v)\cdot\nabla w\,dx. \quad (38)$$

Since

$$\begin{aligned}&|\nabla u + \lambda_k\nabla v|^{p(x)-2}(\nabla u + \lambda_k\nabla v)\cdot\nabla w\\ &\longrightarrow |\nabla u + \lambda\nabla v|^{p(x)-2}(\nabla u + \lambda\nabla v)\cdot\nabla w,\end{aligned} \quad (39)$$

for a.e. $x \in \Omega$ and

$$\begin{aligned}&|\nabla u + \lambda_k\nabla v|^{p(x)-1}|\nabla w|\\ &\leq C\left(|\nabla u|^{p(x)-1}|\nabla w| + |\nabla v|^{p(x)-1}|\nabla w|\right) \in L^1(\Omega),\end{aligned} \quad (40)$$

where $C = \max\{2^{p_2-2}, 2^{p_2-2}(1 + |M|^{p_2-1})\} > 0$, then, by the classical dominated convergence theorem,

$$g(\lambda_k) \longrightarrow g(\lambda) \quad \text{as } k \longrightarrow \infty. \quad (41)$$

Hence, A is hemicontinuous.

Therefore, conditions of Proposition 7 are satisfied and problem (P) has a unique solution. □

4. Numerical Study

In this section, we present some numerical results and applications of the problem:

$$\begin{aligned}&u_t - \operatorname{div}\left(|\nabla u|^{p(x,y)-2}\nabla u\right) = 0, \quad \text{in } Q = \Omega\times(0,T)\\ &u = 0, \quad \text{on } \partial Q = \partial\Omega\times(0,T)\\ &u(x,y,0) = u_0(x,y), \quad \text{in } \Omega,\end{aligned} \quad (P_*)$$

which is a well-posed problem due to Theorem 8. Our objective is to provide a numerical verification of the following decay result:

Proposition 9 (see [13]). *Assume that (1) and (3) hold. Then the solution of (P_*) satisfies the following:*

(i) *If $p_2 = 2$, then there exists a constant $c_2 > 0$ such that*

$$\|u(t)\|_2 \leq \|u_0\|_2 e^{-c_2 t}, \quad \forall t \geq 0. \quad (42)$$

(ii) *If $p_2 > 2$, there exists a constant $c_1 > 0$ and $t_1 \geq 0$ such that*

$$\|u(t)\|_2 \leq \|u_0\|_2 (1 + c_1 t)^{-1/(p_2-2)}, \quad \forall t \geq t_1. \quad (43)$$

We consider two applications to illustrate numerically an exponential decay for the case $p(x,y) = 2$ and a polynomial decay for an exponent function $p(\cdot)$ satisfying conditions (1)–(3).

For this purpose, we introduce a numerical scheme for (P_*), prove its convergence in Section 4.1, and show the decay results in Section 4.2.

4.1. Numerical Method. In this part, we present a linearized numerical scheme to obtain the numerical results of the system (P_*) and confirm the decay results. The system is fully discretized through a finite difference method for the time variable and a finite element Galerkin method for the space variable. Useful background about the numerical and error analysis of these methods is found in [18]. More interestingly in [19], Li and Wang introduced a numerical scheme to solve strongly nonlinear parabolic systems and proved unconditional error estimates of the scheme. Our problem (P_*) is highly nonlinear due to the presence of the gradient and nonlinear exponent in the diffusivity coefficient, which can be zero inside the spatial domain. Below, we introduce our numerical scheme for the purpose of confirming the decay results.

The parabolic equation

$$u_t - \operatorname{div}\left(|\nabla u|^{p(x,y)-2} \nabla u\right) = 0, \quad \text{in } Q = \Omega \times (0,T) \tag{44}$$

is discretized using finite differences for the time derivative and a finite element method for the $p(\cdot)$–Laplacian term. For this, we divide the time interval $[0,T]$ into N equal subintervals by

$$t_n = n\tau, \quad \tau = \frac{T}{N} \tag{45}$$

and denote by

$$u^n(x,y) := u(x,y,t_n), \quad n = 0,1,\ldots,N. \tag{46}$$

The term u_t is approximated using the first-order forward finite difference formula:

$$u_t^{n+1} := \frac{u^{n+1} - u^n}{\tau}. \tag{47}$$

Semidiscrete Problem. A linear semidiscrete formulation of (P_*) takes the following form: given p and u_0, find $\{u^1, u^2, \ldots, u^{n+1}\}$ such that

$$\begin{aligned} u_t^{n+1} - \operatorname{div}\left(|\nabla u^n|^{p(x,y)-2} \nabla u^{n+1}\right) &= 0, \quad \text{in } \Omega \\ u^{n+1} &= 0, \quad \text{on } \partial\Omega \\ u^0 &= u_0(x,y), \quad \text{in } \Omega. \end{aligned} \tag{48}$$

This problem is elliptic and admits a unique solution [20], for every $n = 0,1,\ldots,N$. Also, the Rothe approximation $u^{(n)}$ to the exact solution u, given by

$$\begin{aligned} u^{(n)}(x,y,t) &:= u^{i-1}(x,y) + (t - t_{i-1})\frac{u^i - u^{i-1}}{\tau}, \\ & t \in [t_{i-1}, t_i], \; i = 1,2,\ldots,n, \end{aligned} \tag{49}$$

is well defined and $u^{(n)} \to u$ in $L^2(\Omega)$ as $\tau \to 0$, see [1].

Full-Discrete Problem. The variable u^{n+1} is discretized in space by a finite element method. For this, let Ω_h be a triangulation of Ω with a maximal element size h. Let also v_h be a test function in the linear Lagrangian space $P_1(\Omega_h)$ such that $v_h = 0$ on $\partial\Omega_h$.

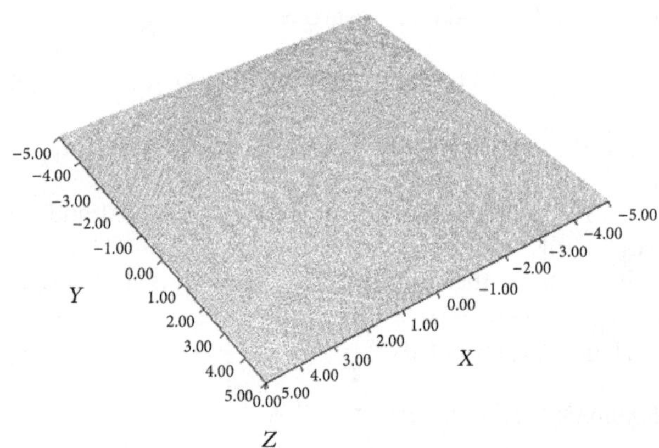

FIGURE 1: Mesh of the domain $[-5,5] \times [-5,5]$.

The semidiscrete problem is then written in a weak form to define the full-discrete problem: given p_h, $u^n \in P_1(\Omega_h)$, find $u_h^{n+1} \in P_1(\Omega_h)$ such that

$$\begin{aligned} \int_{\Omega_h} \left(\frac{u_h^{n+1} - u_h^n}{\tau} v_h + |\nabla u_h^n|^{p_h(x,y)-2} \nabla u_h^{n+1} \cdot \nabla v_h \right) d\Omega_h \\ = 0, \quad \forall v_h \in P_1(\Omega_h). \end{aligned} \tag{50}$$

For $p_h \geq 2$, the above problem has a unique solution $u_h^{n+1} \in H_0^1(\Omega_h)$ for every nontrivial $u_h^n \in H^1(\Omega_h)$. This follows from the Lax-Milgram Lemma, and the Galerkin approximation u_h^{n+1} converges to u^{n+1} in $H_0^1(\Omega_h)$ as $h \to 0$; see [18].

4.2. Numerical Results. In this subsection, we present the following numerical applications of (P_*):

(1) Exponential decay: for $p_h(x,y) = 2$, we show, for some $c > 0$, that

$$g_1(t) := \frac{\|u_h^n\|}{\|u_h^0\|} \leq e^{-ct}, \quad \forall t \geq 0. \tag{51}$$

(2) Polynomial decay: for $p_h(x,y) = (1/5)\lceil x \rceil^2 + 2.5$, we show, for some $c > 0$ and $t_0 > 0$, that

$$g_2(t) := \frac{\|u_h^n\|}{\|u_h^0\|} \leq (1 + ct)^{-1/(p_2-2)}, \quad \forall t \geq t_0. \tag{52}$$

Here, $\lceil \cdot \rceil$ denote the greatest integer function.

In both applications, we set the following parameters:

$$\begin{aligned} T &= 100, \\ \tau &= 0.1, \\ h &= 0.1, \\ \Omega_h &= [-5,5] \times [-5,5]. \end{aligned} \tag{53}$$

Figure 1 shows the mesh used for Ω_h, which involves 23702 triangles and 12052 vertices.

1
0.95
0.9
0.85
0.8
0.75
0.7
0.65
0.6
0.55
0.5
0.45
0.4
0.35
0.3
0.25
0.2
0.15
0.1
0.05
1.3888e – 011

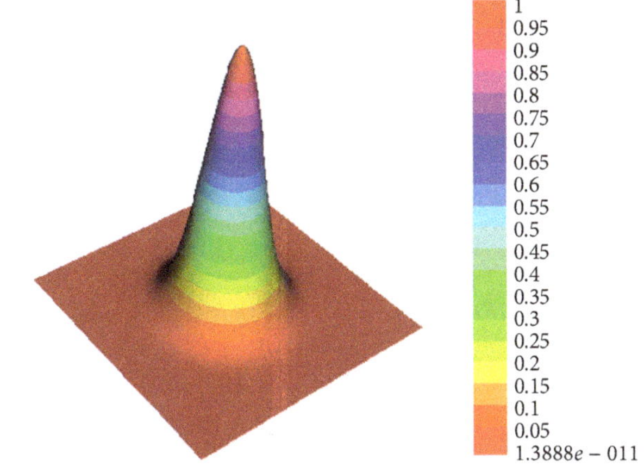

FIGURE 2: Initial condition: $u_h^0(x, y) = e^{-0.5(x^2+y^2)}$.

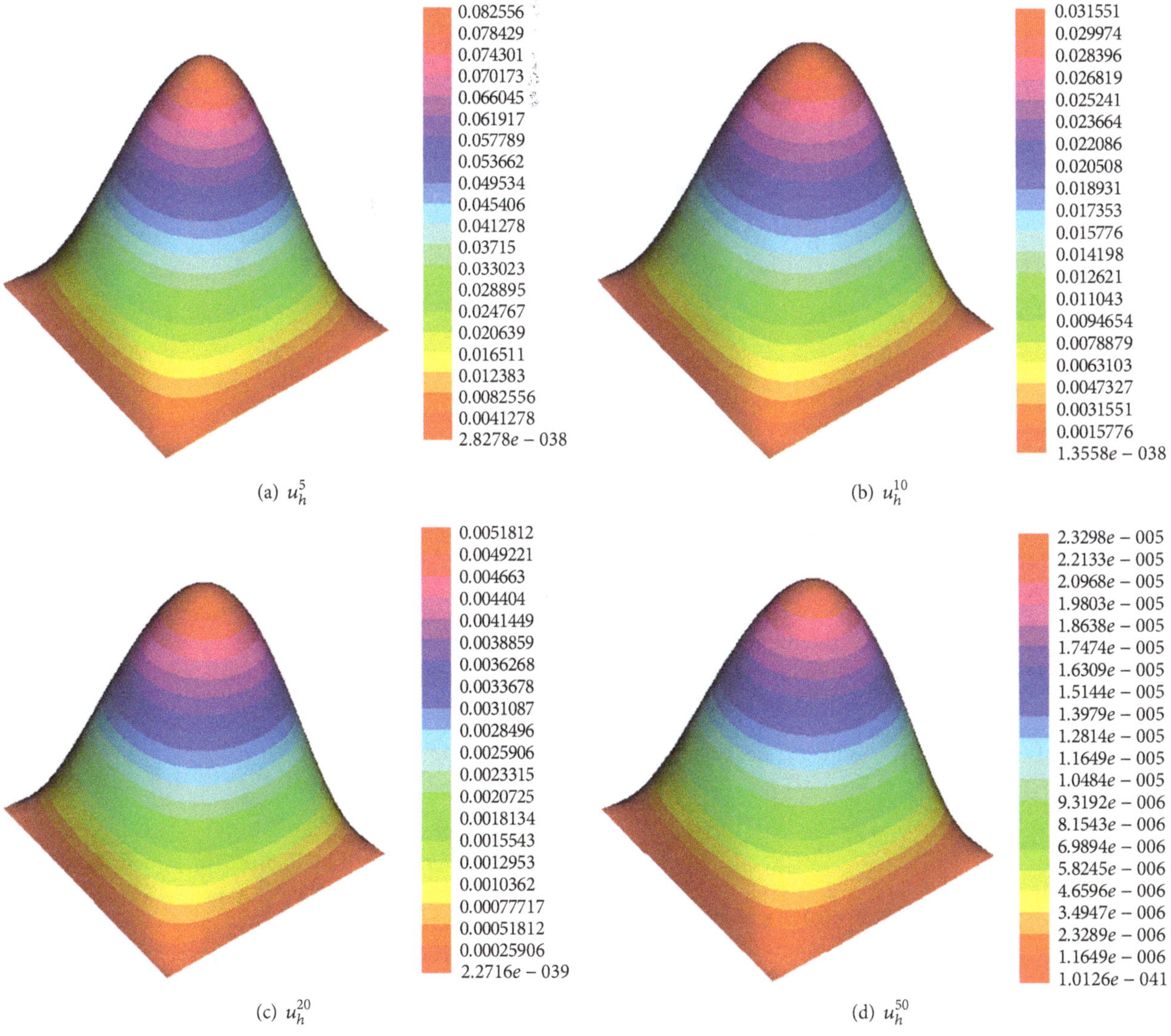

0.082556
0.078429
0.074301
0.070173
0.066045
0.061917
0.057789
0.053662
0.049534
0.045406
0.041278
0.03715
0.033023
0.028895
0.024767
0.020639
0.016511
0.012383
0.0082556
0.0041278
2.8278e – 038
(a) u_h^5
0.031551
0.029974
0.028396
0.026819
0.025241
0.023664
0.022086
0.020508
0.018931
0.017353
0.015776
0.014198
0.012621
0.011043
0.0094654
0.0078879
0.0063103
0.0047327
0.0031551
0.0015776
1.3558e – 038
(b) u_h^{10}
0.0051812
0.0049221
0.004663
0.004404
0.0041449
0.0038859
0.0036268
0.0033678
0.0031087
0.0028496
0.0025906
0.0023315
0.0020725
0.0018134
0.0015543
0.0012953
0.0010362
0.00077717
0.00051812
0.00025906
2.2716e – 039
(c) u_h^{20}
2.3298e – 005
2.2133e – 005
2.0968e – 005
1.9803e – 005
1.8638e – 005
1.7474e – 005
1.6309e – 005
1.5144e – 005
1.3979e – 005
1.2814e – 005
1.1649e – 005
1.0484e – 005
9.3192e – 006
8.1543e – 006
6.9894e – 006
5.8245e – 006
4.6596e – 006
3.4947e – 006
2.3289e – 006
1.1649e – 006
1.0126e – 041
(d) u_h^{50}

FIGURE 3: Numerical solutions for Application 1.

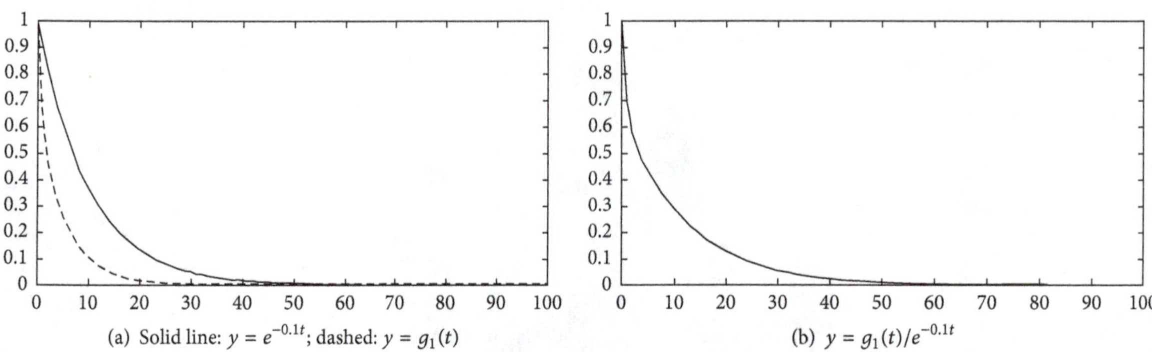

FIGURE 4: Exponential decay.

FIGURE 5: $p_h(x, y) = (1/5)\lceil x \rceil^2 + 2.5$.

The initial condition is taken to be $u_h^0(x, y) = e^{-0.5(x^2+y^2)}$ and projected into $P_1(\Omega_h)$; see Figure 2.

The numerical results are obtained using the noncommercial software, FreeFem++ [21].

Application 1. $p_h(x, y) = 2$ satisfies the required conditions (1)–(3). Figure 3 shows the numerical solutions for $t = 5$, $t = 10$, $t = 20$, and $t = 50$.

With $c = 0.1$, Figure 4(a) shows that u_h^n decays exponentially as

$$g_1(t) = \frac{\|u_h^n\|}{\|u_h^0\|} \leq e^{-0.1t}, \quad 0 \leq t \leq T. \tag{54}$$

This is also confirmed by Figure 4(b) that shows the ratio $y = g_1(t)/e^{-0.1t}$ is less than one and remains decreasing for a large value of T.

Application 2. The exponent function $p_h(x, y) = (1/5)\lceil x \rceil^2 + 2.5$, in Figure 5, satisfies the required conditions (1)–(3) as

(i) $p_1 = 2.5$, $p_2 = 7.5$;

(ii) $|p(x, y) - p(x_0, y_0)| = (1/5)|\lceil x \rceil^2 - \lceil x_0 \rceil^2| \leq -(20\sqrt{2}\log(1/\delta))/\log|x - y|$ for $|x - y| < \delta$ with $0 < \delta < 1$.

Figure 6 shows the numerical solutions for $t = 5$, $t = 10$, $t = 20$, and $t = 50$.

In this case, the solution u_h^n has a polynomial decay. With $c = 1$, Figure 7(a) shows that

$$g_2(t) = \frac{\|u_h^n\|}{\|u_h^0\|} \leq (1 + t)^{-2/11}, \quad 0 \leq t \leq T. \tag{55}$$

This is also confirmed by Figure 7(b), which shows that the ratio $y = g_2(t)/(1 + t)^{-2/11}$ remains less than one and decreasing until T.

We conclude that the numerical results in above applications verify Proposition 9.

Conflicts of Interest

The authors declare that they have no conflicts of interest.

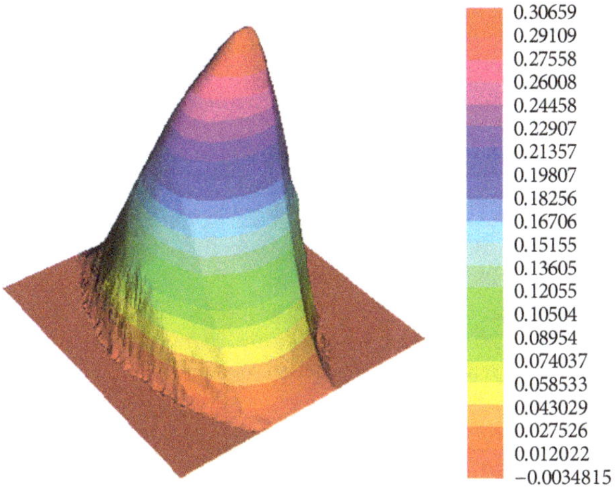

(a) u_h^5

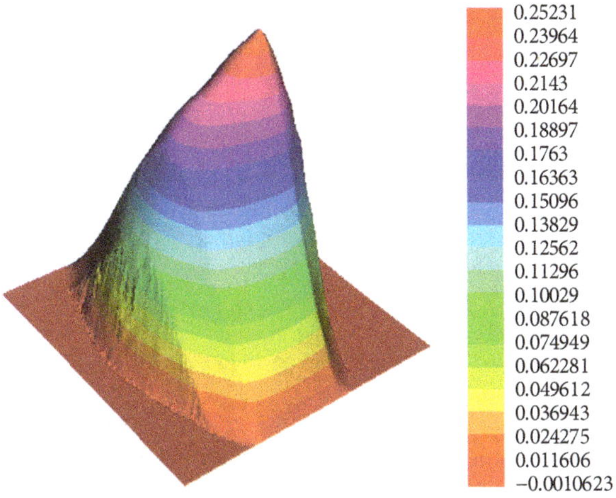

(b) u_h^{10}

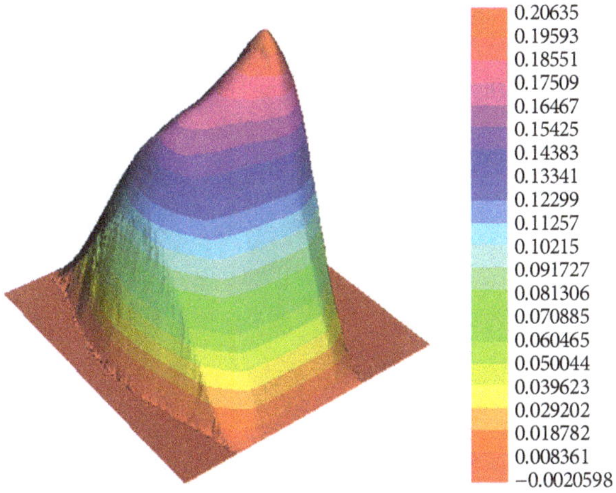

(c) u_h^{20}

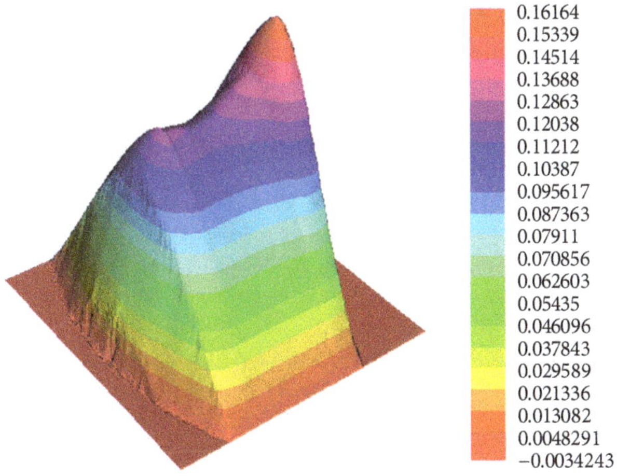

(d) u_h^{50}

FIGURE 6: Numerical solutions for Application 2.

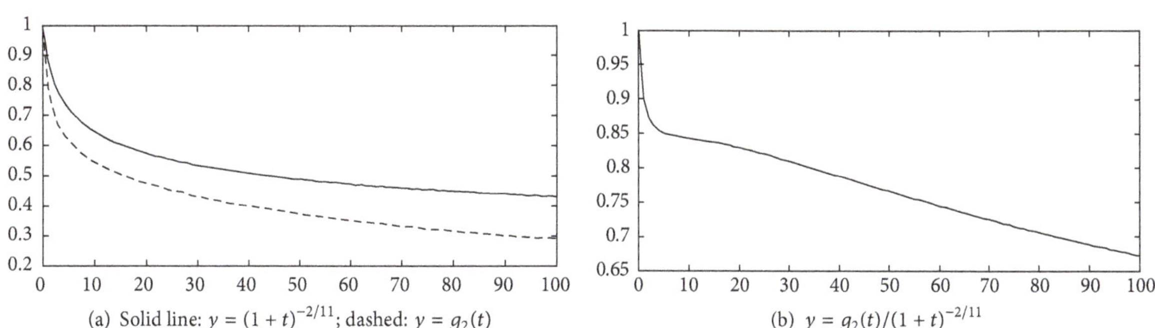

(a) Solid line: $y = (1+t)^{-2/11}$; dashed: $y = g_2(t)$

(b) $y = g_2(t)/(1+t)^{-2/11}$

FIGURE 7: Polynomial decay.

Acknowledgments

The authors acknowledge King Fahd University of Petroleum and Minerals for its support. This work is sponsored by KFUPM under Project no. FT 161004.

References

[1] M. K. Alaoui, S. A. Messaoudi, and H. B. Khenous, "A blow-up result for nonlinear generalized heat equation," *Computers & Mathematics with Applications. An International Journal*, vol. 68, no. 12, part A, pp. 1723–1732, 2014.

[2] J. P. Pinasco, "Blow-up for parabolic and hyperbolic problems with variable exponents," *Nonlinear Analysis. Theory, Methods & Applications. An International Multidisciplinary Journal*, vol. 71, no. 3-4, pp. 1094–1099, 2009.

[3] L. Shangerganesh, A. Gurusamy, and K. Balachandran, "Weak solutions for nonlinear parabolic equations with variable exponents," *Communications in Mathematics*, vol. 25, no. 1, pp. 55–70, 2017.

[4] E. Acerbi and G. Mingione, "Regularity results for a class of functionals with non-standard growth," *Archive for Rational Mechanics and Analysis*, vol. 156, no. 2, pp. 121–140, 2001.

[5] E. Acerbi and G. Mingione, "Regularity results for stationary electro-rheological fluids," *Archive for Rational Mechanics and Analysis*, vol. 164, no. 3, pp. 213–259, 2002.

[6] Y. Chen, S. Levine, and M. Rao, "Variable exponent, linear growth functionals in image restoration," *SIAM Journal on Applied Mathematics*, vol. 66, no. 4, pp. 1383–1406, 2006.

[7] L. Diening, F. Ettwein, and M. Ruzicka, "C1,α-regularity for electrorheological fluids in two dimensions," *Nonlinear Differential Equations and Applications NoDEA*, vol. 14, no. 1-2, pp. 207–217, 2007.

[8] F. Ettwein and M. Ruzicka, "Existence of local strong solutions for motions of electrorheological fluids in three dimensions," *Computers & Mathematics with Applications. An International Journal*, vol. 53, no. 3-4, pp. 595–604, 2007.

[9] Z. Guo, Q. Liu, J. Sun, and B. Wu, "Reaction-diffusion systems with $p(x)$-growth for image denoising," *Nonlinear Analysis: Real World Applications*, vol. 12, no. 5, pp. 2904–2918, 2011.

[10] L. Diening, P. Harjulehto, P. Hästö, and M. Růžička, "Lebesgue and Sobolev spaces with variable exponents," *Lecture Notes in Mathematics*, vol. 2017, pp. 1–518, 2011.

[11] M. Ružička, *Electrorheological Fluids: Modeling and Mathematical Theory*, vol. 1748 of *Lecture Notes in Mathematics*, Springer, Berlin, Germany, 2000.

[12] M. Bendahmane, P. Wittbold, and A. Zimmermann, "Renormalized solutions for a nonlinear parabolic equation with variable exponents and L^1-data," *Journal of Differential Equations*, vol. 249, no. 6, pp. 1483–1515, 2010.

[13] G. Akagi and K. Matsuura, "Well-posedness and large-time behaviors of solutions for a parabolic equation involving $p(x)$-Laplacian," *Discrete and Continuous Dynamical Systems - Series A, Dynamical systems, differential equations and applications. 8th AIMS Conference. Suppl. Vol. I*, pp. 22–31, 2011.

[14] D. E. Edmunds and J. Rakosnik, "Sobolev embeddings with variable exponent," *Studia Mathematica*, vol. 143, no. 3, pp. 267–293, 2000.

[15] D. E. Edmunds and J. Rakosnik, "Sobolev embeddings with variable exponent, II," *Mathematische Nachrichten*, vol. 246/247, pp. 53–67, 2002.

[16] X. Fan and D. Zhao, "On the spaces $L^{p(x)}(\Omega)$ and $W^{m,p(x)}(\Omega)$," *Journal of Mathematical Analysis and Applications*, vol. 263, no. 2, pp. 424–446, 2001.

[17] L. Simon, *Application of Monotone Type Operators to Nonlinear PDEs*, Prime Rate Kft., Budapest, Hungary, 2013.

[18] A. Quarteroni and A. Valli, *Numerical approximation of partial differential equations*, vol. 23 of *Springer Series in Computational Mathematics*, Springer-Verlag, Berlin, Germany, 1994.

[19] D. Li and J. Wang, "Unconditionally optimal error analysis of Crank-Nicolson Galerkin FEMs for a strongly nonlinear parabolic system," *Journal of Scientific Computing*, vol. 72, no. 2, pp. 892–915, 2017.

[20] J. Leray and J.-L. Lions, "Quelques résulatats de Visick sur les problémes elliptiques nonlinéaires par les méthodes de Minty-Browder," *Bulletin de la Société Mathématique de France*, vol. 93, pp. 97–107, 1965.

[21] F. Hecht, Freefem++ manual, http, http://www.freefem.org.

18

Dynamics of a Fractional Order HIV Infection Model with Specific Functional Response and Cure Rate

Adnane Boukhouima,[1] Khalid Hattaf,[1,2] and Noura Yousfi[1]

[1]*Laboratory of Analysis, Modeling and Simulation (LAMS), Faculty of Sciences Ben M'sik, Hassan II University, P.O. Box 7955, Sidi Othmane, Casablanca, Morocco*
[2]*Centre Régional des Métiers de l'Education et de la Formation (CRMEF), 20340 Derb Ghallef, Casablanca, Morocco*

Correspondence should be addressed to Khalid Hattaf; k.hattaf@yahoo.fr

Academic Editor: Yuji Liu

We propose a fractional order model in this paper to describe the dynamics of human immunodeficiency virus (HIV) infection. In the model, the infection transmission process is modeled by a specific functional response. First, we show that the model is mathematically and biologically well posed. Second, the local and global stabilities of the equilibria are investigated. Finally, some numerical simulations are presented in order to illustrate our theoretical results.

1. Introduction

Fractional order differential equations (FDEs) are a generalization of ordinary differential equations (ODEs) and they have many applications in various fields such as mechanics, image processing, viscoelasticity, bioengineering, finance, psychology, and control theory [1–7]. In addition, it has been deduced that the membranes of cells of biological organisms have fractional order electrical conductance [8].

Modeling by FDEs has more advantages to describe the dynamics of phenomena with memory which exists in most biological systems, because fractional order derivatives depend not only on local conditions but also on the past. More precisely, calculating the time-fractional derivative of a function $f(t)$ at some time $t = t_1$ requires all the previous history, that is, all $f(t)$ from $t = 0$ to $t = t_1$. In addition, the region of stability of FDEs is larger than that of ODEs. Moreover, some previous study compared between the results of the fractional order model, the results of the integer model, and the measured real data obtained from 10 patients during primary HIV infection [9]. This study proved that the results of the fractional order model give predictions to the plasma virus load of the patients better than those of the integer model.

From the above biological and mathematical reasons, we propose a fractional order model to describe the dynamics of HIV infection that is given by

$$\begin{aligned} D^{\alpha}T(t) &= \lambda - dT - \frac{\beta TV}{1+\alpha_1 T+\alpha_2 V+\alpha_3 TV} + \rho I, \\ D^{\alpha}I(t) &= \frac{\beta TV}{1+\alpha_1 T+\alpha_2 V+\alpha_3 TV} - (a+\rho)I, \\ D^{\alpha}V(t) &= kI - \mu V, \end{aligned} \tag{1}$$

where $T(t)$, $I(t)$, and $V(t)$ represent the concentrations of uninfected CD4$^+$ T-cells, infected cells, and free virus particles at time t, respectively. Uninfected cells are assumed to be produced at a constant rate λ, die at the rate dT, and become infected by a virus at the rate $\beta TV/(1+\alpha_1 T+\alpha_2 V+\alpha_3 TV)$, where $\alpha_1, \alpha_2, \alpha_3 \geq 0$ are the saturation factors measuring the psychological or inhibitory effect. Infected cells die at the rate aI and return to the uninfected state by loss of all covalently closed circular DNA (cccDNA) from their nucleus at the rate ρI. Free virus particles are produced from infected cells at the rate kI and cleared at the rate μV.

The fractional order derivative used in system (1) is in the sense of Caputo. We use this Caputo fractional derivative for two reasons: the first reason is that the fractional derivative of a constant is zero and the second reason is that the initial value problems depend on the integer order derivative only. In addition, we choose $0 < \alpha \leq 1$ in order to have the same initial conditions as ODE systems.

On the other hand, system (1) generalizes many special cases existing in the literature. For example, when $\alpha_1 = \alpha_2 = \alpha_3 = 0$, we get the model of Arafa et al. [10]. Further, we obtain the model of Liu et al. [11] when $\alpha_3 = 0$. It is very important to note that when $\alpha = 1$, system (1) becomes a model with an ordinary derivative which is the generalization of the ODE models presented in [12–15].

The rest of the paper is organized as follows. In the next section, we give some preliminary results. In Section 3, equilibria and their local stability are investigated. In Section 4, the global stability of the two equilibria is established. Numerical simulations of our theoretical results are presented in Section 5. Finally, the paper ends with conclusion in Section 6.

2. Preliminary Results

We first recall the definitions of the fractional order integral, Caputo fractional derivative, and Mittag-Leffler function that are given in [16].

Definition 1. The fractional integral of order $\alpha > 0$ of a function $f : \mathbb{R}_+ \to \mathbb{R}$ is defined as follows:

$$I^{\alpha} f(t) = \frac{1}{\Gamma(\alpha)} \int_0^t (t-x)^{\alpha-1} f(x)\, dx, \tag{2}$$

where $\Gamma(\cdot)$ is the Gamma function.

Definition 2. The Caputo fractional derivative of order $\alpha > 0$ of a continuous function $f : \mathbb{R}_+ \to \mathbb{R}$ is given by

$$D^{\alpha} f(t) = I^{n-\alpha} D^{n} f(t), \tag{3}$$

where $D = d/dt$ and $n-1 < \alpha \leq n$, $n \in \mathbb{N}$.

In particular, when $0 < \alpha \leq 1$, we have

$$D^{\alpha} f(t) = \frac{1}{\Gamma(1-\alpha)} \int_0^t \frac{f'(x)}{(t-x)^{\alpha}} dx. \tag{4}$$

Definition 3. Let $\alpha > 0$. The function E_α, defined by

$$E_{\alpha}(z) = \sum_{j=0}^{\infty} \frac{z^j}{\Gamma(\alpha j + 1)}, \tag{5}$$

is called the Mittag-Leffler function of parameter α.

Let $f : \mathbb{R}^n \to \mathbb{R}^n$ with $n \geq 1$. Consider the fractional order system

$$\begin{aligned} D^{\alpha} x(t) &= f(x), \\ x(t_0) &= x_0, \end{aligned} \tag{6}$$

with $0 < \alpha \leq 1$, $t_0 \in \mathbb{R}$, and $x_0 \in \mathbb{R}^n$. For the global existence of solution of system (6), we need the following lemma.

Lemma 4. *Assume that f satisfies the following conditions:*

(i) *$f(x)$ and $(\partial f/\partial x)(x)$ are continuous for all $x \in \mathbb{R}^n$.*

(ii) *$\|f(x)\| \leq \omega + \lambda \|x\|$ for all $x \in \mathbb{R}^n$, where ω and λ are two positive constants.*

Then, system (6) has a unique solution on $[t_0, +\infty)$.

The proof of this lemma follows immediately from [17]. For biological reasons, we assume that the initial conditions of system (1) satisfy

$$\begin{aligned} T(0) &= \phi_1(0) \geq 0, \\ I(0) &= \phi_2(0) \geq 0, \\ V(0) &= \phi_3(0) \geq 0. \end{aligned} \tag{7}$$

In order to establish the nonnegativity of solutions with initial conditions (7), we need also the following lemmas.

Lemma 5 (see [18]). *Suppose that $g(t) \in C[a,b]$ and $D^{\alpha} g(t) \in C[a,b]$ for $0 < \alpha \leq 1$; then, one has*

$$\begin{aligned} g(t) = g(a) + \frac{1}{\Gamma(\alpha)} D^{\alpha} g(\xi)(t-a)^{\alpha}, \\ a < \xi < t,\ \forall t \in (a,b]. \end{aligned} \tag{8}$$

Lemma 6 (see [18]). *Suppose that $g(t) \in C[a,b]$ and $D^{\alpha} g(t) \in C[a,b]$ for $0 < \alpha \leq 1$. If $D^{\alpha} g(t) \geq 0$ $\forall t \in [a,b]$, then $g(t)$ is nondecreasing for each $t \in [a,b]$. If $D^{\alpha} g(t) \leq 0$ $\forall t \in [a,b]$, then $g(t)$ is nonincreasing for each $t \in [a,b]$.*

Theorem 7. *For any initial conditions satisfying (7), system (1) has a unique solution on $[0,+\infty)$. Moreover, this solution remains nonnegative and bounded for all $t \geq 0$. In addition, one has*

(i) *$N(t) \leq N(0) + \lambda/\delta$,*

(ii) *$V(t) \leq V(0) + (k/\mu)\|I\|_{\infty}$,*

where $N(t) = T(t) + I(t)$ and $\delta = \min\{a, d\}$.

Proof. It is easy to see that the vector function of system (1) satisfies the first condition of Lemma 4. It remains to prove the second condition. Let

$$\begin{aligned} X(t) &= \begin{pmatrix} T(t) \\ I(t) \\ V(t) \end{pmatrix}, \\ \zeta &= \begin{pmatrix} \lambda \\ 0 \\ 0 \end{pmatrix}. \end{aligned} \tag{9}$$

To this end, we discuss four cases:

(i) If $\alpha_1 \neq 0$, then system (1) can be written as follows:

$$D^{\alpha} X(t) = \zeta + A_1 X + \frac{\alpha_1 T}{1 + \alpha_1 T + \alpha_2 V + \alpha_3 TV} A_2 X, \tag{10}$$

where

$$A_1 = \begin{pmatrix} -d & \rho & 0 \\ 0 & -(a+\rho) & 0 \\ 0 & k & -\mu \end{pmatrix},$$

$$A_2 = \begin{pmatrix} 0 & 0 & -\dfrac{\beta}{\alpha_1} \\ 0 & 0 & \dfrac{\beta}{\alpha_1} \\ 0 & 0 & 0 \end{pmatrix}. \quad (11)$$

Moreover, we have

$$\|D^\alpha X(t)\| \le \|\zeta\| + (\|A_1\| + \|A_2\|)\|X\|. \quad (12)$$

(ii) If $\alpha_2 \neq 0$, we have

$$D^\alpha X(t) = \zeta + A_1 X + \frac{\alpha_2 T}{1 + \alpha_1 T + \alpha_2 V + \alpha_3 TV} A_3 X, \quad (13)$$

where

$$A_3 = \begin{pmatrix} 0 & 0 & -\dfrac{\beta}{\alpha_2} \\ 0 & 0 & \dfrac{\beta}{\alpha_2} \\ 0 & 0 & 0 \end{pmatrix}. \quad (14)$$

Then,

$$\|D^\alpha X(t)\| \le \|\zeta\| + (\|A_1\| + \|A_3\|)\|X\|. \quad (15)$$

(iii) If $\alpha_3 \neq 0$, we have

$$D^\alpha X(t) = \zeta + A_1 X + \frac{\alpha_3 TV}{1 + \alpha_1 T + \alpha_2 V + \alpha_3 TV} A_4, \quad (16)$$

where

$$A_4 = \begin{pmatrix} -\dfrac{\beta}{\alpha_3} \\ \dfrac{\beta}{\alpha_3} \\ 0 \end{pmatrix}. \quad (17)$$

Then,

$$\|D^\alpha X(t)\| \le (\|\zeta\| + \|A_4\|) + \|A_1\|\|X\|. \quad (18)$$

(iv) If $\alpha_1 = \alpha_2 = \alpha_3 = 0$, we have

$$D^\alpha X(t) = \zeta + A_1 X + V A_5 X, \quad (19)$$

where

$$A_5 = \begin{pmatrix} -\beta & 0 & 0 \\ \beta & 0 & 0 \\ 0 & 0 & 0 \end{pmatrix}. \quad (20)$$

Then,

$$\|D^\alpha X(t)\| \le \|\zeta\| + (\|V\|\|A_5\| + \|A_1\|)\|X\|. \quad (21)$$

Thus, the second condition of Lemma 4 is satisfied. Then, system (1) has a unique solution on $[0, +\infty)$. Next, we show that this solution is nonnegative. From (1), we have

$$\begin{aligned} D^\alpha T(t)\big|_{T=0} &= \lambda + \rho I \ge 0, \\ D^\alpha I(t)\big|_{I=0} &= \frac{\beta TV}{1 + \alpha_1 T + \alpha_2 V + \alpha_3 TV} \ge 0, \\ D^\alpha V(t)\big|_{V=0} &= kI \ge 0. \end{aligned} \quad (22)$$

According to Lemmas 5 and 6, we deduce that the solution of (1) is nonnegative.

Finally, we prove that the solution is bounded. By adding the first two equations of system (1), we get

$$D^\alpha N(t) \le \lambda - \delta N(t). \quad (23)$$

Hence,

$$N(t) \le N(0) E_\alpha(-\delta t^\alpha) + \frac{\lambda}{\delta}\left[1 - E_\alpha(-\delta t^\alpha)\right]. \quad (24)$$

Since $0 \le E_\alpha(-\delta t^\alpha) \le 1$, we have

$$N(t) \le N(0) + \frac{\lambda}{\delta}. \quad (25)$$

The third equation of system (1) implies that

$$\begin{aligned} V(t) &= V(0) E_\alpha(-\mu t^\alpha) \\ &\quad + k \int_0^t I(s)\, \alpha (t-s)^{\alpha-1} \frac{dE_\alpha}{ds}(-\mu (t-s)^\alpha)\, ds. \end{aligned} \quad (26)$$

Then,

$$V(t) \le V(0) E_\alpha(-\mu t^\alpha) + \frac{k\|I\|_\infty}{\mu}(1 - E_\alpha(-\mu t^\alpha)). \quad (27)$$

Consequently,

$$V(t) \le V(0) + \frac{k\|I\|_\infty}{\mu}. \quad (28)$$

This completes the proof. □

3. Equilibria and Their Local Stability

It is easy to see that system (1) always has a disease-free equilibrium $E_0(\lambda/d, 0, 0)$. Therefore, the basic reproduction number of our system (1) is given by

$$R_0 = \frac{k\beta\lambda}{\mu(a+\rho)(d+\lambda\alpha_1)}. \quad (29)$$

Biologically, this basic reproduction number represents the average number of secondary infections produced by one infected cell during the period of infection when all cells are uninfected. Further, it is not hard to get the following result.

Theorem 8. *(i) If $R_0 \leq 1$, system (1) has a unique disease-free equilibrium of the form $E_0(T_0, 0, 0)$, where $T_0 = \lambda/d$. (ii) If $R_0 > 1$, the disease-free equilibrium is still present and system (1) has a unique chronic infection equilibrium of the form $E_1(T_1, (\lambda - dT_1)/a,\ k(\lambda - dT_1)/a\mu)$, where $T_1 = 2(a+\rho)(a\mu+\alpha_2\lambda k)/(ak\beta+(a+\rho)(\alpha_2 dk-\alpha_1 a\mu-\alpha_3 k\lambda)+\sqrt{\overline{\delta}})$ with*

$$\begin{aligned}\overline{\delta} &= (ak\beta + (a+\rho)(\alpha_2 dk - \alpha_1 a\mu - \alpha_3 k\lambda))^2 \\ &\quad + 4\alpha_3 kd(a+\rho)^2(a\mu + \alpha_2\lambda k).\end{aligned} \tag{30}$$

Next, we investigate the local stability of equilibria. Let $E_e(T, I, V)$ be an arbitrary equilibrium of system (1). Then, the characteristic equation at E_e is given by

$$\begin{vmatrix} -d - V\dfrac{\partial f}{\partial T} - \xi & \rho & -V\dfrac{\partial f}{\partial V} - f(T,V) \\ V\dfrac{\partial f}{\partial T} & -(a+\rho) - \xi & V\dfrac{\partial f}{\partial V} + f(T,V) \\ 0 & k & -\mu - \xi \end{vmatrix} = 0, \tag{31}$$

where

$$f(T,V) = \frac{\beta T}{1 + \alpha_1 T + \alpha_2 V + \alpha_3 TV}. \tag{32}$$

We recall that the equilibrium E_e is locally asymptotically stable if all roots ξ_i of (31) satisfy the following condition [19]:

$$|\arg(\xi_i)| > \frac{\alpha\pi}{2}. \tag{33}$$

Theorem 9. *(i) If $R_0 < 1$, then E_0 is locally asymptotically stable. (ii) If $R_0 > 1$, then E_0 is unstable.*

Proof. Evaluating (31) at E_0, we have

$$(d+\xi)\left[\xi^2 + (a+\rho+\mu)\xi + \mu(a+\rho)(1-R_0)\right] = 0. \tag{34}$$

Obviously, the roots of (34) are

$$\begin{aligned}\xi_1 &= -d, \\ \xi_2 &= \frac{-(a+\rho+\mu) - \sqrt{(a+\rho+\mu)^2 - 4\mu(a+\rho)(1-R_0)}}{2}, \\ \xi_3 &= \frac{-(a+\rho+\mu) + \sqrt{(a+\rho+\mu)^2 - 4u(a+\rho)(1-R_0)}}{2}.\end{aligned} \tag{35}$$

It is clear that ξ_1 and ξ_2 are negative. However, ξ_3 is negative if $R_0 < 1$ and it is positive if $R_0 > 1$. Therefore, E_0 is locally asymptotically stable if $R_0 < 1$ and unstable if $R_0 > 1$. □

Now, we focus on the local stability of the chronic infection equilibrium E_1. It follows from (31) that the characteristic equation at E_1 is given by

$$P(\xi) := \xi^3 + a_1\xi^2 + a_2\xi + a_3 = 0, \tag{36}$$

where

$$\begin{aligned}a_1 &= \mu + d + a + \rho + \frac{\beta V_1(1+\alpha_2 V_1)}{(1+\alpha_1 T_1 + \alpha_2 V_1 + \alpha_3 T_1 V_1)^2}, \\ a_2 &= d(\mu + a + \rho) \\ &\quad + \frac{\beta V_1[(a+\mu)(1+\alpha_2 V_1) + kT_1(\alpha_2 + \alpha_3 T_1)]}{(1+\alpha_1 T_1 + \alpha_2 V_1 + \alpha_3 T_1 V_1)^2}, \\ a_3 &= \frac{\beta V_1[a\mu(1+\alpha_2 V_1) + kdT_1(\alpha_2 + \alpha_3 T_1)]}{(1+\alpha_1 T_1 + \alpha_2 V_1 + \alpha_3 T_1 V_1)^2}.\end{aligned} \tag{37}$$

It is obvious that $a_1 > 0$, $a_2 > 0$, and $a_3 > 0$. Further, we have

$$\begin{aligned}a_1 a_2 - a_3 &= a\Bigg(d(\mu + a + \rho) \\ &\quad + \frac{\beta V_1[a(1+\alpha_2 V_1) + kT_1(\alpha_2 + \alpha_3 T_1)]}{(1+\alpha_1 T_1 + \alpha_2 V_1 + \alpha_3 T_1 V_1)^2}\Bigg) \\ &\quad + d\Bigg(d(\mu + a + \rho) \\ &\quad + \frac{(a+\mu)\beta V_1(1+\alpha_2 V_1)}{(1+\alpha_1 T_1 + \alpha_2 V_1 + \alpha_3 T_1 V_1)^2}\Bigg) + \Bigg(\mu + \rho \\ &\quad + \frac{\beta V_1(1+\alpha_2 V_1)}{(1+\alpha_1 T_1 + \alpha_2 V_1 + \alpha_3 T_1 V_1)^2}\Bigg) h_2 > 0.\end{aligned} \tag{38}$$

So, Routh–Hurwitz conditions are satisfied. Let $D(P)$ denote the discriminant of the polynomial P given by (36); then,

$$D(P) = 18a_1a_2a_3 + (a_1a_2)^2 - 4a_3a_1^3 - 4a_2^3 - 27a_3^2. \tag{39}$$

Using the results in [19], we easily obtain the following result.

Theorem 10. *Assume that $R_0 > 1$.*

(i) *If $D(P) > 0$, then E_1 is locally asymptotically stable for all $\alpha \in (0, 1]$.*

(ii) *If $D(P) < 0$ and $\alpha < 2/3$, then E_1 is locally asymptotically stable.*

4. Global Stability

In this section, we study the global stability of the disease-free equilibrium E_0 and the chronic infection equilibrium E_1.

Theorem 11. *If $R_0 \leq 1$, then the disease-free equilibrium E_0 is globally asymptotically stable.*

Proof. Define Lyapunov functional $L_0(t)$ as follows:

$$L_0(t) = \frac{T_0}{1+\alpha_1 T_0}\Phi\left(\frac{T}{T_0}\right) + \frac{\rho}{2(1+\alpha_1 T_0)(a+d)T_0}(T - T_0 + I)^2 + \frac{a+\rho}{k}V, \tag{40}$$

where $\Phi(x) = x - 1 - \ln(x)$, $x > 0$. Calculating the derivative of $L_0(t)$ along solutions of system (1) and using the results in [20], we get

$$D^\alpha L_0(t) \le \frac{1}{1+\alpha_1 T_0}\left(1 - \frac{T_0}{T}\right)D^\alpha T + D^\alpha I \cdot \frac{\rho}{(1+\alpha_1 T_0)(a+d)T_0}(T - T_0 + I) \cdot (D^\alpha T + D^\alpha I) + \frac{a+\rho}{k}D^\alpha V. \tag{41}$$

Using $\lambda = dT_0$, we obtain

$$\begin{aligned} D^\alpha L_0(t) &\le -\frac{d(T-T_0)^2}{(1+\alpha_1 T_0)T} - \frac{1}{1+\alpha_1 T_0}\left(1-\frac{T_0}{T}\right)f(T,V)V \\ &+ \frac{\rho}{1+\alpha_1 T_0}\left(1-\frac{T_0}{T}\right)I + f(T,V)V \\ &- \frac{d\rho(T-T_0)^2}{(a+d)T_0(1+\alpha_1 T_0)} \\ &- \frac{a\rho I^2}{(a+d)T_0(1+\alpha_1 T_0)} \\ &+ \frac{\rho}{T_0(1+\alpha_1 T_0)}I(T_0 - T) - \frac{(a+\rho)\mu}{k}V, \end{aligned} \tag{42}$$

$$\begin{aligned} D^\alpha L_0(t) &\le -\left(\frac{1}{T} + \frac{\rho}{(a+d)T_0}\right)\frac{d(T-T_0)^2}{1+\alpha_1 T_0} \\ &- \frac{\rho I(T-T_0)^2}{TT_0(1+\alpha_1 T_0)} - \frac{a\rho I^2}{(a+d)T_0(1+\alpha_1 T_0)} \\ &+ \frac{(a+\rho)\mu}{k}(R_0 - 1)V \\ &- \frac{\beta T_0(\alpha_2 + \alpha_3 T)}{(1+\alpha_1 T_0)(1+\alpha_1 T + \alpha_2 V + \alpha_3 TV)}V^2. \end{aligned}$$

Hence, if $R_0 \le 1$, then $D^\alpha L_0(t) \le 0$. Furthermore, it is clear that the largest invariant set of $\{(T,I,V) \in D : D^\alpha L_0(t) = 0\}$ is the singleton $\{E_0\}$. Therefore, by LaSalle's invariance principle [21], E_0 is globally asymptotically stable. □

Theorem 12. *The chronic infection equilibrium E_1 is globally asymptotically stable if $R_0 > 1$ and*

$$R_0 \le 1 + \frac{d(a+\rho)(\lambda\alpha_2 k + a\mu) + \rho\lambda^2\alpha_3 k}{\rho\mu(a+\rho)(d+\lambda\alpha_1)}. \tag{43}$$

Proof. Define Lyapunov functional $L_1(t)$ as follows:

$$\begin{aligned} L_1(t) &= \frac{1+\alpha_2 V_1}{1+\alpha_1 T_1 + \alpha_2 V_1 + \alpha_3 T_1 V_1}T_1\Phi\left(\frac{T}{T_1}\right) \\ &+ I_1\Phi\left(\frac{I}{I_1}\right) + \frac{a+\rho}{k}V_1\Phi\left(\frac{V}{V_1}\right) \\ &+ \frac{\rho(1+\alpha_2 V_1)}{2T_1(a+d)(1+\alpha_1 T_1 + \alpha_2 V_1 + \alpha_3 T_1 V_1)}(T - T_1 + I - I_1)^2. \end{aligned} \tag{44}$$

Then, we have

$$\begin{aligned} D^\alpha L_1(t) &\le \frac{1+\alpha_2 V_1}{1+\alpha_1 T_1 + \alpha_2 V_1 + \alpha_3 T_1 V_1}\left(1-\frac{T_1}{T}\right) \cdot D^\alpha T + \left(1 - \frac{I_1}{I}\right)D^\alpha I \\ &\cdot \frac{\rho(1+\alpha_2 V_1)}{T_1(a+d)(1+\alpha_1 T_1 + \alpha_2 V_1 + \alpha_3 T_1 V_1)}(T - T_1 + I - I_1)(D^\alpha T + D^\alpha I) + \frac{a+\rho}{k}\left(1-\frac{V_1}{V}\right)D^\alpha V. \end{aligned} \tag{45}$$

Using $\lambda = dT_1 + aI_1$, $f(T_1,V_1)V_1 = (a+\rho)I_1$, $\mu/k = I_1/V_1$, and $1 - f(T_1,V_1)/f(T,V_1) = ((1+\alpha_2 V_1)/(1+\alpha_1 T_1 + \alpha_2 V_1 + \alpha_3 T_1 V_1))(1 - T_1/T)$, we get

$$\begin{aligned} D^\alpha L_1(t) &\le d\left(1 - \frac{f(T_1,V_1)}{f(T,V_1)}\right)(T_1 - T) + (a+\rho) \\ &\cdot I_1\left(4 - \frac{f(T_1,V_1)}{f(T,V_1)} - \frac{I_1}{I}\frac{V}{V_1}\frac{f(T,V)}{f(T_1,V_1)} - \frac{I}{I_1}\frac{V_1}{V} - \frac{f(T,V_1)}{f(T,V)}\right) + (a+\rho)I_1\left(-1 - \frac{V}{V_1} + \frac{f(T,V_1)}{f(T,V)} + \frac{V}{V_1}\frac{f(T,V)}{f(T,V_1)}\right) \\ &- \frac{d\rho(1+\alpha_2 V_1)}{T_1(a+d)(1+\alpha_1 T_1 + \alpha_2 V_1 + \alpha_3 T_1 V_1)}(T - T_1)^2 \\ &- \frac{a\rho(1+\alpha_2 V_1)}{T_1(a+d)(1+\alpha_1 T_1 + \alpha_2 V_1 + \alpha_3 T_1 V_1)}(I - I_1)^2 \\ &- \frac{\rho(1+\alpha_2 V_1)}{TT_1(1+\alpha_1 T_1 + \alpha_2 V_1 + \alpha_3 T_1 V_1)}(T - T_1)^2(I - I_1). \end{aligned} \tag{46}$$

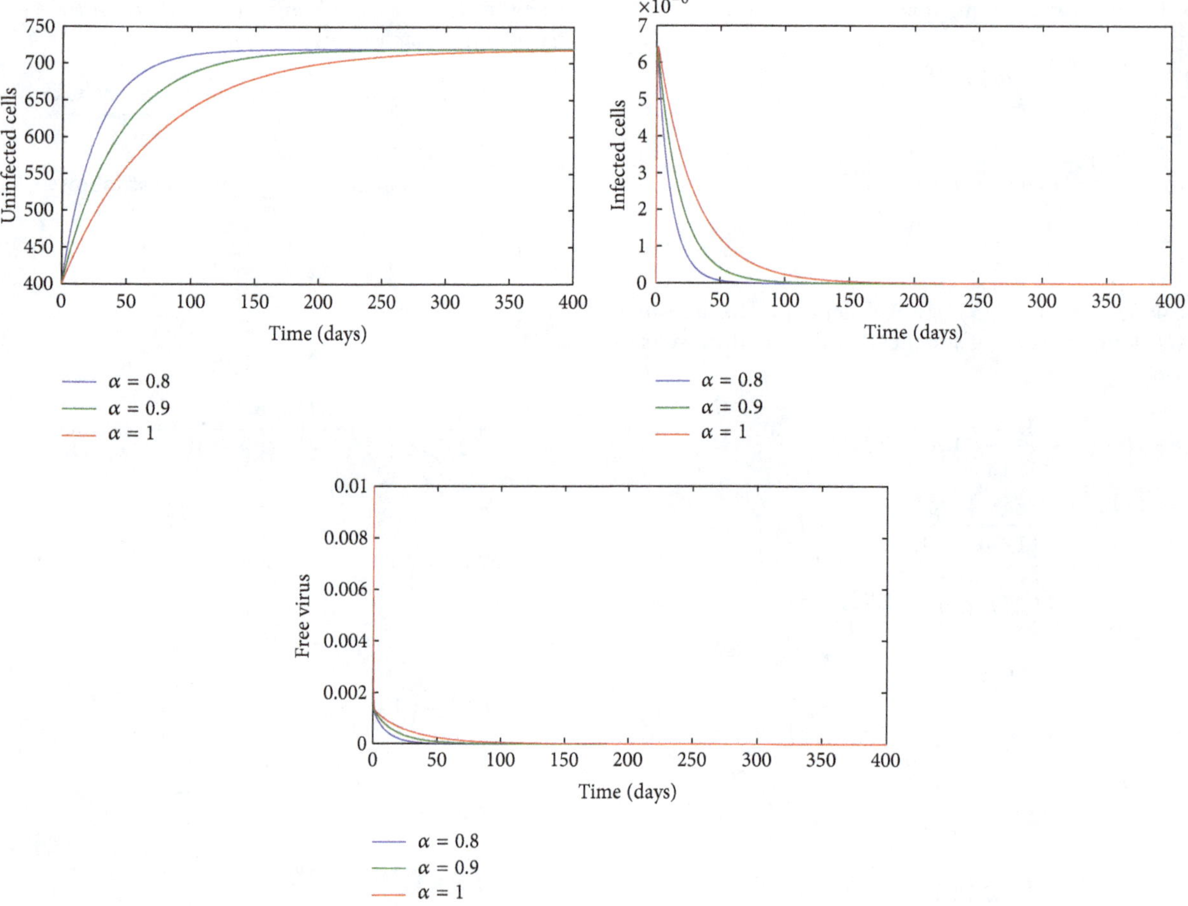

FIGURE 1: Stability of the disease-free equilibrium E_0.

Thus,

$$\begin{aligned}
D^{\alpha}L_1(t) &\le -\frac{(1+\alpha_2 V_1)(T-T_1)^2}{TT_1(1+\alpha_1 T_1+\alpha_2 V_1+\alpha_3 T_1 V_1)}\Bigg((dT_1-\rho I_1) \\
&+\frac{d\rho T}{d+a}+\rho I\Bigg)-(a+\rho)I_1 \\
&\cdot\frac{(1+\alpha_1 T)(\alpha_2+\alpha_3 T)(V-V_1)^2}{V_1(1+\alpha_1 T+\alpha_2 V+\alpha_3 TV)(1+\alpha_1 T+\alpha_2 V_1+\alpha_3 TV_1)} \\
&-(a+\rho)I_1\left(\Phi\left(\frac{f(T_1,V_1)}{f(T,V_1)}\right)+\Phi\left(\frac{I_1}{I}\frac{V}{V_1}\frac{f(T,V)}{f(T_1,V_1)}\right)\right. \\
&\left.+\Phi\left(\frac{I}{I_1}\frac{V_1}{V}\right)+\Phi\left(\frac{f(T,V_1)}{f(T,V)}\right)\right) \\
&-\frac{a\rho(1+\alpha_2 V_1)}{T_1(a+d)(1+\alpha_1 T_1+\alpha_2 V_1+\alpha_3 T_1 V_1)}(I-I_1)^2.
\end{aligned} \tag{47}$$

It is clear that $\Phi(x) \ge 0$. Consequently, $D^{\alpha}L_1(t) \le 0$ if $dT_1 \ge \rho I_1$. In addition, it is easy to see that this condition is equivalent to (43). Further, the largest invariant set of $\{(T,I,V) \in D : D^{\alpha}L_1(t) = 0\}$ is the singleton $\{E_1\}$. By LaSalle's invariance principle, E_1 is globally asymptotically stable. □

It is important to see that

$$\lim_{\rho\to 0}\frac{d(a+\rho)(\lambda\alpha_2 k+a\mu)+\rho\lambda^2\alpha_3 k}{\rho\mu(a+\rho)(d+\lambda\alpha_1)} = +\infty. \tag{48}$$

According to Theorem 12, we obtain the following result.

Corollary 13. *The chronic infection equilibrium E_1 is globally asymptotically stable when $R_0 > 1$ and ρ is sufficiently small.*

5. Numerical Simulations

In this section, we give some numerical simulations in order to illustrate our theoretical results. We discretize system (1) by using fractional Euler's method presented in [22]. Firstly, we take the parameter values as shown in Table 1.

By calculation, we have $R_0 = 0.9283 < 1$. Then, system (1) has a disease-free equilibrium $E_0(719.4245, 0, 0)$. By Theorem 11, the solution of (1) converges to E_0 (see Figure 1). Consequently, the virus is cleared and the infection dies out.

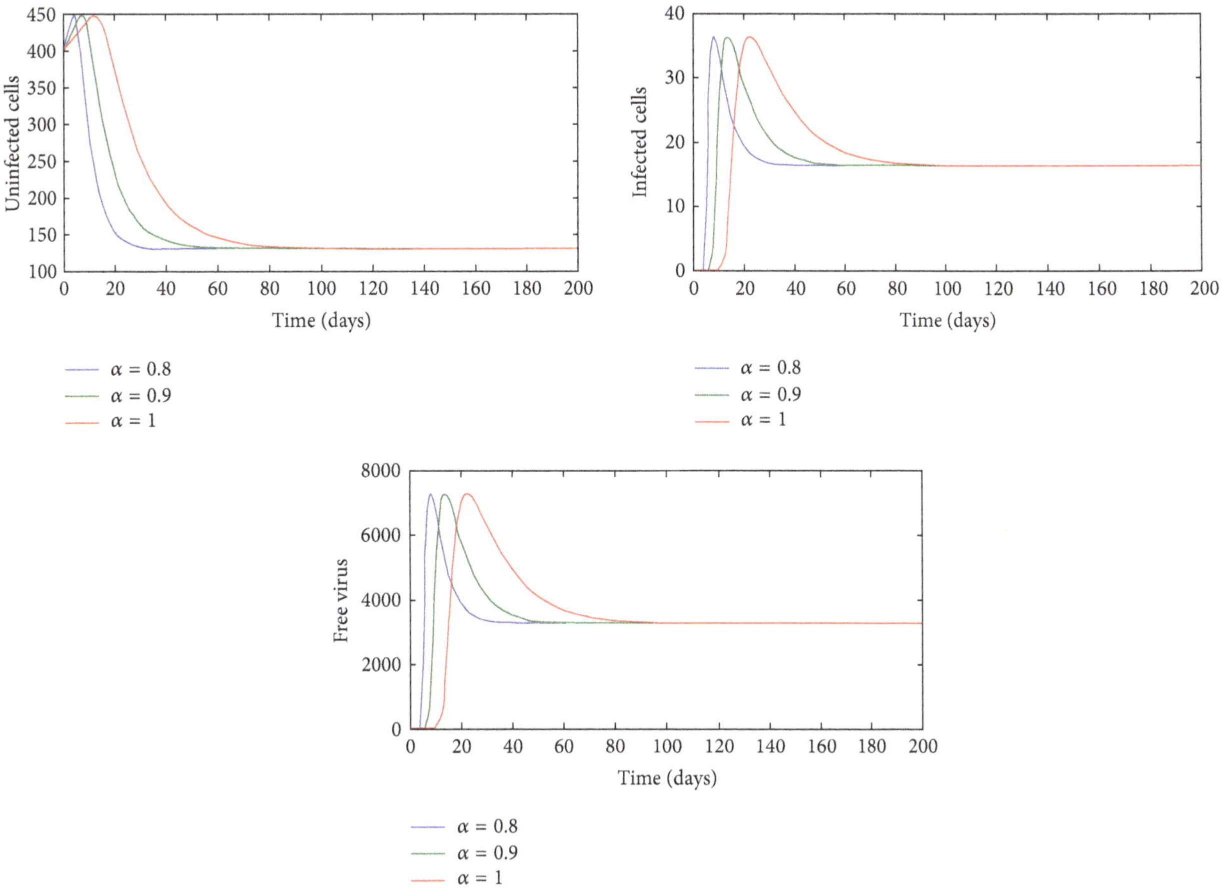

FIGURE 2: Stability of the chronic infection equilibrium E_1.

TABLE 1: Parameter values of system (1).

Parameters	Values
λ	10
d	0.0139
β	0.00024
ρ	0.01
a	0.5
k	600
u	3
α_1	0.1
α_2	0.01
α_3	0.00001

Now, we choose $\beta = 0.001$ and we keep the other parameter values. In this case, $R_0 = 3.8678$ and

$$1 + \frac{d(a+\rho)(\lambda\alpha_2 k + a\mu) + \rho\lambda^2\alpha_3 k}{\rho\mu(a+\rho)(d+\lambda\alpha_1)} = 415.885. \quad (49)$$

Hence, condition (43) is satisfied. Therefore, the chronic infection equilibrium $E_1(130.1613, 16.3815, 3276.3)$ is globally asymptotically stable. Figure 2 demonstrates this result.

6. Conclusion

In this paper, we have proposed a fractional order model of HIV infection with specific functional response and cure rate. This functional response covers the most functional responses used by several authors such as the saturated incidence rate, the Beddington-DeAngelis functional response, and the Crowley-Martin functional response. We have shown that the proposed model has a bounded and nonnegative solution as desired in any population dynamics. By using stability analysis of fractional order system, we have proved that if the basic reproduction number $R_0 \leq 1$, the disease-free equilibrium E_0 is globally asymptotically stable for all $\alpha \in (0, 1]$, which means that the virus is cleared and the infection dies out. However, when $R_0 > 1$, the disease-free equilibrium E_0 becomes unstable and there exists another biological equilibrium, namely, chronic infection equilibrium E_1, that is globally asymptotically stable provided that condition (43) is satisfied. In this case, the HIV virus persists in the host and the infection becomes chronic. Furthermore, we have remarked that if the cure rate ρ is equal to zero or is sufficiently small, condition (43) is satisfied and the global stability of E_1 is only characterized by $R_0 > 1$.

According to the above theoretical analysis, we deduce that the global dynamics of the model are fully determined by the basic reproduction number R_0. In addition, we see that the fractional order parameter α has no effect on the global dynamics of our model, but it can affect the time for arriving at both steady states (see Figures 1 and 2). Moreover, the fractional order model and main results presented by Liu et al. in [11] are generalized and improved.

Conflicts of Interest

The authors declare that there are no conflicts of interest regarding the publication of this paper.

References

[1] Y. A. Rossikhin and M. V. Shitikova, "Applications of fractional calculus to dynamic problems of linear and nonlinear hereditary mechanics of solids," *Applied Mechanics Reviews*, vol. 50, no. 1, pp. 15–67, 1997.

[2] R. J. Marks and M. W. Hall, "Differintegral Interpolation from a Bandlimited Signal's Samples," *IEEE Transactions on Acoustics, Speech, and Signal Processing*, vol. 29, no. 4, pp. 872–877, 1981.

[3] G. L. Jia and Y. X. Ming, "Study on the viscoelasticity of cancellous bone based on higher-order fractional models," in *Proceedings of the 2nd International Conference on Bioinformatics and Biomedical Engineering, iCBBE 2008*, pp. 1733–1736, chn, May 2008.

[4] R. Magin, *Fractional Calculus in Bioengineering, Cretical Reviews in Biomedical Engineering 32*, vol. 32, 2004.

[5] E. Scalas, R. Gorenflo, and F. Mainardi, "Fractional calculus and continuous-time finance," *Physica A. Statistical Mechanics and its Applications*, vol. 284, no. 1-4, pp. 376–384, 2000.

[6] L. Song, S. Xu, and J. Yang, "Dynamical models of happiness with fractional order," *Communications in Nonlinear Science and Numerical Simulation*, vol. 15, no. 3, pp. 616–628, 2010.

[7] R. Capponetto, G. Dongola, L. Fortuna, and I. Petras, "Fractional order systems: Modelling and control applications," in *World Scientific Series in Nonlinear Science*, vol. 72, Series A, Singapore, 2010.

[8] K. S. Cole, "Electric conductance of biological systems," *Cold Spring Harbor Symposia on Quantitative Biology*, vol. 1, pp. 107–116, 1933.

[9] A. A. Arafa, S. Z. Rida, and M. Khalil, "A fractional-order model of HIV infection: numerical solution and comparisons with data of patients," *International Journal of Biomathematics*, vol. 7, no. 4, Article ID 1450036, 1450036, 11 pages, 2014.

[10] A. A. M. Arafa, S. Z. Rida, and M. Khalil, "Fractional modeling dynamics of HIV and CD4 + T-cells during primary infection," *Nonlinear Biomedical Physics*, vol. 6, no. 1, article 1, 2012.

[11] Y. Liu, J. Xiong, C. Hu, and C. Wu, "Stability analysis for fractional differential equations of an HIV infection model with cure rate," in *Proceedings of the 2016 IEEE International Conference on Information and Automation, IEEE ICIA 2016*, pp. 707–711, August 2016.

[12] X. Zhou and J. Cui, "Global stability of the viral dynamics with Crowley-Martin functional response," *Bulletin of the Korean Mathematical Society*, vol. 48, no. 3, pp. 555–574, 2011.

[13] G. Huang, W. Ma, and Y. Takeuchi, "Global properties for virus dynamics model with Beddington-DeAngelis functional response," *Applied Mathematics Letters. An International Journal of Rapid Publication*, vol. 22, no. 11, pp. 1690–1693, 2009.

[14] M. A. Nowak and C. R. M. Bangham, "Population dynamics of immune responses to persistent viruses," *Science*, vol. 272, no. 5258, pp. 74–79, 1996.

[15] P. K. Srivastava and P. Chandra, "Modeling the dynamics of HIV and $CD4^+$ T cells during primary infection," *Nonlinear Analysis. Real World Applications. An International Multidisciplinary Journal*, vol. 11, no. 2, pp. 612–618, 2010.

[16] I. Podlubny, *Fractional Differential Equations*, vol. 198 of *Mathematics in Science and Engineering*, Academic Press, San Diego, Calif, USA, 1999.

[17] W. Lin, "Global existence theory and chaos control of fractional differential equations," *Journal of Mathematical Analysis and Applications*, vol. 332, no. 1, pp. 709–726, 2007.

[18] Z. M. Odibat and N. T. Shawagfeh, "Generalized Taylor's formula," *Applied Mathematics and Computation*, vol. 186, no. 1, pp. 286–293, 2007.

[19] E. Ahmed, A. M. El-Sayed, and H. A. El-Saka, "On some Routh-Hurwitz conditions for fractional order differential equations and their applications in Lorenz, Rossler, Chua and CHEn systems," *Physics Letters. A*, vol. 358, no. 1, pp. 1–4, 2006.

[20] C. V. De-Leon, "Volterra-type Lyapunov functions for fractional-order epidemic systems," *Communications in Nonlinear Science and Numerical Simulation*, vol. 24, no. 1-3, pp. 75–85, 2015.

[21] J. Huo, H. Zhao, and L. Zhu, "The effect of vaccines on backward bifurcation in a fractional order HIV model," *Nonlinear Analysis. Real World Applications*, vol. 26, pp. 289–305, 2015.

[22] Z. Odibat and S. Momani, "An algorithm for the numerical solution of differential equations of fractional order," *Applied Mathematics & Information*, vol. 26, no. 1, pp. 15–27, 2008.

19

A Trigonometrically Fitted Block Method for Solving Oscillatory Second-Order Initial Value Problems and Hamiltonian Systems

F. F. Ngwane[1] and S. N. Jator[2]

[1]*Department of Mathematics, University of South Carolina, Salkehatchie, Walterboro, SC 29488, USA*
[2]*Department of Mathematics and Statistics, Austin Peay State University, Clarksville, TN 37044, USA*

Correspondence should be addressed to F. F. Ngwane; fifonge@yahoo.com

Academic Editor: Julio D. Rossi

In this paper, we present a block hybrid trigonometrically fitted Runge-Kutta-Nyström method (BHTRKNM), whose coefficients are functions of the frequency and the step-size for directly solving general second-order initial value problems (IVPs), including Hamiltonian systems such as the energy conserving equations and systems arising from the semidiscretization of partial differential equations (PDEs). Four discrete hybrid formulas used to formulate the BHTRKNM are provided by a continuous one-step hybrid trigonometrically fitted method with an off-grid point. We implement BHTRKNM in a block-by-block fashion; in this way, the method does not suffer from the disadvantages of requiring starting values and predictors which are inherent in predictor-corrector methods. The stability property of the BHTRKNM is discussed and the performance of the method is demonstrated on some numerical examples to show accuracy and efficiency advantages.

1. Introduction

In what follows, we consider the numerical solution of the general second-order IVPs of the form

$$\begin{aligned} y'' &= f\left(x, y, y'\right), \\ y\left(x_0\right) &= y_0, \\ y'\left(x_0\right) &= y'_0, \end{aligned} \qquad x \in \left[x_0, x_N\right], \tag{1}$$

where $f : \mathbb{R} \times \mathbb{R}^{2m} \rightarrow \mathbb{R}^{2m}$, $N > 0$ is an integer, and m is the dimension of the system. Problems of form (1) frequently arise in several areas of science and engineering such as classical mechanics, celestial mechanics, quantum mechanics, control theory, circuit theory, astrophysics, and biological sciences. Equation (1) is traditionally solved by reducing it into a system of first-order IVPs of double dimension and then solved using the various methods that are available for solving systems of first-order IVPs (see Lambert [1, 2], Hairer and Wanner in [3], Hairer [4], and Brugnano and Trigiante [5, 6]).

Nevertheless, there are numerous methods for directly solving the special second-order IVPs in which the first derivative does not appear explicitly and it has been shown that these methods have the advantages of requiring less storage space and fewer number of function evaluations (see Hairer [4], Hairer et al. [7], Simos [8], Lambert and Watson, and [9], Twizell and Khaliq [10]). Fewer methods have been proposed for directly solving second-order IVPs in which the first derivative appears explicitly (see Vigo-Aguiar and Ramos [11], Awoyemi [12], Chawla and Sharma [13], Mahmoud and Osman [14], Franco [15], and Jator [16, 17]). It is also the case that some of these IVPs possess solutions with special properties that may be known in advance and take advantage of when designing numerical methods. In this light, several methods have been presented in the literature which take advantage of the special properties of the solution that may be known in advance (see Coleman and Duxbury [18], Coleman and Ixaru [19], Simos [20], Vanden Berghe et al. [21], Vigo-Aguiar and Ramos [11], Fang et al. [22], Nguyen et al. [23],

Ramos and Vigo-Aguiar [24], Franco and Gómez [25], and Ozawa [26]). However, most of these methods are restricted to solving special second-order IVPs in a predictor-corrector mode.

Our objective is to present a BHTRKNM that is implemented in a block-by-block fashion; in this way, the method does not suffer from the disadvantages of requiring starting values and predictors which are inherent in predictor-corrector methods (see Jator et al. [27], Jator [16], and Ngwane and Jator [28]). We note that multiderivative trigonometrically fitted block methods for $y'' = f(x, y, y')$ have been proposed in Jator [29] and Jator [16]. However, the BHTRKNM proposed in this paper avoids the computation of higher order derivatives which have the potential to increase computational cost, especially, when applied to nonlinear systems. In this paper, we propose a BHTRKNM which is of order 3 and its application is extended to solving oscillatory systems, PDEs, and Hamiltonian systems including the energy conserving equation.

The organization of this article is as follows. In Section 2, we derive the BHTRKNM for solving (1). The analysis and implementation of the BHTRKNM are discussed in Section 3. Numerical examples are given in Section 4 to show the accuracy and efficiency of the BHTRKNM. Finally, the conclusion of the paper is given in Section 5.

2. Development of the BHTRKNM

In order to numerical integrate (1) we define the BHTRKNM as consisting of the following four discrete formulas:

$$\begin{aligned} y_{n+1} &= y_n + hy'_n + h^2\left(\sum_{j=0}^{1}\beta_j f_{n+j} + \beta_{n+v} f_{n+v}\right), \\ y_{n+v} &= y_n + hy'_n + h^2\left(\sum_{j=0}^{1}\gamma_j f_{n+j} + \gamma_{n+v} f_{n+v}\right), \\ hy'_{n+1} &= hy'_n + h^2\left(\sum_{j=0}^{1}\beta'_j f_{n+j} + \beta'_{n+v} f_{n+v}\right), \\ hy'_{n+v} &= hy'_n + h^2\left(\sum_{j=0}^{1}\gamma'_j f_{n+j} + \gamma'_{n+v} f_{n+v}\right), \end{aligned} \tag{2}$$

where β_j, β'_j, γ_j, and γ'_j are coefficients that depend on the step-length h and frequency w. In general, the frequency w is chosen near the exact frequency of the true solution (see [30]). The coefficients of the method are chosen so that the method integrates the IVP (1) exactly where the solutions are members of the linear space $\langle 1, x, x^2, \sin(wx), \cos(wx)\rangle$.

The main method has the form

$$y_{n+1} = \alpha_0 y_n + \delta_0 hy'_n + h^2\left(\sum_{j=0}^{1}\beta_j f_{n+j} + \beta_{n+v} f_{n+v}\right), \tag{3}$$

where α_0, δ_0, and β_0, β_v, and β_1 are to be determined coefficient functions of the frequency and step-size. In order to derive the main method and additional methods we initially seek a continuous local approximation $\Pi(x)$ on the interval $[x_n, x_{n+1}]$ of the form

$$\Pi(x) = \alpha_0(x) y_n + \delta_0(x) hy'_n + h^2\left(\sum_{j=0}^{1}\beta_j(x) f_{n+j} + \beta_{n+v}(x) f_{n+v}\right), \tag{4}$$

where $\alpha_0(x)$, $\delta_0(x)$, and $\beta_j(x)$, $j = 0, v, 1$, are continuous coefficients. The first derivative of (4) is given by

$$\Pi'(x) = \frac{d}{dx}\Pi(x). \tag{5}$$

We assume that $y_{n+j} = \Pi(x_{n+j})$ is the numerical approximation to the analytical solution $y(x_{n+j})$, $y'_{n+j} = \Pi'(x_{n+j})$ is the numerical approximation to $y'(x_{n+j})$, and $f_{n+j} = \Pi''(x_{n+j})$ is an approximation to $y'(x_{n+j})$, $j = 0, v, 1$.

The following theorem shows how the continuous method (4) is constructed. This is done by requiring that on the interval from x_n to $x_{n+1} = x_n + h$ the exact solution is locally approximated by function (4) with (5) obtained as a consequence.

Theorem 1. *Let $F_i(x) = x^i$, $i = 0, 1, 2$, $F_3(x) = \sin wx$, and $F_4(x) = \cos wx$ be basis functions and let $V = (y_n, y'_n, f_n, f_{n+v}, f_{n+1})^T$ be a vector, where T is the transpose. Define the matrix G by*

$$G = \begin{pmatrix} F_0(x_n) & \cdots & F_4(x_n) \\ F'_0(x_n) & \cdots & F'_4(x_n) \\ F''_0(x_n) & \cdots & F''_4(x_n) \\ F''_0(x_{n+v}) & \cdots & F''_4(x_{n+v}) \\ F''_0(x_{n+1}) & \cdots & F''_4(x_{n+1}) \end{pmatrix} \tag{6}$$

and G_i is obtained by replacing the ith column of G by the vector V. Let the following conditions be satisfied:

$$\begin{aligned} \Pi(x_n) &= y_n, \\ \Pi'(x_n) &= y'_n, \\ \Pi''(x_n + j) &= f_{n+j}, \\ & j = 0, v, 1; \end{aligned} \tag{7}$$

then the continuous representations (4) and (5) are equivalent to the following:

$$\Pi(x) = \sum_{i=0}^{4}\frac{\det(G_i)}{\det(G)}F_i(x), \tag{8}$$

$$\Pi'(x) = \frac{d}{dx}\left(\sum_{i=0}^{4}\frac{\det(G_i)}{\det(G)}F_i(x)\right). \tag{9}$$

Proof. To prove this theorem, we use the approach given in Jator [17] with appropriate notational modification. We start by requiring that the method (4) be defined by the assumed basis functions

$$\alpha_0(x) = \sum_{i=0}^{4} \alpha_{i+1,0} F_i(x),$$
$$h\delta_0(x) = \sum_{i=0}^{4} h\delta_{i+1,0} F_i(x), \quad (10)$$
$$h^2\beta_j(x) = \sum_{i=0}^{4} h^2\beta_{i+1,j} F_i(x),$$

where $\alpha_{i+1,0}$, $h\delta_{i+1,0}$, and $h^2\beta_{i+1,j}$ are coefficients to be determined. Substituting (10) into (4) we get

$$\Pi(x) = \sum_{i=0}^{4} \alpha_{i+1,0} F_i(x)\, y_n + \sum_{i=0}^{4} h\delta_{i+1,0} F_i(x)\, y'_n + \sum_{j=0}^{1}\sum_{i=0}^{4} h^2\beta_{i+1,j} F_i(x)\, f_{n+j} \quad (11)$$

which is simplified to

$$\Pi(x) = \sum_{i=0}^{4} \left(\alpha_{i+1,0} F_i(x)\, y_n + h\delta_{i+1,0} F_i(x)\, y'_n + \sum_{j=0}^{1} h^2\beta_{i+1,j} F_i(x)\, f_{n+j} \right) \quad (12)$$

and expressed as

$$\Pi(x) = \sum_{i=0}^{4} \ell_i F_i(x), \quad (13)$$

where

$$\ell_i = \alpha_{i+1,0} y_n + h\delta_{i+1,0} y'_n + \sum_{j=0}^{1} h^2\beta_{i+1,j} f_{n+j}. \quad (14)$$

By imposing conditions (7) on (13), we obtain a system of five equations which can be expressed as

$$GL = V, \quad (15)$$

where $L = (\ell_0, \ell_1, \ldots, \ell_4)^T$ is a vector whose coefficients are determined via Cramer's rule as

$$\ell_i = \frac{\det(G_i)}{\det(G)}, \quad i = 0, 1, \ldots, 4, \quad (16)$$

where G_i is obtained by replacing the ith column of G by V. In order to obtain the continuous approximation, we use the elements of L to rewrite (13) as

$$\Pi(x) = \sum_{i=0}^{4} \frac{\det(G_i)}{\det(G)} F_i(x), \quad (17)$$

whose first derivative is given by

$$\Pi'(x) = \frac{d}{dx}\left(\sum_{i=0}^{4} \frac{\det(G_i)}{\det(G)} F_i(x) \right). \quad (18)$$

□

Remark 2. We note that, in the derivation of the BHTRKNM, the basis functions $F_i(x) = x^i$, $i = 0, 1, 2$, $F_3(x) = \sin wx$, and $F_4(x) = \cos wx$ are chosen because they are simple to analyze. Nevertheless, other possible bases are possible (see Nguyen et al. [23]).

2.1. Specification of the Method. The continuous methods (8) and (9) which are equivalent to forms (4) and (5) are used to generate two discrete methods and two additional methods. The discrete and additional methods are then applied as a BHTRKNM for solving (1). We choose $v = 1/2$ and evaluating (8) at $x = x_{n+v}$ and $x = x_{n+1}$, respectively, gives the two discrete methods $y_{n+v} = \Pi(x_n + vh)$ and $y_{n+1} = \Pi(x_n + h)$ which takes the form of the main method. Evaluating (9) at $x = x_{n+v}$ and $x = x_{n+1}$, respectively, gives the additional methods $y'_{n+v} = \Pi'(x_n+vh)$ and $y'_{n+1} = \Pi'(x_n+h)$. The coefficients and their corresponding Taylor series equivalence of y_{n+v}, y_{n+1}, hy'_{n+v}, and hy'_{n+1} are, respectively, given as follows:

$$\begin{aligned}
\alpha_{v,0} &= 1,\\
\delta_{v,0} &= \frac{1}{2},\\
\beta_{v,0} &= \frac{\csc(u/4)\csc(u/2)\left(\left(8+u^2\right)\cos(u/4) - 4\left(2\cos(3u/4) + u\sin(3u/4)\right)\right)}{16u^2}\\
&= \frac{7}{96} + \frac{7u^2}{7680} + \frac{71u^4}{3870720} + \frac{53u^6}{123863040} + \frac{23u^8}{2179989504} + \cdots,\\
\beta_{v,v} &= -\frac{\csc(u/4)^2\left(8 + \left(-8+u^2\right)\cos(u/2) - 4u\sin(u/2)\right)}{16u^2} = \frac{1}{16} - \frac{u^2}{2304} - \frac{u^4}{276480} - \frac{u^6}{34406400} - \frac{u^8}{4459069440} + \cdots,\\
\beta_{v,1} &= \frac{\csc(u/4)^2\left(u + 4\cot(u/2) - 4\csc(u/2)\right)}{32u} = -\frac{1}{96} - \frac{11u^2}{23040} - \frac{19u^4}{1290240} - \frac{247u^6}{619315200} - \frac{1013u^8}{98099527680} + \cdots,
\end{aligned} \quad (19)$$

$$\alpha_{1,0} = 1,$$

$$\delta_{1,0} = 1,$$

$$\beta_{1,0} = \frac{\csc(u/4)\csc(u/2)\left(\left(2+u^2\right)\cos(u/4) - 2\left(\cos(3u/4) + u\sin(3u/4)\right)\right)}{4u^2}$$

$$= \frac{1}{6} + \frac{u^2}{480} + \frac{19u^4}{483840} + \frac{17u^6}{19353600} + \frac{29u^8}{1362493440} + \cdots, \quad (20)$$

$$\beta_{1,v} = -\frac{\csc(u/4)^2\csc(u/2)\left(-2+2\cos u + u\sin u\right)}{8u} = \frac{1}{3} - \frac{u^2}{720} - \frac{u^4}{80640} - \frac{u^6}{9676800} - \frac{u^8}{1226244096} + \cdots,$$

$$\beta_{1,1} = \frac{\csc(u/4)^2\left(-4+u^2+4\cos(u/2)+2u\cot(u/2)-2u\csc(u/2)\right)}{8u^2}$$

$$= -\frac{u^2}{1440} - \frac{13u^4}{483840} - \frac{u^6}{1290240} - \frac{251u^8}{12262440960} + \cdots,$$

$$\alpha_{v,0} = 0,$$

$$\delta_{v,0} = 1,$$

$$\gamma'_{v,0} = -\frac{\cot(u/2)}{u} + \frac{1}{8}\csc\left(\frac{u}{4}\right)^2 = \frac{5}{24} + \frac{19u^2}{5760} + \frac{23u^4}{322560} + \frac{263u^6}{154828800} + \frac{1033u^8}{24524881920} + \cdots, \quad (21)$$

$$\gamma'_{v,v} = -\frac{\csc(u/4)^2\csc(u/2)\left(-2+2\cos(u)+u\sin(u)\right)}{8u} = \frac{1}{3} - \frac{u^2}{720} - \frac{u^4}{80640} - \frac{u^6}{9676800} - \frac{u^8}{1226244096} + \cdots,$$

$$\gamma'_{v,1} = \frac{\csc(u/4)^2\left(u+4\cot(u/2)-4\csc(u/2)\right)}{8u} = -\frac{1}{24} - \frac{11u^2}{5760} - \frac{19u^4}{322560} - \frac{247u^6}{154828800} - \frac{1013u^8}{24524881920} + \cdots,$$

$$\alpha_{1,0} = 0,$$

$$\delta_{1,0} = 1,$$

$$\beta'_{1,0} = \frac{\csc(u/4)^2\left(u-2\sin(u/2)\right)}{4u} = \frac{1}{6} + \frac{u^2}{720} + \frac{u^4}{80640} + \frac{u^6}{9676800} + \frac{u^8}{1226244096} + \cdots,$$

$$\beta'_{1,v} = \frac{\csc(u/4)^2\left(u-2\sin(u/2)\right)}{4u} = \frac{2}{3} - \frac{u^2}{360} - \frac{u^4}{40320} - \frac{u^6}{4838400} - \frac{u^8}{613122048} + \cdots, \quad (22)$$

$$\beta'_{1,1} = \frac{\csc(u/4)^2\left(-4+u^2+4\cos(u/2)+2u\cot(u/2)-2u\csc(u/2)\right)}{8u^2}$$

$$= \frac{1}{6} + \frac{u^2}{720} + \frac{u^4}{80640} + \frac{u^6}{9676800} + \frac{u^8}{1226244096} + \cdots.$$

Remark 3. We note that the Taylor series expansions in (19) through (22) must be used when $u \to 0$ because the corresponding trigonometric coefficients given in these equations are vulnerable to heavy cancelations (see [8]).

2.2. Block Form. BHTRKNM is formulated from the four discrete hybrid formulas stated in (2) which are provided by the continuous one-step hybrid trigonometrically fitted method with one off-grid point given by (4) and its first derivative (5). We define the following vectors:

$$Y_{\mu+1} = \left[y_{n+v}, y_{n+1}, hy'_{n+v}, hy'_{n+1}\right]^T,$$

$$Y_{\mu} = \left[y_{n-v}, y_n, hy'_{n-v}, hy'_n\right]^T,$$

$$F_{\mu+1} = \left[f_{n+v}, f_{n+1}, hf'_{n+v}, hf'_{n+1}\right]^T,$$
$$F_{\mu} = \left[f_{n-v}, f_{n}, hf'_{n-v}, hf'_{n}\right]^T, \tag{23}$$

where $\mu = 0,\ldots,N$, $n = 0,\ldots,N$. The methods in (2) specified by the coefficients (19)–(22) are combined to give the BHTRKNM, which is expressed as

$$A_1 Y_{\mu+1} = A_0 Y_{\mu} + h^2\left(B_0 F_{\mu} + B_1 F_{\mu+1}\right), \tag{24}$$

where A_0, A_1, B_0, and B_1 are matrices of dimension four whose elements characterize the method and are given by the coefficients of (2).

3. Error Analysis and Stability

3.1. Local Truncation Error (LTE). We define the local truncation error of (24) as

$$L\left[(Z(x);h)\right] = Z_{\mu+1} - \left[AZ_{\mu} + h^2 B\overline{F}_{\mu} + h^2 C\overline{F}_{\mu+1}\right], \tag{25}$$

where

$$Z_{\mu+1} = \left[y(x_{n+v}), y(x_{n+1}), hy'(x_{n+v}), hy'(x_{n+1})\right]^T,$$
$$Z_{\mu} = \left[y(x_{n-v}), y(x_{n}), hy'(x_{n-v}), hy'(x_{n})\right]^T,$$
$$\overline{F}_{\mu+1} = \left[f(x_{n+v}, y_{n+v}), f(x_{n+1}, y_{n+1}), hf'(x_{n+v}, y_{n+v}), hf'(x_{n+1}, y_{n+1})\right]^T, \tag{26}$$
$$\overline{F}_{\mu} = \left[f(x_{n-v}, y_{n-v}), f(x_{n}, y_{n}), hf'(x_{n-v}, y_{n-v}), hf'(x_{n}, y_{n})\right]^T,$$

and $L[(Z(x);h)] = [L_1[z(x);h], L_2[z(x);h],\ldots,L_4[z(x);h]]^T$ is linear different operator.

Suppose that $Z(x)$ is sufficiently differentiable. Then, a Taylor series expansion of the terms in (25) about the point x gives the following expression for local truncation error:

$$L\left[Z(x);h\right] = C_0 Z(x) + C_1 hZ'(x) + \cdots + C_q h^q Z^q(x) + \cdots, \tag{27}$$

where C_i, $i = 0, 1, \ldots$, are constant coefficients (see [17]).

Definition 4. The block method (24) has algebraic order at least $p \geq 1$ provided there exists a constant $C_{p+2} \neq 0$ such that the local truncation error E_{μ} satisfies $\|E_{\mu}\| = C_{p+2}h^{p+2} + O(h^{p+3})$, where $\|\cdot\|$ is the maximum norm.

Remark 5. (i) The local truncation error constants $(\overline{C}_{p+2})$ of $(y_{n+v}, y_{n+1}, hy'_{n+v}, hy'_{n+1})^T$ of the block method (24) are given, respectively, by $\overline{C}_5 = (1/1440, 1/720, 1/384, 0)^T$, where $\overline{C}_0 = \overline{C}_1 = \overline{C}_2 = \overline{C}_3 = \overline{C}_4 = 0$.

(ii) From the local truncation error constant computation, it follows that the method (24) has order p at least three.

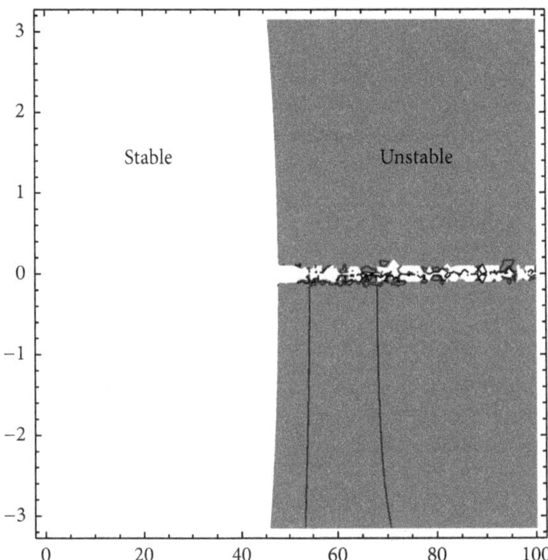

Figure 1: The stability region plotted in the (q, u)-plane.

3.2. Stability. The linear-stability of the BHTRKNM is discussed by applying the method to the test equation $y'' = -\lambda^2 y$, where λ is a real constant (see [18]). Letting $\Upsilon = \lambda h$, it is easily shown as in [19] that the application of (24) to the test equation yields

$$Y_{\mu+1} = M\left(\Upsilon^2; u\right) Y_{\mu},$$
$$M\left(\Upsilon^2; u\right) := \left(A_1 - \Upsilon^2 B_1\right)^{-1}\left(A_0 + \Upsilon^2 B_0\right), \tag{28}$$

where the matrix $M(\Upsilon^2; u)$ is the amplification matrix which determines the stability of the method. In the spirit of [22], the spectral radius of $\rho(M(\Upsilon^2; u))$ can be obtained from the characteristics equation

$$\rho^2 - 2\Gamma\left(\Upsilon^2; u\right)\rho + \Theta\left(\Upsilon^2; u\right) = 0, \tag{29}$$

where $u = \omega h$, $\Gamma(\Upsilon^2; u) = \operatorname{trace} M(\Upsilon^2; u)$, and $\Theta(\Upsilon^2; u) = \det M(\Upsilon^2; u)$ are rational functions. We let $q = \lambda h$ in the following definition.

Definition 6. A region of stability is a region in the q-u plane, throughout which $\rho(M(\Upsilon^2; u)) \leq 1$ and any closed curve given by $\rho(M(\Upsilon^2; u)) = 1$ defines the stability boundary of the method (see [22]). We note that the plot for the stability region of the BHTRKNM method is given in Figure 1.

Remark 7. It is observed that, in the q-u plane, the BHTRKNM is stable for $q \in [0, 47.96]$ and $u \in [-\pi, \pi]$ (see Figure 1).

3.3. Implementation. The main method and the additional methods specified by (19)–(22) are combined to form the block method BHTRKNM (24), which is used to solve (1) without requiring starting values and predictors. BHTRKNM is implemented in a block-by-block fashion using a Mathematica 10.0 code, enhanced by the feature *NSolve*[ ] for linear problems while nonlinear problems were solved by Newton's method enhanced by the feature *FindRoot*[ ] (see Keiper and Gear [33]). Mathematica can symbolically compute derivatives and so the entries of the Jacobian matrix which involve partial derivatives are automatically generated. In what follows, we summarize how BHTRKNM is applied.

Step 1. Choose N, $h = (b-a)/N$, and the number of blocks $\Gamma = N$. Using (24), $n = 0$, and $\mu = 0$, the values of $(y_{1/2}, y_1, y'_{1/2}, y'_1)^T$ are simultaneously obtained over the subinterval $[x_0, x_1]$, as y_0 and y'_0 are known from the IVP (1).

Step 2. For $n = 1$ and $\mu = 1$, the values of $(y_{3/2}, y_2, y'_{3/2}, y'_2)^T$ are simultaneously obtained over the subinterval $[x_1, x_2]$, as y_1 and y'_1 are known from the previous block.

Step 3. The process is continued for $n = 2, \ldots, N-1$ and $\mu = 2, \ldots, \Gamma - 1$ to obtain the numerical solution to (1) on the subintervals $[x_0, x_1], [x_1, x_2], \ldots, [x_{N-1}, x_N]$.

In order to illustrate the efficiency of our method, we solved a variety of problems including oscillatory systems, PDEs such as the Telegraph equation, and Hamiltonian systems. The following methods are selected for comparison:

(i) BHTRKNM given in this paper.

(ii) ARKN: adapted Runge-Kutta-Nyström method in [34] which has order five.

(iii) (DS3.12): difference scheme (3.12) in [32].

(iv) ESDIRK: explicit singly diagonally implicit Runge-Kutta method in [26].

(v) FESDIRK: functionally fitted ESDIRK in [26].

(vi) EFRK: exponentially fitted Runge-Kutta method (Method (b)) in Simos [8].

(vii) N4: fourth-order standard Runge-Kutta-Nyström method in [35].

4. Numerical Examples

In this section, numerical experiments are performed using a code in Mathematica 10.0 to illustrate the accuracy and efficiency of the method.

Example 1. We consider the following inhomogeneous IVP by Simos [8].

$$y'' = -100y + 99\sin(x),$$
$$y(0) = 1, \qquad (30)$$
$$y'(0) = 11,$$
$$x \in [0, 1000],$$

where the analytical solution is given by

$$\text{Exact: } y(x) = \cos(10x) + \sin(10x) + \sin(x). \qquad (31)$$

Table 1: Results, with $\omega = 10$, for Example 1.

Our method		Simos [8]	
N	Err	N	Err
1000	2.14×10^{-3}	1000	1.4×10^{-1}
2000	5.98×10^{-5}	2000	3.5×10^{-2}
4000	2.06×10^{-5}	4000	1.1×10^{-3}
8000	1.26×10^{-6}	8000	8.4×10^{-5}
16000	7.79×10^{-8}	16000	5.5×10^{-6}
32000	4.67×10^{-9}	32000	3.5×10^{-7}

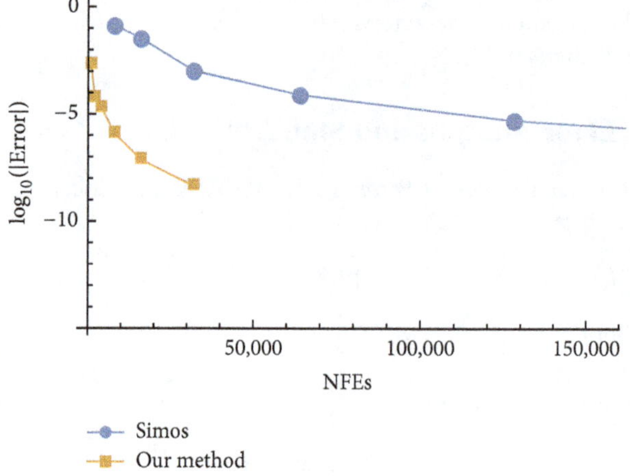

Figure 2: Efficiency curve for Example 1.

This example was solved using the order 3 BHTRKNM and the endpoint errors (Err $= |y(x_N) - y_N|$) obtained were compared to the order 4 exponentially fitted method given in Simos [8]. In Table 1 it is shown that BHTRKNM is more efficient than the method in Simos [8]. We also compare the computational efficiency of the two methods in Figure 2 by considering the FNEs (number of function evaluations) over N integration steps for each method. This example illustrates that the BHTRKNM performs better.

Example 2. We consider the nonlinear Duffing equation which was also solved by Simos [8] and Ixaru and Vanden Berghe [31]:

$$y'' + y + y^3 = B\cos(\Omega x),$$
$$y(0) = C_0, \qquad (32)$$
$$y'(0) = 0.$$

The analytical solution is given by

$$\text{Exact: } y(x) = C_1\cos(\Omega x) + C_2\cos(3\Omega x) + C_3\cos(5\Omega x) + C_4\cos(7\Omega x), \qquad (33)$$

where $\Omega = 1.01$, $B = 0.002$, $C_0 = 0.200426728069$, $C_1 = 0.200179477536$, $C_2 = 0.246946143 \times 10^{-3}$, $C_3 = 0.304016 \times 10^{-6}$, and $C_4 = 0.374 \times 10^{-9}$. We choose $\omega = 1.01$.

TABLE 2: Results, with $\omega = 1.01$, for Example 2.

Our method		Simos [8]		Ixaru and Vanden Berghe [31]	
N	Err	N	Err	N	Err
300	7.52×10^{-5}	300	1.7×10^{-3}	300	1.1×10^{-3}
600	2.47×10^{-6}	600	1.9×10^{-4}	600	5.4×10^{-5}
1200	1.34×10^{-7}	1200	1.4×10^{-5}	1200	1.9×10^{-6}
2400	8.11×10^{-9}	2400	8.7×10^{-7}	2400	6.2×10^{-8}

TABLE 3: Steps and absolute errors, with $\omega = 1$, for Example 3 $[0, 50\pi]$.

Our method		FESDIRK4(3) [26]		ESDIRK4(3) [26]	
Steps	Err	Steps	Err	Steps	Err
200	4.42×10^{-4}	381	1.40×10^{-3}	884	9.36×10^{-3}
300	3.2×10^{-5}	680	1.69×10^{-4}	1573	6.20×10^{-4}
400	5.39×10^{-8}	1207	1.85×10^{-5}	2796	4.42×10^{-5}
600	4.25×10^{-7}	2144	1.94×10^{-6}	4970	3.41×10^{-6}
1000	1.06×10^{-8}	3806	1.99×10^{-7}	8833	2.85×10^{-7}
1200	1.76×10^{-9}	6762	2.02×10^{-8}	15706	2.53×10^{-8}

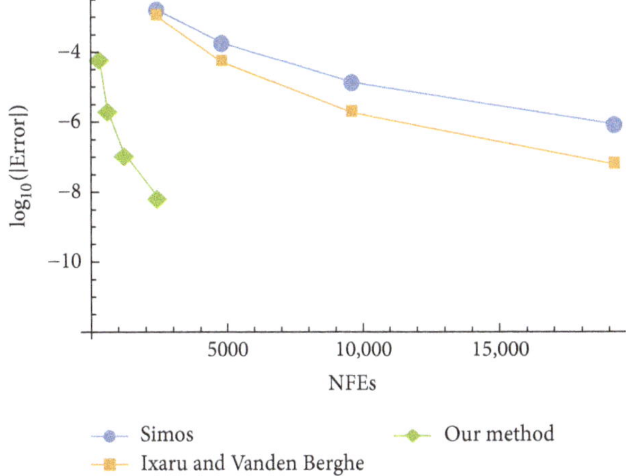

FIGURE 3: Efficiency curves for Example 2.

We compare the endpoint global errors for our method with those of Simos [8] and Ixaru and Vanden Berghe [31]. We see from Table 2 that the results produced by our method are competitive to those given in Simos [8] and Ixaru and Vanden Berghe [31]. Hence our method is more accurate and efficient as demonstrated in Figure 3.

Example 3. We consider the following two-body problem which was solved by Ozawa [26] on $[0, 50\pi]$:

$$y_1'' = -\frac{y_1}{r^3},$$
$$y_2'' = -\frac{y_2}{r^3},$$
$$r = \sqrt{y_1^2 + y_2^2},$$
$$y_1(0) = 1 - e,$$
$$y_1'(0) = 0,$$
$$y_2(0) = 0,$$
$$y_2'(0) = \sqrt{\frac{1+e}{1-e}}, \tag{34}$$

where e $(0 \le e < 1)$ is an eccentricity. The exact solution of this problem is

$$\text{Exact: } y_1(x) = \cos(k) - e, \quad y_2(x) = \sqrt{1-e^2}\sin(k), \tag{35}$$

where k is the solution of Kepler's equation $k = x + e\sin(k)$. We choose $\omega = 1$.

We show in Table 3 that the results obtained using the BHTRKNM method are more accurate than the explicit singly diagonally implicit Runge-Kutta (ESDIRK) and the functionally fitted ESDIRK (FESDIRK) methods given in Ozawa [26]. In Figure 4, we also illustrate the efficiency advantage of the BHTRKNM method over those in Ozawa [26].

Example 4. We consider the stiff second-order IVP (see [16] and references herein)

$$y_1'' = (\epsilon - 2) y_1 + (2\epsilon - 2) y_2,$$
$$y_2'' = (1 - \epsilon) y_1 + (1 - 2\epsilon) y_2,$$
$$y_1(0) = 2,$$
$$y_1'(0) = 0,$$

Table 4: Results for Example 5.

	Our method			ARKN	
N	Error ($\delta = 10^{-6}$)	Error ($\delta = 10^{-10}$)	N	Error ($\delta = 10^{-6}$)	Error ($\delta = 10^{-10}$)
2000	1.82×10^{-8}	2.00×10^{-12}	2000	9.05×10^{-8}	9.00×10^{-12}
4000	1.14×10^{-9}	8.32×10^{-14}	4000	5.43×10^{-9}	7.06×10^{-13}
8000	7.13×10^{-11}	3.50×10^{-13}	8000	2.03×10^{-10}	2.87×10^{-13}
16000	4.33×10^{-12}	1.17×10^{-13}	16000	7.25×10^{-12}	3.56×10^{-13}
32000	5.22×10^{-14}	1.59×10^{-12}	32000	3.45×10^{-13}	5.91×10^{-13}

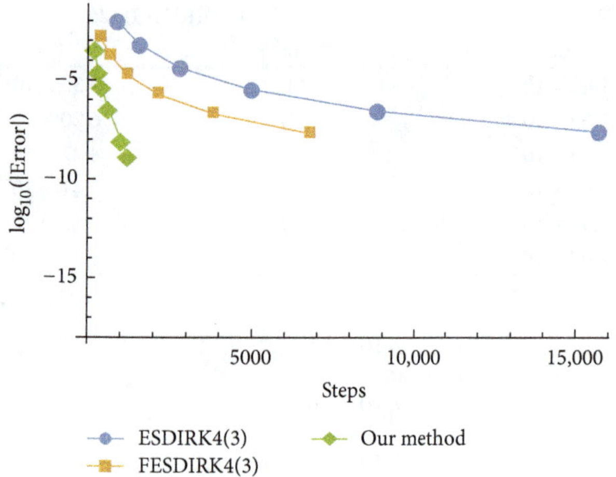

Figure 4: Efficiency curves for Example 3.

$$
\begin{aligned}
y_2(0) &= -1, \\
y_2'(0) &= 0, \\
\epsilon &= 2500, \\
w &= 1, \\
x &\in [0, 100].
\end{aligned}
\tag{36}
$$

$y_1(x) = 2\cos x$; $y_2(x) = -\cos x$ where ϵ is an arbitrary parameter.

This problem was chosen to demonstrate the stability of the BHTRKNM (Figure 5). As mentioned in Remark 7, the method is stable when $q \in [0, 47.06]$ and $u \in [-\pi, \pi]$.

4.1. Problems Where y' Appears Explicitly

Example 5. We consider the harmonic oscillator with frequency Ω and small perturbation δ that was solved in Franco [15] and Guo and Yan [34]:

$$
\begin{aligned}
y'' + \delta y' + \Omega^2 y &= 0, \\
y(0) &= 0,
\end{aligned}
$$

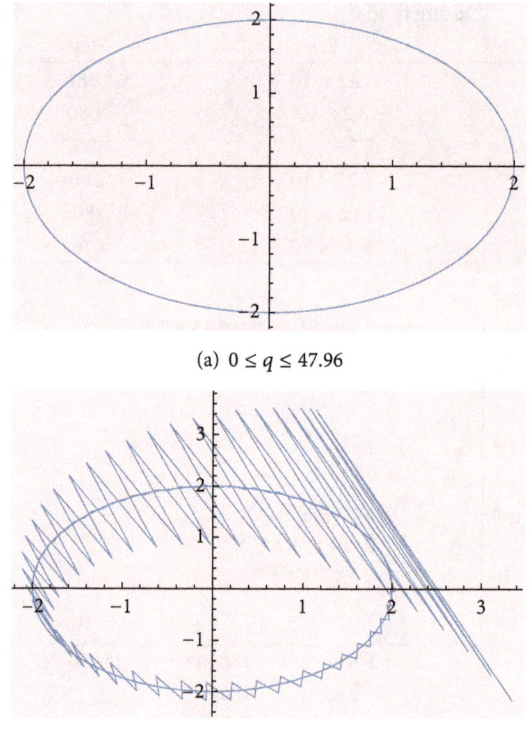

(a) $0 \le q \le 47.96$

(b) $q > 47.96$

Figure 5: These figures illustrate the stability of the BHTRKNM applied to Example 4. In (a) the method is stable with $N = 722$, $q \in [0, 47.96]$, and the global error is 1.7×10^{-10}, whereas in (b) the method is unstable with $N = 721$, $q > 47.96$, and the global error is 7005.78.

$$
\begin{aligned}
y'(0) &= -\frac{\delta}{2}, \\
x &\in [0, 1000],
\end{aligned}
\tag{37}
$$

where the analytical solution is given by

$$
\text{Exact: } y(x) = e^{(\delta/2)x} \cos\left(\Omega^2 - \frac{\delta^2}{4}\right), \tag{38}
$$

where $\Omega = 1$, $\delta = 10^{-6}$, and $\delta = 10^{-10}$. Guo and Yan [34] solved this problem using ARKN method. The results in Table 4 show that the BHTRKNM is competitive with the order 5 Runge-Kutta-Nyström method.

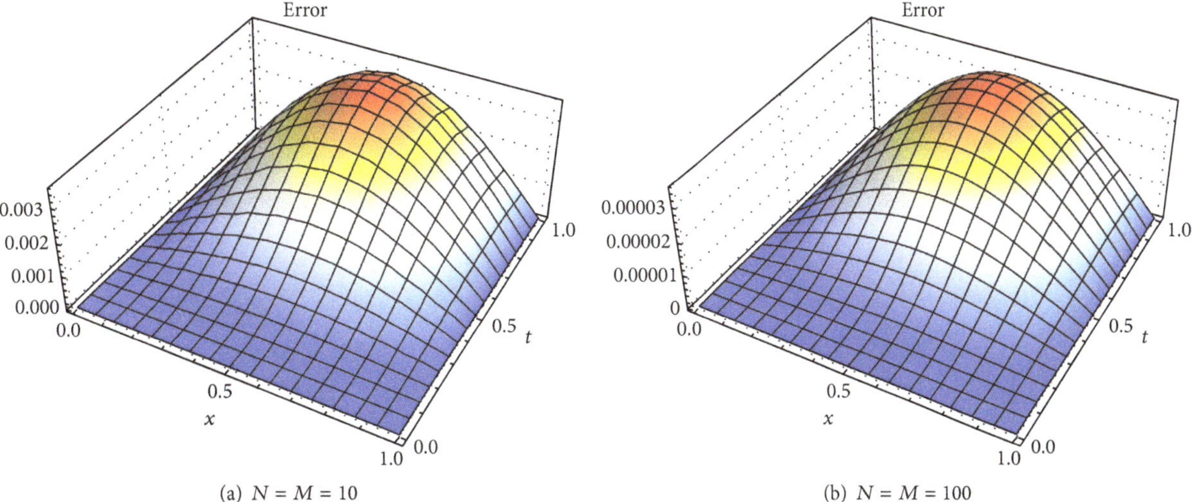

(a) $N = M = 10$

(b) $N = M = 100$

FIGURE 6: Absolute errors for Example 6.

4.2. Hyperbolic PDE

Example 6. We consider the given Telegraph equation (see Ding et al. [32]).

$$\frac{\partial^2 u}{\partial t^2} + 2\pi\frac{\partial u}{\partial t} + \pi^2 u = \frac{\partial^2 u}{\partial x^2} + \pi^2 \sin(\pi x)(\sin(\pi t) + \cos(\pi t)) \quad (39)$$

$$0 \le x \le 1, \quad 0 \le t \le 1.$$

The exact solution is given by $u(x, y) = \sin(\pi x)\sin(\pi t)$.

In order to solve this PDE using the BHTRKNM, we carry out the semidiscretization of the spatial variable x using the second-order finite difference method to obtain the following second-order system in the second variable t.

$$\begin{aligned} &\frac{\partial^2 u_m}{\partial t^2} + 2\pi\frac{\partial u_m}{\partial t} + \pi^2 u_m - \frac{(u_{m+1} - 2u_m + u_{m-1})}{(\Delta x)^2} \\ &\quad = g_m, \quad 0 < t < 1, \ m = 1, \ldots, M-1, \\ &u(x_m, 0) = u_m, \\ &u_t(x_m, 0) = u'_m, \end{aligned} \quad (40)$$

where $\Delta x = (b-a)/M$, $x_m = a + m\Delta x$, $m = 0, 1, \ldots, M$, $\mathbf{u} = [u_1(t), \ldots, u_M(t)]^T$, $\mathbf{g} = [g_1(t), \ldots, g_m(t)]^T$, $u_m(t) \approx u(x_m, t)$, and $g_m(t) \approx g(x_m, t) = \pi^2 \sin(\pi x_m)(\sin(\pi t) + \cos(\pi t))$, which can be written in the form

$$\mathbf{u}'' = \mathbf{f}(t, \mathbf{u}, \mathbf{u}'), \quad (41)$$

subject to the boundary conditions $\mathbf{u}(t_0) = \mathbf{u}_0$ and $\mathbf{u}'(t_0) = \mathbf{u}'_0$ where $\mathbf{f}(t,\mathbf{u},\mathbf{u}') = \mathbf{A}\mathbf{u} + \mathbf{g}$, and $\mathbf{A}$ is $(M-1)\times(M-1)$, matrix arising from the semidiscretized system, and $\mathbf{g}$ is a vector of constants.

TABLE 5: Results, with $\omega = \pi$, for Example 6.

	Our method	Ding et al. [32]
x	Err	Err
0.2	2.46×10^{-10}	9.62×10^{-10}
0.4	3.96×10^{-10}	1.56×10^{-9}
0.6	3.98×10^{-10}	1.56×10^{-9}
0.8	2.46×10^{-10}	9.62×10^{-10}

The boundary conditions are chosen accordingly. This example was chosen to demonstrate that the BHTRKNM can be used to solve the Telegraph equation. In Table 5, the results produced by the BHTRKNM using $\Delta t = 1/100$ and space step $\Delta x = 1/100$ are compared to scheme (3.12) ($\lambda_1 = 1/12$ and $\lambda_2 = 5/6$), time step $\Delta t = 1/200$, and space step $\Delta x = 1/100$, given in Ding et al. [32]. It is obvious from Table 5 that the BHTRKNM is more accurate than the method given in [32]. Moreover, the errors produced by BHTRKNM method using $\Delta t = 1/100$ and space step $\Delta x = 1/100$ are given in Figure 6.

4.3. Hamiltonian Systems and Energy Conservation. In this section we present additional examples to show that the BHTRKNM preserves energy. To do so we consider Hamiltonian systems of the form

$$\begin{aligned} p' &= -\nabla_q H(p, q), \\ q' &= -\nabla_p H(p, q), \end{aligned} \quad (42)$$

where $H(p, q)$ is an arbitrary scalar function of the variables (p, q). Let M be a positive definite matrix and let $U(q)$ be a potential and the total energy H expressed as the sum of the kinetic and potential energy namely in the form

$$H(p, q) = \frac{1}{2}p^T M^{-1} p + U(q); \quad (43)$$

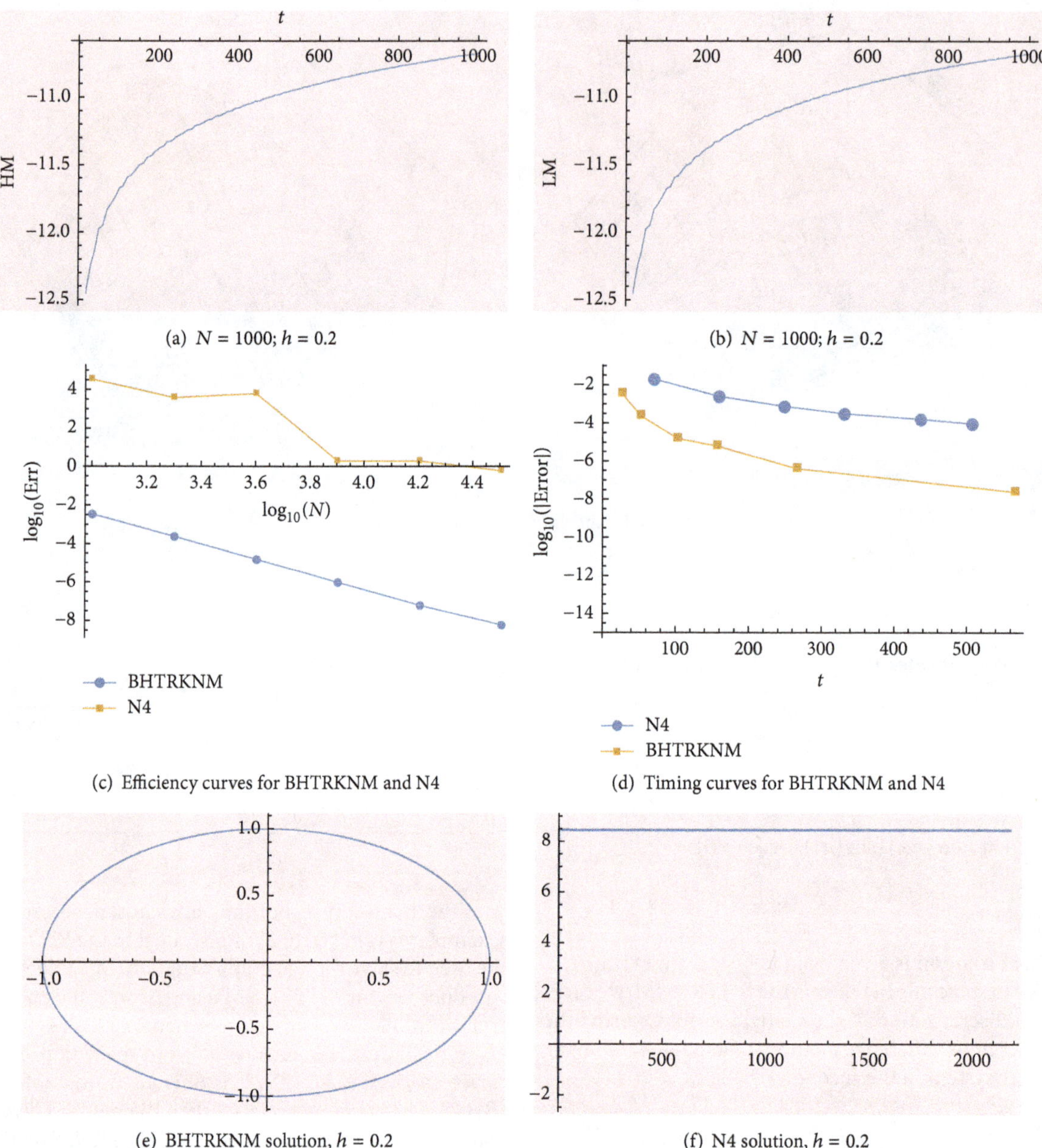

(a) $N = 1000$; $h = 0.2$

(b) $N = 1000$; $h = 0.2$

(c) Efficiency curves for BHTRKNM and N4

(d) Timing curves for BHTRKNM and N4

(e) BHTRKNM solution, $h = 0.2$

(f) N4 solution, $h = 0.2$

FIGURE 7: Perturbed Kepler problem: the logarithm of the global error of the Hamiltonian EH $= |H_n - H_0|$ against t for $h = 0.2$ and the momentum EL $= |L_n - L_0|$ are presented in (a) and (b), respectively. In (c) we compare the efficiency curves for the BHTRKNM and N4. Timing comparison is provided in (d). It is clear from the timing curves that BHTRKNM is very efficient.

then systems (42) can be written as a system of first-order differential equations

$$q' = v, \quad v' = f(q), \tag{44}$$

where the momentra $p = Mv$ is in terms of the velocities and $f(q) = -M^{-1}\nabla U(q)$ is in terms of the negative gradient of a potential. See [36–39] and references therein for further details. The Hamiltonian function, $H(y)$, defined by $H(y) = H(p, q)$ is a polynomial in the variables p and q. The Hamiltonian function conserves energy if

$$H(y_{n+1}) = H(y_n), \quad \forall n, \ h > 0. \tag{45}$$

Example 7. We consider the perturbed Kepler's problem in [40] given by

$$q_1'' = -\frac{q_1}{(q_1^2 + q_2^2)^{3/2}} - \frac{(2\epsilon + \epsilon^2) q_1}{(q_1^2 + q_2^2)^{5/2}},$$

$$q_1(0) = 1,$$

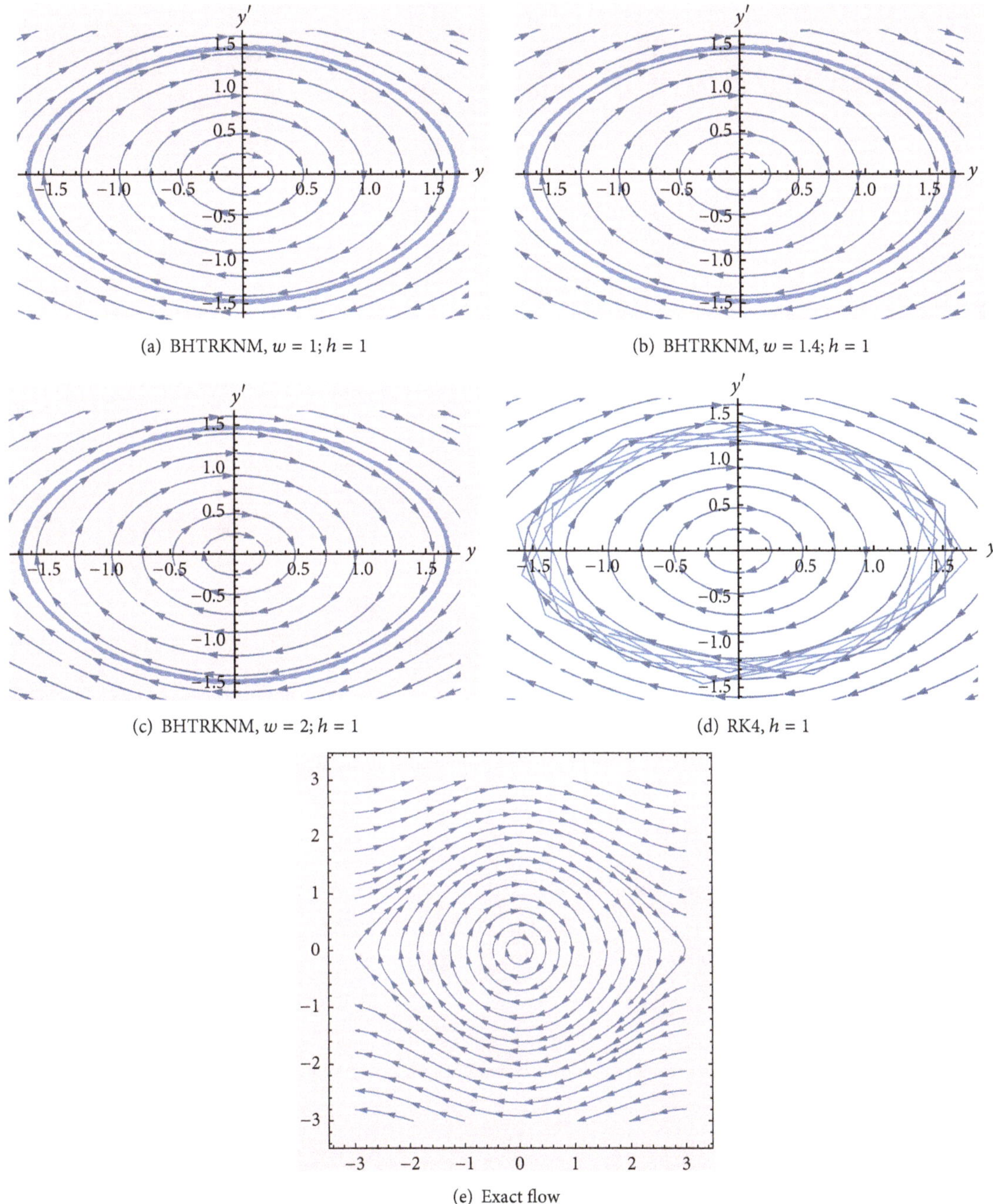

(a) BHTRKNM, $w = 1; h = 1$

(b) BHTRKNM, $w = 1.4; h = 1$

(c) BHTRKNM, $w = 2; h = 1$

(d) RK4, $h = 1$

(e) Exact flow

FIGURE 8: The pendulum problem: phase diagrams for BHTRKNM with $w = 1, 1.4, 2$ are presented in (a), (b), and (c), respectively. (d) illustrates a distortion in the flow for the RK4, while (e) shows exact flow of the pendulum problem. In the diagrams, $y = q$; $y' = q'$.

$$
\begin{aligned}
q_1'(0) &= 0,\\
q_2'' &= -\frac{q_2}{\left(q_1^2+q_2^2\right)^{3/2}} - \frac{\left(2\epsilon+\epsilon^2\right)q_2}{\left(q_1^2+q_2^2\right)^{5/2}},\\
q_2(0) &= 0,\\
q_2'(0) &= 1+\epsilon.
\end{aligned} \tag{46}
$$

The exact solution of this problem is

$$
\begin{aligned}
q_1(t) &= \cos(t+\epsilon t),\\
q_2(t) &= \sin(t+\epsilon t).
\end{aligned} \tag{47}
$$

The Hamiltonian is

$$
H = \frac{1}{2}\left(q_1'^2 + q_2'^2\right) - \frac{1}{\sqrt{q_1^2+q_2^2}} - \frac{\left(2\epsilon+\epsilon^2\right)}{3\left(q_1^2+q_2^2\right)^{3/2}}. \tag{48}
$$

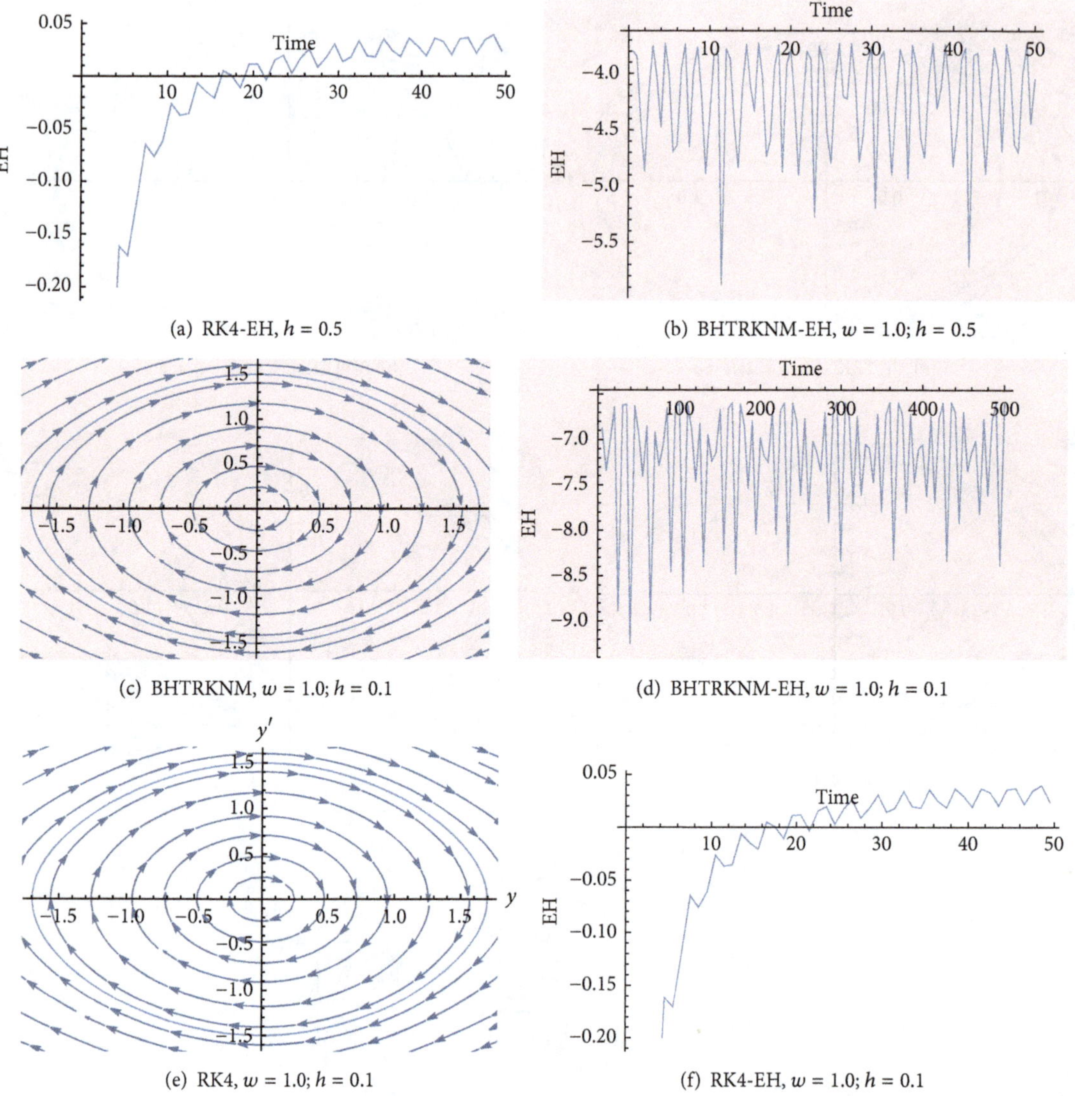

Figure 9: The pendulum problem: Hamiltonian error EH $= |H_n - H_0|$ using RK4 with $h = 0.5$ is presented in (a), while (b) shows EH for BHTRKNM. We also solve the problem on a larger interval of integration $[0, 500]$, using the BHTRNM in (c) and (d) and the RK4 in (e) and (f), respectively.

The system also has the angular momentum $L = q_1 q_2' - q_2 q_1'$ as a first integral. We take the parameter value $\epsilon = 10^{-3}$.

This problem is solved in the interval $[0, 1000]$ using the BHTRKNM for various values of $h = 0.1/2^{i-1}$, $i = 0, 1, 2, 3, 4$. The BHTRKNM preserves the Hamiltonian energy and to demonstrate this, we plot the logarithm of the global error of the Hamiltonian EH $= |H_n - H_0|$ and the momentum EL $= |L_n - L_0|$ as given in Figures 7(a) and 7(b), respectively. The problem was also solved using N4 given in Sommeijer [35] and in Figure 7(c), the efficiency curves for the BHTRKNM and N4 are compared showing that the BHTRKNM is superior.

Example 8. We consider the pendulum oscillator in [36] (and references herein) given by

$$q'' = -\sin q \tag{49}$$

with initial conditions

$$\begin{aligned} q(0) &= 0, \\ q'(0) &= 1.5 \end{aligned} \tag{50}$$

and Hamiltonian

$$H = \frac{1}{2} q'^2 - \cos q. \tag{51}$$

This problem is solved using the BHTRKNM on the interval $[0, 50]$ for $h = 1$ and $w = 1, 1.4, 2$ and the results for the phase diagrams produced by the BHTRKNM in the q-q' plane are presented in Figures 8(a), 8(b), and 8(c), respectively. We observe that the BHTRKNM gives good results for all the values of w, since all the diagrams follow the exact flow of the pendulum problem as given in Figure 8(e). As illustrated in these Figures, the numerical solutions are periodic and in accordance with the fact that the pendulum equation has a periodic solution. We note that Van Daele and Vanden Berghe in [36] obtained similar results for a smaller step-size $h = 0.5$ and $w = 1$ using S/V method including other versions of the S/V method and it was observed that the $SE1_{EF}$ method in [36] produced better numerical results for $w = 1$ than for $w = 1.4$. The problem was also solved using the fourth-order Runge-Kutta method (RK4) and the results presented in Figure 8(d) show a distortion in the flow diagram for the RK4; hence the BHTRKNM is superior. The pendulum problem was also presented in Figure 9.

5. Conclusion

This paper presents a BHTRKNM whose coefficients are functions of the frequency and the step-size for directly solving general second-order initial value problems (IVPs), oscillatory systems, and Hamiltonian systems, as well as systems arising from the semidiscretization of hyperbolic PDEs, such as the Telegraph equation. We implement the BHTRKNM in a block-by-block fashion; thus the method does not need starting values and predictors which are inherent in predictor-corrector methods. Numerical experiments presented in this paper clearly demonstrate that our method has a reasonably wide stability region and enjoys accuracy and efficiency advantages when compared to existing methods in the literature. Technique for accurately estimating the frequency as suggested in [30, 41] as well as implementing the method in a variable step mode will be considered in future.

Competing Interests

The authors declare that there is no conflict of interests regarding the publication of this paper.

Acknowledgments

This work was supported by the University of South Carolina RISE Grant no. 17660-15-38959.

References

[1] J. D. Lambert, *Numerical Methods for Ordinary Differential Systems*, John Wiley & Sons, New York, NY, USA, 1991.

[2] J. D. Lambert, *Computational Methods in Ordinary Differential Equations*, John Wiley & Sons, New York, NY, USA, 1973.

[3] E. Hairer and G. Wanner, *Solving Ordinary Differential Equations II*, vol. 14 of *Springer Series in Computational Mathematics*, Springer-Verlag, Berlin, Germany, 2nd edition, 1996.

[4] E. Hairer, "A one-step method of order 10 for $y'' = f(x, y)$," *IMA Journal of Numerical Analysis*, vol. 2, pp. 83–94, 1982.

[5] L. Brugnano and D. Trigiante, *Solving ODEs By Multistep Initial and Boundary Value Methods*, Gordon & Breach, Amsterdam, The Netherlands, 1998.

[6] L. Brugnano and D. Trigiante, "Block implicit methods for ODEs," in *Recent Trends in Numerical Analysis*, D. Trigiante, Ed., pp. 81–105, Nova Science Publishers, New York, NY, USA, 2001.

[7] E. Hairer, S. P. Nørsett, and G. Wanner, *Solving ordinary differential equations I*, vol. 8 of *Springer Series in Computational Mathematics*, Springer, Berlin, Germany, Second edition, 1993.

[8] T. E. Simos, "An exponentially-fitted Runge-Kutta method for the numerical integration of initial-value problems with periodic or oscillating solutions," *Computer Physics Communications*, vol. 115, no. 1, pp. 1–8, 1998.

[9] J. D. Lambert and I. A. Watson, "Symmetric multistep methods for periodic initial value problems," *Journal of the Institute of Mathematics and Its Applications*, vol. 18, no. 2, pp. 189–202, 1976.

[10] E. H. Twizell and A. Q. Khaliq, "Multiderivative methods for periodic initial value problems," *SIAM Journal on Numerical Analysis*, vol. 21, no. 1, pp. 111–122, 1984.

[11] J. Vigo-Aguiar and H. Ramos, "Dissipative Chebyshev exponential-fitted methods for numerical solution of second-order differential equations," *Journal of Computational and Applied Mathematics*, vol. 158, no. 1, pp. 187–211, 2003.

[12] D. O. Awoyemi, "A new sixth-order algorithm for general second order ordinary differential equations," *International Journal of Computer Mathematics*, vol. 77, no. 1, pp. 117–124, 2001.

[13] M. M. Chawla and S. R. Sharma, "Families of three-stage third order Runge-Kutta-Nyström methods for $y'' = f(x, y, y')$," *Australian Mathematical Society. Journal. Series B. Applied Mathematics*, vol. 26, no. 3, pp. 375–386, 1985.

[14] S. M. Mahmoud and M. S. Osman, "On a class of spline-collocation methods for solving second-order initial-value problems," *International Journal of Computer Mathematics*, vol. 86, no. 4, pp. 616–630, 2009.

[15] J. M. Franco, "Runge–Kutta–Nyström methods adapted to the numerical integration of perturbed oscillators," *Computer Physics Communications. An International Journal and Program Library for Computational Physics and Physical Chemistry*, vol. 147, no. 3, pp. 770–787, 2002.

[16] S. N. Jator, "Implicit third derivative Runge-Kutta-Nyström method with trigonometric coefficients," *Numerical Algorithms*, vol. 70, no. 1, pp. 133–150, 2015.

[17] S. N. Jator, "A continuous two-step method of order 8 with a block extension for $y'' = f(x, y, y')$," *Applied Mathematics and Computation*, vol. 219, pp. 781–791, 2012.

[18] J. P. Coleman and S. C. Duxbury, "Mixed collocation methods for $y'' = f(x,y)$," *Journal of Computational and Applied Mathematics*, vol. 126, no. 1-2, pp. 47–75, 2000.

[19] J. P. Coleman and L. G. Ixaru, "P-stability and exponential-fitting methods for $y'' = f(x, y)$," *IMA Journal of Numerical Analysis*, vol. 16, pp. 179–199, 1996.

[20] T. E. Simos, "Dissipative trigonometrically-fitted methods for second order IVPs with oscillating solution," *International Journal of Modern Physics C. Computational Physics and Physical Computation*, vol. 13, no. 10, 2002.

[21] G. Vanden Berghe, L. G. Ixaru, and M. Van Daele, "Optimal implicit exponentially-fitted Runge-Kutta methods," *Computer Physics Communications*, vol. 140, no. 3, pp. 346–357, 2001.

[22] Y. Fang, Y. Song, and X. Wu, "A robust trigonometrically fitted embedded pair for perturbed oscillators," *Journal of Computational and Applied Mathematics*, vol. 225, no. 2, pp. 347–355, 2009.

[23] H. S. Nguyen, R. B. Sidje, and N. H. Cong, "Analysis of trigonometric implicit Runge-Kutta methods," *Journal of Computational and Applied Mathematics*, vol. 198, no. 1, pp. 187–207, 2007.

[24] H. Ramos and J. Vigo-Aguiar, "A trigonometrically-fitted method with two frequencies, one for the solution and another one for the derivative," *Computer Physics Communications*, vol. 185, no. 4, pp. 1230–1236, 2014.

[25] J. M. Franco and I. Gómez, "Trigonometrically fitted nonlinear two-step methods for solving second order oscillatory IVPs," *Applied Mathematics and Computation*, vol. 232, pp. 643–657, 2014.

[26] K. Ozawa, "A functionally fitted three-stage explicit singly diagonally implicit Runge-Kutta method," *Japan Journal of Industrial and Applied Mathematics*, vol. 22, no. 3, pp. 403–427, 2005.

[27] S. N. Jator, S. Swindell, and R. French, "Trigonometrically fitted block Numerov type method for $y'' = f(x, y, y')$," *Numerical Algorithms*, vol. 62, no. 1, pp. 13–26, 2013.

[28] F. F. Ngwane and S. N. Jator, "Block hybrid method using trigonometric basis for initial value problems with oscillating solutions," *Numerical Algorithms*, vol. 63, no. 4, pp. 713–725, 2013.

[29] S. N. Jator, "Block third derivative method based on trigonometric polynomials for periodic initial-value problems," *Afrika Matematika*, vol. 27, no. 3-4, pp. 365–377, 2016.

[30] H. Ramos and J. Vigo-Aguiar, "On the frequency choice in trigonometrically fitted methods," *Applied Mathematics Letters*, vol. 23, no. 11, pp. 1378–1381, 2010.

[31] L. G. Ixaru and G. Vanden Berghe, *Exponential Fitting*, vol. 568 of *Mathematics and Its Applications*, Springer Netherlands, 2004.

[32] H.-f. Ding, Y.-x. Zhang, J.-x. Cao, and J.-h. Tian, "A class of difference scheme for solving telegraph equation by new non-polynomial spline methods," *Applied Mathematics and Computation*, vol. 218, no. 9, pp. 4671–4683, 2012.

[33] J. B. Keiper and C. W. Gear, "The analysis of generalized backward difference formula methods applied to Hessenberg form differential-algebraic equations," *SIAM Journal on Numerical Analysis*, vol. 28, no. 3, pp. 833–858, 1991.

[34] B.-y. Guo and J.-p. Yan, "Legendre-Gauss collocation method for initial value problems of second order ordinary differential equations," *Applied Numerical Mathematics*, vol. 59, no. 6, pp. 1386–1408, 2009.

[35] B. P. Sommeijer, "Explicit high-order Runge-Kutta-Nyström methods for parallel computers," *Applied Numerical Mathematics*, vol. 13, no. 1–3, pp. 221–240, 1993.

[36] M. Van Daele and G. Vanden Berghe, "Geometric numerical integration by means of exponentially-fitted methods," *Applied Numerical Mathematics*, vol. 57, no. 4, pp. 415–435, 2007.

[37] A. A. Kosti and Z. A. Anastassi, "Explicit almost P-stable Runge–Kutta–Nyström methods for the numerical solution of the two-body problem," *Computational & Applied Mathematics*, vol. 34, no. 2, pp. 647–659, 2015.

[38] Z. A. Anastassi and A. A. Kosti, "A 6(4) optimized embedded Runge-Kutta-Nyström pair for the numerical solution of periodic problems," *Journal of Computational and Applied Mathematics*, vol. 275, pp. 311–320, 2015.

[39] G. A. Panopoulos, Z. A. Anastassi, and T. E. Simos, "A new eight-step symmetric embedded predictor-corrector method (EPCM) for orbital problems and related IVPs with oscillatory solutions," *The Astronomical Journal*, vol. 145, no. 3, article 75, 2013.

[40] B. Wang, A. Iserles, and X. Wu, "Arbitrary-order trigonometric Fourier collocation methods for multi-frequency oscillatory systems," *Foundations of Computational Mathematics*, vol. 16, no. 1, pp. 151–181, 2016.

[41] G. Vanden Berghe, L. G. Ixaru, and H. De Meyer, "Frequency determination and step-length control for exponentially-fitted Runge-Kutta methods," *Journal of Computational and Applied Mathematics*, vol. 132, no. 1, pp. 95–105, 2001.

20

Identifying Initial Condition in Degenerate Parabolic Equation with Singular Potential

K. Atifi, Y. Balouki, El-H. Essoufi, and B. Khouiti

Laboratoire de Mathématiques, Informatique et Sciences de l'Ingénieur (MISI), Université Hassan 1er, 26000 Settat, Morocco

Correspondence should be addressed to B. Khouiti; khouiti_b@yahoo.fr

Academic Editor: Jen-Chih Yao

A hybrid algorithm and regularization method are proposed, for the first time, to solve the one-dimensional degenerate inverse heat conduction problem to estimate the initial temperature distribution from point measurements. The evolution of the heat is given by a degenerate parabolic equation with singular potential. This problem can be formulated in a least-squares framework, an iterative procedure which minimizes the difference between the given measurements and the value at sensor locations of a reconstructed field. The mathematical model leads to a nonconvex minimization problem. To solve it, we prove the existence of at least one solution of problem and we propose two approaches: the first is based on a Tikhonov regularization, while the second approach is based on a hybrid genetic algorithm (married genetic with descent method type gradient). Some numerical experiments are given.

1. Introduction

The inverse problem is expressed when the PDE solution is measured or specified, and we are interested in determining some properties: coefficients, forcing term, boundary, or initial condition from the partial knowledge of the system in a limited time interval (see [1, 2]).

In the last recent years, an increasing interest has been devoted to degenerate parabolic equations. Indeed, many problems coming from physics (boundary layer models in [3], models of Kolmogorov type in [4], etc.), biology (Wright-Fisher models in [5] and Fleming-Viot models in [6]), and economics (Black-Merton-Scholes equations in [7]) are described by degenerate parabolic equations [8].

The identification of the initial state of nondegenerate parabolic problems is well studied in the literature (see [9–11]). However, as far as we know, the degenerate case has not been analysed in the literature.

In this paper, we are interested in estimating the initial condition by the variational method in data assimilation of degenerate/singular parabolic equation:

$$\partial_t \psi - \partial_x \left(a(x)\partial_x \psi(x)\right) - \frac{\lambda}{x^\beta}\psi = f, \quad a(x) = x^\alpha, \qquad (1)$$

where $\Omega =]0,1[$, $\alpha \in]0,1[$, $\beta \in]0, 2-\alpha[$, $\lambda \leqslant 0$, and $f \in L^2(\Omega\times]0,T[)$. With initial and boundary conditions

$$\psi(x,0) = \psi_0, \quad \psi|_{x=0} = \psi|_{x=1} = 0. \qquad (2)$$

The mathematical model leads to a nonconvex minimization problem

$$\begin{aligned} &\text{find} \quad \psi_0 \in A_{\text{ad}} \\ &\text{such that} \quad J(\psi_0) = \min_{u\in A_{\text{ad}}} J(u), \end{aligned} \qquad (3)$$

where the functional J is defined as follows:

$$J(u) = \frac{1}{2T}\int_0^T \left\| C\psi(t) - \psi^{\text{obs}}(t) \right\|^2_{L^2(\Omega)}, \qquad (4)$$

subject to ψ being the weak solution of the parabolic problem (1) with initial state u, ψ^{obs} an observation of ψ in $\Omega\times]0,T[$, and C the observation operator. The space A_{ad} is the set of admissible initial states.

Problem (3) is ill-posed in the sense of Hadamard. To solve this problem, we propose two approaches.

The first approach is based on regularization, for the first time, applied to solve a degenerate inverse problem. The problem thus consists of minimizing a functional of the form

$$J_T(u) = \frac{1}{2T}\int_0^T \left\|C\psi(t) - \psi^{\text{obs}}(t)\right\|_{L^2(\Omega)}^2 dt + \frac{\varepsilon}{2}\left\|u - \psi^b\right\|_{L^2(\Omega)}^2. \quad (5)$$

Here, the last term in (5) stands for the so-called Tikhonov-type regularization ([12, 13]), ε being a small regularizing coefficient that provides extra convexity to the functional J_T and ψ^b a priori (background) knowledge of the true state ψ_0^{exact} (the state to estimate). We consider that the values of ψ^b are given in each point of analysis grid-points.

The second approach is applied when there is a partial knowledge of values of ψ^b (example 20%); the regularization parameter is very difficult to determine. To overcome this problem, we propose a new approach, based on a hybrid genetic algorithm (married genetic with descent method gradient type). Finally, we make a comparison between the two mentioned approaches (with 20% of ψ^b).

First of all, we prove that problem (3) has at least one solution. The gradient of the functional J is calculated with the adjoint method. Numerical experiments are presented to show the performance of our approaches.

2. Problem Statement and Main Result

Consider the following problem:

$$\begin{aligned} &\partial_t\psi + A(\psi) = f \\ &\qquad \psi(0,t) = \psi(1,t) = 0 \quad \forall t \in]0,T[\\ &\qquad \psi(x,0) = \psi_0(x) \quad \forall x \in \Omega, \end{aligned} \quad (6)$$

where $\Omega =]0,1[$, $f \in L^2(\Omega\times]0,T[)$,and A is the operator defined as

$$A(\psi) = -\partial_x(a(x)\partial_x\psi(x)) - \frac{\lambda}{x^\beta}\psi, \quad a(x) = x^\alpha, \quad (7)$$

with $\alpha \in]0,1[$, $\beta \in]0,2-\alpha[$, and $\lambda \leqslant 0$.

We want to estimate ψ_0 thanks to an observation $\psi^{\text{obs}}(x,t)$ of $\psi(x,t)$ in $\Omega\times]0,T[$. The minimization problem associated with this problem is

$$\begin{aligned} &\text{find} \quad \psi_0 \in A_{\text{ad}} \\ &\text{such that} \quad J(\psi_0) = \min_{u\in A_{\text{ad}}} J(u), \end{aligned} \quad (8)$$

where the functional J is as follows:

$$J(u) = \frac{1}{2T}\int_0^T \left\|C\psi(t) - \psi^{\text{obs}}(t)\right\|_{L^2(\Omega)}^2, \quad (9)$$

subject to ψ being the weak solution of the parabolic problem (6) with initial state u, ψ^b the background state, and C the observation operator. The space A_{ad} is the set of admissible initial states (will be defined later).

We now specify some notations we shall use. Let us introduce the following functional spaces (see [14–16]):

$$\begin{aligned} V &= \{u \in L^2(\Omega),\ u \text{ absolutely continuous on } [0,1]\}, \\ S &= \{u \in L^2(\Omega),\ \sqrt{a}u_x \in L^2(\Omega),\ u(0) = u(1) = 0\}, \\ H_a^1(\Omega) &= V \cap S, \\ H_a^2(\Omega) &= \{u \in H_a^1(\Omega),\ au_x \in H^1(\Omega)\}, \\ H_{\alpha,0}^1 &= \{u \in H_\alpha^1 \mid u(0) = u(1) = 0\}, \\ H_\alpha^1 &= \{u \in L^2(\Omega) \cap H_{\text{Loc}}^1(]0,1]) \mid x^{\alpha/2}u_x \in L^2(\Omega)\}, \end{aligned} \quad (10)$$

with

$$\begin{aligned} \|u\|_{H_a^1(\Omega)}^2 &= \|u\|_{L^2(\Omega)}^2 + \left\|\sqrt{a}u_x\right\|_{L^2(\Omega)}^2, \\ \|u\|_{H_a^2(\Omega)}^2 &= \|u\|_{H_a^1(\Omega)}^2 + \left\|(au_x)_x\right\|_{L^2(\Omega)}^2, \\ \langle u,v\rangle_{H_\alpha^1} &= \int_\Omega (uv + x^\alpha u_x v_x)\,dx. \end{aligned} \quad (11)$$

We recall that (see [16]) H_a^1 is an Hilbert space and it is the closure of $C_c^\infty(0,1)$ for the norm $\|\cdot\|_{H_a^1}$. If $1/\sqrt{a} \in L^1(\Omega)$ then the injections

$$\begin{aligned} &H_a^1(\Omega) \hookrightarrow L^2(\Omega), \\ &H_a^2(\Omega) \hookrightarrow H_a^1(\Omega), \\ &H^1\left(0,T;L^2(\Omega)\right) \cap L^2(0,T;D(A)) \\ &\quad \hookrightarrow L^2\left(0,T;H_a^1\right) \cap C\left(0,T;L^2(\Omega)\right) \end{aligned} \quad (12)$$

are compacts.

Firstly, we prove that problem (6) is well-posed, the functional J is continuous, and J is G-derivable in A_{ad}.

The weak formulation of problem (6) is

$$\int_\Omega \partial_t\psi v\,dx + \int_\Omega \left(a(x)\partial_x\psi\partial_x v - \frac{\lambda}{x^\beta}\psi v\right)dx = \int_\Omega fv\,dx, \quad \forall v \in H_0^1(\Omega). \quad (13)$$

Let

$$B[\psi,v] = \int_\Omega \left(a(x)\partial_x\psi\partial_x v - \frac{\lambda}{x^\beta}\psi v\right)dx. \quad (14)$$

We discuss the following cases.

(1) Noncoercive Case (see [14], $\lambda = 0$). In this case, the bilinear form B becomes

$$B[\psi,v] = \int_\Omega (a(x)\partial_x\psi\partial_x v)\,dx. \quad (15)$$

We have $a(x) = 0$ at $x = 0$, from where the bilinear form B will be noncoercive.

Let

$$A_{\text{ad}} = \left\{u \in H_a^1(\Omega);\ \|u\|_{H_a^1(\Omega)} \leqslant r\right\}, \tag{16}$$

where r is a real strictly positive constant.

We recall the following theorem.

Theorem 1 (see [14, 17, 18]). *For all $f \in L^2(\Omega\times]0,T[)$ and $\psi_0 \in L^2(0,1)$, there exists a unique weak solution which solves problem (6) such that*

$$\psi \in C^0\left([0,T];L^2(\Omega)\right) \cap L^2\left(0,T;H_a^1\right) \tag{17}$$

and there is a constant C_T such that for any solution of (6)

$$\begin{aligned}\sup_{t\in[0,T]} \|\psi(t)\|_{L^2(\Omega)}^2 + \int_0^T \|\sqrt{a}\psi_x(t)\|_{L^2(\Omega)}^2\, dt \\ \leq C_T\left(\|\psi_0\|_{L^2(\Omega)}^2 + \|f\|_{L^2(\Omega\times]0,T[)}^2\right);\end{aligned} \tag{18}$$

if more $\psi_0 \in H_a^1(\Omega)$ then

$$\begin{aligned}\psi \in C^0\left([0;T],H_a^1\right) \cap L^2\left(0,T;H_a^2\right) \\ \cap H^1\left(0,T;L^2(\Omega)\right)\end{aligned} \tag{19}$$

and there is a constant C_T such that

$$\begin{aligned}&\sup_{t\in[0,T]} \|\psi(t)\|_{H_a^1}^2 \\ &+ \int_0^T \left(\|\psi_t\|_{L^2(\Omega)}^2 + \|(a\psi_x)_x(t)\|_{L^2(\Omega)}^2\right) dt \\ &\leqslant C_T\left(\|\psi_0\|_{H_a^1}^2 + \|f\|_{L^2(\Omega\times]0,T[)}^2\right).\end{aligned} \tag{20}$$

Theorem 2. *Let ψ be the weak solution of (6) with initial state ψ_0. In noncoercive case, the function*

$$\begin{aligned}\varphi : H_a^1(\Omega) \longrightarrow C^0\left([0,T];H_a^1(\Omega)\right) \\ \cap L^2\left(0,T;H_a^2(\Omega)\right) \\ \cap H^1\left(0,T;L^2(\Omega)\right) \\ \psi_0 \longmapsto \psi\end{aligned} \tag{21}$$

is continuous, and the functional J has at least one minimum in A_{ad}.

Theorem 3. *Let ψ be the weak solution of (6) with initial state ψ_0. The function*

$$\begin{aligned}\varphi : H_a^1(\Omega) \longrightarrow C^0\left([0,T];H_a^1(\Omega)\right) \\ \cap L^2\left(0,T;H_a^2(\Omega)\right) \\ \cap H^1\left(0,T;L^2(\Omega)\right) \\ \psi_0 \longmapsto \psi\end{aligned} \tag{22}$$

is G-derivable in A_{ad}.

(2) Subcritical Potential Case (see [19, 20], $\lambda \neq 0$). Then the bilinear form B becomes

$$B[\psi,v] = \int_\Omega \left(a(x)\partial_x\psi\partial_x v - \frac{\lambda}{x^\beta}\psi v\right) dx. \tag{23}$$

Since $a(x) = 0$ at $x = 0$ and $\lim_{x\to 0}(\lambda/x^\beta) = +\infty$, the bilinear form B is noncoercive and is noncontinuous at $x = 0$.

Consider the not bounded operator $(K, D(K))$ where

$$\begin{aligned}&Ku = (x^\alpha u_x)_x + \frac{\lambda}{x^\beta}u, \quad \forall u \in D(K), \\ &D(k) = \left\{u \in H_{\alpha,0}^1 \cap H_{\text{Loc}}^2(]0,1]) \mid (x^\alpha u_x)_x + \frac{\lambda}{x^\beta}u \right. \\ &\left. \quad \in L^2(\Omega)\right\}.\end{aligned} \tag{24}$$

Let

$$A_{\text{ad}} = \left\{u \in L^2(\Omega);\ \|u\|_{L^2(\Omega)} \leqslant r\right\}, \tag{25}$$

where r is a real strictly positive constant.

We recall the following theorem.

Theorem 4 (see [15, 19]). *If $f = 0$ then, for all $\psi_0 \in L^2(\Omega)$, problem (6) has a unique weak solution*

$$\begin{aligned}\psi \in C^0\left([0,T];L^2(\Omega)\right) \cap C^0(]0,T];D(K)) \\ \cap C^1\left(]0,T];L^2(\Omega)\right);\end{aligned} \tag{26}$$

if more $\psi_0 \in D(K)$ then

$$\psi \in C^0([0,T];D(K)) \cap C^1\left([0,T];L^2(\Omega)\right); \tag{27}$$

if $f \in L^2(]0,1[\times]0,T[)$ then, for all $\psi_0 \in L^2(\Omega)$, problem (6) has a unique solution

$$\psi \in C^0\left([0,T];L^2(\Omega)\right). \tag{28}$$

Theorem 5. *Let ψ be the weak solution of (6) with initial state ψ_0. In subcritical potential case, the function*

$$\begin{aligned}\varphi : L^2(\Omega) \longrightarrow C\left([0,T];L^2(\Omega)\right) \\ \psi_0 \longmapsto \psi\end{aligned} \tag{29}$$

is continuous, and the functional J is continuous in A_{ad}.

Theorem 6. *Let ψ be the weak solution of (6) with initial state ψ_0. The function*

$$\begin{aligned}\varphi : L^2(\Omega) \longrightarrow C\left([0,T];L^2(\Omega)\right) \\ \psi_0 \longmapsto \psi\end{aligned} \tag{30}$$

is G-derivable in A_{ad}.

3. Proof

Proof of Theorem 2. Let $\delta\psi_0 \in H_a^1(\Omega)$ be a small variation such that $\psi_0 + \delta\psi_0 \in A_{\text{ad}}$.

Consider $\delta\psi = \psi^\delta - \psi$, with ψ being the weak solution of (6) with initial state ψ_0 and ψ^δ is the weak solution of (6) with initial state $\psi_0^\delta = \psi_0 + \delta\psi_0$.

Consequently, $\delta\psi$ is the solution of the variational problem:

$$\begin{aligned} &\int_\Omega \partial_t \delta\psi v\, dx + \int_\Omega a(x)\,\partial_x \delta\psi(x)\,\partial_x v\, dx = 0 \\ &\delta\psi(0,t) = \delta\psi(1,t) = 0 \quad \forall t \in]0,T[\\ &\delta\psi(x,0) = \delta\psi_0(x) \quad \forall x \in \Omega. \end{aligned} \tag{31}$$

Hence, $\delta\psi$ is the weak solution of (6) with $f = 0$. We apply the estimate in Theorem 1 with $f = 0$. This gives the following.

There is a constant C_T such that

$$\begin{aligned} &\sup_{t\in[0,T]} \|\delta\psi(t)\|_{H_a^1(\Omega)}^2 \\ &\quad + \int_0^T \left(\|\partial_t \delta\psi\|_{L^2(\Omega)}^2 + \|\partial_x (a\partial_x \delta\psi)(t)\|_{L^2(\Omega)}^2 \right) dt \\ &\leqslant C_T \|\delta\psi_0\|_{H_a^1(\Omega)}^2; \end{aligned} \tag{32}$$

therefore

$$\sup_{t\in[0,T]} \|\delta\psi(t)\|_{H_a^1(\Omega)}^2 \leqslant C_T \|\delta\psi_0\|_{H_a^1(\Omega)}^2 \tag{33}$$

$$\|\delta\psi\|_{C([0,T];H_a^1(\Omega))}^2 \leqslant C_T \|\delta\psi_0\|_{H_a^1(\Omega)}^2. \tag{34}$$

And from (32) we have

$$\begin{aligned} &\|\delta\psi(t)\|_{H_a^1(\Omega)}^2 + \int_0^T \|\partial_x (a\partial_x \delta\psi)(t)\|_{L^2(\Omega)}^2 dt \\ &\quad\leqslant C_T \|\delta\psi_0\|_{H_a^1(\Omega)}^2, \quad \forall t \in [0,T] \\ &\int_0^T \|\delta\psi(t)\|_{H_a^1(\Omega)}^2 dt + T\int_0^T \|\partial_x (a\partial_x \delta\psi)(t)\|_{L^2(\Omega)}^2 dt \\ &\quad\leqslant TC_T \|\delta\psi_0\|_{H_a^1(\Omega)}^2 \\ &\inf(1,T)\left(\int_0^T \|\delta\psi(t)\|_{H_a^1(\Omega)}^2 dt \right. \\ &\quad \left. + \int_0^T \|\partial_x (a\partial_x \delta\psi)(t)\|_{L^2(\Omega)}^2 dt \right) \leqslant TC_T \|\delta\psi_0\|_{H_a^1(\Omega)}^2 \\ &\int_0^T \|\delta\psi(t)\|_{H_a^1(\Omega)}^2 dt + \int_0^T \|\partial_x (a\partial_x \delta\psi)(t)\|_{L^2(\Omega)}^2 dt \\ &\quad\leqslant \frac{TC_T}{\inf(1,T)} \|\delta\psi_0\|_{H_a^1(\Omega)}^2. \end{aligned} \tag{35}$$

Hence,

$$\|\delta\psi\|_{L^2(0,T,H_a^2(\Omega))}^2 \leqslant \frac{TC_T}{\inf(1,T)} \|\delta\psi_0\|_{H_a^1(\Omega)}^2. \tag{36}$$

In addition, from (32) we have

$$\begin{aligned} &\|\delta\psi(t)\|_{H_a^1(\Omega)}^2 \\ &\quad + \int_0^T \|\partial_t \delta\psi(t)\|_{L^2(\Omega)}^2 dt \leqslant C_T \|\delta\psi_0\|_{H_a^1(\Omega)}^2, \\ &\qquad \forall t \in [0,T] \end{aligned} \tag{37}$$

$$\begin{aligned} &\|\delta\psi(t)\|_{L^2(\Omega)}^2 + \|\sqrt{a}\partial_x \delta\psi(t)\|_{L^2(\Omega)}^2 \\ &\quad + \int_0^T \|\partial_t \delta\psi(t)\|_{L^2(\Omega)}^2 dt \leqslant C_T \|\delta\psi_0\|_{H_a^1(\Omega)}^2, \\ &\qquad \forall t \in [0,T] \end{aligned} \tag{38}$$

$$\begin{aligned} &\|\delta\psi(t)\|_{L^2(\Omega)}^2 \\ &\quad + \int_0^T \|\partial_t \delta\psi(t)\|_{L^2(\Omega)}^2 dt \leqslant C_T \|\delta\psi_0\|_{H_a^1(\Omega)}^2, \\ &\qquad \forall t \in [0,T] \end{aligned} \tag{39}$$

$$\begin{aligned} &\int_0^T \|\delta\psi(t)\|_{L^2(\Omega)}^2 dt \\ &\quad + T\int_0^T \|\partial_t \delta\psi(t)\|_{L^2(\Omega)}^2 dt \leqslant TC_T \|\delta\psi_0\|_{H_a^1(\Omega)}^2 \end{aligned} \tag{40}$$

$$\|\delta\psi\|_{H^1(0,T;L^2(\Omega))}^2 \leqslant \frac{TC_T}{\inf(1,T)} \|\delta\psi_0\|_{H_a^1(\Omega)}^2. \tag{41}$$

Equations (34), (36), and (41) imply the continuity of the function

$$\begin{aligned} \varphi : H_a^1(\Omega) \longrightarrow C^0&\left([0,T]; H_a^1(\Omega)\right) \\ &\cap L^2\left(0,T; H_a^2(\Omega)\right) \\ &\cap H^1\left(0,T; L^2(\Omega)\right) \\ \psi_0 \longmapsto \psi&. \end{aligned} \tag{42}$$

Hence, the functional J is continuous in

$$A_{\text{ad}} = \left\{ u \in H_a^1(\Omega);\ \|u\|_{H_a^1(\Omega)} \leqslant r \right\}. \tag{43}$$

We have $1/\sqrt{a(x)} = x^{-\alpha/2} \in L^1(0,1)$, where $\alpha \in]0,1[$, which gives $H_a^1(\Omega) \underset{\text{compact}}{\hookrightarrow} L^2(\Omega)$. Since the set A_{ad} is bounded in $H_a^1(\Omega)$, then A_{ad} is a compact in $L^2(\Omega)$. Therefore, J has at least one minimum in A_{ad}. □

Proof of Theorem 3. Let $\psi_0 \in A_{ad}$ and $\delta\psi_0$ such that $\psi_0+\delta\psi_0 \in A_{ad}$; we define the function

$$\varphi'(\psi_0) : \delta\psi_0 \in A_{ad} \longrightarrow \delta\psi, \tag{44}$$

where $\delta\psi$ is the solution of the variational problem

$$\begin{gathered}\int_\Omega \partial_t(\delta\psi)\, v\, dx + \int_\Omega a(x)\,\partial_x(\delta\psi)\,\partial_x v\, dx = 0 \\ \forall v \in H_0^1(\Omega) \\ \delta\psi(0,t) = \delta\psi(1,t) = 0 \quad \forall t \in]0,T[\\ \delta\psi(x,0) = \delta\psi_0 \quad \forall x \in \Omega\end{gathered} \tag{45}$$

and we pose

$$\phi(\psi_0) = \varphi(\psi_0+\delta\psi_0) - \varphi(\psi_0) - \varphi'(\psi_0)\,\delta\psi_0. \tag{46}$$

We want to show that

$$\phi(\psi_0) = o(\delta\psi_0). \tag{47}$$

We easily verify that the function ϕ is solution of the following variational problem:

$$\begin{gathered}\int_\Omega \partial_t\phi v\, dx + \int_\Omega a(x)\,\partial_x\phi\,\partial_x v\, dx = 0 \quad \forall v \in H_0^1(\Omega) \\ \phi(0,t) = \phi(1,t) = 0 \\ \forall t \in]0,T[\\ \phi(x,0) = \delta\psi_0 - (\delta\psi_0)^2 \\ \forall x \in \Omega.\end{gathered} \tag{48}$$

By the same way as that used in the proof of continuity, we deduce

$$\begin{gathered}\|\phi\|^2_{C([0,T],H_a^1(\Omega))} \leqslant C_T \left\|\delta\psi_0 - (\delta\psi_0)^2\right\|^2_{H_a^1(\Omega)}, \\ \|\phi\|^2_{L^2(0,T,H_a^2(\Omega))} \leqslant \frac{TC_T}{\inf(1,T)} \left\|\delta\psi_0 - (\delta\psi_0)^2\right\|^2_{H_a^1(\Omega)}, \\ \|\phi\|^2_{H^1(0,T;L^2(\Omega))} \leqslant \frac{TC_T}{\inf(1,T)} \left\|\psi_0 - (\delta\psi_0)^2\right\|^2_{H_a^1(\Omega)}.\end{gathered} \tag{49}$$

Hence, the function $\varphi : \psi_0 \to \psi$ is G-derivable in A_{ad} and we deduce the existence of the gradient of the functional J. □

Proof of Theorem 5. Let $\delta\psi_0 \in L^2(\Omega)$ be a small variation such that $\psi_0 + \delta\psi_0 \in A_{ad}$.

Consider $\delta\psi = \psi^\delta - \psi$, with ψ being the weak solution of (6) with initial state ψ_0, and ψ^δ is the weak solution of (6) with initial state $\psi_0^\delta = \psi_0 + \delta\psi_0$.

Consequently, $\delta\psi$ is the solution of variational problem

$$\begin{gathered}\int_\Omega \partial_t\delta\psi v\, dx + \int_\Omega \left(a(x)\,\partial_x\delta\psi\partial_x v - \frac{\lambda}{x^\beta}\delta\psi v\right) dx = 0, \\ \forall v \in H_0^1(\Omega) \\ \delta\psi(0,t) = \delta\psi(1,t) = 0 \quad \forall t \in]0,T[\\ \delta\psi(x,0) = \delta\psi_0(x) \quad \forall x \in \Omega.\end{gathered} \tag{50}$$

Take $v = \delta\psi$; this gives

$$\int_\Omega \partial_t\delta\psi\delta\psi\, dx + \int_\Omega \left(a(x)(\partial_x\delta\psi)^2 - \frac{\lambda}{x^\beta}(\delta\psi)^2\right) dx = 0, \tag{51}$$

since Ω is independent of t, which gives

$$\frac{1}{2}\frac{d}{dt}\int_\Omega (\delta\psi)^2\, dx\, dt + \int_\Omega \left(a(x)(\partial_x\delta\psi)^2 - \frac{\lambda}{x^\beta}(\delta\psi)^2\right) dx = 0. \tag{52}$$

By integrating between 0 and t with $t \in [0,T]$ we obtain

$$\frac{1}{2}\|\delta\psi(t)\|^2_{L^2(\Omega)} + \int_0^t\int_\Omega \left(a(x)(\partial_x\delta\psi)^2 - \frac{\lambda}{x^\beta}(\delta\psi)^2\right) dx\, dt = \frac{1}{2}\|\delta\psi(0)\|^2_{L^2(\Omega)} \tag{53}$$

$$\frac{1}{2}\|\delta\psi(t)\|^2_{L^2(\Omega)} + \int_0^t\int_\Omega \left(a(x)(\partial_x\delta\psi)^2 - \frac{\lambda}{x^\beta}(\delta\psi)^2\right) dx\, dt = \frac{1}{2}\|\delta\psi_0\|^2_{L^2(\Omega)}, \tag{54}$$

and since $a(x) \geqslant 0$ and $-\lambda/x^\beta > 0, \forall x \in \Omega$, we obtain

$$\|\delta\psi(t)\|_{L^2(\Omega)} \leqslant \|\delta\psi_0\|_{L^2(\Omega)}; \tag{55}$$

this gives

$$\operatorname*{Sup}_{t\in[0,T]} \|\delta\psi(t)\|_{L^2(\Omega)} \leq \|\delta\psi_0\|_{L^2(\Omega)}. \tag{56}$$

From where

$$\|\delta\psi(t)\|_{C([0;T],L^2(\Omega))} \leq \|\delta\psi_0\|_{L^2(\Omega)}. \tag{57}$$

Which gives the continuity of the function

$$\begin{gathered}\varphi : L^2(\Omega) \longrightarrow C([0,T];L^2(\Omega)) \\ \psi_0 \longmapsto \psi.\end{gathered} \tag{58}$$

Hence, the functional J is continuous in

$$A_{\text{ad}} = \left\{ u \in L^2(\Omega);\ \|u\|_{L^2(\Omega)} \leqslant r \right\}. \tag{59}$$

□

Proof of Theorem 6. Let $\psi_0 \in A_{\text{ad}}$ and $\delta\psi_0$ such that $\psi_0 + \delta\psi_0 \in A_{\text{ad}}$; we define the function

$$\varphi'(\psi_0) : \delta\psi_0 \in A_{\text{ad}} \longrightarrow \delta\psi, \tag{60}$$

where $\delta\psi$ is the solution of the variational problem

$$\begin{aligned} &\int_\Omega \partial_t(\delta\psi)\, v\, dx \\ &\quad + \int_\Omega \left(a(x)\,\partial_x(\delta\psi)\,\partial_x v - \frac{\lambda}{x^\beta}\delta\psi v \right) dx = 0 \\ &\qquad\qquad \forall v \in H_0^1(\Omega) \\ &\delta\psi(0,t) = \delta\psi(1,t) = 0 \quad \forall t \in\,]0,T[\\ &\delta\psi(x,0) = \delta\psi_0 \quad \forall x \in \Omega \end{aligned} \tag{61}$$

and we pose

$$\phi(\psi_0) = \varphi(\psi_0 + \delta\psi_0) - \varphi(\psi_0) - \varphi'(\psi_0)\,\delta\psi_0. \tag{62}$$

We want to show that

$$\phi(\psi_0) = o(\delta\psi_0). \tag{63}$$

We easily verify that the function ϕ is the solution of the following variational problem:

$$\begin{aligned} &\int_\Omega \partial_t \phi v\, dx + \int_\Omega \left(a(x)\,\partial_x\phi\partial_x v - \frac{\lambda}{x^\beta}\phi v \right) dx = 0 \\ &\qquad\qquad \forall v \in H_0^1(\Omega) \\ &\phi(0,t) = \phi(1,t) = 0 \quad \forall t \in\,]0,T[\\ &\phi(x,0) = \delta\psi_0 - (\delta\psi_0)^2 \quad \forall x \in \Omega. \end{aligned} \tag{64}$$

By the same way as that used in the proof of continuity, we deduce

$$\|\phi\|_{C([0,T],L^2(\Omega))} \leqslant \left\| \delta\psi_0 - (\delta\psi_0)^2 \right\|_{L^2(\Omega)}. \tag{65}$$

Hence, in all cases, the function $\varphi : \psi_0 \to \psi$ is G-derivable in A_{ad} and we deduce the existence of the gradient of the functional J. □

Now, we are going to compute the gradient of J with the adjoint state method.

4. Gradient of J

We define the Gâteaux derivative of ψ at ψ_0 in the direction $h \in L^2(\Omega)$, by

$$\widehat{\psi} = \lim_{\varepsilon \to 0} \frac{\psi(\psi_0 + \varepsilon h) - \psi(\psi_0)}{\varepsilon}, \tag{66}$$

where $\psi(\psi_0 + \varepsilon h)$ is the weak solution of (6) with initial state $\psi_0 + \varepsilon h$, and $\psi(\psi_0)$ is the weak solution of (6) with initial state ψ_0.

We compute the Gâteaux (directional) derivative of (6) at ψ_0 in some direction $h \in L^2(\Omega)$, and we get the so-called tangent linear model:

$$\begin{aligned} &\partial_t \widehat{\psi} + A\widehat{\psi} = 0 \\ &\widehat{\psi}(0,t) = \widehat{\psi}(1,t) = 0 \quad \forall t \in\,]0,T[\\ &\widehat{\psi}(x,0) = h \quad \forall x \in \Omega. \end{aligned} \tag{67}$$

We introduce the adjoint variable P, and we integrate

$$\begin{aligned} &\int_0^1 \int_0^T \partial_t \widehat{\psi} P\, dt\, dx + \int_0^1 \int_0^T A\widehat{\psi} P\, dx = 0 \\ &\int_0^1 \left([\widehat{\psi} P]_0^T - \int_0^T \widehat{\psi}\partial_t P\, dt \right) dx \\ &\quad + \int_0^T \langle A\widehat{\psi}, P\rangle_{L^2(\Omega)}\, dt = 0 \end{aligned} \tag{68}$$

$$\begin{aligned} &\int_0^1 \left[\widehat{\psi}(T)\, P(T) - \widehat{\psi}(0)\, P(0) \right] dx \\ &\quad - \int_0^T \langle \widehat{\psi}, \partial_t P\rangle_{L^2(\Omega)}\, dt + \int_0^T \langle A\widehat{\psi}, P\rangle_{L^2(\Omega)}\, dt = 0. \end{aligned} \tag{69}$$

Let us take $P(x=0) = P(x=1) = 0$; then we may write $\langle \widehat{\psi}, AP\rangle_{L^2(\Omega)} = \langle A\widehat{\psi}, P\rangle_{L^2(\Omega)}$.

And with $P(T) = 0$ we may now rewrite (69) as

$$\int_0^1 \widehat{\psi}(0)\, P(0)\, dx + \int_0^T \langle \widehat{\psi}, \partial_t P - AP\rangle_{L^2(\Omega)}\, dt = 0; \tag{70}$$

this gives

$$\begin{aligned} &\int_0^T \langle \widehat{\psi}, \partial_t P - AP\rangle_{L^2(\Omega)}\, dt = \langle -P(0), h\rangle_{L^2(\Omega)} \\ &\qquad P(x=0) = P(x=1) = 0, \\ &\qquad P(T) = 0. \end{aligned} \tag{71}$$

The Gâteaux derivative of the functional

$$J(\psi_0) = \frac{1}{2T} \int_0^T \left\| C\psi(t) - \psi^{\text{obs}}(t) \right\|_{L^2(\Omega)}^2 dt \tag{72}$$

at ψ_0 in the direction $h \in L^2(\Omega)$ is given by

$$\widehat{J}(h) = \lim_{\varepsilon \to 0} \frac{J(\psi_0 + \varepsilon h) - J(\psi_0)}{\varepsilon}. \tag{73}$$

After some calculations, we arrive at

$$\widehat{J}(h) = \frac{1}{T}\int_0^T \left\langle C^*\left(C\psi - \psi^{\text{obs}}\right), \widehat{\psi}\right\rangle_{L^2(\Omega)} dt. \tag{74}$$

The adjoint model is

$$\begin{aligned} &\partial_t P - AP = \frac{1}{T}C^*\left(C\psi - \psi^{\text{obs}}\right) \\ &P(x=0) = P(x=1) = 0 \quad \forall t \in]0,T[\\ &P(T) = 0. \end{aligned} \tag{75}$$

Problem (75) is retrograde; we make the change of variable $t \leftrightarrow T-t$, which gives

$$\begin{aligned} &\partial_t P + AP = \frac{1}{T}C^*\left(\widetilde{\psi}^{\text{obs}} - C\widetilde{\psi}\right) \\ &P(x=0) = P(x=1) = 0 \quad \forall t \in]0,T[\\ &P(0) = 0, \end{aligned} \tag{76}$$

with $\widetilde{\psi}(t) = \psi(T-t)$.

From (71), (74), and (75) the gradient of J is given by

$$\frac{\partial J}{\partial \psi_0} = -P(0). \tag{77}$$

With the change of variable $t \leftrightarrow T-t$, the gradient becomes

$$\frac{\partial J}{\partial \psi_0} = -P(T). \tag{78}$$

To calculate a gradient of J, we solve two problems: (6) and (76). The result solution of (6) is used in the second member of problem (76).

5. Discretization of Problem

Step 1 (full discretization). To resolve problem (6) and (76), we use the method θ-schema in time. This method is unconditionally stable for $1 > \theta \geq 1/2$.

Let h be the steps in space and Δt the steps in time.

Let

$$\begin{aligned} &x_i = ih, \quad i \in \{0,1,2,\ldots,N+1\}, \\ &c(x_i) = a(x_i), \\ &t_j = j\Delta t, \quad j \in \{0,1,2,\ldots,M+1\}, \\ &f_i^j = f(x_i, t_j); \end{aligned} \tag{79}$$

we put

$$\psi_i^j = \psi(x_i, t_j). \tag{80}$$

Let

$$\begin{aligned} da(x_i) &= \frac{c(x_{i+1}) - c(x_i)}{h}, \\ b(x) &= -\frac{\lambda}{x^\beta}. \end{aligned} \tag{81}$$

Therefore,

$$\partial_t \psi + A\psi = f \tag{82}$$

is approximated by

$$\begin{aligned} &-\frac{\theta\Delta t}{h^2}c(x_i)\psi_{i-1}^{j+1} + \left(1 + \frac{2\theta\Delta t}{h^2}c(x_i) + da(x_i)\frac{\theta\Delta t}{h}\right. \\ &\left. + b(x_i)\theta\Delta t\right)\psi_i^{j+1} - \left(\frac{\theta\Delta t}{h^2}c(x_i) + da(x_i)\frac{\theta\Delta t}{h}\right) \\ &\cdot \psi_{i+1}^{j+1} = \frac{(1-\theta)\Delta t}{h^2}c(x_i)\psi_{i-1}^j + \left(1\right. \\ &- \frac{(1-\theta)\Delta t}{h}da(x_i) - \frac{2(1-\theta)\Delta t}{h^2}c(x_i) \\ &\left. - (1-\theta)b(x_i)\Delta t\right)\psi_i^j + \left(\frac{(1-\theta)\Delta t}{h}da(x_i)\right. \\ &\left. + \frac{(1-\theta)\Delta t}{h^2}c(x_i)\right)\psi_{i+1}^j + \Delta t \cdot \left[(1-\theta)f_i^j\right. \\ &\left. + \theta f_i^{j+1}\right]. \end{aligned} \tag{83}$$

Let us define

$$\begin{aligned} g_1(x_i) &= -\frac{\theta\Delta t}{h^2}c(x_i), \\ g_2(x_i) &= 1 + \frac{2\theta\Delta t}{h^2}c(x_i) + da(x_i)\frac{\theta\Delta t}{h} + b(x_i)\theta\Delta t, \\ g_3(x_i) &= -\frac{\theta\Delta t}{h^2}c(x_i) - da(x_i)\frac{\theta\Delta t}{h}, \\ k_1(x_i) &= \frac{(1-\theta)\Delta t}{h^2}c(x_i), \\ k_2(x_i) &= 1 - \frac{(1-\theta)\Delta t}{h}da(x_i) - \frac{2(1-\theta)\Delta t}{h^2}c(x_i) \\ &\quad - (1-\theta)b(x_i)\Delta t, \\ k_3(x_i) &= \frac{(1-\theta)\Delta t}{h}da(x_i) + \frac{(1-\theta)\Delta t}{h^2}c(x_i). \end{aligned} \tag{84}$$

Letting $\psi^j = (\psi_i^j)_{i\in\{1,2,\ldots,N\}}$, finally we get

$$\begin{aligned} &D\psi^{j+1} = B\psi^j + V^j \text{ avec } j \in \{1,2,\ldots,M\} \\ &\psi^0 = (\psi_0(ih))_{i\in\{1,2,\ldots,N\}}, \end{aligned} \tag{85}$$

where

$$D=\begin{bmatrix} g_2(x_1) & g_3(x_1) & 0 & & & & & 0 \\ g_1(x_2) & g_2(x_2) & g_3(x_2) & 0 & & & & \\ 0 & g_1(x_3) & g_2(x_3) & g_3(x_3) & 0 & & & \\ & 0 & g_1(x_4) & g_2(x_4) & g_3(x_4) & 0 & & \\ & & 0 & \cdot & \cdot & \cdot & 0 & \\ & & & \cdot & \cdot & \cdot & \cdot & 0 \\ & & & & 0 & g_1(x_{N-1}) & g_2(x_{N-1}) & g_3(x_{N-1}) \\ 0 & & & & & 0 & g_1(x_N) & g_2(x_N) \end{bmatrix}$$

$$B=\begin{bmatrix} k_2(x_1) & k_3(x_1) & 0 & & & & & 0 \\ k_1(x_2) & k_2(x_2) & k_3(x_2) & 0 & & & & \\ 0 & k_1(x_3) & k_2(x_3) & k_3(x_3) & 0 & & & \\ & 0 & k_1(x_4) & k_2(x_4) & k_3(x_4) & 0 & & \\ & & 0 & \cdot & \cdot & \cdot & 0 & \\ & & & \cdot & \cdot & \cdot & \cdot & 0 \\ & & & & 0 & k_1(x_{N-1}) & k_2(x_{N-1}) & k_3(x_{N-1}) \\ 0 & & & & & 0 & k_1(x_N) & k_2(x_N) \end{bmatrix} \quad (86)$$

$$V^j=\begin{bmatrix} \Delta t\cdot\left[(1-\theta)f\left(x_1,t_j\right)+\theta f\left(x_1,t_j+\Delta t\right)\right] \\ \Delta t\cdot\left[(1-\theta)f\left(x_2,t_j\right)+\theta f\left(x_2,t_j+\Delta t\right)\right] \\ \cdot \\ \cdot \\ \cdot \\ \cdot \\ \Delta t\cdot\left[(1-\theta)f\left(x_{N-1},t_j\right)+\theta f\left(x_{N-1},t_j+\Delta t\right)\right] \\ \Delta t\cdot\left[(1-\theta)f\left(x_N,t_j\right)+\theta f\left(x_N,t_j+\Delta t\right)\right] \end{bmatrix}.$$

Step 2 (discretization of the functional one has).

$$J(u)=\frac{1}{2T}\int_0^T\int_0^1\left(C\psi(x,t)-\psi^{\text{obs}}(x,t)\right)^2 dx\,dt. \quad (87)$$

We recall that the method of Thomas Simpson to calculate an integral is

$$\int_a^b f(x)\,dx \simeq \frac{h}{2}\left[f(x_0)+2\sum_{i=1}^{(N+1)/2-1} f(x_{2i}) + 4\sum_{i=1}^{(N+1)/2} f(x_{2i+1})+f(x_{N+1})\right], \quad (88)$$

with $x_0=a$, $x_{N+1}=b$, $x_i=a+ih$, $i\in\{1,\ldots,N+1\}$.

Let the functions

$$\varphi(x,t)=\left(C\psi(x,t)-\psi^{\text{obs}}(x,t)\right)^2 \quad \forall t\in[0,T],\ x\in\Omega, \quad (89)$$

$$\varnothing(t)=\int_0^1\varphi(x,t)\,dx.$$

This gives

$$\varnothing(t)\simeq\frac{h}{2}\left[\varphi(0,t)+2\sum_{i=1}^{(N+1)/2-1}\varphi(x_{2i},t) + 4\sum_{i=1}^{(N+1)/2}\varphi(x_{2i+1},t)+\varphi(1,t)\right], \quad (90)$$

where

$$\int_0^T \int_0^1 \left(C\psi(x,t) - \psi^{\text{obs}}(x,t)\right)^2 dx\,dt \simeq \int_0^T \varnothing(t)\,dt$$
$$\simeq \frac{dt}{2}\left[\varnothing(0) + 2\sum_{j=1}^{(M+1)/2-1} \varnothing(t_{2j}) + 4\sum_{j=1}^{(M+1)/2} \varnothing(t_{2j+1}) + \varnothing(T)\right], \quad (91)$$

with $t_0 = 0, t_{M+1} = T, t_j = jdt, j \in \{1, \ldots, M+1\}$.

Therefore,

$$J(u) \simeq \frac{1}{2T}\frac{dt}{2}\left[\varnothing(0) + 2\sum_{j=1}^{(M+1)/2-1} \varnothing(t_{2j}) + 4\sum_{j=1}^{(M+1)/2} \varnothing(t_{2j+1}) + \varnothing(T)\right]. \quad (92)$$

Step 3 (discretization of ∇J). The adjoint problem (76) is discretized as (85), so,

$$\nabla J \simeq -P^{M+1}. \quad (93)$$

6. Numerical Experiments and Results

In this section, we discuss two cases:

In case we have a priori knowledge ψ^b of ψ_0^{exact} in each point of analysis grid-points, we apply the Tikhonov approach to solve the minimization problem (8). The data ψ^b is assumed to be corrupted by measurement errors, which we will refer to as noise. In particular, we suppose that $\psi^b = \psi_0^{\text{exact}} + e$. Here, we study the impact of err (err $= \|e\|_2$) on the construction of the solution.

In case we have a partial knowledge of values of ψ^b (example 20%): firstly, we apply the hybrid approach to rebuild the initial state. Secondly, we make a comparison between both hybrid and Tikhonov approaches.

The tests have been performed in Matlab 2012A, on a Windows 7 platform.

6.1. Regularization Approach. The differentiability and continuity in A_{ad} of the functional,

$$J_T(\psi_0) = \frac{1}{2T}\int_0^T \left\|C\psi(t) - \psi^{\text{obs}}(t)\right\|_{L^2(\Omega)}^2 dt + \frac{\varepsilon}{2}\left\|\psi_0 - \psi^b\right\|_{L^2(\Omega)}^2, \quad (94)$$

is deduced from the differentiability and continuity of the functional J, and we have

$$\frac{\partial J_T}{\partial \psi_0} = -P(T) + \varepsilon\left(\psi_0 - \psi^b\right), \quad (95)$$

where P is the solution of (76).

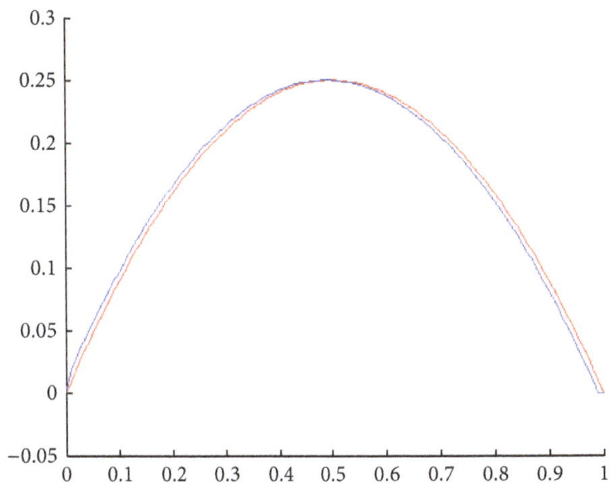

FIGURE 1: Initial temperature. This figure shows that we can rebuild the initial state.

The main steps for descent method at each iteration are the following:

(i) Calculate ψ^k solution of (6) with initial condition ψ_0.

(ii) Calculate P^k solution of (76).

(iii) Calculate the descent direction $d_k = -\nabla J_T(\psi_0)$.

(iv) Find $t_k = \text{argmin}_{t>0} J_T(\psi_0 + td_k)$

(v) Update the variable $\psi_0 = \psi_0 + t_k d_k$.

The algorithm ends when $|J_T(\psi_0)| < \mu$, where μ is a given small precision.

t_k is chosen by the inaccurate linear search by Rule Armijo-Goldstein as follows:

let $\alpha_i, \beta \in [0, 1[$ and $\alpha > 0$

if $J_T(\psi_0^k + \alpha_i d_k) \leqslant J_T(\psi_0^k) + \beta\alpha_i d_k^T d_k$

$t_k = \alpha_i$ and stop

if not

$\alpha_i = \alpha\alpha_i$.

We do all the tests on Pc with the following configurations: Intel Core i3 CPU 2.27 GHz; RAM = 4 GB (2.93 usable).

In all figures, the observed function is drawn in red and built function in blue.

Let N be number of points in space and M number of points in time.

6.1.1. The Noncoercive Case. Let $\alpha = 1/2$, $\lambda = 0$ and parameters $N = 100$, $M = 100$.

(i) Tests with err $= 0$. See Figures 1, 2, 3, and 4.

(ii) Tests with err $\neq 0$. In Figures 5, 6, 7, and 8, ψ_0^{exact} is drawn in red and ψ_0 (rebuilt initial condition) in blue.

6.1.2. Sub Critical Potential Case. Let $\alpha = 1/2, \lambda = -(1-\alpha)^2/4$, $\beta = 3/4$ and the parameters $N = 100$, $M = 100$.

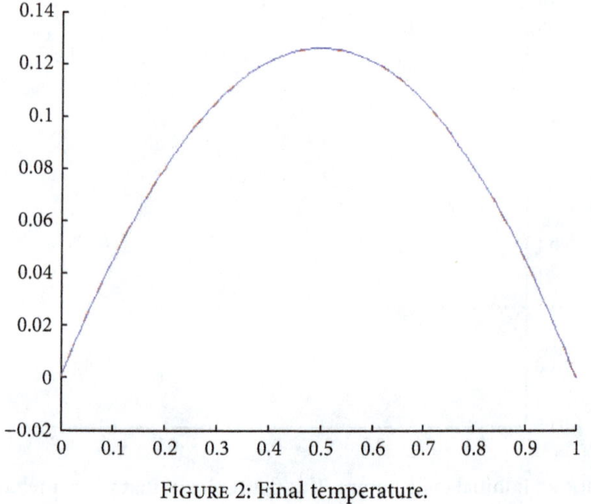

Figure 2: Final temperature.

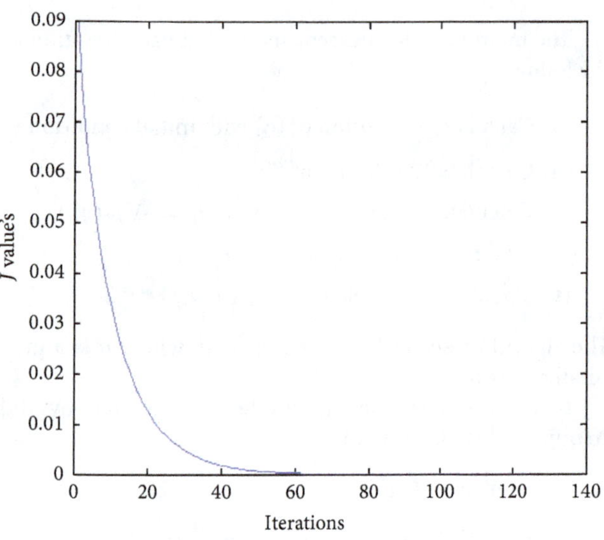

Figure 3: Graph of J.

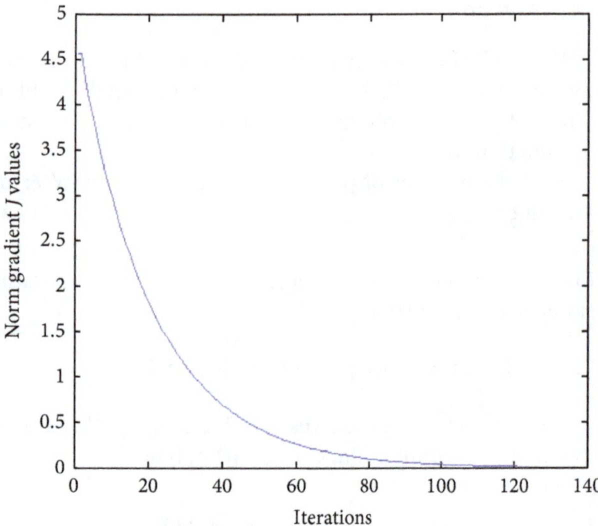

Figure 4: Norm of gradient of J.

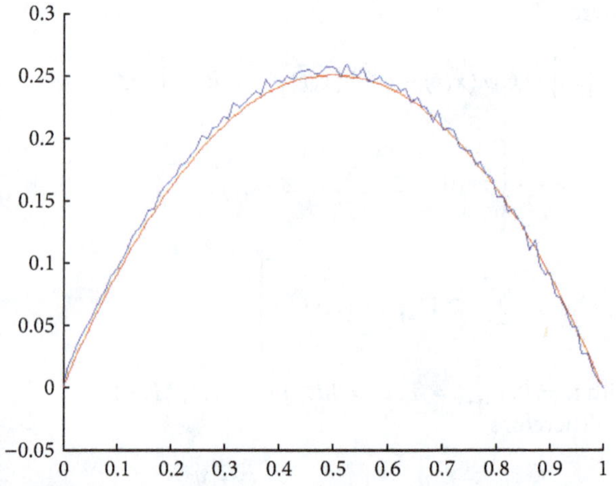

Figure 5: Initial temperature in err = 2% $\|\psi_0^{\text{exact}}\|_2$ case. This figure shows that we can rebuild the true state.

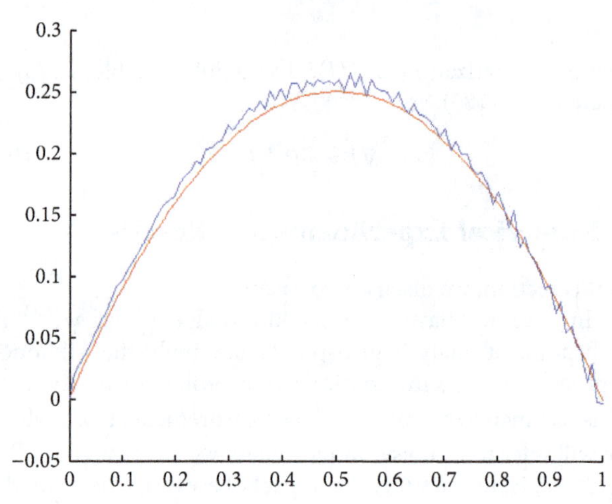

Figure 6: Initial temperature in err = 5% $\|\psi_0^{\text{exact}}\|_2$ case. The reconstructed initial condition is not far from the true state.

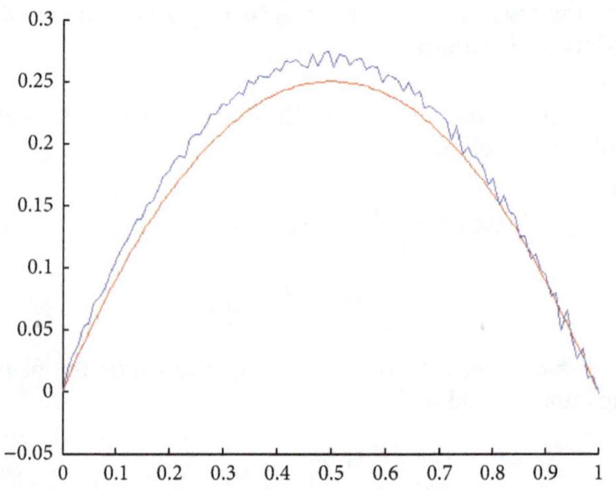

Figure 7: Initial temperature in err = 10% $\|\psi_0^{\text{exact}}\|_2$ case. The reconstructed initial condition begins to move away from the true state.

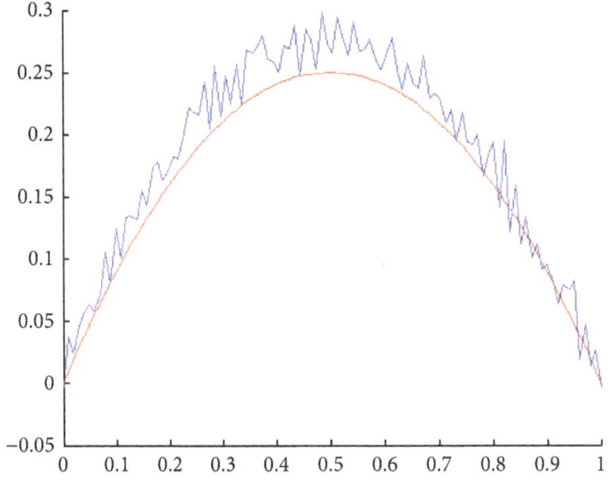

FIGURE 8: Initial temperature in err = 20% $\|\psi_0^{\text{exact}}\|_2$ case. This figure shows that we cannot rebuild the true state.

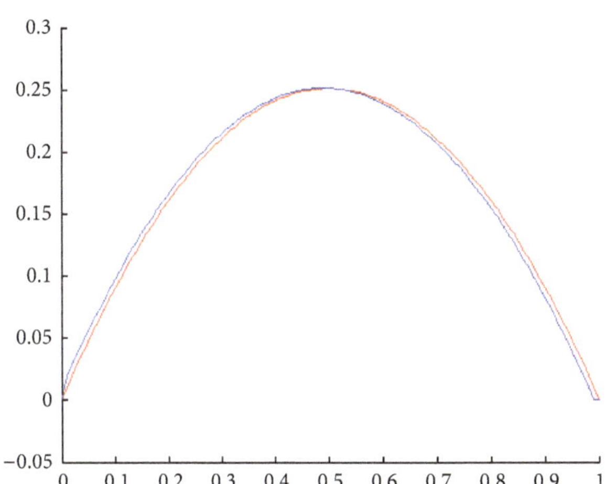

FIGURE 9: Initial temperature. This figure shows that we can rebuild the initial state.

(i) Tests with err $= 0$. See Figures 9, 10, 11, and 12.

(ii) Tests with err $\neq 0$. See Figures 13, 14, 15, and 16.

6.2. Hybrid Algorithm. The genetic algorithms (GA) are adaptive search and optimization methods that are based on the genetic processes of biological organisms. Their principles have been first laid down by Holland. The aim of GA is to optimize a problem-defined function, called the fitness function. To do this, GA maintain a population of individuals (suitably represented candidate solutions) and evolve this population over time. At each iteration, called generation, the new population is created by the process of selecting individuals according to their level of fitness in the problem domain and breeding them together using operators borrowed from natural genetics, as, for instance, crossover and mutation. As the population evolves, the individuals in general tend toward

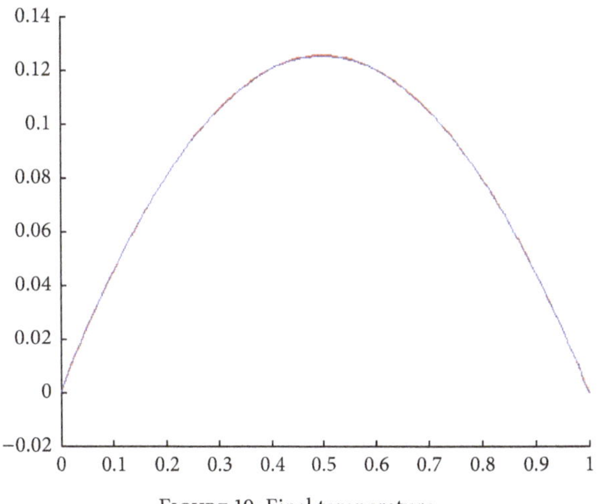

FIGURE 10: Final temperature.

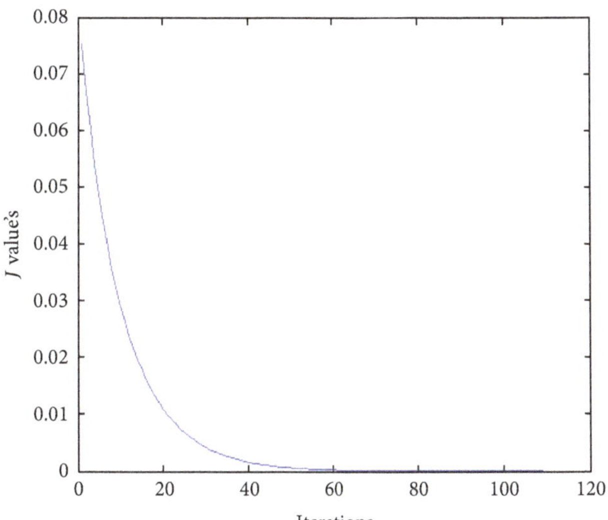

FIGURE 11: Graph of J.

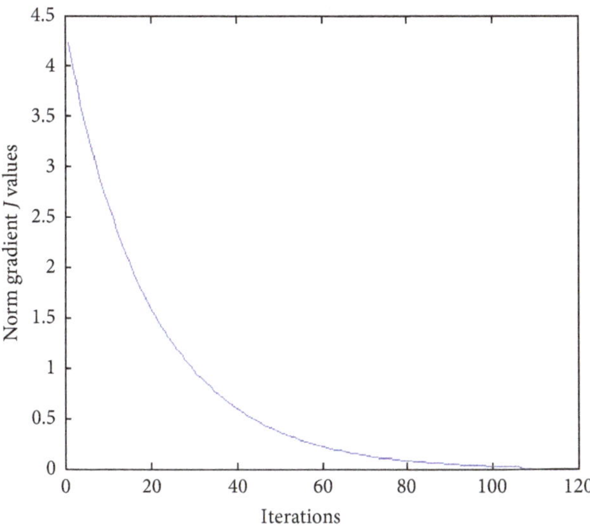

FIGURE 12: Norm of gradient of J.

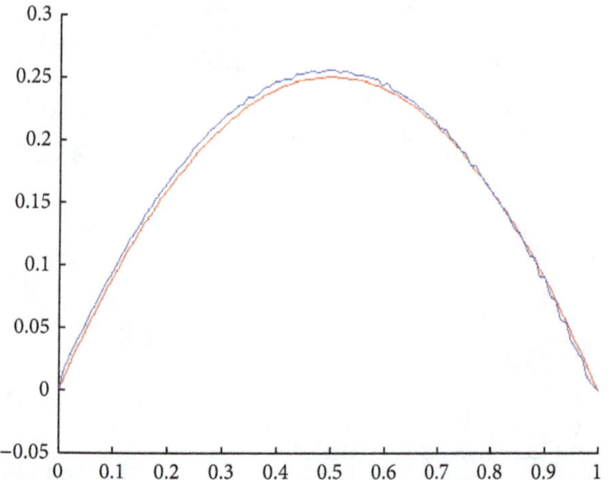

FIGURE 13: Initial temperature in err = 2% $\|\psi_0^{\text{exact}}\|_2$ case. This figure shows that we can rebuild the true state.

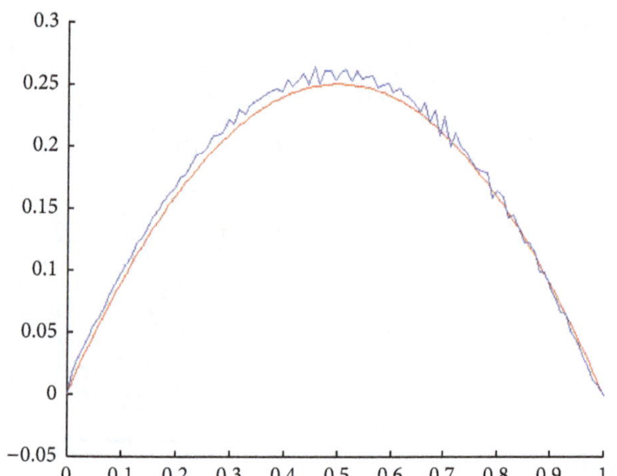

FIGURE 14: Initial temperature in err = 5% $\|\psi_0^{\text{exact}}\|_2$ case. The reconstructed initial condition is not far from the true state.

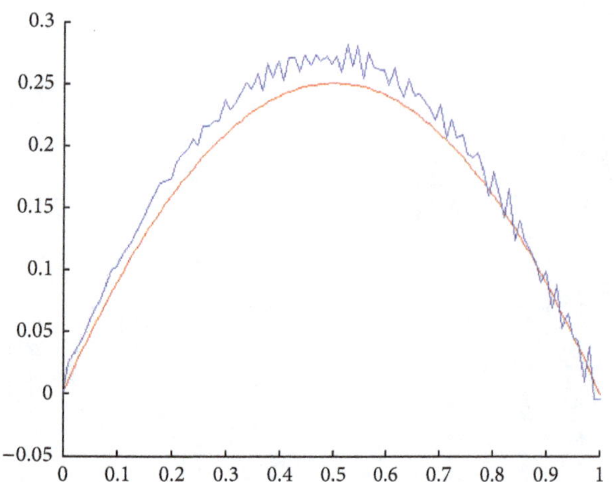

FIGURE 15: Initial temperature in err = 10% $\|\psi_0^{\text{exact}}\|_2$ case. The reconstructed initial condition began to move away from the true state.

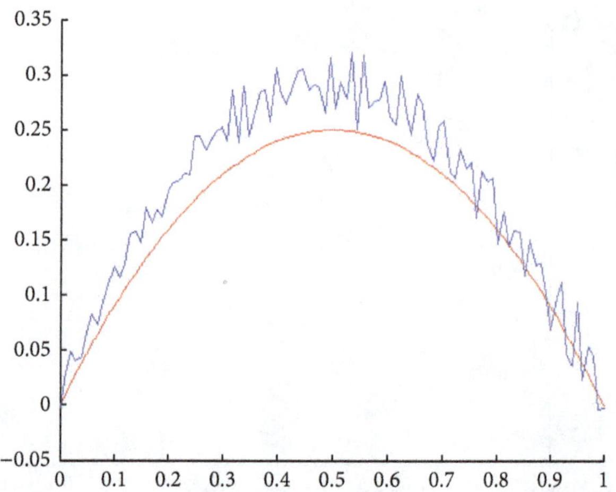

FIGURE 16: Initial temperature in err = 20% $\|\psi_0^{\text{exact}}\|_2$ case. This figure shows that we cannot rebuild the true state.

the optimal solution [21–24]. The basic structure of a GA is the following:

(1) Initialize a population of individuals;

(2) Evaluate each individual in the population;

(3) While the stop criterion is not reached do

{

(4) Select individuals for the next population;

(5) Apply genetic operators (crossover, mutation) to produce new individuals;

(6) Evaluate the new individuals;

}

(7) return the best individual.

The hybrid methods combine principles from genetic algorithms and other optimization methods. In this approach, we will combine the genetic algorithm with method descent (steepest descent algorithm (FP)).

We assume that we have a partial knowledge of background state ψ^b at certain points $(x_i)_{i\in I}$, $I \subset \{1, 2, \ldots, N+1\}$.

We assume the individual is a vector ψ_0; the population is a set of individuals.

The initialization of individual is as follows:

$$\begin{aligned}
&\text{for } i = 1 \text{ to } N+1 \\
&\quad \text{if } i \in I \\
&\qquad \psi_0(x_i) \text{ is chosen in the vicinity of } \psi^b(x_i) \\
&\quad \text{else} \\
&\qquad \psi_0(x_i) \text{ is chosen randomly} \\
&\quad \text{end if} \\
&\text{end for.}
\end{aligned} \tag{96}$$

Starting by initial population, we apply genetic operators (crossover, mutation) to produce a new population in which

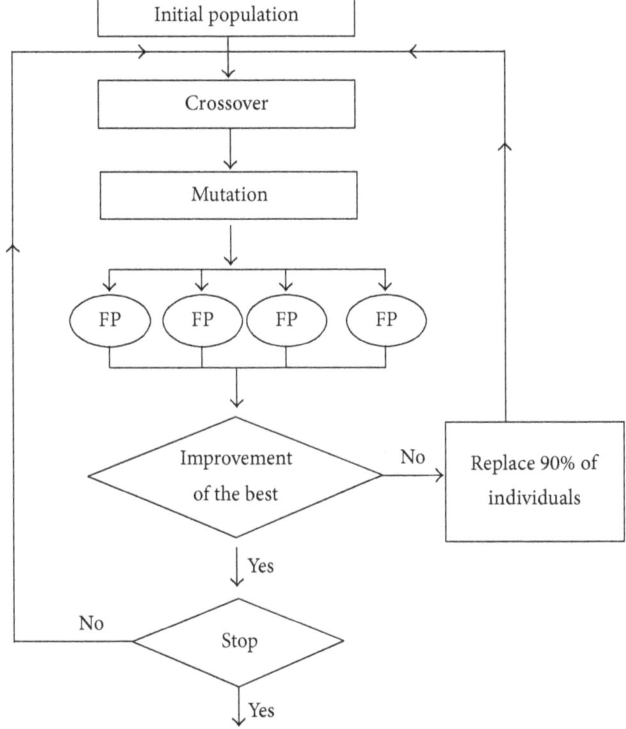

FIGURE 17: Hybrid algorithm.

each individual is an initial point for the descent method (FP). When a specific number of generations is reached without improvement of the best individual, only the fittest individuals (e.g., the first 10% fittest individuals in the population) survive. The remaining die and their place is occupied by new individuals with new genetic (45% are chosen randomly; the other 45% are chosen as (96)). At each generation we keep the best. The algorithm ends when $|J(\psi_0)| < \mu$ or generation $\geqslant$ Maxgen, where μ is a given precision (see Figure 17).

The main steps for descent method (FP) at each iteration are the following:

(i) Calculate ψ^k solution of (6) with initial condition ψ_0.

(ii) Calculate P^k solution of (76).

(iii) Calculate the descent direction $d_k = -\nabla J(\psi_0)$.

(iv) Find $t_k = \text{argmin}_{t>0} J(\psi_0 + t d_k)$.

(v) Update the variable $\psi_0 = \psi_0 + t_k d_k$.

The algorithm ends when $|J(\psi_0)| < \mu$, where μ is a given small precision.

t_k is chosen by the inaccurate linear search by Rule Armijo-Goldstein as follows:

let $\alpha_i, \beta \in [0, 1[$ and $\alpha > 0$.

if $J(\psi_0^k + \alpha_i d_k) \leqslant J(\psi_0^k) + \beta \alpha_i d_k^T d_k$

$t_k = \alpha_i$ and stop

if not

$\alpha_i = \alpha \alpha_i$.

TABLE 1: Results on the Tikhonov approach. Comparison between different values of regularizing coefficient ε. The smallest value of J is reached when $\varepsilon = 10^{-06}$.

ε	Minimum value of J	Elapsed time (seconds)
10^{-08}	$1.106317 \cdot 10^{-02}$	8.38
10^{-07}	$1.092014 \cdot 10^{-02}$	24.47
10^{-06}	$6.630517 \cdot 10^{-03}$	23.11
10^{-05}	$7.752620 \cdot 10^{-03}$	21.50
10^{-04}	$7.857129 \cdot 10^{-03}$	22.64
10^{-03}	$8.510799 \cdot 10^{-03}$	18.95
10^{-02}	$8.733989 \cdot 10^{-03}$	15.01
10^{-01}	$1.018406 \cdot 10^{-02}$	17.33
1	$1.552344 \cdot 10^{-02}$	6.04

TABLE 2: Results on the hybrid method.

Minimum value of J	Elapsed time
$6.581908 \cdot 10^{-03}$	1 min
$5.850810 \cdot 10^{-03}$	2 min
$3.362100 \cdot 10^{-04}$	7 min
$1.071378 \cdot 10^{-04}$	11 min
$8.739839 \cdot 10^{-05}$	23 min
$5.958016 \cdot 10^{-05}$	6 h 43 min
$6.175260 \cdot 10^{-06}$	11 h 20 min

Consider we know 20% of values of background state (ψ^b), in this test we try to build the solution with the hybrid method.

In the figures below, the observed function is drawn in red and built function in blue.

Let N be number of points in space and M number of points in time.

6.2.1. The Noncoercive Case. Let $\alpha = 1/2$, $\lambda = 0$ and parameters $N = 100$, $M = 100$, number of individuals = 30, and number of generations = 2000.

The test with simple descent gives Figure 18.

The test with genetic algorithm gives Figure 19.

Now we turn the hybrid algorithm. This gives Figure 20.

6.2.2. Subcritical Potential Case. Let $\alpha = 1/2$, $\lambda = -(1-\alpha)^2/4$, $\beta = 3/4$ and parameters $N = 100$, $M = 100$, number of individuals = 30, and number of generations = 2000.

The test with simple descent gives Figure 21.

The test with genetic algorithm gives Figure 22.

Now we turn the hybrid algorithm. This gives Figure 23.

6.3. Comparison between Hybrid Approach and Tikhonov Approach. Here, we assume that we know 20% of values of background state (ψ^b).

(i) Noncoercive Case. see Tables 1 and 2.

The minimum value of J reached by the Tikhonov algorithm was $6.630517 \cdot 10^{-03}$, whereas with the hybrid

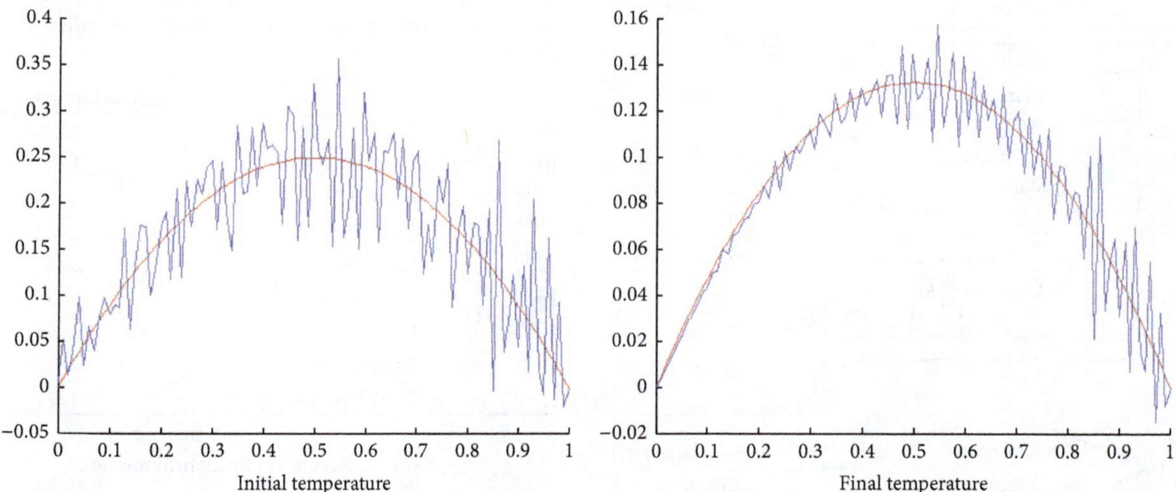

FIGURE 18: This figure shows that we cannot rebuild the solution.

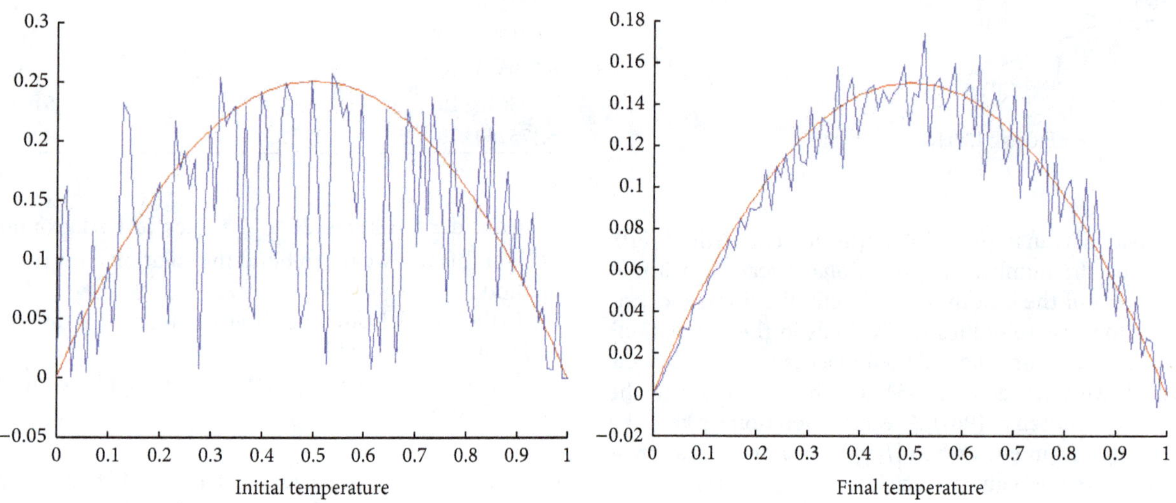

FIGURE 19: This figure shows that we cannot rebuild the solution.

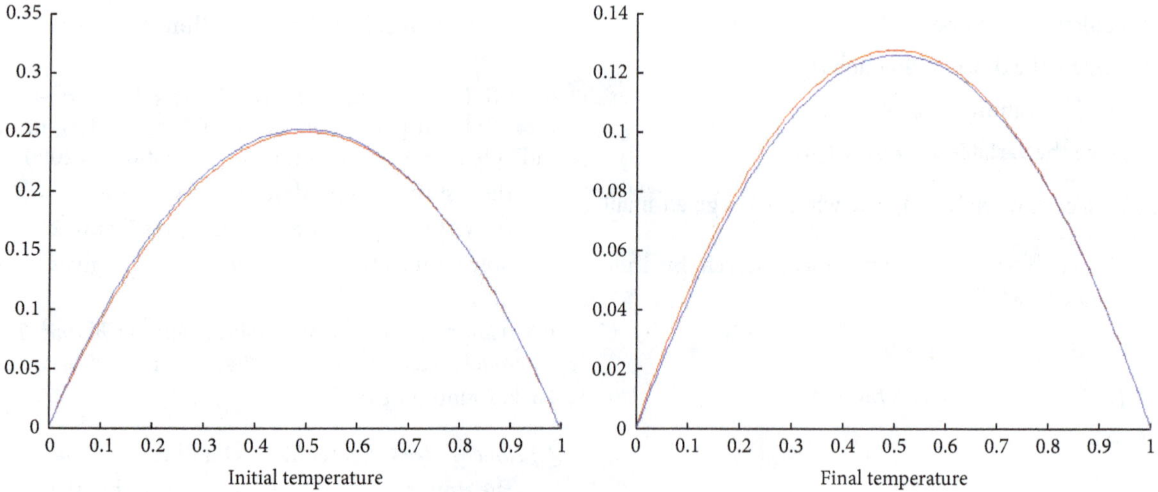

FIGURE 20: This figure shows that we can rebuild the solution.

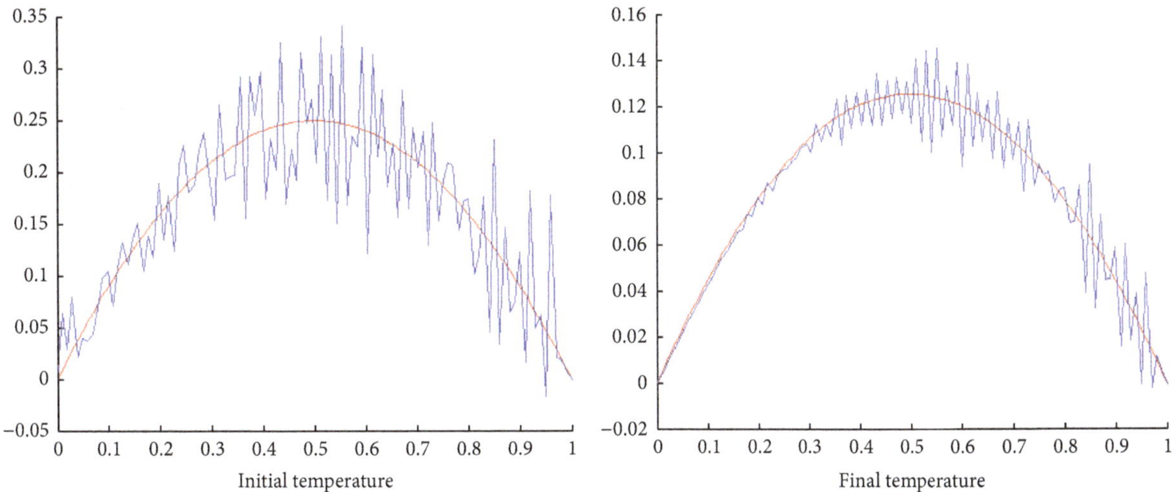

FIGURE 21: This figure shows that we cannot rebuild the solution.

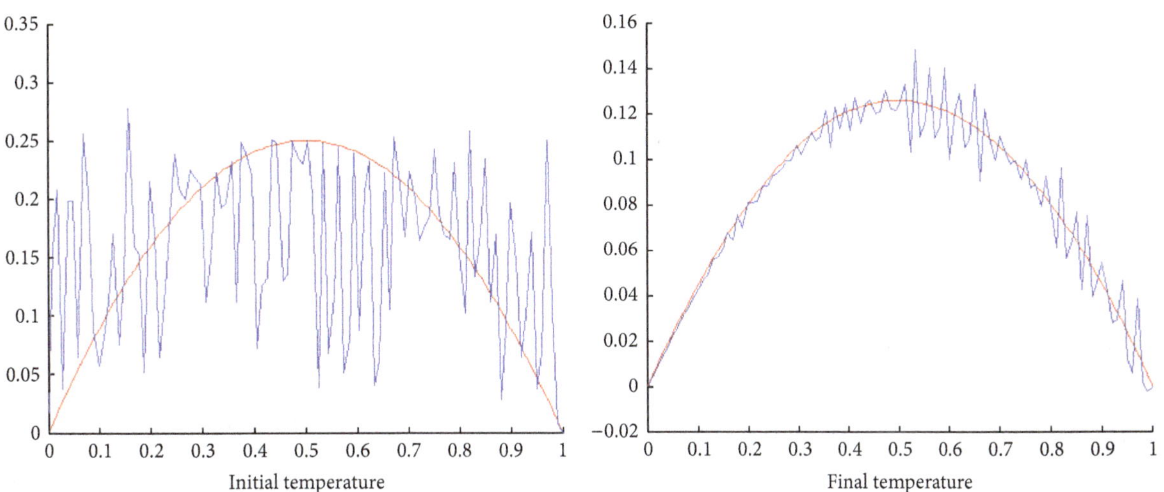

FIGURE 22: This figure shows that we cannot rebuild the solution.

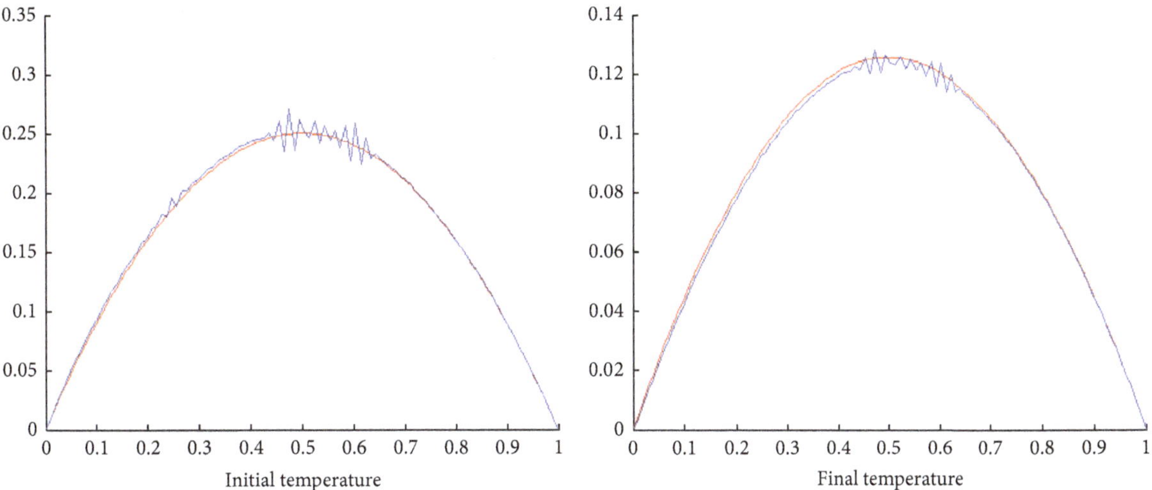

FIGURE 23: This figure shows that we can rebuild the solution.

TABLE 3: Results on the Tikhonov aporoach. Comparison between different values of regularizing coefficient ε. The smallest value of J is reached when $\varepsilon = 10^{-03}$.

ε	Minimum value of J	Elapsed time (seconds)
10^{-08}	$1.113538 \cdot 10^{-02}$	10.91
10^{-07}	$1.188092 \cdot 10^{-02}$	8.68
10^{-06}	$1.099187 \cdot 10^{-02}$	8.75
10^{-05}	$1.267204 \cdot 10^{-02}$	7.91
10^{-04}	$9.648060 \cdot 10^{-03}$	11.17
10^{-03}	$7.320995 \cdot 10^{-03}$	11.61
10^{-02}	$9.188603 \cdot 10^{-03}$	10.20
10^{-01}	$9.799159 \cdot 10^{-03}$	9.57
1	$1.042463 \cdot 10^{-02}$	9.66

TABLE 4: Results on the hybrid method.

Minimum value of J	Elapsed time
$7.605018 \cdot 10^{-03}$	1 min
$7.505279 \cdot 10^{-03}$	2 min
$1.762564 \cdot 10^{-03}$	4 min
$9.407809 \cdot 10^{-04}$	43 min
$2.981666 \cdot 10^{-05}$	1 h 56 min
$1.378356 \cdot 10^{-05}$	6 h 43 min
$8.203546 \cdot 10^{-06}$	13 h 40 min

algorithm it was possible to reach the value $6.175260 \cdot 10^{-06}$ in 11 h and 20 min with knowledge of 20% of ψ^b; if we take more than 20%, we got less than elapsed time.

(ii) Subcritical Potential Case. see Tables 3 and 4.

The minimum value of J reached by the Tikhonov algorithm was $7.320995 \cdot 10^{-03}$, whereas with the hybrid algorithm it was possible to reach the value $8.203546 \cdot 10^{-06}$ in 13 h and 40 min with knowledge of 20% of ψ^b; if we take more than 20%, we got less than elapsed time.

7. Conclusion

In this paper, we have presented the regularization method and the hybrid method which are applied to determine an initial state from the point of measurements of parabolic degenerate/singular problem. These methods have proven efficiency to rebuild the solution. The proposed reconstruction algorithms are easily implanted.

The elapsed time of the hybrid method is long enough. To reduce it, in our coming work we will use the multiprogramming to run two approaches of parallelism.

Conflicts of Interest

The authors declare that there are no conflicts of interest regarding the publication of this paper.

References

[1] V. Isakov, *Inverse Problems for Partial Differential Equations*, Springer, New York, NY, USA, 1998.

[2] A. Kirsch, *An Introduction to the Mathematical Theory of Inverse Problems*, Springer, New York, NY, USA, 1999.

[3] J.-M. Buchot and J.-P. Raymond, "A linearized model for boundary layer equations, Optimal control of complex structures (Oberwolfach, 2000)," in *Internat. Ser. Numer. Math*, vol. 139, pp. 31–42, Birkhäuser, Basel, Swizerland, 2002.

[4] K. Beauchard and E. Zuazua, "Some controllability results for the 2D Kolmogorov equation," *Annales de l'Institut Henri Poincare (C) Non Linear Analysis*, vol. 26, no. 5, pp. 1793–1815, 2009.

[5] N. Shimakura, *Partial differential operators of elliptic type*, vol. 99 of *Translations of Mathematical Monographs*, American Mathematical Society, Providence, RI, USA, 1992.

[6] W. H. Fleming and M. Viot, "Some measure-valued Markov processes in population genetics theory," *Indiana University Mathematics Journal*, vol. 28, no. 5, pp. 817–843, 1979.

[7] H. Emamirad, G. R. Goldstein, and J. A. Goldstein, "Chaotic solution for the Black-Scholes equation," *Proceedings of the American Mathematical Society*, vol. 140, no. 6, pp. 2043–2052, 2012.

[8] G. Fragnelli and D. Mugnai, "Carleman estimates for singular parabolic equations with interior degeneracy and non smooth coefficients," *Advances in Nonlinear Analysis*, 2016.

[9] F. Bourquin and A. Nassiopoulos, "Assimilation thermique 1D par méthode adjointe libérée," in *Problèmes Inverses*, Collection Recherche du LCPC, 2006.

[10] T. Min, B. Geng, and J. Ren, "Inverse estimation of the initial condition for the heat equation," *International Journal of Pure and Applied Mathematics*, vol. 82, no. 4, pp. 581–593, 2013.

[11] L. B. L. Santos, L. D. Chiwiacowsky, and H. F. Campos-Velho, "Genetic algorithm and variational method to identify initial conditions: worked example in hyperbolic heat transfer," *TEMA. Tendências em Matemática Aplicada e Computacional*, vol. 14, no. 2, pp. 265–276, 2013.

[12] M. Bonnet, *Ecole Centrale de Paris Mention Matière, Structures, Fluides, Rayonnement Spécialité Dynamique des Structures et Systèmes Couplés [M.S. thesis]*, 2008, bonnet@lms.polytechnique.fr.

[13] F. Jens, *Generalized Tikhonov regularization, Basic theory and comprehensive results on convergence rates [Dissertation]*, Fakultät für Mathematik Technische Universität Chemnitz, Chemnitz, 2011.

[14] F. Alabau-Boussouira, P. Cannarsa, and G. Fragnelli, "Carleman estimates for degenerate parabolic operators with applications to null controllability," *Journal of Evolution Equations*, vol. 6, no. 2, pp. 161–204, 2006.

[15] E. M. Ait Ben Hassi, F. Ammar Khodja, A. Hajjaj, and L. Maniar, "Carleman estimates and null controllability of coupled degenerate systems," *Evolution Equations and Control Theory*, vol. 2, no. 3, pp. 441–459, 2013.

[16] P. Cannarsa and G. Fragnelli, "Null controllability of semilinear degenerate parabolic equations in bounded domains," *Electronic Journal of Differential Equations*, vol. 136, pp. 1–20, 2006.

[17] P. Cannarsa, P. Martinez, and J. Vancostenoble, "Carleman estimates for a class of degenerate parabolic operators," *SIAM Journal on Control and Optimization*, vol. 47, no. 1, article no. 119, 2008.

[18] P. Cannarsa, P. Martinez, and J. Vancostenoble, "Null controllability of degenerate heat equations," *Advances in Differential Equations*, vol. 10, no. 2, pp. 153–190, 2005.

[19] J. Vancostenoble, "Improved Hardy-Poincaré inequalities and sharp Carleman estimates for degenerate/singular parabolic problems," *Discrete and Continuous Dynamical Systems*, vol. 4, no. 3, Article ID 761790, 2011.

[20] E. M. Ait Ben Hassi, F. Ammar Khodja, A. Hajjaj, and L. Maniar, "Null controllability of degenerate cascade systems," *Journal of Evolution Equations and Control Theory*, vol. 2, no. 3, pp. 441–459, 2013.

[21] Z. Michalewicz, *Genetic Algorithms + Data Structures = Evolution Programs*, Springer, New York, NY, USA, 1996.

[22] D. E. Goldberg, *Genetic Algorithms in Search, Optimisation and Machine Learning*, 1989.

[23] J. H. Holland, *Adaptation in Natural and Artificial Systems*, University of Michigan Press, Oxford, UK, 1975.

[24] J. R. Koza, *Genetic Programming*, MIT Press, 1992.

Permissions

All chapters in this book were first published in IJDE, by Hindawi Publishing Corporation; hereby published with permission under the Creative Commons Attribution License or equivalent. Every chapter published in this book has been scrutinized by our experts. Their significance has been extensively debated. The topics covered herein carry significant findings which will fuel the growth of the discipline. They may even be implemented as practical applications or may be referred to as a beginning point for another development.

The contributors of this book come from diverse backgrounds, making this book a truly international effort. This book will bring forth new frontiers with its revolutionizing research information and detailed analysis of the nascent developments around the world.

We would like to thank all the contributing authors for lending their expertise to make the book truly unique. They have played a crucial role in the development of this book. Without their invaluable contributions this book wouldn't have been possible. They have made vital efforts to compile up to date information on the varied aspects of this subject to make this book a valuable addition to the collection of many professionals and students.

This book was conceptualized with the vision of imparting up-to-date information and advanced data in this field. To ensure the same, a matchless editorial board was set up. Every individual on the board went through rigorous rounds of assessment to prove their worth. After which they invested a large part of their time researching and compiling the most relevant data for our readers.

The editorial board has been involved in producing this book since its inception. They have spent rigorous hours researching and exploring the diverse topics which have resulted in the successful publishing of this book. They have passed on their knowledge of decades through this book. To expedite this challenging task, the publisher supported the team at every step. A small team of assistant editors was also appointed to further simplify the editing procedure and attain best results for the readers.

Apart from the editorial board, the designing team has also invested a significant amount of their time in understanding the subject and creating the most relevant covers. They scrutinized every image to scout for the most suitable representation of the subject and create an appropriate cover for the book.

The publishing team has been an ardent support to the editorial, designing and production team. Their endless efforts to recruit the best for this project, has resulted in the accomplishment of this book. They are a veteran in the field of academics and their pool of knowledge is as vast as their experience in printing. Their expertise and guidance has proved useful at every step. Their uncompromising quality standards have made this book an exceptional effort. Their encouragement from time to time has been an inspiration for everyone.

The publisher and the editorial board hope that this book will prove to be a valuable piece of knowledge for researchers, students, practitioners and scholars across the globe.

List of Contributors

Fengyan Wu, Xiujun Cheng and Xiaoli Chen
Center for Mathematical Sciences, Huazhong University of Science and Technology,Wuhan 430074, China
School of Mathematics and Statistics, Huazhong University of Science and Technology,Wuhan 430074, China

Qiong Wang
School of Mathematics and Statistics, Huazhong University of Science and Technology,Wuhan 430074, China

Cheng-Min Su, Jian-Ping Sun and Ya-Hong Zhao
Department of Applied Mathematics, Lanzhou University of Technology, Lanzhou 730050, China

Oluwaseun Adeyeye and Zurni Omar
Mathematics Department, School of Quantitative Sciences, Universiti Utara Malaysia, Sintok, Kedah, Malaysia

Brajesh Kumar Singh and Pramod Kumar
Department of Applied Mathematics, School for Physical Sciences, Babasaheb Bhimrao Ambedkar University, Lucknow 226 025, India

A. Acosta and H. Leiva
School of Mathematical Sciences and Information Technology, Department of Mathematics, Yachay Tech, Urcuqui, Ecuador

P. García
Facultad de Ingeniería en Ciencias Aplicadas, Universidad Técnica del Norte, Ibarra, Ecuador

A. Merlitti
Departamento de Estadística, Facultad de Ciencias Económicas y Sociales, Universidad Central de Venezuela, Caracas, Venezuela

Rezvan Ghoochani-Shirvan
Department of Applied Mathematics, Ferdowsi University of Mashhad, International Campus, Mashhad, Iran

Jafar Saberi-Nadjafi and Morteza Gachpazan
Department of Applied Mathematics, School of Mathematical Sciences, Ferdowsi University of Mashhad, Mashhad, Iran

Tetsutaro Shibata
Laboratory of Mathematics, Graduate School of Engineering, Hiroshima University, Higashi-Hiroshima, 739-8527, Japan

Seda İğret Araz
Siirt University, Department of Elementary Mathematics Education, Siirt 56100, Turkey

Murat Subaşi
Atatürk University, Department of Mathematic, Erzurum 25240, Turkey

Quanxiang Wang
College of Engineering, Nanjing Agricultural University, Nanjing 210031, China

Tengjin Zhao and Zhiyue Zhang
Jiangsu Key Laboratory for NSLSCS, School of Mathematical Sciences, Nanjing Normal University, Nanjing 210023, China

Ling De Su
Institute of Mathematics and Information Science, North-Eastern Federal University, Russia

Tong Song Jiang
Department of Mathematics, Heze University, Shandong, China

Aziz Bouhlal and Hamad Talibi Alaoui
Labo Math Applied, Faculty of Sciences, B. P. 20, El Jadida, Morocco

Abderrahmane El Hachimi
Laboratory of Mathematical Analysis and Applications, Mohammed V University, Faculty of Sciences, Rabat, Morocco

Jaouad Igbida
Labo DGTIC, department of Mathematics, CRMEF El Jadida, Morocco

El Mostafa Sadek
Laboratory Lab SIPE, ENSA, d'EL Jadida, University Chouaib Doukkali, Morocco

Meng-Rong Li
Department of Mathematical Sciences, National Chengchi University, Taipei 116, Taiwan

Tsung-Jui Chiang-Lin
Graduate Institute of Finance, National Taiwan University of Science and Technology, Taipei 106, Taiwan

Yong-Shiuan Lee
Department of Statistics, National Chengchi University, Taipei 116, Taiwan

Kwanpaka Naboonmee
Department of Mathematics, Faculty of Science, Naresuan University, Phitsanulok 65000, Thailand

Supaporn Suksern
Department of Mathematics, Faculty of Science, Naresuan University, Phitsanulok 65000, Thailand
Research Center for Academic Excellence in Mathematics, Naresuan University, Phitsanulok 65000, Thailand

Loredana Caso, Patrizia Di Gironimo, Sara Monsurrò and Maria Transirico
University of Salerno, Via Giovanni Paolo II,No. 132, 84084 Fisciano, Italy

Fernando Bernal-Vílchis
Instituto de Matemáticas, UNAM, Campus Morelia, AP 61-3 Xangari, 58089 Morelia, MICH, Mexico

Nakao Hayashi
Department of Mathematics, Graduate School of Science, Osaka University, Osaka, Toyonaka 560-0043, Japan

Pavel I. Naumkin
Centro de Ciencias Matemáticas, UNAM, Campus Morelia, AP 61-3 Xangari, 58089 Morelia, MICH, Mexico

Jeffry Kusuma, Agustinus Ribal and Andi Galsan Mahie
Department of Mathematics, Hasanuddin University, Makassar, Indonesia

Jamal H. Al-Smail, Salim A. Messaoudi and Ala A. Talahmeh
Department of Mathematics and Statistics, King Fahd University of Petroleum and Minerals, Dhahran 31261, Saudi Arabia

Adnane Boukhouima and Noura Yousfi
Laboratory of Analysis, Modeling and Simulation (LAMS), Faculty of Sciences Ben M'sik, Hassan II University, Sidi Othmane, Casablanca, Morocco

Khalid Hattaf
Laboratory of Analysis, Modeling and Simulation (LAMS), Faculty of Sciences Ben M'sik, Hassan II University, Sidi Othmane, Casablanca, Morocco
Centre Régional des Métiers de l'Education et de la Formation (CRMEF), 20340 Derb Ghallef, Casablanca, Morocco

F. F. Ngwane
Department of Mathematics, University of South Carolina, Salkehatchie, Walterboro, SC 29488, USA

S. N. Jator
Department of Mathematics and Statistics, Austin Peay State University, Clarksville, TN 37044, USA

K. Atifi, Y. Balouki, El-H. Essoufi and B. Khouiti
Laboratoire de Math´ematiques, Informatique et Sciences de l'Ing´enieur (MISI), Universit´e Hassan 1er, 26000 Settat, Morocco

Index

www.ingramcontent.com/pod-product-compliance
Lightning Source LLC
Chambersburg PA
CBHW080634200326
41458CB00013B/4626

* 9 7 8 1 6 8 2 8 5 8 2 4 0 *